Coagulation and Flocculation in Water and Wastewater Treatment

Coagulation and Flocculation in Water and Wastewater Treatment

Third Edition

John Bratby

IWA PUBLISHING

Published by **IWA Publishing**
Alliance House
12 Caxton Street
London SW1H 0QS, UK
Telephone: +44 (0)20 7654 5500
Fax: +44 (0)20 7654 5555
Email: publications@iwap.co.uk
Web: www.iwapublishing.com

First published 2016

British Library Cataloguing in Publication Data
A CIP catalogue record for this book is available from the British Library

ISBN: 9781780407494 (Hardback)
ISBN: 9781780407500 (eBook)

Contents

Preface

> "Fryday, May 9, 1851 ... Made 22 miles to day, Good forage for the animals. Not much wood. The water of the plat is very good when settled, which we do by throwing a little alumn in and let it stand awhile."
>
> *From the diary of Amelia Hardy in Covered wagon woman, diaries and letters from the western trails, 1840–1890.*

A number of new developments have been included in this third edition. Most of these were included by expanding existing sections, or new sections were added. However, the general organization of the second edition has been maintained in this book. Topics that have been included as new sections include algae removal and harvesting; dissolved organic nitrogen (DON) removal; wastewater treatment by coagulation and chemically enhanced primary treatment (CEPT); activated sludge bulking and foaming control and enhanced bioflocculation; inorganics removal; and turbidity and its measurement, included as an Appendix. In some cases, topics have been separated into specific treatment using metal coagulants (Chapter 4) or organic polymers (Chapter 5) to conform with the general layout of the book. These cases include algae removal and harvesting; wastewater treatment by coagulation and CEPT; and activated sludge bulking and foaming control and enhanced bioflocculation.

Despite the additions and modifications to the book, most of the comments in the Preface to the second edition are still relevant. Coagulation and flocculation play a dominant role in many water and wastewater treatment schemes, including those incorporating membrane treatment. Because of the complex interdependence of numerous factors inherent in the coagulation and flocculation processes, a thorough understanding of the phenomena involved is essential.

There are several excellent books on water treatment that include chapters and sections on coagulation and flocculation. However, current literature often does not present an in-depth account of coagulation and flocculation in a single publication. Moreover, a given publication may emphasize theoretical aspects with little indication of the practical significance of the phenomena, or largely ignores the theoretical development and presents essentially practical material only. In many cases the uninitiated reader is left without an overall picture of the processes as a whole and, without further extensive literature searching, succeeds in achieving only a superficial understanding. The intention of this book is to help remedy these inadequacies.

An attempt has been made to present the subject matter logically and sequentially from theoretical principles to practical applications. Successive chapters deal with, in turn, properties of materials present in waters and wastewaters; characteristics and types of coagulants commonly in use; mechanisms and practical implications of destabilization of waterborne material using metal coagulants and polyelectrolytes; considerations related to coagulant addition at the *rapid mixing* stage; theoretical and practical considerations of flocculation; and details of experimental procedures for assessing primary coagulants, flocculant aids, sludge conditioners, and flocculation parameters. Numerous examples are included as appropriate.

Treatment and disposal of sludges resulting from coagulation-flocculation related operations is dealt with in an Appendix. This important topic has been separated from the main text to avoid disturbing the continuum of the presentation.

The book in its present form evolved from material offered to undergraduate and postgraduate university students. Although it has been progressively modified and increased in scope to cater for the requirements of practicing chemists, chemical and civil engineers involved with water and wastewater treatment, the book still retains a didactic nature. Therefore, it is hoped that the book will serve three functions:

(1) A readable and useful presentation for the water scientist and engineer.
(2) A convenient reference handbook in the form of numerous examples and appended information.
(3) Course material for under- and postgraduate students.

Chapter 1
Introduction

1.1 GENERAL

Coagulation and flocculation are essential processes in a number of diverse disciplines, including biochemistry, cheese manufacturing, rubber manufacturing, and in water and wastewater treatment. Many people are probably familiar with the general concept of coagulation from the well known chemically caused clotting of blood when exposed to air. Another well known example of coagulation is the hard-boiling of an egg, in which the protein coagulates and hardens by heat.

In water and wastewater treatment, coagulation and flocculation phenomena are extremely important. In the area of potable water treatment, clarification of water with coagulating agents has been practiced from ancient times, using a variety of substances – most notably crushed seeds. The Egyptians as early as 2000 BC used almonds smeared around a vessel to clarify river water. Flocculation was achieved by plunging an arm into the vessel. The early Romans were also familiar with alum, although perhaps not for water treatment. However, its use as a coagulant by the Romans was mentioned in ca. 77 AD. By 1757, alum was used for coagulation in water treatment in England, and more formally for the treatment of public water supplies in 1881 (Faust & Aly, 1998).

In modern water treatment, coagulation and flocculation are still essential components of the overall suite of treatment processes. From an engineering survey of the quality of water treatment at over twenty operating water treatment plants, Cleasby *et al.* (1989) concluded: *Chemical pretreatment prior to filtration is more critical to success than the physical facilities at the plant.*

The need for a profound understanding of coagulation-flocculation processes is as important today as it has been in the past – even more so since many waters used for public supplies are less than pristine, and the requirements for the removal of particulates and organics has become increasingly stringent.

For example, since 1989, the regulatory limit in the United States for treated water turbidity has progressively reduced from 1.0 NTU in 1989, to 0.3 NTU today.

The committed goal of many water utilities is to consistently produce treated water turbidities of less than 0.1 NTU.

Several utilities have adopted micro- or ultrafiltration membranes for particulate removal in water treatment. In some cases, the raw water quality is such that efficient particulate removal is sufficient to produce acceptable treated water quality. However, in the majority of cases this is not the case, and natural organic material or other substances in solution must also be removed. In these cases, coagulation and flocculation play an important role for both treatment and for mitigating fouling on the membranes (Wiesner *et al.* 1989; Guigui *et al.* 2001; Kimura *et al.* 2005; Jang *et al.* 2005). The membrane processes are used for final separation of the precipitated and flocculated particles.

In some uncommon cases tighter membranes are used for the direct removal of macromolecules and material in solution. These membranes are nanofiltration and reverse osmosis membranes. However, because of the potential for fouling these membranes, careful pretreatment is required. Coagulation and flocculation are useful in this regard because the flocculated particles can reduce fouling by producing more permeable deposits (Gregory, 2006).

In the areas of wastewater treatment, coagulation and flocculation operations can be important. For very low phosphorus concentrations in effluents discharged to sensitive catchments, chemical precipitation of the residual phosphorus is often employed. Coagulation can achieve final concentrations in the nanogram per litre range. (Chemical phosphorus removal is included in Chapter 4).

Flocculation is also important for a number of other wastewater treatment processes. For example, the degree of agitation imposed upon biological flocs in activated sludge systems, and the relative degrees of floc formation and of floc breakup are important influences on suspended solids concentrations in secondary effluents. The concepts of floc aggregation and floc breakup influenced the development of flocculating feedwells in secondary clarifiers to optimize performance (Parker *et al.* 1971, 1983).

In overloaded wastewater treatment plants, one way of alleviating downstream processes is the practice of chemically enhanced primary treatment to reduce suspended solids and organic loads, using primary clarifiers. Here, coagulant dosage control and initial mixing into the volume of incoming wastewater are also important issues. (Wastewater treatment by coagulation is included in Chapter 4; and Chapter 6 is devoted to rapid mixing).

1.2 STABILITY AND DESTABILIZATION

The material in waters and wastewaters arise from land erosion, the dissolution of minerals, the decay of vegetation, and domestic and industrial waste discharges. For a given water or wastewater, such material may comprise suspended and/or dissolved organic and/or inorganic matter and numerous biological forms such as bacteria, algae, and viruses. As shown in Table 1.1, much of the

Table 1.1 Classification of particle sizes.

Particle Size (mm)	Classification	Examples	Total Surface Area (m^2/cm^3)	Time Required to Settle 100 mm if Specific Gravity = 2.65
10	Coarse dispersion	Gravel, coarse sand, mineral	6×10^{-4}	0.1 second
1	(visible to naked eye)	substances, precipitated and	6×10^{-3}	1 second
10^{-1}		flocculated particles, silt, macroplankton	6×10^{-2}	13 seconds
10^{-2}	Fine particulate	Mineral substances, precipitated	0.6	11 minutes
10^{-3}	dispersion (visible	and flocculated particles, silt,	6	20 hours
10^{-4}	under microscope)	bacteria, plankton and other organisms	60	80 days
10^{-5}	Colloidal dispersion	Mineral substances, hydrolysis	600	2 years
10^{-6}	(submicroscopic)	and precipitated products, macromolecules, biopolymers, viruses	6,000	20 years
$<10^{-6}$	Solution	Inorganic simple and complex ions, molecules and polymeric species, polyelectrolytes, organic molecules, undissociated solutes		

suspended material present in waters and wastewaters are in the microscopic to submicroscopic size range. Particles smaller than approximately 10^{-5} mm are referred to as colloids.

Colloidal material includes mineral substances, small aggregates of precipitated and flocculated matter, silt, bacteria, plankton, viruses, biopolymers, and macromolecules. Material smaller than approximately 10^{-6} mm are referred to as solutions. Such material includes inorganic simple and complex ions, molecules and polymeric species, polyelectrolytes, organic molecules, undissociated solutes, and small aggregates.

Particles of colloidal or lesser dimensions are able to retain a dispersed state because of certain inherent characteristics which promote their stability. The term stability refers to the capacity of such particles to remain as independent entities within a given dispersion.

All particles in a liquid medium possess certain properties associated with interfacial phenomena. Such phenomena (discussed in detail in Chapter 2) include the effects of the surface charge carried by particles and the degree of hydration (or solvation) of the particles' surface layers.

It is due to the overriding influence of surface phenomena that colloids, which possess a colossal surface area to mass ratio, have the ability to exist as stable dispersions. For illustrative purposes, Table 1.1 presents the total surface area for an original particle of diameter 10 mm split into (reformed) spheres of progressively smaller diameters. As the size of particles become progressively smaller, the total surface area becomes extremely large for a given total particle mass. Hence, it is apparent that for a given total mass, the smaller the particles the more predominant become the influences of phenomena associated with interfaces. Moreover, the lesser become the influences of gravity effects associated with mass.

With particulate material in the colloidal size range, it is apparent from Table 1.1 that, considering hydrodynamic effects alone, time scales of up to several years may be required for colloidal material to settle through a significant distance. Furthermore, because of a number of phenomena to which such small particles are subjected, a change in position arising from gravity effects could, in a statistical sense, involve geological time spans. Such phenomena, which include thermal convection currents within the dispersion medium, and molecular and ionic bombardment, serve to maintain particles in effectively permanent suspension.

It is feasible that colloidal particles may be removed from dispersions by methods other than those relying on gravity effects. Such methods include adsorption and physical straining. Figure 1.1 shows various filter types and the size spectrum of dispersed material to which they may be applicable. In the engineering sense, considering the volumes involved in water and wastewater treatment, direct adsorption or straining by means of a given filter type is not generally economically feasible, although exceptions may exist in certain cases.

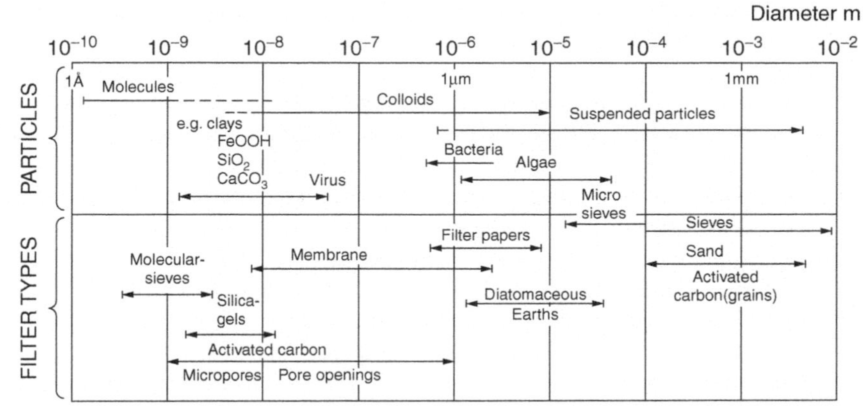

Figure 1.1 Size spectrum of waterborne particles and filter pores (after Stumm, 1977).

From the preceding comments, the value of processes which convert fine particulate, colloidal or dissolved material into a form whereby separation from the dispersion or solvent is practicable is self apparent. Such processes could either:

(a) Alter the surface properties of particulate material thereby increasing the adsorptivity of the particles to a given filter medium, or generating a tendency for aggregation of smaller particles into larger units;
(b) Precipitate dissolved material thereby creating particulate material for which separation by sedimentation and/or filtration is feasible.

Such a conversion of the stable state of a given dispersion or solution to an unstable state is termed destabilization.

The processes which are able to accomplish destabilization of waterborne particulate or dissolved material are coagulation and flocculation.

1.3 DEFINITIONS

Ideally, this section should commence with widely accepted definitions of coagulation and flocculation so as not to perpetrate confusion. However, when the abundant literature on the subject is consulted it becomes evident that a number of definitions have been variously adopted. Furthermore, it appears that at times the terms coagulation and flocculation have been used interchangeably. It is apparent that some of the confusion has been contributed by the water treatment industry itself, which has used all manner of definitions of its own. There have been attempts to remedy the lack of a universal set of definitions, (such as in La Mer *et al.* 1963) but in certain respects there still remained a certain unsatisfactoriness.

In an attempt to avoid ambiguity, and to follow *generally* accepted definitions, this book adopts the following definitions:

- Coagulation is the process whereby destabilization of a given suspension or solution is effected. That is, the function of coagulation is to overcome those factors which promote the stability of a given system.
- Flocculation is the process whereby destabilized particles, or particles formed as a result of destabilization, are induced to come together, make contact and thereby form large(r) agglomerates.
- Primary coagulant refers to that chemical or substance added to a given suspension or solution to effect destabilization.
- Flocculant (or filter) aids are those chemicals or substances added to a destabilized suspension or solution to accelerate the rate of flocculation or to strengthen flocs formed during flocculation.
- Sludge conditioners are those chemicals or substances added to a suspension to promote and strengthen flocs prior to thickening or dewatering processes.

Hence, for the case of a metal coagulant (for example aluminum sulphate or ferric chloride) added to, say, a clay suspension, the metal coagulant is being employed as a primary coagulant. If a cationic polyelectrolyte were to be added in place of the metal coagulant then one would refer to the polyelectrolyte as the primary coagulant. In both cases, the primary coagulant is added to destabilize the clay dispersion. If, in the case of metal coagulant addition to the clay suspension, an anionic polyelectrolyte, say, was subsequently added, in this instance the polyelectrolyte would be employed as a flocculant aid. (Chapter 3 is devoted to the types and characteristics of metal coagulants and polyelectrolytes; Chapter 4 to treatment with metal coagulants; and Chapter 5 to treatment with polymers).

Note that from the instant that the primary coagulant is added, flocculation begins immediately as a result of molecular bombardment. This flocculation stage, as will be explained further in Chapter 7, is termed perikinetic flocculation. However, for appreciable floc formation, some form of induced turbulence in the suspending medium is required. This stage, termed orthokinetic flocculation is effected by mechanical agitation or by inducing a tortuous flow pattern to the suspending medium by passage through granular filter media or around baffles installed in a flocculation tank.

After coagulation (during which destabilization has occurred) and flocculation (during which flocs of required size and characteristics have formed) the destabilized and flocculated stream is passed to a solid-liquid separation stage where flocs are either separated by gravity or are filtered from the liquid. The accumulated material may then be further treated to facilitate ultimate disposal. Polyelectrolytes are often added at this stage to enhance the dewatering characteristics of the material. In this case one refers to the polyelectrolyte as a sludge conditioner.

1.4 PERFORMANCE CRITERIA

Although the areas of application of the combined coagulation and flocculation processes vary widely, including those in the industrial and municipal wastewater treatment fields, by far the most extensive application has been in the area of potable water treatment.

With few exceptions, surface waters require some form of treatment before distribution as a potable supply complying with health standards. Such standards include physical and chemical quality criteria, and bacteriological, protozoan and virological criteria.

When evaluating a potential potable water supply, a thorough analysis is usually carried out to identify those parameters particularly requiring treatment. In general, the initial phase of the appropriate treatment study is to assess the degree of removal of the pertinent parameters by coagulation and flocculation. Here, a number of separate primary coagulant types, various combinations of coagulants, different flocculant aids and so on, will be added to the water under study through a range of imposed experimental conditions. Testing and control of coagulation and flocculation operations are dealt with in Chapter 8.

The criteria for the viability and/or optimum conditions for coagulation and flocculation applied to a potential or existing water supply, depend on the water constituents. In the majority of cases several constituents are of concern and, essentially, analysis of the water after coagulation and flocculation will be based on the removal of each. However, there will often be a particular constituent that may be taken as representative and which will be used as the criterion of process efficiency, especially for the initial phases of the viability study of a potential supply, or during routine control of an existing supply. In many instances, such constituents comprise those giving rise to turbidity, color, or UV absorption.

1.5 SUMMARY

Coagulation is the process whereby a given system may be transformed from a stable to an unstable state. The manifestation of destabilization depends on the particular system; For example, in the case of disperse suspensions or solutions, visible floc or precipitate formation occurs; In the case of highly concentrated suspensions, dewatering of the sludge mass occurs, and so on.

Flocculation is the process whereby the manifestation of destabilization is realized in practicable terms; In effect, flocculation accelerates floc formation, influences the physical characteristics of flocs formed (for example, their strength, size and density) and governs the final number concentration of destabilized particles. Flocculation, by whatever means it occurs, is a necessary adjunct to destabilization.

Considering the wide variety of material of which coagulation and flocculation effect removal in the area of water treatment alone, one may appreciate why these processes are veritable work-horses of whichever overall treatment scheme is applied.

The diversity of material present in a given water source, ranging from clay particles through various biological forms to substances in solution, each with a particular set of conditions for optimal removal, and the potential for relative variations in concentration, present considerable difficulties during both optimization and control of the coagulation and flocculation processes.

Optimization is influenced by the properties of the material present, characteristics of the different coagulants which may be applied, and the methods by which coagulants are added to the water. With a particular combination of such factors, the characteristics of flocs formed for a given set of flocculation parameters, including the magnitude of induced velocity gradients, their duration, and so on, also influence optimization. Control is influenced largely by changes in water quality and, in cases where a range of materials are present, the relative concentration of each at a given time.

The complications inherent in coagulation and flocculation, outlined above, are most effectively overcome with the support of a rational understanding of both the various phenomena involved and their interdependence. It is hoped that the following chapters contribute to making such a level of understanding more widespread.

1.6 REFERENCES

Cleasby J. L., Dharmarajah A. H., Sindt G. L. and Baumann E. R. (1989). Design and Operation Guidelines for Optimization of the High-Rate Filtration Process: Plant Survey Results. AWWA Research Foundation, September, Denver, CO, U.S.A.

Faust S. D. and Aly O. M. (1998). Chemistry of Water Treatment. Ann Arbor Press, MI, U.S.A.

Gregory J. (2006). Particles in Water. Properties and Processes. IWA Publishing/CRC Press, London, U.K.

Guigui C., Bonnelye V., Durand-Bourlier L., Rouch J. C. and Aptel P. (2001). Combination of coagualtion and ultrafiltration for drinking water production: impact of process configuration and module design. *Wat. Sci. Tech.: Water Supply*, **1**(5–6), 107–118.

Jang N.-Y., Watanabe Y. and Minegishi S. (2005). Performance of ultrafiltration membrane process combined with coagulation/sedimentation. *Wat. Sci. Tech.*, **51**(6–7), 209–219.

Kimura K., Hane Y. and Watanabe Y. (2005). Effect of pre-coagualtion on mitigating irreversible fouling during ultrafiltration of a surface water. *Wat. Sci. Tech.*, **51**(6–7), 93–100.

La Mer V. K. and Healy T. W. (1963). The role of filtration in investigating flocculation and re-dispersion of colloidal dispersions. *J. Phys. Chem.*, **67**, 2417.

Parker D. S., Kaufman W. J. and Jenkins D. (1971). Physical conditioning of activated sludge floc. *Jour. WPCF*, **43**(9), 1817–1833.

Parker D. S. (1983). Assessment of secondary clarification design concepts. *Jour. WPCF*, **55**(4), 349–359.

Stumm W. (1977). Chemical interactions in particle separation. *Env. Sci. and Tech.*, **11**(12), 1066–1070.

Wiesner M. R., Clark M. M. and Mallevialle J. (1989). Membrane filtration of coagulated suspensions. *Jour. Env. Eng., ASCE*, **115**(1), 20–40.

Chapter 2

Colloids and interfaces

2.1 INTRODUCTION

In the introductory chapter of this book it was shown that the substances in water and wastewater treatment to which the coagulation-flocculation process is applied may generally be described as being within or less than the colloidal size range. Coagulation is the phase in the overall process whereby the constituents of a given water are destabilized; and flocculation is the phase whereby destabilized particles, or particles formed during destabilization, are induced to collect into aggregates.

With colloids, the term *stability* describes the ability of individual particles to remain as separate entities or, in other words, to maintain a dispersed state. The stability of colloidal material arises from the predominance of forces associated with the solid-liquid interface. Principally, the interfacial forces promoting the stability of a colloidal dispersion arise from:

(1) the presence of a surface charge at the interface between colloid and liquid and
(2) hydration of surface layers of the colloid.

There are two general classes of colloids, which are given the names lyophobic and lyophilic. In water and wastewater treatment the terms are made more specific with the names hydrophobic and hydrophilic. Hydro- refers to the water phase in which the colloids exist and -phobic and -philic refer to the *degree* of affinity of the colloids for the liquid phase. That is, hydrophobic infers water repellency and hydrophilic infers a strong affinity for water molecules in the surface layers of the colloid. The amount of water bound by hydrophilic particles may in some cases be three to ten times their own dry mass. Examples of hydrophobic colloids are gold sols, silver halogenides, and nonhydrated metal oxides. Hydrophilic colloids include gelatin, starch, gums, proteins, and all biocolloids, for example, viruses and bacteria.

In water and wastewater treatment, it is usually difficult to classify a particular water as being either a hydrophobic or hydrophilic colloid dispersion; Both types may co-exist in a particular system and there may be a continuous transition from one state to the other, as occurs during treatment for example. A further difficulty in classification is that in some instances both hydrophobic and hydrophilic areas may exist on the colloids together.

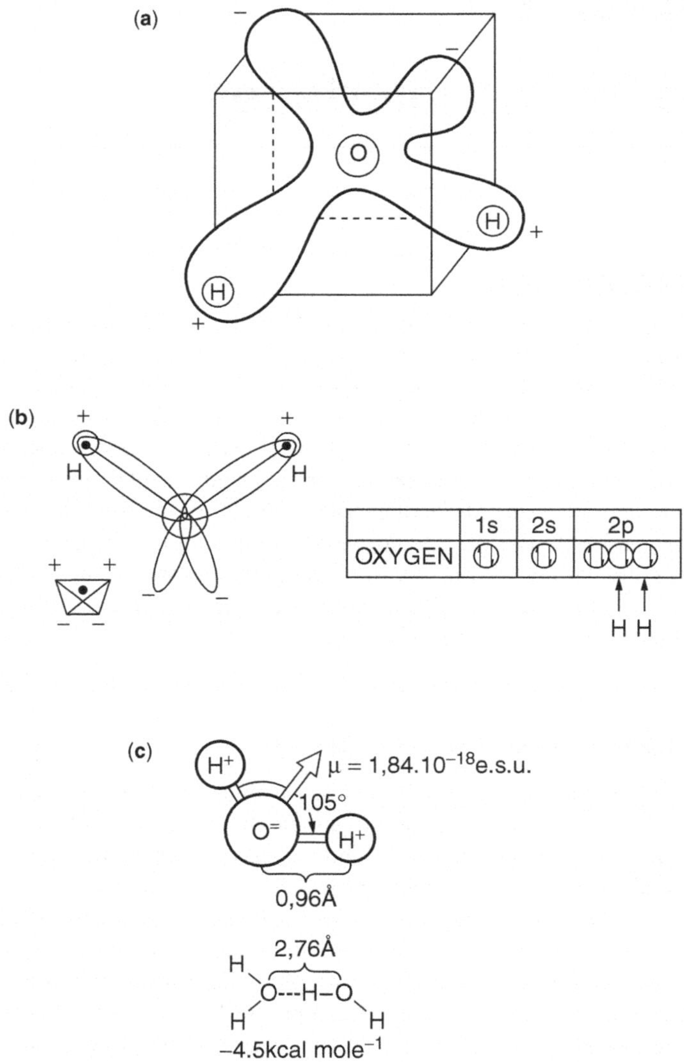

Figure 2.1 The water molecule: (a) Electronic cloud; (b) Electronic orbitals; (c) Structural features and the hydrogen bond. (Adapted from Horne, 1969).

Although the term hydrophobic infers a definite solid-water boundary, there is in reality a layer of water molecules, perhaps in some cases only a molecule thick, bound strongly enough to the colloid surface such that as the colloid moves in the water, the plane of shear is a water-water boundary. As shown in Figure 2.1, the water molecule exists as a dipole with a positive end resulting from the two bonded hydrogen atoms and a negative end from the two free pairs of electrons of the oxygen atom. The result of this charge distribution is a strong dipole moment of 1.84×10^{-18} esu. In the conditions encountered in water and wastewater treatment, colloidal material almost always possesses a net negative surface charge. The effects of these two factors: the dipolarity of the water molecule and the surface charge carried by the colloid are to:

(1) bind water molecules to the solid-liquid interface;
(2) orient water molecules, or alter the way in which molecules are arranged, in the vicinity of the charged surface.

The influence of the charged surface on the molecular arrangement of water is reflected in the different dielectric constants of successive water layers with distance from the charged surface, as shown in Figure 2.2.

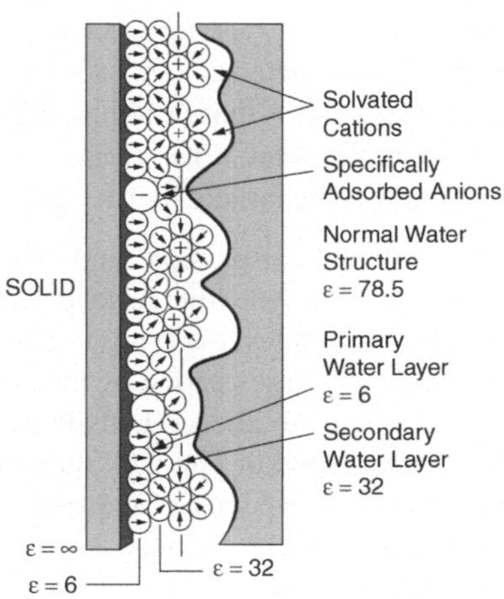

Figure 2.2 Position and orientation of water molecules in the electrical double layer. (Adapted from Bockris *et al.* 1963).

The mechanism of destabilization of hydrophobic colloids is often considered in terms of the following: a reduction of the effective surface charge; a reduction of the zone of influence of the surface charge; and a reduction of the number of adsorbed water molecules or extent of zone of ordered water structure. The effect of these factors is that adjacent colloid particles, heretofore remaining apart principally due to electrostatic repulsion and to a certain extent to the ordered water structure around them, are now able to approach close enough together so that London-Van der Waals forces of attraction can hold them together.

These latter forces are of an electrostatic character and arise from at least four components, of which by far the most dominant results from the interaction of two non-overlapping atoms; The electron cloud of each atom is attracted to the nucleus of the other with an intensity greater than the repulsions due to adjacent electron clouds and nuclei, such that the net result is an attractive force of short range.

Although the above mechanism certainly does occur and may in some instances be dominant during destabilization, there are other phenomena of as great an importance, such as adsorption of polymeric ions or molecules. Here, although electrostatic considerations may play a role during destabilization, in certain cases the dominant mechanism may be one of bridging by the adsorbed polymeric species between adjacent particles. A further mechanism, of importance with removal of dissolved substances from water during coagulation, is the formation of precipitates. This mechanism is probably also highly significant during destabilization of hydrophilic colloid dispersions. Again, electrostatic considerations arising from the surface charge carried by the colloids may play a part during destabilization, but it is likely that the extensive hydration of surface layers of hydrophilic particles has the effect of a physical barrier. Destabilization is brought about principally by dehydration and coordinative chemical reactions between the applied coagulant and functional groups on the colloid surface.

From the above discussion, it is seen that destabilization reactions of colloids in aqueous dispersion are complex and arise from several mechanisms. The predominant mechanism depends very largely on two factors:

(1) the nature of the colloidal dispersion and whether hydrophobic or hydrophilic particles are prevalent; the surface nature of the colloid, for example with hydrophilic particles, the types of functional groups present; and the intensity of surface charge carried by the colloid, and
(2) the type of coagulant added to the colloid dispersion, that is, whether coagulant species are uncharged or charged; the intensity of charge in the latter instance; the adsorptive capacity of the species; and the capacity for bridge formation between adjacent colloids.

For an understanding of destabilization phenomena of colloid dispersions with coagulant addition, the first steps are to consider the surface charge carried by hydrophobic colloids, the effect or zone of influence of the surface charge, the effect of simple coagulants on both the effective surface charge and its zone of influence, and destabilization of hydrophobic colloids with simple coagulants. These aspects are the subject of the present chapter.

Subsequent chapters will deal with, in turn, more complex coagulants with enhanced adsorptive capacity and facility for bridge formation; the destabilization of hydrophobic colloids by such coagulants; and the destabilization of hydrophilic colloids and substances in solution.

2.2 ORIGIN OF SURFACE CHARGE

Before discussing what effect the surface charge of a colloid has on its environment, it is necessary to state the origin of this surface charge.

There are three principal ways in which the surface charge may originate (Stumm, 1962):

The charge could arise from chemical reactions at the surface. Many solid surfaces contain functional groups which are readily ionizable, such as –OH; –COOH; and –OPO_3H_2. For example, with a bacterium, the overall charge may be visualized as resulting from ionization of functional amino and carboxyl groups:

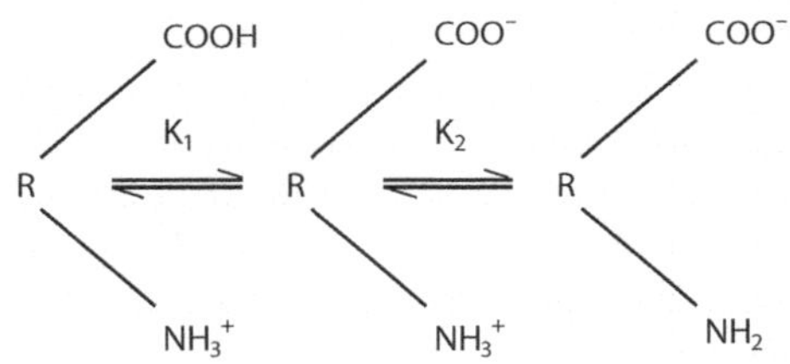

The charge of particles with such surfaces becomes dependent on the degree of ionization (proton transfer) and consequently on the pH of the surrounding liquid. At low pH values, a positively charged surface prevails while at higher pH, a negatively charged surface prevails. At some intermediate pH (the isoelectric point, or the zero point of charge, pH_{zpc}), the charge will be zero.

Surface charges can also originate by processes in which solutes become coordinatively bound to solid surfaces.

(1) Surface charge could be caused by lattice imperfections at the solid surface. For example, if in an array of solid SiO_2 tetrahedra a Si atom is replaced by an Al atom (Al having one electron less than Si), a negatively charged framework is established, viz:

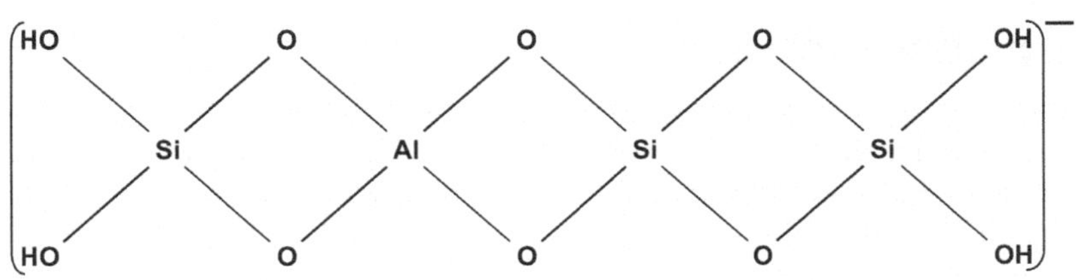

Clays are examples of such atomic substitutions. Surface charges of this type are independent of the pH of the liquid.

(2) Surface charges could also be established by the adsorption of ions. Adsorption of one type of ion on a surface can arise from London-van der Waals forces, and from hydrogen bonding.

2.3 EFFECT OF SURFACE CHARGE

The surface charge influences the distribution of nearby ions in the liquid. Ions of opposite charge (counter-ions) are attracted towards the surface and ions of like charge (co-ions) are repelled away from the surface. This, together with the mixing tendency of thermal motion and mutual ionic repulsion or attraction, leads to the formation of an *electrical double layer* made up of the charged surface and a neutralizing excess of counter-ions over co-ions distributed in a diffuse manner in the nearby liquid. The theory of the electrical double layer (see Sections 2.4, 2.5 and 2.6) deals with this distribution of ions and hence with the magnitude of the electric potentials which occur in the locality of the charged surface.

Electric potential is analogous to gravitational potential, but now one thinks of points in an electric field. For example, in the field around a negative charge, another negative charge moves from points near the charge to points further away. Therefore, points around the charge have an electric potential whose magnitude depends on their relative proximity to the charge.

The electrical double layer can be regarded generally as consisting of two regions: an inner region which probably includes water molecules and adsorbed (hydrated) ions, and a diffuse region in which ions are distributed according to the influence of electrical forces and random thermal motion.

Historically, the electrical double layer was first considered as a charged surface and a diffused region of ions around that surface. This led to a model developed by Gouy and Chapman. In addition to the diffuse region of ions, the second development was to consider a region in which ions are adsorbed and held to the surface. This led to the model developed by Stern.

The development of the electrical double layer presented here will first deal with Stern's consideration of an adsorbed layer of ions, then the treatment for the diffuse part of the double layer (due to Gouy and Chapman) and finally Stern's model for the complete double layer.

2.4 ADSORPTION

Since the acknowledgement of adsorption is inherent in Stern's model of the electrical double layer, it is worthwhile to briefly consider what is meant by it. Adsorption of a molecule (or ion) from solution on the surface of a solid involves a sequence of events (Stumm, 1962):

(1) Removal of the molecule from solution
(2) Removal of solvent from the solid surface
(3) Attachment of the molecule to the surface of the solid.

Because of thermal agitation, molecules and ions in solution are in a continuous state of movement. As a consequence, colloidal particles are haphazardly bombarded by molecules which gives rise to a random motion. This continuous random movement of colloidal particles is known as Brownian motion (after the British botanist Robert Brown who showed in 1827 that such motion was not confined to living matter, as originally supposed). Because of Brownian motion there will frequently be collisions and contact made between molecules and ions in solution and solid surfaces. When such contact occurs, three events are possible:

(1) The molecule or ion could be held to the solid surface by ionic, covalent, hydrogen, or dipolar bonding. These types of bonds may be classified in terms of bond energies, expressed in kcal/mole, as follows: ionic crystal bonds have bond energies greater than 150; covalent bonds between 50 and 100; hydrogen bonds 1 to 10; and polar less than 5. Adsorption arising from such mechanisms is referred to as chemical adsorption. Adsorption could also arise from (and is always enhanced by) London-van der Waals forces of attraction. Adsorption arising from this mechanism is referred to as physical adsorption.
(2) The molecule could be repelled from the surface by means of electrostatic repulsion before event (1) can take place.
(3) Even if conditions are favourable for attachment, the molecule could once more be installed in solution because of natural Brownian movement.

The affinity of the solid surface for some solutes (those molecules and ions in solution) may be less than the solid's affinity for the solvent itself (for example, water molecules, in the case of an aqueous solvent). Organic dipoles and large organic ions are preferentially accumulated at the solid surface primarily because their hydrocarbon parts have little affinity for water. Simple inorganic ions such as Na^+; Ca^{2+}; and Cl^- may remain in solution, even if they are specifically attracted to the solid surface, because they are readily hydrated in aqueous solutions. Less hydrated ions, such as Cs^+; $CuOH^+$; and many anions tend to seek positions at the solid surface to a larger extent than easily hydrated ions.

If adsorption does take place in solution, it follows that there is a net release of energy since the energy of attraction between the solid surface and the adsorbate must be greater than the energy of repulsion (if it is present) due to electrostatic effects. Furthermore, since adsorption involves the removal of a certain amount of solvent from the solid surface, then it follows that the energy required to remove the solvent molecules must be less than the energy released by the attachment of the adsorbate (see Figure 2.3).

This energy released as a consequence of adsorption is termed the *overall standard free energy of adsorption*, $\Delta\bar{G}^\circ$. It is equal to the sum of the energy released by the satisfaction of whatever forces of attraction prevail (such as chemical bonding and London-van der Waals forces) and either the energy released by the satisfaction of electrostatic forces of attraction, or the energy which has to be overcome because of electrostatic forces of repulsion.

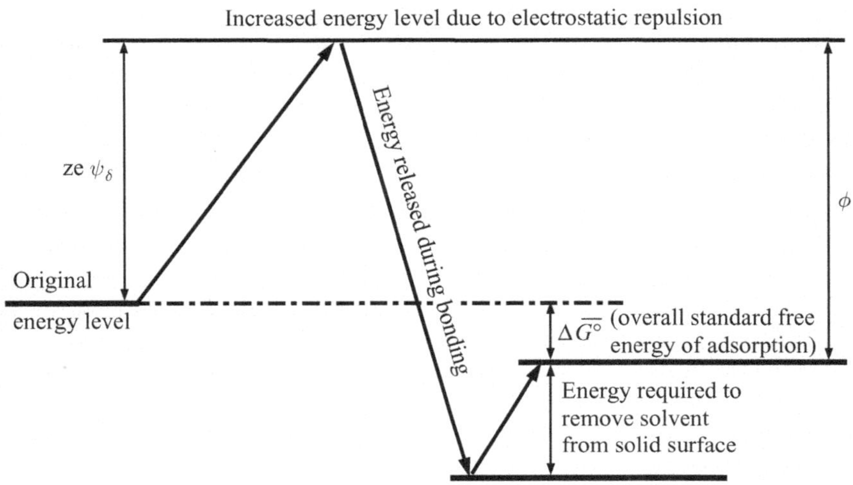

Figure 2.3 Scheme of energy changes during adsorption.

Hence, the overall standard free energy of adsorption is given by:

$$\Delta\bar{G}^{\circ} = \phi \pm ze\psi_{\delta} \tag{2.1}$$

where ϕ is the energy released during bonding and $ze\psi_{\delta}$ is the energy either released or adsorbed during electrostatic interactions (see later). If $ze\psi_{\delta}$ is of opposite sign to ϕ yet is smaller in magnitude, then adsorption takes place quite readily.

Experimentally, it is difficult to separate the energy of adsorption into its chemical and electrostatic components, but it is known that many organic ions of similar charge to the surface are readily adsorbed, indicating that the chemical adsorption energy released can easily be greater than the electrostatic energy of repulsion. Stumm and O'Melia (1968) present a numerical example illustrating the above points:

For the adsorption of a monovalent organic ion to a surface of similar charge and against a potential drop (ψ_{δ}) of 100 mV, the electrostatic term in Equation 2.1, $ze\psi_{\delta} = 2.3$ kcal/mole. The standard chemical adsorption energy for typical monovalent organic ions is within the range −2 to −8 kcal/mole, which demonstrates that the electrostatic contribution to adsorption can easily be less than the chemical contribution. From this, it is at least qualitatively appreciated that the addition of suitable counter-ions not only decreases the effective surface charge, but may produce a reversal of surface charge at higher concentrations.

The adsorptivity of a particular system is described by empirical expressions, known as isotherms, which relate the mass of solute adsorbed per unit mass of adsorbent to the concentration of solute remaining after adsorption. Two widely used isotherms are the Langmuir and Freundlich isotherms.

For systems exhibiting Langmuirian adsorption behaviour, the relevant parameters fit the equation:

$$\frac{C}{x/m} = a + \frac{C}{b} \tag{2.2}$$

where

C = concentration of solute (those molecules and ions in solution) which remains after adsorption (mg/l);
(x/m) = mass of solute adsorbed per unit mass of adsorbent;
a, b = constants for a particular system.

Systems of interest here which exhibit Langmuirian adsorption behaviour are the destabilization characteristics often observed with hydrolyzed iron (FeIII) coagulants, and with polyelectrolytes (Stumm & O'Melia, 1968; Black *et al.* 1965; Committee Report, 1971). For the latter case, a modified Langmuir type isotherm, known as the Simha-Frisch-Eirich (SFE) isotherm, has been used to describe polyelectrolyte adsorption behaviour. However, in many cases the Langmuir isotherm has been found to provide adequate representation of the phenomena.

For systems exhibiting Freundlichian adsorption behaviour, the appropriate equation is as follows:

$$\log\left(\frac{x}{m}\right) = \left(\frac{1}{a}\right)\log C + \log b \tag{2.3}$$

Freundlich adsorption behaviour has been reported for the destabilization characteristics of dilute clay suspensions by hydrolyzed aluminum sulphate (Black & Chen, 1967). The difference between the adsorption behaviour of aluminum coagulants compared to ferric coagulants and polyelectrolytes serves as a further indication of the complexity of the aluminum hydrolysis products formed.

2.5 INNER PART OF ELECTRICAL DOUBLE LAYER

The treatment of the diffuse part of the double layer proposed by Gouy and Chapman (see later) is based on an assumption of point charges in the electrolyte solution. However, Stern recognized that the finite size of the ions will limit the inner boundary of the diffuse part of the double layer since the centre of an ion can only approach the surface to within its (hydrated) radius without becoming specifically adsorbed. Stern proposed a model in which the double layer is divided into two parts separated by a plane (the Stern Plane) located at about a hydrated ion radius from the surface (see Figures 2.4 and 2.5).

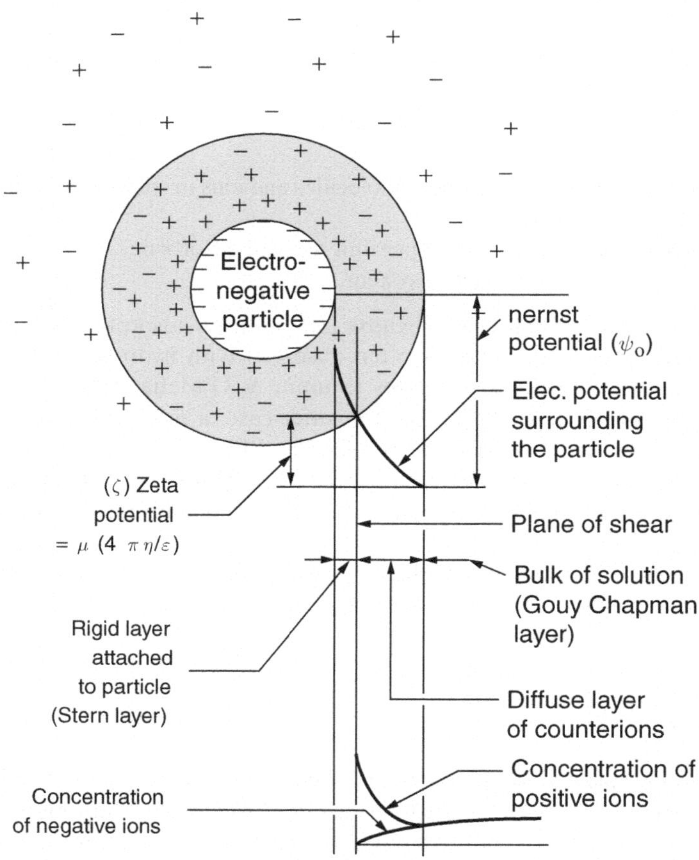

Figure 2.4 Conceptual representation of the electrical double layer (from Priesing, 1962).

The adsorbed ions may be dehydrated in the direction of the solid surface so their centres will lie somewhere between the solid surface and the Stern plane. Ions whose centres are located beyond the Stern plane are considered to be part of the diffuse layer for which the Gouy-Chapman treatment (see next section) is considered applicable. The electric potential at the surface of the particle is given by ψ_o (or the Nernst potential) and that of the Stern plane (the Stern potential) is given by ψ_δ. The potential in the diffuse layer decays exponentially from ψ_δ at the Stern plane to zero at infinity (see later).

When specific adsorption takes place (that is, adsorption arising from electrostatic and/or van der Waals forces) counter-ion adsorption generally predominates over co-ion adsorption; such a situation is depicted in Figure 2.5.

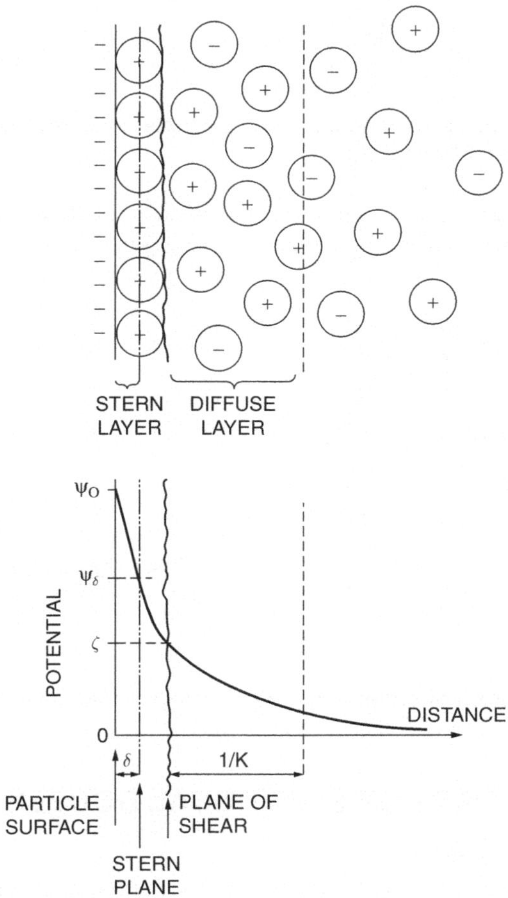

Figure 2.5 Counter-ion adsorption generally predominates over co-ion adsorption.

It is possible, especially with polyvalent or surface active counter-ions for reversal of charge to take place within the Stern layer, that is, for ψ_o and ψ_δ to have opposite signs (see Figure 2.6).

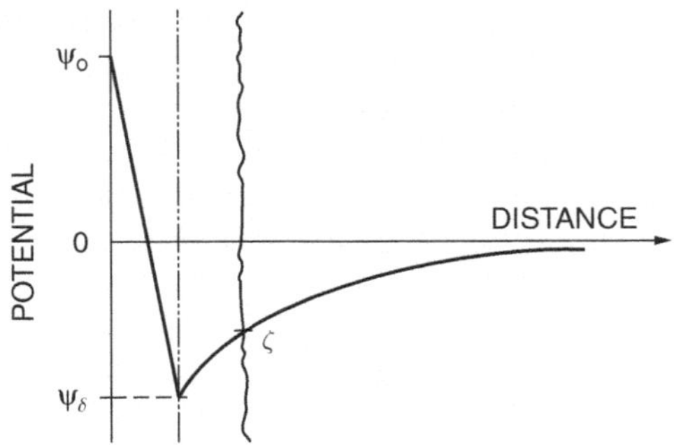

Figure 2.6 With polyvalent or surface-active counter-ions, reversal of charge may take place within the Stern layer.

Adsorption of surface active co-ions could create a situation in which ψ_o has the same sign as ψ_δ and is greater in magnitude (see Figure 2.7).

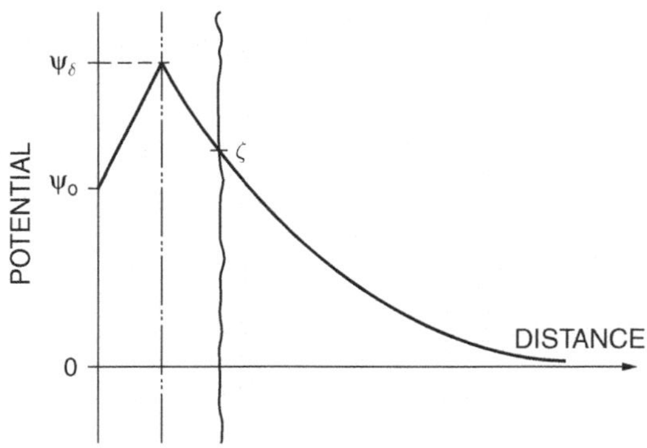

Figure 2.7 Adsorption of surface-active co-ions could increase effective surface charge.

2.6 DIFFUSE PART OF ELECTRICAL DOUBLE LAYER

2.6.1 Assumptions

The Gouy-Chapman treatment for the diffuse part of the double layer is based on the following assumptions (Shaw, 1970):

- The surface is assumed to be flat, of infinite extent, and uniformly charged;
- The ions in the diffuse part of the double layer are assumed to be point charges distributed according to the Maxwell-Boltzmann distribution. This distribution is based on probability theory and leads to an expression for the number of molecules (ions) in a gas (liquid) at equilibrium possessing

a certain specified energy. It considers a system made up of n_o similar molecules which do not attract or repel each other.

The original expression for the distribution was:

$$n_i = n_o \cdot \exp\left(-\frac{E_i}{kT}\right) \tag{2.4}$$

where

n_i is the number of molecules with energy E_i at absolute temperature, T.
kT is Boltzmann's constant times absolute temperature and is a measure of the thermal kinetic energy of the particles. It is derived from entropy and probability considerations where the entropy of a system, S is related to the probability of that system, W by the equation:

$$S = k \cdot \ln W$$

or

$$W = \exp\left(\frac{S}{k}\right) \tag{2.5}$$

W was defined by Boltzmann to be the ratio of the probability of an actual state to one having the same total energy and volume, in which the molecules are completely ordered.

The liquid is assumed to influence the double layer only through its permittivity, which is assumed to have the same value throughout the diffuse part of the double layer. The permittivity may be conceptually visualized by considering it to be the inverse of the resistance to electron flow; that is, the lower the permittivity, the greater is the resistance to flow experienced by electrons.

2.6.2 Distribution of potential with distance from the charged surface

Let ψ_δ be the electric potential at the Stern plane, and ψ the electric potential at a distance x in the solution from the Stern plane.

Taking the solid surface to be negatively charged and that, even with counter-ion adsorption, charge reversal has not taken place (which is the usual case), and an excess of counter-ions over co-ions exists in the diffuse layer, the numbers of positive and negative ions are determined by applying the Maxwell-Boltzmann distribution:

$$n_+ = n_o \cdot \exp\left(\frac{ze\psi}{kT}\right) \tag{2.6}$$

and

$$n_- = n_o \cdot \exp\left(-\frac{ze\psi}{kT}\right) \tag{2.7}$$

Where n_+ and n_- are the respective numbers of positive and negative ions per unit volume at points where the potential is ψ, and n_o is the concentration of ions at infinity (that is, at infinity, $n_+ = n_- = n_o$, as shown schematically in Figure 2.5);

z = valency of ion;
e = charge of electron;
hence, ze = charge of ion.

The potential ψ at a point in an electric field (or in the diffuse layer) is defined to be the work done in moving a unit charge (of the same sign as that of the surface) from infinity to that point. If a charge of ze is moved from infinity to that point it follows that the amount of work done on it is $ze\psi$. That is, the charge is capable of performing $ze\psi$ units of work, which is the potential energy possessed by the charge at that point.

The charge density ρ, at points where the potential is ψ, is given by:

$$\rho = ze(n_+ - n_-)$$

$$= zen_0\left[\exp\left(\frac{ze\psi}{kT}\right) - \exp\left(-\frac{ze\psi}{kT}\right)\right] \tag{2.8}$$

$$\therefore \rho = +2zen_0 \cdot \sinh\left(\frac{ze\psi}{kT}\right)$$

where $\sinh x = \left(\dfrac{e^x - e^{-x}}{2}\right)$.

ρ is related to ψ by Poisson's equation and, for a *flat* double layer takes the form:

$$\frac{d^2\psi}{dx^2} = -\frac{\rho}{\varepsilon} \tag{2.9}$$

where ε = permittivity.

(Poisson's equation, derived from vector analysis, is based on Coulomb's law).

From equations (2.8) and (2.9):

$$\frac{d^2\psi}{dx^2} = -2\left(\frac{zen_0}{\varepsilon}\right) \cdot \sinh\left(\frac{ze\psi}{kT}\right) \tag{2.10}$$

The boundary conditions for the latter equation are:

when $x = 0$: $\psi = \psi_\delta$

and

when $x = \infty$: $\psi = 0$ and $\dfrac{d\psi}{dx} = 0.$

The solution to the above equation is:

$$\psi = \frac{2kT}{ze} \cdot \ln\left[\frac{1 + \gamma \cdot \exp(-Kx)}{1 - \gamma \cdot \exp(-Kx)}\right] \tag{2.11}$$

where

$$\gamma = \frac{\exp\left(\dfrac{ze\psi_\delta}{2kT}\right) - 1}{\exp\left(\dfrac{ze\psi_\delta}{2kT}\right) + 1} \tag{2.11a}$$

and

$$K = \sqrt{\frac{2z^2e^2n_0}{\varepsilon kT}} \tag{2.11b}$$

At low potentials, where $\psi << 25\,\text{mV}$ $\left(\text{and } \dfrac{ze\psi}{2kT} << 1\right)$ one may use the approximation that $e^x = 1 + x$ $\left(\text{or } \exp\left(\dfrac{ze\psi}{2kT}\right) = 1 + \dfrac{ze\psi}{2kT}\right)$.

Equation (2.11) thus simplifies to:

$$\psi = \psi_\delta \cdot \exp(-Kx) \tag{2.12}$$

Equation (2.12) shows that at low potentials, the potential decreases exponentially with distance from the Stern plane. Close to the charged surface where the potential is likely to be relatively high, the above approximation cannot be used and the potential increases at a rate greater than exponential.

2.6.3 *Thickness* of double layer

The Stern potential ψ_δ can be related to the net space charge density σ in the diffuse part of the double layer as follows:

$$\sigma = \int_0^\infty \rho dx$$

Substituting equation (2.8):

$$\sigma = \int_0^\infty 2zen_0 \cdot \sinh\frac{ze\psi}{kT} dx \tag{2.13}$$

and using the approximation for low potentials:

$$\sigma = \frac{2z^2e^2n_0}{kT}\int_0^\infty \psi dx \tag{2.14}$$

Substituting from equations (2.12) and (2.11b):

$$\sigma = K\varepsilon\psi_\delta \tag{2.15}$$

or

$$\psi_\delta = \frac{\sigma}{K\varepsilon} \tag{2.16}$$

From equation (2.16) the surface potential ψ_δ (at low potentials) is dependent on the net space charge density, σ; the permittivity of the diffuse layer, ε; and a quantity $1/K$ whose units are those of length. The significance of the quantity $1/K$ is conceptually demonstrated as follows:

Consider a parallel plate condenser whose plates are of unit area and separated by a distance d. The potential difference between the plates is V and the space between the plates is of permittivity ε, as shown in Figure 2.8.

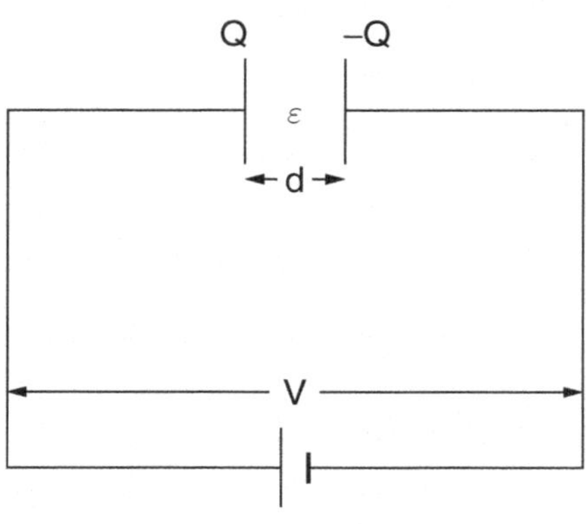

Figure 2.8 Parallel plate condenser analogy with electrical double layer.

The capacitance C of the condenser is defined to be the ratio of the charge on either plate to the potential difference V between them: $C = Q/V$.

The permittivity is defined as:

$$\varepsilon = \frac{\text{Electric flux density}}{\text{Electric force}} = \frac{Q/1}{V/d} = \frac{Q}{V}\cdot\frac{d}{1} = C\cdot d \quad \text{(for plates of unit area)}$$

Hence, for a parallel plate condenser,

$$C = \frac{\varepsilon}{d} \tag{2.17}$$

Consider now the diffuse part of the double layer where the potential ranges from ψ_δ at the surface to zero at infinity:

The capacitance per unit area of the diffuse layer can be written as:

$$C = \frac{\text{'}Q\text{'}}{V} = \frac{\sigma \cdot 1}{\psi_\delta - 0} = \frac{\sigma}{\psi_\delta}$$

From equation (2.16), one can write that:

$$\frac{\sigma}{\psi_\delta}(=\text{capacitance}) = K \cdot \varepsilon \left(\text{or} = \frac{\varepsilon}{(1/K)} \right) \tag{2.18}$$

Comparing equations (2.17) and (2.18) leads one to consider the diffuse layer as analogous to a parallel plate condenser with plates separated by a distance of $1/K$. For this reason, the quantity $1/K$, which is known as the Debye-Huckel length, is often referred to as the *thickness* of the double layer. However, it should be noted that in reality, the diffuse layer extends to infinity.

2.6.4 Effect of ionic strength on double layer

As shown below, the thickness of the double layer is dependent on the ionic strength of the electrolyte solution.

From equation (2.11b):

$$K = \sqrt{\frac{2z^2e^2n_0}{\varepsilon kT}}$$

Noting that $n_0 = N \cdot C_0/10^3$

where

N = Avogadro's Number

= number of ions per mole of electrolyte solution (=6.03 × 10^{23}/mole)

and C_0 = concentration of electrolyte solution, moles/l

$$\text{then } K = \sqrt{\frac{2z^2e^2}{\varepsilon kT} \cdot \frac{NC_0}{10^3}} \tag{2.19}$$

Noting that the ionic strength, I is given by $I = \dfrac{C_0z^2}{2}$

$$\text{then } K = \sqrt{\frac{4e^2N \cdot I}{\varepsilon kT \times 10^3}} \quad \text{or} \quad \frac{1}{K} = \sqrt{\frac{\varepsilon kT \times 10^3}{4e^2N \cdot I}} \tag{2.20}$$

From equation (2.20) it is seen that increasing the ionic strength results in a decrease in the thickness of the double layer.

For illustrative purposes, the thickness of the diffuse layer is quantified as follows, for water at 20°C and with ionic strength I:

$$\frac{1}{K} = \frac{2.0 \times 10^{-8}}{\sqrt{I}} \tag{2.21}$$

where the following values were assumed:

e = charge of electron = 1.6×10^{-19} coulombs;
ε = permittivity = 7.08×10^{-12} coulombs/V/cm
kT = Boltzmann constant × absolute temperature = 0.41×10^{-20}V coulombs at 20°C.

2.6.5 Effect of nature of counter ions

The Gouy-Chapman model also shows that multivalent counter-ions are concentrated in the double layer to a much greater extent than monovalent ions. This is demonstrated in the expressions below:

From equation (2.6):

$$n_+ = n_0 \cdot \exp\left(\frac{ze\psi}{kT}\right)$$

For monovalent ions, the expression can be written as:

$$n_+ = n_0 e^A$$

For divalent ions:

$$n_+ = n_0 e^{2A}$$

For trivalent ions:

$$n_+ = n_0 e^{3A}$$

2.7 STERN'S MODEL OF COMPLETE DOUBLE LAYER

Stern's complete model for the electrical double layer is based on considerations of electrical neutrality for the whole system.

If σ_0 is the charge density at the particle surface, then treating the Stern layer as a condenser of thickness δ and with permittivity ε':

$$\sigma_0 = \frac{\varepsilon'}{\delta} \cdot (\psi_0 - \psi_\delta) \tag{2.22}$$

Stern assumed that a Langmuir type adsorption isotherm could be used to describe the equilibrium between ions adsorbed in the Stern layer and those in the diffuse part of the double layer. Considering only the adsorption of counter-ions, the charge density of the Stern layer is given by the expression:

$$\sigma_\delta = \frac{\sigma_m}{1 + \left(\frac{N}{n_0 V_m}\right) \cdot \exp\left(\frac{ze\psi_\delta + \phi}{kT}\right)} \tag{2.23}$$

where

σ_δ = the charge density of the Stern layer;
σ_m = the charge density of the Stern layer corresponding to a complete monolayer of counter-ions;
N = Avogadro's Number;
V_m = the molar volume of the solvent;
ϕ = the energy (per mole) of adsorption due to chemical forces;
$ze\psi_\delta$ = the energy (per mole) of adsorption due to electrostatic forces.

If σ is the charge density of the diffuse part of the double layer, then:

$\sigma = K\varepsilon\psi_\delta$ (see equation 2.15).

For overall electrical neutrality throughout the double layer, then:

$$\sigma_0 + \sigma_\delta + \sigma = 0$$

and

$$\frac{\varepsilon'}{\delta} \cdot (\psi_0 - \psi_\delta) + \frac{\sigma_m}{1 + \left(\frac{N}{n_0 V_m}\right) \cdot \exp\left(\frac{ze\psi_\delta + \phi}{kT}\right)} + K\varepsilon\psi_\delta = 0 \tag{2.24}$$

Equation (2.24) is the Stern model for the double layer. It contains a number of unknown quantities of which the most significant is the Stern potential ψ_δ. However, although ψ_δ cannot be directly measured, a close approximation may be obtained by determining the potential at the plane of shear, designated the zeta potential, by electrokinetic measurements. (See Chapter 8).

2.8 COLLOID STABILITY IN TERMS OF THE DOUBLE LAYER

2.8.1 Energy of interaction between particles

Verwey and Overbeek, and Derjaguin and Landau independently developed a quantitative theory based on the Gouy-Chapman model in which the stability of hydrophobic colloids is treated in terms of the energy changes which take place when particles approach one another. The theory (known as the VODL theory) involves estimations of the energy of attraction (London-van der Waals forces) and the energy of repulsion (overlapping of electrical double layers) in terms of the distance between particles.

The general expression developed for the repulsive energy V_R is very complex. However, for the condition of low ψ_δ and low values of $1/e^{KH}$ (defined below) a relatively simple expression was developed by Verwey and Overbeek (1948) which gives a good approximation for V_R:

$$V_R = \frac{B\varepsilon k^2 T^2 a \gamma^2}{z^2 e^{KH}} \tag{2.25}$$

where

B = a constant equal to $\dfrac{4.36 \times 10^{20}}{A^2 S^2}$

A = Hamaker's constant (depends on the number of atoms per unit volume and the polarisability of the material of the particles)

$$S = \frac{R}{a}$$

R = distance between centres of spheres

a = radius of (assumed) spherically shaped particles

ε = the permittivity (=7.08 × 10^{-12} coulombs/V/cm)

kT = Boltzmann constant times absolute temperature (=0.417.08 × 10^{-20} V coulombs at 20°C)

$$\gamma = \frac{\exp\left(\dfrac{ze\psi_\delta}{2kT}\right) - 1}{\exp\left(\dfrac{ze\psi_\delta}{2kT}\right) + 1}$$

z = valency of counter-ions

$\dfrac{1}{K}$ = thickness of diffuse part of double layer

H = shortest distance between spheres.

The London-van der Waals attraction energy between two molecules is of very short range and is assumed to vary inversely with the sixth power of the intermolecular distance (Hamaker, 1936, 1937, and 1937; Overbeek, 1977). For an assembly of molecules, these forces are approximately additive and the attractive energy between two particles can be computed by summing the attractions between all pairs of molecules within the particles.

For the case of two identical spheres of radius a with the shortest distance between them equal to H (and $H \ll a$) then the London-van der Waals energy of attraction is approximated by (Overbeek, 1977):

$$V_A = \frac{-Aa}{12H} \tag{2.26}$$

There are certain difficulties involved in computing V_A, of which the greatest difficulty is in the measurement of the Hamaker constant, A.

The total energy of interaction between two charged particles is obtained by summation of the attraction and repulsive energies as shown in Figure 2.9.

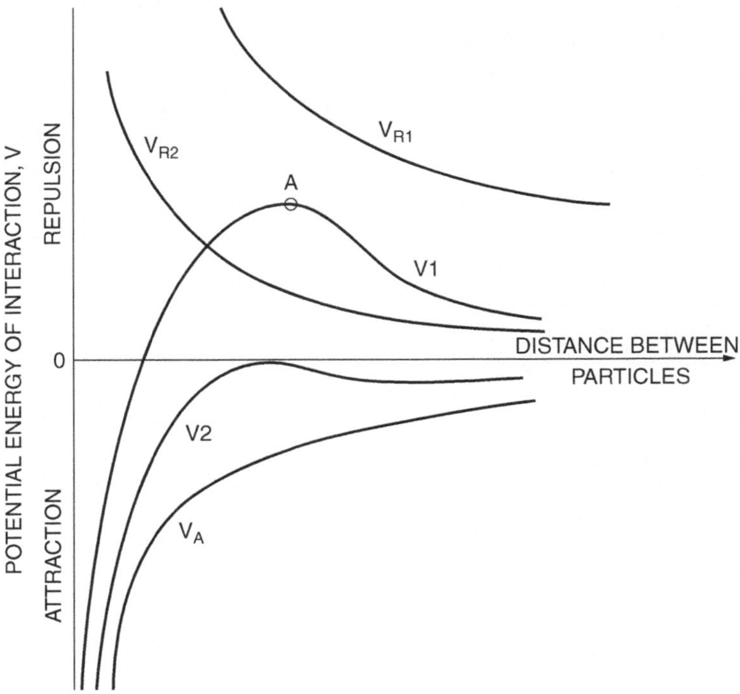

Figure 2.9 Potential energy of interaction between two particles.

The general characteristics of the resulting potential energy-distance curve can be deduced from the properties of the two forces. The repulsion energy is an exponential function of the distance between the particles with a range of the order of the thickness of the electrical double layer, $1/K$. The attraction energy decreases as an inverse power of the distance between the particles. Consequently London-van der Waals attraction will predominate at small distances. At intermediate distances, double layer repulsion may predominate, depending on the relative values of V_A and V_R. Note that, although not shown on the potential energy curve, at very small distances there will be repulsion due to overlapping of electron clouds, known as the *Born* repulsion.

If the potential energy maximum (point A on Figure 2.9) is large compared with the thermal kinetic energy of the particles, kT, the system will be stable. If the kinetic energy of the particles is large enough to surmount the potential energy *hump*, then the particles will coalesce. The height of this energy barrier depends upon the magnitude of ψ_δ and upon the range of the repulsive forces, which is represented by $1/K$.

2.8.2 Theoretical optimal concentration of electrolyte required for destabilization

An expression for the concentration of an indifferent (non-adsorbing) electrolyte required to destabilize a dispersion can be derived by assuming that a potential energy curve such as V_2 in Figure 2.9 can be taken to represent the conditions for optimum destabilization:

From the curve there is a certain distance, H at which $V = 0$ and $dV/dH = 0$. At this point, $V = V_R + V_A = 0$;

therefore, from equations (2.25) and (2.26):

$$V = \frac{B\varepsilon k^2T^2a\gamma^2}{z^2e^{KH}} - \frac{Aa}{12H} = 0$$

also, $$\frac{dV}{dH} = \frac{dV_R}{dH} + \frac{dV_A}{dH} = -KV_R - \frac{V_A}{H} = 0$$

from which $KHV_R + V_A = 0$.

Since $V_R + V_A = 0$

then $KH = 1$

giving $$\frac{B\varepsilon k^2T^2a\gamma^2}{z^2e^1} - \frac{AaK}{12} = 0.$$

and $$K = 4.415\frac{B\varepsilon(kT\gamma)^2}{Az^2}.$$

Substituting the expression found previously for K in equation (2.19):

$$K = \sqrt{\frac{2z^2e^2}{\varepsilon kT} \cdot \frac{NC_0}{10^3}}$$

and rearranging, one can deduce that C_0, the molar concentration of electrolyte solution required to destabilize a colloidal dispersion, is given by:

$$C_0 = \frac{9.75 \times 10^3 B^2 \varepsilon^3 (kT)^5 \gamma^4}{(Ae)^2 N z^6} = \frac{\text{'constant'}}{z^6} \tag{2.27}$$

2.8.3 Schulze-Hardy rule

An interesting point is illustrated by equation (2.27): Optimum concentrations of *indifferent* electrolytes yielding counter-ions with valency, $z = 1$, 2, and 3, for example, should be in the ratio $1/1^6 : 1/2^6 : 1/3^6$ or approximately 800 : 12 : 1. In fact, this is in reasonable agreement with experimental evidence on the coagulation of hydrophobic colloids by non-adsorbable ions. The phenomenon is known as the Schulze-Hardy Rule after the work of H.J. Schulze and W.B. Hardy in 1883 and 1900, respectively.

It should be noted that the above expression for the optimal concentration of indifferent electrolyte relies on a model for double layer depression according to the Gouy-Chapman treatment. It does not take into account the extensive adsorption likely during most instances of colloid destabilization with metal coagulants. For this reason, and because of the simplifications required in developing equations (2.25) through (2.27), the $1/z^6$ rule is inapplicable in most cases.

However, there are instances where the Schulze-Hardy rule is observed. Such cases would include those where highly charged polynuclear coagulant hydrolysis species are present at low concentrations in solution (see Chapter 3). At sufficiently low concentrations, adsorption may be negligible and the destabilization mechanism is one of double layer depression analogous to that of non-adsorbable ions. At higher coagulant concentrations, adsorption increases and the Schulze-Hardy rule becomes inapplicable (Matijevic, 1973).

2.9 ELECTROKINETIC MEASUREMENTS

It was stated in Section 2.6 that a difficulty in quantifying a model of the electrical double layer is measurement of the Stern potential ψ_δ. However, an approximation may be obtained by measuring the potential at the plane of shear between moving particle and surrounding liquid. This potential is known as the zeta potential. The exact location of the shear plane is not known since it depends on the adsorbed ions in the Stern layer and the degree of hydration. It is assumed that the shear plane is located at a small distance further out from the surface than the Stern plane and that the zeta potential is, in general, marginally smaller in magnitude than ψ_δ.

Zeta potential measurements are useful, not only in the testing of double layer theories (particularly in relation to the stability of lyophobic sols) but also in the following applications:

- *Colloid stability*: This involves calculation of the zeta potential and thereby an approximation of the Stern potential, ψ_δ;
- *Ion adsorption studies*;
- *Characterization of particle surfaces*: For example, studies of the effects of pH, surface active agents and enzymes on zeta potential yield valuable information regarding the nature of particle surfaces. This technique has proved useful in characterizing the surfaces of bacterial cells.

The usefulness of zeta potential measurements in the control of coagulation – flocculation processes is largely dependent on whether the destabilization mechanism is predominantly due to charge effects, in which case control to a given zeta potential may be feasible. This is discussed further in Chapter 8.

2.10 REFERENCES

Black A. P., Birkner F. B. and Morgan J. J. (1965). Destabilization of dilute clay suspensions with labelled polymers. *Jour. AWWA*, **57**, 1547–1560.

Black A. P. and Chen C. (1967). Electrokinetic behaviour of aluminum species in dilute dispersed kaolinite systems. *Jour. AWWA*, **59**, 1173–1183.

Bockris J. O., Devanathan M. A. V. and Muller K. (1963). *Proc. Royal Soc. (London)*, **274A**, 55.

Committee Report (1971). State of the art of coagulation. *Jour. AWWA*, **63**, 99–108.

Hamaker H. C. (1936). *Rec. Trav. Chim.*, **55**, 1015.

Hamaker H. C. (1937). *Rec. Trav. Chim.*, **56**(3), 727; *Physica*, **4**, 1058.

Hamaker H. C. (1937). *Physica*, **4**, 1058.

Horne R. A. (1969). Marine Chemistry. Wiley-Interscience, J. Wiley and Sons Inc.

Matijevic E. (1973). Colloid stability and complex chemistry. *Jour. Coll. Interface Sci.*, **43**(2), 217–245.

Overbeek J. Th. G. (1977). Recent developments in the understanding of colloid stability. *Jour. Coll. Interface Sci.*, **58**(2), 408–422.

Priesing C. P. (1962). A theory of coagulation useful for design. *Ind. Eng. Chem.*, **54**(8), 38–45.

Shaw D. J. (1970). Introduction to Colloid and Surface Chemistry, Butterworths.

Stumm W. and Morgan J. J. (1962). Chemical aspects of coagulation. *Jour. AWWA*, **54**(8), 971–991.

Stumm W. and O'Melia C. R. (1968). Stoichiometry of coagulation. *Jour. AWWA*, **60**, 514–539.

TeKippe R. J. and Ham R. K. (1970). Coagulation testing: a comparison of techniques – Part 1, *Jour. AWWA*, **62**, 594–602; – Part 2, *Jour. AWWA*, **62**, 620–628.

Verwey E. J. W. and Overbeek J. Th. G. (1948). Theory of the Stability of Lyophobic Colloids. Elsevier, Amsterdam.

Chapter 3

Coagulants

3.1 INTRODUCTION

In the previous chapter, properties of colloids in suspension were dealt with: The effect of the surface charge carried by the colloids was presented and the stability of hydrophobic colloids in terms of the electrical double layer produced by the charged interfaces was discussed. Destabilization of such particles by simple electrolytes was dealt with in terms of a model describing repression of the electrical double layer. A very important point emphasized in Chapter 2 was that there are other mechanisms besides charge neutralization involved in colloid destabilization which, in some cases, may be dominant during destabilization. These mechanisms, to be described in later chapters, are promoted by coagulants other than those which may be described as simple electrolytes. Such coagulants include those given the general names *metal coagulants* and *polymers*.

The objective of this chapter is to present a description of commonly used coagulants and polymers and, for each, to discuss the reactions occurring from the instant of coagulant addition to the suspension, to the onset of destabilization. Mechanisms of destabilization effected by such coagulants, for different substances in suspension or solution, will be presented in later chapters.

3.2 METAL COAGULANTS

3.2.1 Commonly used metal coagulants

The commonly used metal coagulants fall into two general categories: those based on aluminum and those based on iron. The aluminum coagulants include aluminum sulfate, aluminum chloride, sodium aluminate, aluminum chlorohydrate, polyaluminum chloride, polyaluminum sulfate chloride, polyaluminum silicate chloride, and forms of polyaluminum chloride with organic polymers. The iron coagulants include ferric sulfate, ferrous sulfate, ferric chloride, ferric chloride sulfate, polyferric sulfate, and ferric salts with organic polymers. Other chemicals used as coagulants include hydrated lime and magnesium carbonate.

The popularity of aluminum and iron coagulants arises not only from their effectiveness as coagulants but also from their ready availability and relatively low cost. The efficacy of these coagulants arises principally from their ability to form multi-charged polynuclear complexes in solution with enhanced adsorption characteristics. The nature of the complexes formed may be controlled by the pH of the system.

The following presents a summary of some of the relevant chemical and physical properties of aluminum and iron coagulants. It should be noted that the chemical equations presented are only intended to portray a general idea of the reactions which occur. The hydrolysis reactions of iron and aluminum salts are far more complicated – as will be seen in later sections – than the following suggests. Furthermore, in many cases of destabilization, the reactions do not necessarily proceed to equilibrium, as shown here, before destabilization has occurred.

3.2.1.1 Aluminum sulfate

This is probably the most widely used coagulant, and has been in use for water treatment for several centuries. It is manufactured from the digestion of bauxite ores with sulfuric acid. The quantities of bauxite are just over the stoichiometric amounts needed to combine with the acid, so that in the final product no free acid is present. Evaporation of water in the process results in the dry product having the approximate formula $Al_2(SO_4)_3 \cdot 14H_2O$ and with an aluminum content ranging from 7.4% to 9.5% (usually close to 9% as Al) by mass. (Note: traditionally, alum strength and purity were expressed in terms of equivalent alumina, or aluminum oxide, Al_2O_3 content; 9% as Al is equivalent to 17% as Al_2O_3).

Liquid alum is usually supplied in 20 tonne tanker loads or, where appropriate, by railroad tanker. It is produced by withdrawal before evaporation proceeds as far as the dry product, and is standardized at approximately 4.2% as Al, which is approximately equivalent to a 50% solution of the dry product.

The dry product may be stored in concrete, iron, or steel tanks. Liquid aluminum sulfate, or solutions of the dry product are corrosive and need to be stored in lead, rubber, plastic or epoxy lined steel, or 316-stainless steel tanks.

The specific gravity of liquid alum at 4.2% Al varies from approximately 1.32 at 15°C to 1.33 at 40°C. The corresponding viscosity varies from approximately 0.011 $N \cdot s/m^2$ (11 centipoise) at 40°C to 0.028 $N \cdot s/m^2$ (28 centipoise) at 15°C.

In general, storage of liquid alum at temperatures below 10°C (50°F) is not recommended, to avoid the effects of crystallization.

Aluminum sulfate solutions show pH values similar to those of equimolar acetic acid solutions. pH values of various concentrations of aluminum sulfate solutions are shown in Figure 3.1.

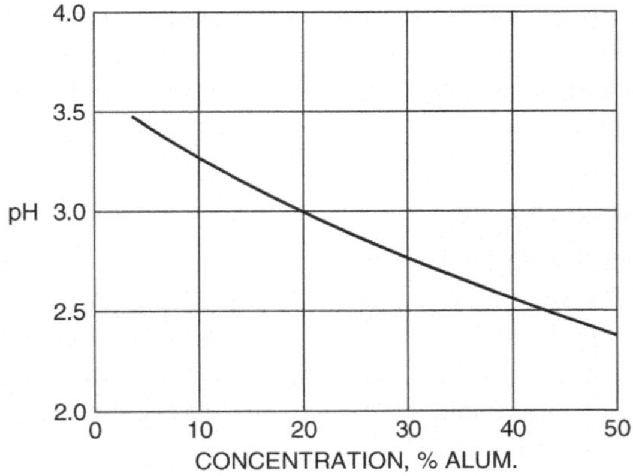

Figure 3.1 pH of solutions of commercial dry aluminum sulfate (9% as Al in dry alum).

Assuming that reactions of aluminum sulfate in water proceed to the electroneutral precipitate, $Al(OH)_3$ (which, as stated earlier, is not necessarily the case) a general idea of the reactions of aluminum sulfate with common alkaline reagents is given below:

$$Al_2(SO_4)_3 \cdot 14H_2O + 6OH^- = 2Al(OH)_3 + 3SO_4^{2-} + 14H_2O$$

$$Al_2(SO_4)_3 \cdot 14H_2O + 6Ca(HCO_3)_2 = 2Al(OH)_3 + 3CaSO_4 + 6CO_2 + 14H_2O$$

$$Al_2(SO_4)_3 \cdot 14H_2O + 6NaHCO_3 = 2Al(OH)_3 + 3Na_2SO_4 + 6CO_2 + 14H_2O$$

$$Al_2(SO_4)_3 \cdot 14H_2O + 3Na_2CO_3 = 2Al(OH)_3 + 3Na_2SO_4 + 3CO_2 + 14H_2O$$

$$Al_2(SO_4)_3 \cdot 14H_2O + 6NaOH = 2Al(OH)_3 + 3Na_2SO_4 + 14H_2O$$

$$Al_2(SO_4)_3 \cdot 14H_2O + 3Ca(OH)_2 = 2Al(OH)_3 + 3CaSO_4 + 14H_2O$$

3.2.1.2 Acidified aluminum sulfate (Acid alum)

This coagulant is prepared from aluminum sulfate and sulfuric acid. The components vary from 26.5 to 43.5% aluminum sulfate and 3 to 14% sulfuric acid. The chemical is supplied as a liquid with pH varying from 1.7 to 2.3.

Acid alum is used in the paper making industry, as well as in enhanced coagulation applications. In the latter case, it may avoid the separate dosing of acid to bring the coagulation pH down to the optimal range.

3.2.1.3 Aluminum chloride

This coagulant ($AlCl_3 \cdot 6H_2O$) is normally supplied in solution form containing 10.5% as Al with a pH and density of approximately 2.5 and 1,300 kg/m^3, respectively. It has been widely used for sludge conditioning and has often been described as a good general purpose conditioner. Because of hydrochloric acid (HCl) released on hydrolysis, solutions need to be stored under similar conditions to ferric chloride. (See later).

3.2.1.4 Sodium aluminate

Sodium aluminate ($NaAlO_2$) is usually supplied as a viscous, strongly alkaline and corrosive liquid. The solution strength is usually 13% as Al. The solution is kept stable by maintaining the molar ratio of Na_2O to Al at approximately 2.5:1.

Crystallization of this product can occur at temperatures around –15°C. The viscosity at 20°C is approximately 0.47 N · s/m^2 (470 centipoise), and increases to 1.5 N · s/m^2 (1,500 centipoise) at 10°C.

Handling and storage considerations are similar to caustic soda solutions. Storage materials can be stainless steel, rubber-lined mild steel, glass, plastics or mild steel. Because of a tendency for some insoluble matter to deposit in the tanks, outlet lines are often designed with a small upstand within the tank, with provisions for draining and removing the deposited material as required.

Because of the high viscosity at low temperatures (up to 30 N · s/m^2, or 30,000 centipoise, at –10°C), provisions for tank heating and tracing of feed lines is required for exposure to severe weather conditions.

This coagulant differs from alum in that it is alkaline rather than acidic in its reactions. It is rarely used alone, but generally with alum to obtain some special result. For example, in the coagulation of highly colored waters, alum (plus acid usually) is added to the water to coagulate the color at its requisite low pH. This, however, results in undesirable concentrations of soluble aluminum compounds; alkaline sodium aluminate is then added – also lime if required – to increase the pH to at least 6.0 which causes the soluble aluminum to precipitate out in a secondary settling basin.

An example of staged coagulation using alum and sodium aluminate is the treatment of the highly colored waters from the Table Mountain catchment area of South Africa, where the sequence of 10 mg/l sodium aluminate, followed by lime, followed by 50 mg/l aluminum sulfate was used. The dosage of lime was typically adjusted to achieve a pH after alum addition of 5.5. The raw water pH of these waters is typically between 4 and 5 (Bratby & Marais, 1977).

$NaAlO_2$ has also been used in the lime-soda softening process as an aid in flocculating the fine precipitates of calcium carbonate and magnesium hydroxide resulting from the softening reactions.

The reactions of $NaAlO_2$ with $Al_2(SO_4)_3 \cdot 14H_2O$ and with free CO_2 produce insoluble aluminum compounds:

$$6NaAlO_2 + Al_2(SO_4)_3 \cdot 14H_2O = 8Al(OH)_3 + 3Na_2SO_4 + 2H_2O$$

$$2NaAlO_2 + CO_2 + 3H_2O = Na_2CO_3 + 2Al(OH)_3$$

3.2.1.5 Ferric sulfate

This coagulant ($Fe_2(SO_4)_3 \cdot 8H_2O$) is available in solid or liquid form. In the solid form the material is granular and free flowing with the following typical specifications: 72 to 75% $Fe_2(SO_4)_3$ and 20 to 21% Fe^{3+}, by mass. In the liquid form, typical specifications are 40 to 42% $Fe_2(SO_4)_3$ and 11.5% Fe^{3+}, by mass. Lower purity liquid grades are available with 30% by mass $Fe_2(SO_4)_3$ content.

Recommended handling and storage materials include PVC; polyethylene; polypropylene; rubber lined mild steel; stainless steel; and glass; etc. There has been evidence of swelling of epoxy coated mild steel so care should be taken with this option.

In general, all ferric coagulants are used over a wide range of pH from 4.0 to 11.0. Ferric sulfate is particularly useful when used for color removal at low pH values and also at high pH values, where it is used for iron and manganese removal and in the softening process. For the latter uses the insolubility of the ferric hydroxides at high pH values makes the iron coagulants in general preferable to alum.

The comments on the chemical equations relating to alum are also applicable to all the following equations:

$$Fe_2(SO_4)_3 + 3Ca(HCO_3)_2 = 2Fe(OH)_3 + 3CaSO_4 + 6CO_2$$

$$Fe_2(SO_4)_3 + 6NaHCO_3 = 2Fe(OH)_3 + 3Na_2SO_4 + 6CO_2$$

$$Fe_2(SO_4)_3 + 3Na_2CO_3 + 3H_2O = 2Fe(OH)_3 + 3Na_2SO_4 + 3CO_2$$

$$Fe_2(SO_4)_3 + 6NaOH = 2Fe(OH)_3 + 3Na_2SO_4$$

$$Fe_2(SO_4)_3 + 3Ca(OH)_2 = 2Fe(OH)_3 + 3CaSO_4$$

3.2.1.6 Ferrous sulfate

Ferrous sulfate ($FeSO_4 \cdot 7H_2O$) is also known as copperas (although the latter term has virtually fallen into disuse). It is available either as crystals or granules containing 20% Fe, both of which are readily soluble in water. Ferrous sulfate reacts either with natural alkalinity or added alkalinity to form ferrous hydroxide, $Fe(OH)_2$, but since ferrous hydroxide is relatively soluble, it must be oxidized to ferric hydroxide in order to be useful. At pH values above 8.5 oxidation may be accomplished by aeration, by the dissolved oxygen in the water, or by adding chlorine. With the exception of chlorine, lime must be added to obtain sufficient alkalinity.

The important reactions for ferrous sulfate are:

$$FeSO_4 \cdot 7H_2O + Ca(OH)_2 = Fe(OH)_2 + CaSO_4 + 7H_2O$$

$$4Fe(OH)_2 + O_2 + 2H_2O = 4Fe(OH)_3$$

Ferrous sulfate and lime find their greatest use at high pH values; for example, the lime-soda softening process and in iron and manganese removal.

3.2.1.7 Chlorinated ferrous sulfate

In contrast to the high pH values required for oxidation by oxygen, chlorine will react over a wide pH range to oxidize the ferrous hydroxide obtained from ferrous sulfate.

In practice, ferrous sulfate and chlorine are fed separately and are generally mixed just prior to entry into the coagulation system. The reaction with chlorine produces ferric sulfate and ferric chloride and each mg/l of ferrous sulfate theoretically requires 0.13 mg/l of chlorine, although an excess of chlorine is generally added to ensure complete reaction and to provide chlorine for disinfection purposes. The reaction is as follows:

$$6FeSO_4 \cdot 7H_2O + 3Cl_2 = 2Fe_2(SO_4)_3 + 2FeCl_3 + 7H_2O$$

Coagulation with chlorinated copperas (or ferrous sulfate) is especially useful where pre-chlorination is required. It has the further advantage over ferrous sulfate in that coagulation may be obtained over a wide range of pH values: from 4.0 to 11.0.

3.2.1.8 Ferric chloride

This coagulant ($FeCl_3$) is available commercially in the liquid, crystal, or anhydrous forms, although the liquid form is by far the most common. The liquid and crystal forms are extremely corrosive and must be handled in a similar fashion to hydrochloric acid. Handling and storage materials should be steel lined with fiberglass reinforced polyester; PVC; epoxy resin; polyethylene; polypropylene; rubber; glass; etc. Polyamide type plastics are not suitable.

The liquid grade is supplied as a 40 to 43% by mass $FeCl_3$ dark brown syrupy solution. Physical properties of a 43% commercial solution are summarized as follows:

- $FeCl_3$ content: 43% m/m (approx. 628 g $FeCl_3$/l)
- Specific gravity at 20°C: 1.45 (1.45 kg/l; 12 lb/U.S. gal)
- Viscosity: 10°C: 0.0169 $N \cdot s/m^2$ (16.9 centipoise)
20°C: 0.0137 $N \cdot s/m^2$ (13.7 centipoise)
- pH of 43% solution: 1.0

The first ionization constant (K_1) of the aquo-ferric ion is similar to that of phosphoric acid, such that a 10^{-3} solution of $FeCl_3$ has a pH of about 3.2.

The anhydrous form may be stored in steel tanks but only if moisture is excluded. Because ferric chloride is very hygroscopic, drums must remain sealed until needed, and their entire contents must be dissolved at one time. The most popular commercial form for coagulation operations is liquid.

The reactions of ferric chloride with natural or added alkalinity may be written as follows:

$$2FeCl_3 + 3Ca(HCO_3)_2 = 2Fe(OH)_3 + 3CaCl_2 + 6CO_2$$

$$2FeCl_3 + 3Ca(OH)_2 = 2Fe(OH)_3 + 3CaCl_2$$

3.2.1.9 Prepolymerized aluminum and iron coagulants

As discussed in more detail later, when metal coagulants are added to water, the metal ions (Al and Fe) hydrolyze rapidly, but in a somewhat uncontrolled manner, forming a series of metal hydrolysis species. The efficiency of rapid mixing, the pH, and the coagulant dosage determine which hydrolysis species is effective for treatment.

There has been considerable interest in the development of pre-hydrolyzed inorganic coagulants, based on both aluminum and iron, to produce the correct hydrolysis species independent of the process conditions during treatment. The principal advantages of pre-polymerized inorganic coagulants are that they are able to function efficiently over wide ranges of pH and raw water temperatures. Pre-polymerization effectively enhances charge interactions between the coagulant and colloids by slowing down the hydrolysis reactions of the metal coagulant after it is added to water. In general, these coagulants are less sensitive to low water temperatures; lower dosages are required to achieve water treatment goals; less chemical residuals are produced; lower chloride or sulfate residuals are produced, resulting in lower final TDS; and they produce lower metal residuals.

Pre-polymerized inorganic coagulants are prepared with varying basicity ratios, base concentration, base addition rate, initial metal concentrations, aging time, and aging temperature.

The basicity ratio, r is defined as the molar ratio of hydroxide ions bound per mole of metal ion. Gillberg *et al.* (2003) compared required flocculation times for turbidity removal using alum and polyaluminum chloride compounds with basicity ratios of 1.1 and 2.2, respectively. The required flocculation times were 19; 10; and 7 minutes, respectively. These results parallel those of Gregory and Dupont (2001) who found that flocs formed with polyaluminum chloride were larger, stronger and settled faster than flocs formed with alum.

The higher the basicity of a coagulant, the lower is the alkalinity consumption, and the lesser is the pH reduction of the treated water.

Another important parameter describing the makeup of polymerized inorganic coagulants is the *relative basicity*, normally expressed as a percentage. This is essentially described by the molar ratio of negative to positive charges of the compound.

For example, the stable species $Al_{13}O_4(OH)_{24}(H_2O)_{12}^{7+}$ is a major component of many polyaluminum chloride preparations. It has a total of ($13 \times 3 = 39$ positive charges) and a total of ($4 \times 2 + 24 \times 1 = 32$ negative charges). Therefore, the relative basicity of this compound is $(32/39) \times 100 = 82\%$.

Gillberg *et al.* (2003) pointed out that in general, for turbidity removal, polyaluminum chloride products with relative basicities close to 65% provide the best removal efficiencies. However, because of the highly specific nature of these products, the best formulation for a particular water is case specific, and needs to be determined by jar testing (see Chapter 8). For example, Polasek and Mutl (2002) found inferior performance on the whole when comparing the polyaluminum chloride product tested by them, and alum. In terms of turbidity results were similar. However, alum was superior in terms of organics removal.

Gillberg *et al.* also showed that the efficiency of orthophosphate removal in wastewater treatment decreased with increasing basicity. In general, alum was superior to the polyaluminum chloride coagulants tested.

Polymerized aluminum coagulants consist of species such as $Al_2(OH)_2^{4+}$; $Al_3(OH)_4^{5+}$; $Al_8(OH)_{20}^{4+}$; $AlO_4[Al(OH)_2]_{12}^{7+}$; $Al_{13}O_4(OH)_{24}(H_2O)_{12}^{7+}$; and $Al_3(OH)_4^{5+}$. The presence and distribution of these species in preparations of aluminum chlorohydrate, polyaluminum chloride, polyaluminum sulfate chloride, polyaluminum silicate chloride, and polyaluminum chlorides with organic polymers, depend on the type of coagulant, the total aluminum concentration, the quantity of sodium hydroxide or sodium bicarbonate added during preparation, the rate of addition of sodium hydroxide, the intensity of mixing, and the degree of aging (Fan *et al.* 2003).

Polymerized coagulants can be manufactured with various substitutions that can be beneficial for treatment of certain waters. For example, some products have 1 to 3-percent sulfate substituting for chloride, to produce polyaluminum chlorosulfates (PACS). There are several commercial products with stated formulae such as $Al_{13}(OH)_{27.5}(SO_4)_2Cl_{9.5}$ and $Al_{13}(OH)_{20}(SO_4)_2Cl_{15}$. Other substitutions include phosphate and silicate. Some formulations contain additives such as cationic polyelectrolytes to enhance performance and reduce sludge production, in some cases.

Gray *et al.* (1995) and Wang *et al.* (2003) pointed out the difficulty in preparing stable cationic polymerized iron species by partial neutralization of ferric chloride solutions, when compared with the corresponding formation of stable polymerized aluminum compounds. Fan *et al.* (2002) found that the amount of sulfuric acid should be 51.7% by weight of the stoichiometric requirement for poly-ferric sulfate produced at 55°C, to produce a stable product that can be stored for 6 months.

There are various commercial aluminum-iron blends, including blends of ferric sulfate and polyaluminum silicate sulfate; ferric sulfate and aluminum sulfate; and ferric chloride and aluminum sulfate.

3.2.1.9.1 Aluminum chlorohydrate

This is probably the first of the pre-polymerized coagulants, and is a high density polyaluminum chloride. It was developed in the U.K. in the early 1950s primarily as a sludge conditioner. The theoretical formula is $Al_2(OH)_5Cl$, with a relative basicity of approximately 83%. The commercial product contains approximately 13% Al (compared with alum at approximately 4% Al content). Aluminum chlorohydrate is effective as a sludge conditioner, as well as primary coagulant.

Tests carried out on Colorado River water, comparing the use of alum and aluminum chlorohydrate as primary coagulants, demonstrated overall advantages with aluminum chlorohydrate (Jones, 1990). In general, the chemical was effective when raw water turbidities were less than 30 NTU. Chemical sludge volumes, backwash water volumes, and finished water residual aluminum were significantly less with aluminum chlorohydrate, compared with alum. However, for this water, final turbidities were lower with alum: on average 0.07 NTU, compared with 0.16 NTU with aluminum chlorohydrate. No pH adjustment chemicals were added for coagulation during the tests. The pH using alum was 7.7, and with aluminum chlorohydrate 8.2.

Ruehl (1999) reported on the advantages of aluminum chlorohydrate compared with ferric sulfate when treating wide turbidity swings in a river water. Treated water turbidities were more consistent and generally lower, particularly during high raw water turbidity events. When raw water turbidities rose from approximately 55 to 220 NTU, the required ferric sulfate dosage increased from 80 to 130 mg/l, whereas aluminum chlorohydrate increased from 37 to 45 mg/l. Corresponding filtered turbidities with ferric sulfate increased from 0.06 to 1.12 NTU, whereas with aluminum chlorohydrate, filtered turbidities remained at 0.05 NTU.

3.2.1.9.2 Polyaluminum chloride

There are numerous formulations of polyaluminum chloride preparations, apart from aluminum chlorohydrate. Most are in liquid form, but some spray dried granular powder products are available. Relative basicities range from 30 to 80% and aluminum contents range from 5 to 12% as Al in the liquid products. Specific gravities of the liquid products vary from 1.2 to 1.4 at 20°C.

3.2.1.9.3 Polyaluminum silicate sulfate and polyaluminum silicate chloride

Hasegawa *et al.* (1991) describe the incorporation of aluminum sulfate in pre-polymerized silicic acid (activated silica) to form a very effective coagulant. Commercial preparations of polyaluminum silicate chloride coagulants are available, and can be effective in wastewater and surface water treatment, although they may be less effective for phosphorus precipitation (Gillberg *et al.* 2003).

Commercial polyaluminum silicate sulfate and polyaluminum silicate chloride products are supplied in liquid form, although spray dried products are available. The liquid products are supplied with 4.4 to 7.5% as Al contents, with relative basicities of 40 to 70%. Specific gravities of the solutions vary but are approximately 1.30.

3.2.1.9.4 Polymerized ferric and blended aluminum-ferric coagulants

Various authors have described the benefits of polymerized ferric coagulants (Leprince *et al.* 1984; Jiang *et al.* 1993, 1996; Fan *et al.* 2002; He *et al.* 2011). Jiang *et al.* described the superior performance of a poly-ferric sulfate coagulant in terms of color, UV-absorbance and DOC removals, compared with ferric sulfate and aluminum sulfate.

Commercial poly-ferric sulfate products are available with 12.5% Fe content and 11% relative basicity. The chemical structure of poly-ferric sulfate is described as $[Fe_2(OH)_n(SO_4)_{(6-n)/2}]_m$ where m is a function of n.

Other commercial products incorporate blends of ferric salts and organic polymers.

Jiang and Graham (2003) describe the development of a hybrid polyalumino-iron sulfate coagulant that was superior to ferric sulfate or aluminum sulfate in removing natural organic matter. Residual metal concentrations also were lower. For preparing a stable polymerized coagulant with a high cationic charge, the basicity ratio should be approximately 0.3; the mole Al: mole Fe ratio between 0.6 and 0.9; the aging temperature 40°C; and the aging time 1 hour.

Gao *et al.* (2002) compared the performance of a polyaluminum ferric chloride preparation, with polyaluminum chloride and poly-ferric sulfate. It was found that blending aluminum with the polymerized ferric provided increased stability of the product, when compared with poly-ferric sulfate. The polyaluminum ferric chloride blend was a more effective coagulant for the removal of turbidity, TSS, color, COD and oil from petrochemical wastewaters.

Wang *et al.* (2002) investigated a polysilicato-iron coagulant preparation. It was found to be more effective when compared with polyaluminum chloride for the removal of phytoplankton, turbidity and DOC from a eutrophic surface water. The silica-to-iron molar ratio was 3:1.

3.2.1.9.5 Handling of pre-polymerized coagulants

In general, reactions with the polymerized coagulants with alum or ferric coagulants produce precipitates. Therefore, before switching between coagulants on a trial or a permanent basis, it is important to thoroughly clean tanks, lines and pumps. Similarly, if one of the pre-polymerized coagulants is to be used in conjunction with alum or one of the ferric coagulants, they should be fed through different lines, and not mixed in-line before the point of application, to avoid precipitation and reduced effectiveness of both coagulants.

In general, polymerized coagulant liquid products should be fed directly to the flash mixer without prior dilution, to avoid precocious precipitation reactions. Similar considerations apply to laboratory jar testing where the products should be used as-is without pre-dilution. When using micro-pipettes or syringes for chemical injection of the small test volumes, normally the tip of the syringe or pipette should not be immersed in the water, to avoid blockages due to hydrolysis of the chemicals. After use the syringe or pipette should be rinsed thoroughly with deionized water to prevent blockages.

The storage life of polymerized coagulants depends on the product and the storage conditions. But as an example, a polyaluminum coagulant with 5.2% aluminum content is stable for 4 to 5 months when stored at less than 50°C.

3.2.1.10 Preparation and feeding of coagulant products

Figure 3.2 shows a schematic of a liquid coagulant feed system. The delivered product is stored in bulk storage tanks and transferred to one or more day tanks

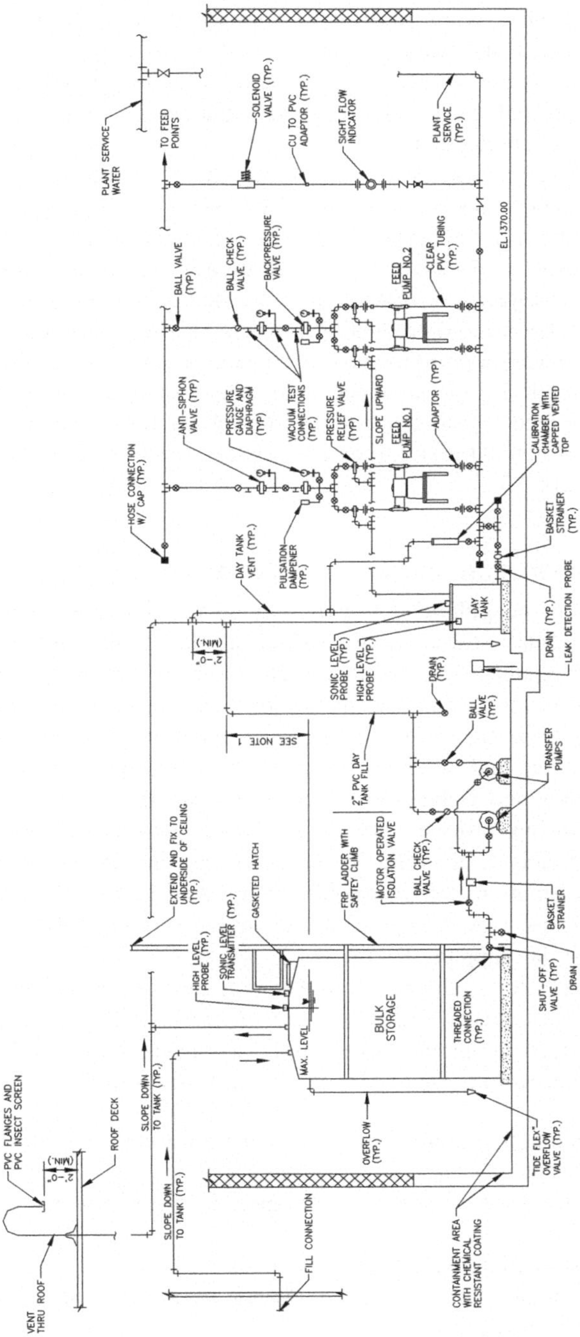

Figure 3.2 Schematic of a liquid chemical feed system.

from which the product is metered to the application point. In-line dilution of the metered product, if it occurs, is often carried out as close as possible to the application point to avoid hydroxide precipitation in the line. However, this is not the case with the pre-polymerized products, as explained previously. These should not be diluted before application.

Figure 3.3 shows a schematic of a batch dry chemical feed system. Lump or ground dry coagulant products may be dissolved in batch fashion, or continuously. With a batch method, the appropriate quantity of chemical (either pre-weighed or a number of bags) is dissolved in a tank, usually with a mechanical mixer, then metered either directly or after dilution to the application point. As before, in-line dilution of the metered product, if it occurs, is often carried out as close as possible to the application point to avoid hydroxide precipitation in the line. With a continuous method, dry granular chemical is fed continuously, by means of a screw or vibrating feeder, from a hopper to a dissolving tank. The dissolving tank is fitted with a mechanical mixer, then the solution is fed in a similar fashion to that shown in Figure 3.3.

3.2.1.11 Contamination of commercial chemicals

Chemicals used for coagulation operations inevitably contain a small amount of contaminants. Usually these are at a sufficiently low level that they are not of concern, even for potable water treatment. However, this is not always the case. The most common causes of contamination of these chemicals can be transport related, typically resulting from improperly or incompletely cleaned vehicles– particularly those that had previously been used for transporting goods such as feed corn, pesticides, and asphalt (Brown & Wierenga, 2005).

The manufacturing process itself has also created issues. For example, crystallization typically occurs if the liquid product is evaporated to concentrations that are too high. Solution strength should be confirmed upon delivery, to verify that the strength is less than 48.5 percent. The inventory should also preferably be kept to below a 30-day supply.

The production of ferric chloride from scrap metals and inadequate storage of the raw material has caused a large amount of debris in the final product that has clogged plant feed equipment. Brown and Wierenga recommend the following safeguards before accepting chemical deliveries at water plants:

- Stricter specifications, particularly requiring filtration of the products before delivery
- Dedicated trucks for deliveries of specific chemicals to avoid cross-contamination
- Random testing of delivered products
- Use of pneumatic off-loading equipment to avoid contamination from mechanical pumps and seals

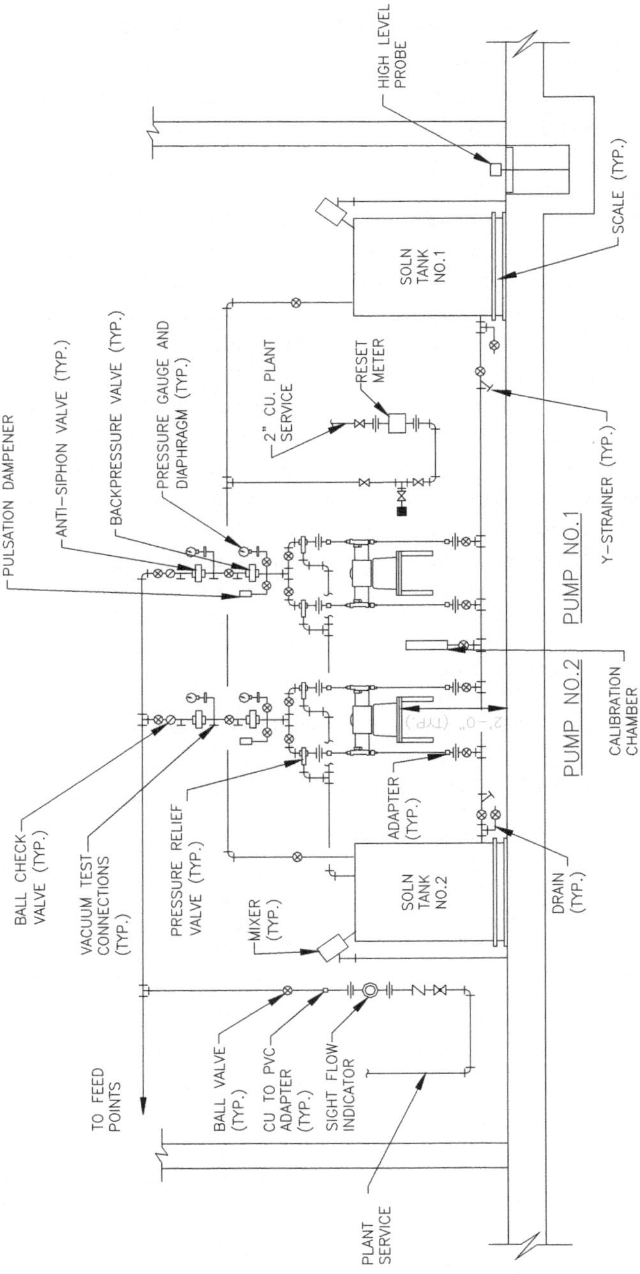

Figure 3.3 Schematic of a batch dry chemical feed system.

3.2.2 Chemistry of metal coagulants

3.2.2.1 Hydration and stepwise substitution reactions

Having identified the common metal coagulants, the following discussions are devoted to the reactions taking place when metal coagulants are introduced into the stream of water to be treated.

Aluminum and ferric salts, when in solution, immediately dissociate to form hydrated reaction products. The metal ions form coordination compounds with water molecules to give $Al(H_2O)_6^{3+}$ and $Fe(H_2O)_6^{3+}$ (Basolo, 1964; Beck, 1970; Orgel, 1970). These species, referred to as the trivalent ions of aluminum and iron, are often presented as Al^{3+} and Fe^{3+} for reasons of convenience in presentation.

There will also be other ligands in solution, and a stepwise substitution of the ligand molecules or ions for the water molecule takes place. The extent of this substitution depends on the concentration of the substituted ligand. For the destabilization of hydrophobic colloids with iron and aluminum coagulants, the complexes of principal interest are those comprising H_2O and OH^- as ligands. The OH^- ligands may be thought of as arising from the dissociation of bound H_2O ligands, or from the replacement of H_2O by OH^- ligands. It can be seen that in either case, the extent to which OH^- ions are bound to the metal complex is dependent primarily on the pH or the concentration of OH^- ions in solution.

The scheme of these stepwise substitution reactions for Al(III) is shown in Table 3.1 (Stumm, 1962). From the first reaction, one can write:

$$Al(H_2O)_6^{3+} \Leftrightarrow [Al(H_2O)_5OH^{2+}] + H^+$$

from which it is seen that complexes of metal ions in water act as weak acids.

Table 3.1 Scheme of stepwise hydrolysis reactions for aluminum.

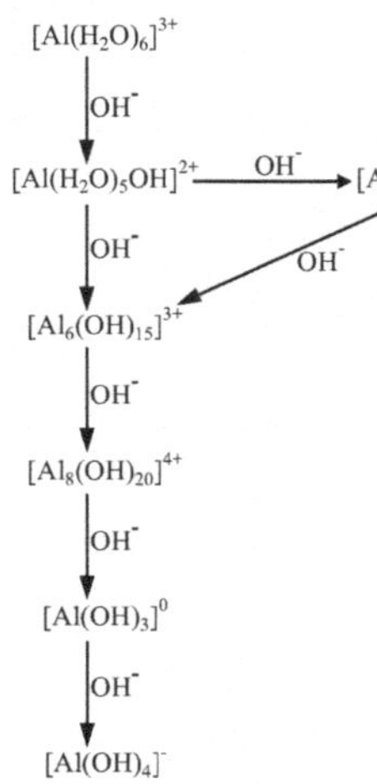

Other polynuclear species are of particular importance with polymerized aluminum based coagulants. These include $Al_2(OH)_2^{4+}$; $Al_3(OH)_4^{5+}$; $Al_8(OH)_{20}^{4+}$; $AlO_4[Al(OH)_2]_{12}^{7+}$; $Al_3(OH)_4^{5+}$; and $Al_{13}O_4(OH)_{24}(H_2O)_{12}^{7+}$. This latter species is considered to be one of the most effective and stable polynuclear species for treatment (Jiang, 2001; Gillberg *et al.* 2003).

The donor capacity of a ligand is frequently not exhausted by being coordinated to a metal ion and there is a possibility of it forming a coordinative bond with another metal ion. The ligand thus acts as a bridge between the two central ions. Both unidentate and multidentate ligands may serve as bridges in polynuclear complexes.

Both iron and aluminum-hydroxo complexes have pronounced tendencies to form polynuclear complexes. The simplest reaction leading to a bi-nuclear iron complex species is:

$$2[Fe(H_2O)_5 \cdot (OH)]^{2+} = [Fe_2(H_2O)_8 \cdot (OH)_2]^{4+} + 2H_2O$$

The bi-nuclear ion so formed has sufficient stability to exist in appreciable concentrations in iron solutions which are more concentrated than 10^{-4}M. The two metal ions are thought to be bound by two hydroxo bridges:

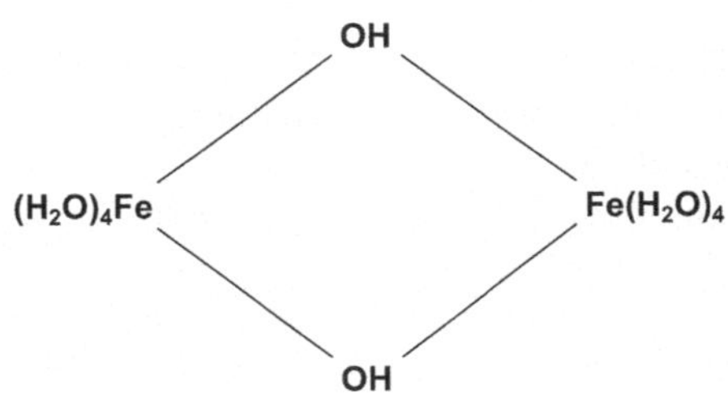

Dousma and DeBruyn (1978) pointed out that hydrolysis proceeds in four recognizable steps:

- hydrolysis to mono- and bi-nuclear species
- reversible rapid growth to small polynuclear species
- formation of slowly reacting large polynuclear species
- precipitation of a solid phase.

3.2.2.2 Alkalinity consumption by metal coagulants

With the exception of sodium aluminate explained previously, the hydrolysis reactions of metal coagulants added to water generally result in a consumption of alkalinity. Supplemental alkalinity addition may be required if the metal coagulant dosage causes a sufficient drop in pH that impedes effective coagulation.

The alkalinity consumption can be calculated by determining the equivalent dosage of alkalinity expressed as calcium carbonate ($CaCO_3$). Table 3.2 lists the alkalinity consumed by a number of coagulants. The following is an example of the scheme of calculation required (also refer to Chapter 8):

Table 3.2 Alkalinity consumption by various coagulants.

Coagulant	Coagulant Dosage (mg/L)	Alkalinity Consumed (mg/L as $CaCO_3$)
Aluminum sulfate	1.0 as $Al_2(SO_4)_3 14H_2O$	0.51
Acid alum (3 to 14% H_2SO_4)	1.0 as $Al_2(SO_4)_3 14H_2O$	0.58 to 0.98
Ferric chloride	1.0 as $FeCl_3$	0.93
Ferric sulfate	1.0 as $Fe_2(SO_4)_3$	0.75
Ferrous sulfate	1.0 as $FeSO_4$	0.66
Polyaluminum chloride (PACl, 50% basicity)	1.0 as Al	2.78
PACl (70% basicity)	1.0 as Al	1.67
Aluminum chlorohydrate (82% basicity)	1.0 as Al	1.00

- Assume coagulant dosage = 1.0 mg/L alum (as $Al_2(SO_4)_3 14H_2O$)
- Aluminum dosage = (54/594) × 1.0 = 0.0909 mg/L as Al
- Molecular weight of Al = 27
- Equivalent weight of Al = 27/valency = 27/3 = 9
- Molecular weight of $CaCO_3$ = 100
- Equivalent weight of $CaCO_3$ = 100/valency = 100/2 = 50
- Alkalinity consumption of 1 mg/L alum = 0.0909 × 50/9 = 0.505 mg/L as $CaCO_3$.

Note that calculations for PACl and aluminum chlorohydrate are modified by taking into consideration the portion of acidity related to the metal coagulant that has been neutralized by hydroxyl ions. For example, a compound with a relative basicity of 50% would consume half the alkalinity of alum at the same metal (Al) concentration.

In the case of acid alum, in which the excess sulphuric acid consumes additional alkalinity, the calculations are modified by including the additional alkalinity consumption of the sulphuric acid component of the compound.

Table 3.3 shows the alkalinity provided by various chemicals used for supplemental alkalinity and pH correction during coagulation. Note that when alkalinity chemicals are used to assist coagulation they should be added before coagulant addition at a point where complete dissolution is achieved before coagulant addition. Note that the chemical dosages in Table 3.3 assume 100% purity. The calculated quantities of chemicals would need to be adjusted by taking into account actual purities.

Table 3.3 Alkalinity provided by various chemicals.

Chemical	Dosage (mg/L)	Alkalinity Assuming 100% Purity (mg/L as $CaCO_3$)
Hydrated lime	1.0 as $Ca(OH)_2$	1.35
Quicklime	1.0 as CaO	1.78
Soda ash	1.0 as Na_2CO_3	0.94
Sodium bicarbonate	1.0 as $NaHCO_3$	0.60
Caustic soda	1.0 as NaOH	1.25

3.2.2.3 Stability of metal-ion hydrolysis species

The stepwise formation of metal-ion hydrolysis species can be described by a set of stability constants. The individual stability constants cannot be directly calculated and to obtain them it is necessary to find a suitable relationship between the individual constants (or their products) and the experimentally measurable parameters, as exemplified for mononuclear species below (Beck, 1970):

The total concentration of the central metal ion $[T_{Me}]$ and of the ligand $[T_L]$ are known.

$[T_{Me}]$ is the sum of the concentrations of the different species containing it:

$$T_{Me} = [Me] + [MeL_1] + \cdots + [MeL_N] = \sum_{i=0}^{N} [MeL_i]$$

where L is the ligand.

The total concentration of the ligand is the weighted sum of the species concentrations containing it:

$$T_L = [L] + [MeL] + 2[MeL_2] + \cdots + N[MeL_N] = [L] + \sum_{i=1}^{N} i[MeL_i]$$

Further:

$$K_1 = \frac{[MeL]}{[Me][L]}; \cdots; K_N = \frac{[MeL_N]}{[MeL_{N-1}][L]}$$

where K is the stability constant, and

$$\beta_1 = K_1;\ \beta_2 = K_1 \cdot K_2;\ \beta_N = K_1 \cdot K_2 \cdots K_N = \frac{[MeL_N]}{[Me][L]^N}$$

where β is known as the overall stability constant.

Values of stability constants, K for some hydrolysis reactions with iron and aluminum are given in Table 3.4. From such stability constants, the concentration (or fraction present) of each species in solution under equilibrium conditions may be computed (see later).

Table 3.4 Stability constants for iron and aluminum (after Stumm & Morgan, 1962).

Reaction	log K (25°C)
$[Fe(H_2O)_6]^{3+} + H_2O = [Fe(H_2O)_5OH]^{2+} + H_3O^+$ (Note: H_2O molecules omitted from subsequent reactions for brevity)	−2.17
$[Fe(OH)]^{2+} + H_2O = [Fe(OH)_2]^+ + H^+$	−4.7
$2[Fe(OH)]^{2+} = [Fe_2(OH)_2]^{4+}$	1.46
$Fe^{3+} + 3H_2O = Fe(OH)_3^0 + 3H^+$	−6.0
$Fe(OH)_3^0 + H_2O = [Fe(OH)_4]^- + H^+$	−18.5
$Al^{3+} + H_2O = [Al(OH)]^{2+} + H^+$	−5.03
$Al^{3+} + 3H_2O = Al(OH)_3^0 + 3H^+$	−9.10
$2Al^{3+} + 2H_2O = [Al_2(OH)_2]^{4+} + 2H^+$	−6.27
$Al(OH)_3^0 + H_2O = [Al(OH)_4]^- + H^+$	−12.74
$6Al^{3+} + 15H_2O = [Al_6(OH)_{15}]^{3+} + 15H^+$	163.0

3.2.2.4 Average coordination number

The extent of complex formation may be characterized by the average coordination number (or formation function) $\bar{n}$, where

$$\bar{n} = \frac{\text{concentration of ligands bound to metal ions}}{\text{total metal ion concentration}}$$

$$= \frac{(T_L - [L])}{T_{Me}},$$

where $[L]$ is the free ligand concentration.

The average coordination number is determined experimentally by alkalimetric titration. The procedure is illustrated by the following example (Stumm, 1962): To an $Al(ClO_4)_3$ (aluminum perchlorate) solution, NaOH (sodium hydroxide) is incrementally added. After each addition of NaOH the pH of the solution is determined with a glass electrode and hence the hydrogen ion concentration of the solution is determined.

It will be assumed that together with free Al^{3+} ions, the mononuclear complexes $Al(OH)^{2+}$ and $Al(OH)_3$ are formed together with polynuclear complexes collectively given by $Al_n(OH)_x^{3n-x}$.

The total aluminum concentration at any given pH is given by:

$$Al_T = [Al^{3+}] + [Al(OH)^{2+}] + \left[Al(OH)_3\right] + \sum n[Al_n(OH)_x^{3n-x}] = \frac{[ClO_4^-]}{3} \tag{3.1}$$

For electroneutrality at any given pH:

$$3[Al^{3+}] + 2[Al(OH)^{2+}] + \sum (3n-x)\cdot[Al_n(OH)_x^{3n-x}] + [Na^+] + H^+ = [ClO_4^-] + [OH^-] \tag{3.2}$$

From the addition of one mole of NaOH, one mole of Na^+ and one mole of OH^- are formed.

Hence,

$$[Na^+] = [OH^-]_{added}$$

Combining equations (3.1) and (3.2):

$$[OH^-]_{added} + [H^+] - [OH^-] = [Al(OH)^{2+}] + 3[Al(OH)_3] + \sum x[Al_n(OH)_x^{3n-x}]$$

Studying the right-hand side of the above equation, one sees that each term gives the moles of OH^- bound to each species. Hence the concentration of OH^- ions bound to aluminum is given by the right-hand side of the above equation–and consequently by the left-hand side. Since all the terms of the left-hand side are known, the $[OH^-]_{bound}$ may be computed.

Hence, for the above case, the average coordination number $\bar{n}$ is given by:

$$\bar{n} = \frac{[OH^-]_{bound}}{[Al_T]} = \frac{[OH^-]_{added} + [H^+] - [OH^-]}{[Al_T]}$$

From this number, the mean charge of the total aluminum species may be determined at any particular pH:

$$\text{Mean charge} = (3 - \bar{n})$$

Hence, if there is an average of, say, three OH^- ions bound per aluminum ion, then the mean charge of all the complex species will be zero. This is the isoelectric point. It can be said that the charge found above has a statistical character. The isoelectric point of the above solution may then be defined as the pH at which equal quantities of the aluminum species migrate to the anode and cathode of an electrophoretic cell.

If curves are drawn of pH versus the average coordination number $\bar{n}$ or the mean charge $(3-\bar{n})$, valuable information may be gleaned concerning the nature of the complexes present at any particular pH. As exemplified in Figure 3.4, one can see that for aluminum between pH 10 and 10.5 for instance, the constant value of $\bar{n} = 4$ indicates that a definite compound is formed, such as aluminate, $[Al(OH)_4]^-$.

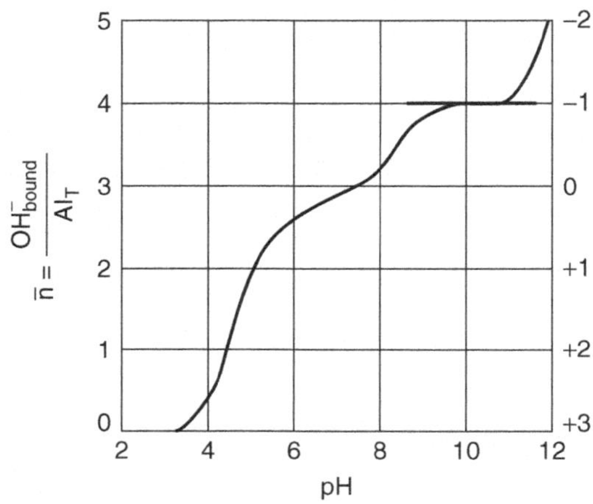

Figure 3.4 Values of average coordination number $\bar{n}$, for different pH values (adapted from Stumm & Morgan, 1962).

3.2.2.5 Species distribution during destabilization

Figures 3.5(a) and (b) (from Stumm & O'Melia, 1968) show the concentrations of species present at different pH values for iron and aluminum solutions, respectively, at equilibrium. The values are derived from stability constants such as those shown in Table 3.4. The cross-hatched areas indicate the range of coagulant dosages which are normally used in water treatment. From these Figures one might be tempted to conclude that the species involved in coagulation are the insoluble metal hydroxide precipitates. Although precipitation reactions are of doubtless importance in certain cases (for example, during organic color removal, phosphorus removal in tertiary wastewater treatment processes, and iron removal–although the reactions in these examples may be more complex, as discussed in later Chapters) destabilization reactions are often other than that of precipitation. The rate of transition from aquo metal ions to metal hydroxide precipitates is dependent on the pH and applied concentration of the solution. For the concentrations used in water treatment, true equilibrium (as shown in Figures 3.5a and b) may in some cases require time spans longer than required to attain destabilization. For this reason the use of stability constants in identifying the species effective during destabilization in water treatment processes may not be valid.

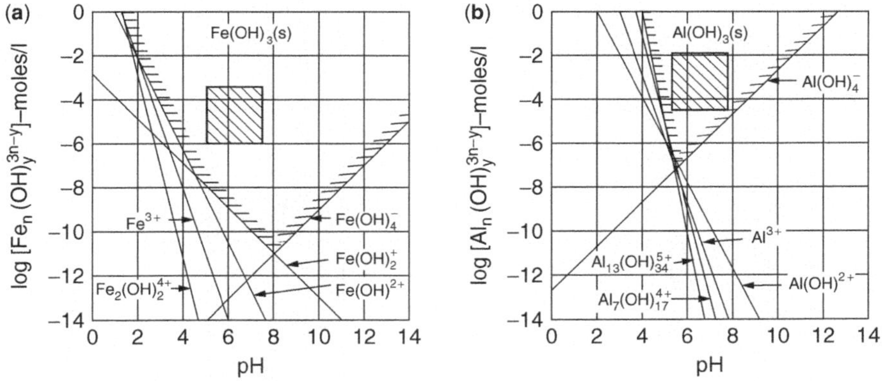

Figure 3.5 Equilibrium–solubility domains of (a) ferric hydroxide and (b) aluminum hydroxide in water. These diagrams have been calculated using representative values for the equlibrium constants for solubility and hydrolysis equilibria. Shaded areas are approximate operating regions in water treatment practice. Coagulation in these systems occurs under conditions of over-saturation with respect to the metal hydroxide (after Stumm & O'Melia, 1968).

Singley and Black (1967), Sullivan and Singley (1968), and Singley and Sullivan (1969) attempted to identify the probable species present during destabilization by using the concept of average coordination number, described above. They recognized that during the time required for alkalimetric titrations a quasi-equilibrium condition was attained for which the mass action relationship is best referred to as an *instantaneous concentration quotient*, Q. For example, in the general reaction

$$Fe^{3+} + nH_2O \Leftrightarrow Fe(OH)_n^{3-n} + nH^+$$

Q may be defined as:

$$Q_n = ([Fe(OH)_n]^{3-n}[H^+]^n)/[Fe^{3+}]$$

That is, Q is equivalent to an instantaneous stability constant.

In their analysis, they considered mono-nuclear species only. For example, in the case of iron, the species present were assumed to be:

$[Fe^{3+}]$; $[Fe(OH)^{2+}]$; $[Fe(OH)_2^+]$; $[Fe(OH)_3]$; $Fe(OH)_4^-$; $[Fe(OH)_5^{2-}]$; and $[Fe(OH)_6^{3-}]$

A maximum coordination number of 6 was assumed for both iron and aluminum. This is probably valid as a first approximation.

By setting up equations for each of the six concentration quotients, Q, and for the average coordination number, $\bar{n}$, it was possible to calculate the species present and their concentration at each pH value. Results for iron are shown in Figure 3.6a, and for aluminum in Figure 3.6b. It is seen that in each case, as the concentration is reduced, the dominance of the electroneutral species $Fe(OH)_3$ and $Al(OH)_3$

reduces. If the concentration of coagulants applied is large enough and the pH of the solution is suitable, then polynuclear precipitates are dominant.

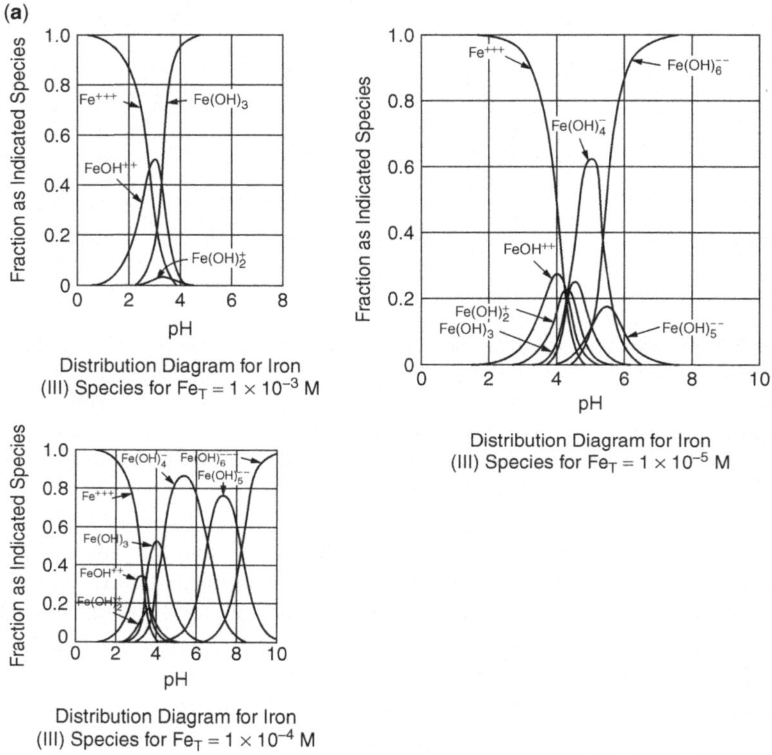

Figure 3.6(a) Species distribution diagrams for iron (III) during destabilization (from Singley & Sullivan, 1969).

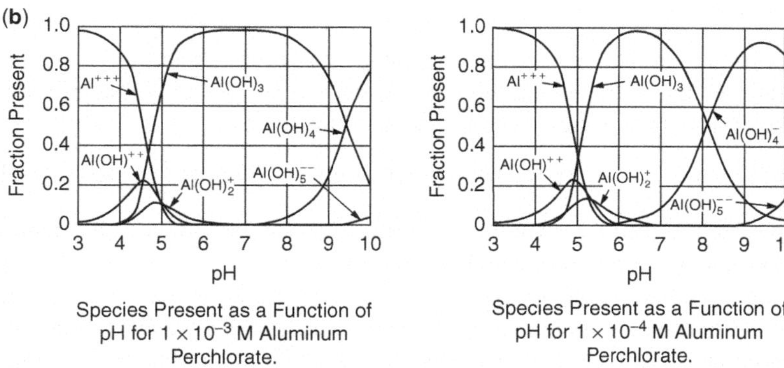

Figure 3.6(b) Species distribution diagrams for aluminum (III) (from Sullivan & Singley, 1968).

It should be noted that the mononuclear species considered above are a simplification, especially for the case of aluminum. When a quantity of aluminum salts are put into solution, then during progressive hydrolysis and substitution reactions, numerous polymeric species are formed, the extent of polymerization becoming more pronounced as the charge of the metal species decreases. As shown in Figure 3.5b, colloidal metal hydroxide polymers and insoluble metal hydroxide precipitates may eventually form, the extent depending on the concentration of coagulant added (Figure 3.6b) and whether destabilization reactions are such as to permit equilibrium conditions to be attained. At slightly acid or neutral pH values, positively charged or electroneutral metal hydroxide polymeric colloids form, whereas at higher pH values, the polymerization of negative complexes takes place and negatively charged polymeric species are formed.

The polymerization action of metal coagulants, described above, is of high importance during destabilization since, as will be expanded upon in Chapter 4, it promotes an enhanced adsorption capacity of metal coagulant species to colloids.

It is seen that the mononuclear species model does not indicate the often reported Al:OH ratio of 1:2.5 for the predominant aluminum hydroxide complex species during destabilization (see, for example, Brosset, 1954 and Matijevic, 1961). Singley and co-workers suggested that in systems where the latter ratio has been reported, ligands other than OH^- were present which would give rise to species more complex than expected yet give an apparent coordination number, $\bar{n}$ of 2.5. In their model the only ligands present were considered to be H_2O and OH^-.

Of interest from Figure 3.6a is that for $Fe_T = 10^{-4}$M and at a pH of approximately 4.0 the predominant species is $Fe(OH)_3$, which is consistent with the conditions for organic color removal. Although the mononuclear model suggested by Singley and co-workers predicted closely their experimental results for iron solutions, this was not the case for aluminum solution concentrations less than 10^{-5} to 10^{-4} M. The reason put forward was the probable formation of complicated polynuclear aluminum hydrolysis species at lower concentrations. The more complicated hydrolysis reactions of aluminum salt solutions is in large part responsible for the narrower pH range often evident for optimum destabilization when compared with iron solutions. Furthermore, the different adsorption characteristics of aluminum hydrolysis species when compared with, say, iron coagulants or polymers is a further indication of the complicated hydrolysis products formed. With iron salts and polymers, adsorption often follows a Langmuir adsorption isotherm, whereas with aluminum, adsorption often follows a Freundlich isotherm.

3.3 POLYMERS

3.3.1 General

Polymers refer to a large variety of natural or synthetic, water soluble, macromolecular compounds which have the ability to destabilize or enhance flocculation of the constituents of a body of water.

A polymer molecule may be described as a series of repeating chemical units held together by covalent bonds (polymer, from the literal Latin translation, means many parts). If the repeating units are of the same molecular structure, the compound is termed a homopolymer. However, if the molecule is formed from more than one type of repeating chemical unit, it is termed a copolymer. The individual repeating units are called monomers and the molecular weight of the polymer molecule is the sum of the molecular weights of the individual monomers. The total number of monomer units is referred to as the degree of polymerization.

Polyelectrolytes are special classes of polymers containing certain functional groups along the polymer backbone which may be ionizable. If present, when the ionizable groups dissociate, the polymer molecules become charged either positively or negatively, depending on the specific functional groups present, and are thus referred to as cationic or anionic polyelectrolytes, respectively. Polymers that possess both positively and negatively charged sites are referred to as ampholytic, whereas those that possess no ionizable functional groups are termed nonionic polymers. All polyelectrolytes are typical hydrophilic colloids. They have molecular weights generally in the range 10^4 to 10^7 and are soluble in water due to hydration of functional groups.

The following presents several types of polyelectrolytes currently in use. It is not intended as a comprehensive list but merely to provide a general picture of polyelectrolyte properties. The mechanisms of destabilization by polyelectrolytes are reserved for Chapter 5.

3.3.2 Activated silica

Activated silica is probably the first polyelectrolyte to be used widely in water clarification. In preparing activated silica (which is an anionic polyelectrolyte) commercial sodium silicate solutions (pH approximately 12) at concentrations in excess of 2×10^{-3} M are neutralized with acid reagent (sulfuric acid, chlorine, aluminum sulfate or sodium bicarbonate, have been used) to a pH of less than 9. Under these conditions, the silica solution is oversaturated with respect to amorphous silica and polymeric silicates are formed. These polymeric species are negatively charged kinetic intermediates in the transition to a silica precipitate. The reaction is stopped prior to precipitate formation by dilution to a concentration less than 2×10^{-3} M (compare this to the intermediate species formed during hydrolysis reactions of metal coagulants, discussed previously).

Activation is usually carried out on-site. Worthington (1978) demonstrated that activation using alum was superior to using sulfuric acid, in terms of performance. However, it was unknown whether the residual alum used for activation assisted in flocculation reactions, or whether the effect was due to better activation. It is generally easier to use sulfuric acid for activation, in terms of avoiding gelling and deposit formation. With this approach, cleaning of equipment is typically required every 2 to 5 years (Gibson, 1994).

The nature of the activation process is to polymerize the silica particles into spherical micelles. Aging is beneficial to the effectiveness of activated silica, although it has been shown that at low dosages, aging does not significantly improve flocculation (Worthington, 1978).

Activated silica is commonly used at dosages of 1/15 to 1/10 the primary metal coagulant dosage and typical dosages are 2 to 5 mg/l (Packham, 1967). A disadvantage with this chemical is that with carelessness in the method of preparation, that is, with the concentration of reagents, degree of mixing, degree of neutralization, and aging, silica gel may form. For this reason, the use of activated silica often leads to blockages in pumps and feed pipes and may give rise to reduced filter runs (Packham, 1967; Robinson, 1974).

Several workers have reported on the enhanced organics removal promoted by activated silica (Batchelor & Yang, 1982; Gibson, 1994). Batchelor and Yang suggested that the negatively charged activated silica polymer better attached to the positively charged sites on natural organic matter coagulated with alum. At higher dosages, restabilization occurred which supported the charge neutralization hypothesis. However, they also found similar removals using activated silica alone, without alum. This suggested an adsorption-bridging mechanism.

Gibson suggested that the enhanced NOM removal by activated silica was due to the interlaced fibrillar structure of the polymer, that serves to sweep pinpont flocs and precipitates. The fact that activated silica also performs well–and sometimes better when it is added before alum, is puzzling, but might suggest an effect of mixing and distribution in the water to be treated.

3.3.3 Natural polyelectrolytes

Natural polymers have long been used as flocculants. Sanskrit literature (*ca.* 2000 BC) mentions the use of the crushed nuts of the Nirmali tree, *Strychnos potatorum Linn* as a means of clarifying water (Cohen, 1958). The Nirmali tree grows in India and Southeast Asia. The extract of the seeds from the tree are an anionic polyelectrolyte. The main functional groups are carboxyl and hydroxyl, mainly contributed by the protein portion of the material. Tripathi *et al.* (1976) analyzed the constituents of the Nirmali seed extract and determined the following proportions: carbohydrates 52.5%; proteins 16.3%; lipids 9%. The balance was made up of other secondary ingredients. Preparation of the Nirmali seed is simple and requires grinding before preparing solutions in water.

There have been numerous naturally derived substances used to clarify waters for human consumption. There is evidence of the use of crushed kernels from almonds, apricots and peaches in Egypt, Sudan, Tunisia, Lesotho, South Africa, and in Potosi, Bolivia (Jahn, 1988). The sap from certain cactus (*Opuntia*) species in Peru and Chile, as well as Haiti produces an anionic polyelectrolyte that has been used to clarify water, but more effectively as a flocculant aid. Kirchmer *et al.* (1975) demonstrated the effectiveness of the anionic polymer extracted from the

Opuntia cactus when used as flocculant aid with alum as the primary coagulant. Results obtained with this natural polymer were similar to those obtained with synthetic polymers.

The seed of the Red Sorrela plant (*Hibiscus sabdariffa*) is widely cultivated in India and was shown to possess coagulating properties (Bulusu & Pathak, 1974). The pods were dried and thrashed to release the seeds. These in turn were cleaned of fibrous material and pulverized. The powder was then sieved and mixed with sodium carbonate in the proportion 9:1 by weight. The mixture was then mixed in water, at a concentration of 2 g/l. The resulting solution was found to be effective for clarifying high turbidity water.

Other natural flocculants include starch derivatives that are pre-gelatinized and water-soluble. They can be natural starches, anionic oxidized starches or amine treated cationic starches. The use of this class of products is particularly important in the paper industry. Other classes include polysaccharides, such as guar gums, tannins, chitosan, and the alginates.

In general, the advantages of natural polyelectrolytes are that they are virtually toxic free; bio-degradable in the environment; and the raw products are often locally available, whereas industrialized coagulants may not be.

However, the natural polymers may not always be effective and need to be assessed on a case-by-case basis, just as any other coagulant product. For example, Patel and Shah (1978) point out that natural polymers such as those prepared from Nirmali, Guar, Fenugreek and Red Sorrela seeds are effective as primary coagulants with very high turbidity waters, but not with lower turbidities below about 300 NTU. However, they are effective as flocculant aids at typical dosages from 2 to 20 mg/l, in conjunction with metal coagulants.

Examples of natural polymer products are as follows:

3.3.3.1 Seeds from the moringa oleifera tree

There are approximately 14 known varieties of *Moringa oleifera* trees around the world, particularly in developing countries. The tree is a native of northern India, but now grows widely throughout tropical countries. Different varieties appear to have differing coagulating properties that depend on the geographical location, climate, altitude, and soil characteristics. The trees grow freely in the wild, as well as in rural areas, where they serve as fence ornamentation, or parts of the tree are used as fodder for domestic animals, or for food preparation, traditional herbal remedies, and for cooking oil (Narasiah *et al.* 2002).

The seeds contain up to 40% by weight of oil. The oil is similar in content to olive oil and is of high edible and market value. Almost all parts of the tree, including the seed pods, leaves and bark are utilized for diverse uses. The tree is easy to cultivate and allows multiple harvests in a single year (Folkard & Sutherland, 2002).

When harvesting the seeds for coagulant production or other uses, the seedpods are often allowed to dry naturally on the tree before harvesting. The seeds are then easily removed from the pods, shelled, crushed, ground, and sieved. The seed powder is then mixed with water to yield water soluble proteins that possess a net positive charge. The cationic polyelectrolyte thus formed has molecular weights of the order 13,000.

The coagulant is usually prepared as a 2 to 5% solution (20 to 50 g/l). After adding water, the suspension is vigorously shaken for at least 5 minutes to fully expand the molecules. The suspension is then filtered. Traditionally this is done through cotton cloth (Jahn, 1988). In the laboratory, two stage filters have been used: 2.5 μm, followed by 1.2 or 0.45 μm filter papers. In hot climates, fresh suspensions should be prepared every day to avoid deterioration. If the suspensions can be refrigerated, they can be kept up to one week without significant deterioration.

Narasiah *et al.* (2002) compared the efficiencies of two *Moringa* seed extracts from Burundi and from Madagascar, on the coagulation of a laboratory prepared kaolin turbid water. Both shelled and non-shelled crushed seeds were used. Initial turbidities during the tests were approximately 100 NTU. The goal was to achieve less than 1 NTU. Neither of the preparations using non-shelled seeds from the two sources was able to reduce turbidity to less than 1 NTU. However, the seeds from Burundi were clearly superior to those from Madagascar. The preparations made from shelled seeds did attain lower than 1 NTU in the treated water. In this case the Burundi seed preparation required a dosage of 1 ml/l of 5% solution, whereas the Madagascar preparation required 4 ml/l.

Narasiah *et al.* showed that when the kaolin suspension was prepared at an initial turbidity of 6 NTU, settled turbidities actually increased with increasing coagulant addition. Note that filtered water turbidities were not determined. Other authors have also reported on the difficulty of treating low turbidity waters with this coagulant. Therefore, it is possible that in general, *Moringa* coagulants are not suitable for low turbidity waters. This may be due to the low molecular weight of the coagulant and the patch mechanism of charge neutralization and floc formation that forms smaller and light flocs.

Folkard and Sutherland (2002) tested a *Moringa* coagulant on a natural river water in Malawi. During the study period the raw water turbidity was approximately 350 NTU. The coagulant dosage was approximately 75 mg/l and final filtered turbidities were below 2 NTU. At turbidities below 50 NTU, direct filtration after flocculation was effective, using a *Moringa* coagulant dosage of 25 mg/l and achieving filtered turbidities of less than 1 NTU.

Apart from oils and flocculant material, *Moringa* seeds contain natural anti-microbial mustard oils. The amount of this antibiotic substance is relatively low and insufficient for water disinfection. Some detrimental effects were detected in some protozoa and on fish (guppies). However, no detrimental effect on rats was detected at dosages up to 500 mg/kg body weight (Jahn, 1988). No mutagenic effects were detected in Ames *Salmonella* mutagenicity assays, and no adverse effects on human consumption have been identified (Grabow *et al.* 1985).

3.3.3.2 Starches

Polymers may be processed from various sources of starches, including potato, corn, cassava (manioc), arrowroot, and yams. Starches are basically highly polymerized carbohydrates. These polymers may be non-ionic, cationic, or anionic, depending on the form of processing and the substitutions. The cationic types usually have quaternary ammonium group substitutions, and the anionic types have carboxylic substitutions. The natural polymer without substitutions is largely nonionic. Molecular weights of the processed polymers are on the order of several millions.

Campos *et al.* (1984) investigated the use of arrowroot starch pre-gelatinized with sodium hydroxide. Full scale tests were carried out at the São Carlos water plant. Using 0.5 mg/l activated arrowroot starch, the alum dosage could be reduced by 20% and achieve better settled and filtered water qualities when compared with the use of alum alone.

Campos *et al.* carried out further investigations using several sources of starch as flocculant aid for potable water treatment. Both sodium hydroxide and heat were tested and compared for pre-gelatinizing and activating the starches. The primary coagulant applied to all the waters tested was aluminum sulfate. The starch flocculants were added to enhance flocculation and sedimentation. Results obtained with four water types are summarized below. From the results, it appeared that selection of the best starch flocculants was principally dependent on raw water turbidity, more so than with alkalinity.

Water	**Characteristics**			**Best Results**	
	Turbidity (NTU)	**Color (mg/l Pt-Co)**	**Alkalinity (mg/l $CaCO_3$)**	**NaOH**	**Heat**
I	6.5 to 8.9	20 to 40	8.8 to 9.0	Arrowroot	Yam; arrowroot
II	110 to 150	70	3.3 to 14.6	Corn; cassava	Yam; arrowroot; cassava; potato
III	15 to 39	70	65 to 83	Potato; corn	Potato; arrowroot
IV	124 to 140	70	75 to 88	Yam; arrowroot	Yam; arrowroot

3.3.3.3 Guar gums

These are neutral (non-ionic) polysaccharides relatively unaffected by pH and ionic strength. They are subject to enzymatic degradation on storage, but this may be prevented by the addition of citric or oxalic acid. Guar gum has been used in uranium ore processing.

3.3.3.4 Tannins

These are complex polysaccharide tannin derivatives that have been used extensively in potable water, wastewater, and industrial effluent treatment applications. They are generally most effective under acidic conditions. Care must be taken on storage as they are subject to degradation reactions if left for lengthy periods. One commercial product (*Floccotan*) is a brown syrupy liquid that is amphoteric, with cationic or anionic characteristics depending on the pH. It is prepared as a partially condensed product of commercial wattle bark tannin extract, treated to produce active amine groups along the polymerized molecule. One such product is blended with synthetic polyDADMAC to produce a cationic polyelectrolyte.

3.3.3.5 Chitosan

Chitin is the skeletal substance of the shells of crustaceans, such as crabs, lobsters and shrimps, and is described as a high-nitrogen-containing linear amino-polysaccharide polymer, with a molecular weight of several hundreds of thousand. It is the most abundant biopolymer in nature, after cellulose. Chitosan is usually prepared in acetic acid where it undergoes hydrolysis reactions (Kawamura, 1991). Other solvents have been used for its preparation, including adipic, formic, malic, propionic and succinic acids. Chitosan is a cationic polyelectrolyte with a molecular weight of approximately 10^6. It is widely used in the food industry, as well as sold as a health food.

Sekine *et al.* (2006) applied 1.5 mg/l of a commercial chitosan solution directly to the river during a river construction project, to reduce the detrimental ecological effects arising from increased turbidity. Chitosan was used in preference to alum to reduce the impacts of chemical addition on the river biota. Some limited effect on fish larvae was observed, but the effect was for a short stretch of river and was less than observed using alum or the untreated turbidity itself.

Vogelsang *et al.* (2004) demonstrated the effectiveness of chitosan on the removal of humic substances from Norwegian surface waters. They observed that the highest charged chitosan molecules tested were the most effective, indicating that charge neutralization was an important mechanism for the coagulation of the humic matter. Bridging or electrostatic patch mechanisms of flocculation may have been of importance, but the absence of a benefit in increased molecular weight of the chitosan molecules did not strongly support this. Adding small amounts of Fe^{3+} improved overall performance and allowed a reduction in chitosan dosage.

Guibal *et al.* (1999) used processed chitosan for metal ion recovery. They produced chitosan gel beads and found uptake capacities for molybdate of 750 to 800 mg/g.

3.3.3.6 Sodium alginate

Sodium alginate is an extract from certain brown seaweeds, or kelp. It is a non-toxic polysaccharide polymer that is widely used as a food additive in ice cream, milk, juice, beer, soup, ketchup, mayonnaise, yogurt, and margarine (Kawamura,

1991). The extracted product forms a negatively charged polyelectrolyte that is effective as a flocculant aid. It has been widely used for this purpose in the U.K. and Japan since the 1950s.

One commercial product contains sodium and calcium carbonates. Sufficient Na_2CO_3 is present to allow a 0.5% solution to be prepared in waters containing up to 500 mg/l calcium (as $CaCO_3$). For higher concentration solutions or higher calcium concentrations, additional Na_2CO_3 may be added to precipitate $CaCO_3$ and thereby prevent the relatively ineffective calcium alginate from forming. When used as a flocculant aid, dosages are normally within the range 0.1 to 0.6 mg/l.

3.3.4 Synthetic polymers

Although natural polyelectrolyte products have the advantage of being virtually toxic free, the use of synthetic polyelectrolytes is more widespread. They are, in general, more effective as flocculants principally due to the possibility of controlling properties such as the number and type of charged units and the molecular weight.

Because of the level of control possible during manufacture, synthetic polyelectrolytes have the potential of being applied almost in a tailor-made fashion. However, in many cases, because of the proprietary nature of these products, detailed information available to the user is often limited (Benedek, 1977; O'Brien, 1977; Packham, 1967). Potential users of polyelectrolytes are faced with a multitude of products which must be evaluated on a trial and error basis to identify the most appropriate for the particular application, although polymer suppliers will often provide various samples for given applications that they believe could be successful.

In general, polymer properties that are potentially important in a given application (although not all the information is always available) include the following (Packham, 1967; Ghosh *et al.* 1985):

- Type and chemical structure of polymer or copolymer
- Concentration of active ingredient
- Concentration of free monomer
- Concentration and nature of other impurities or constituents
- Molecular weight
- Viscosity under specified conditions
- Proportion of ionizable groups
- Charge density
- Amount of insoluble material present.

3.3.4.1 Structure of synthetic polyelectrolytes

Many synthetic polyelectrolytes are based on polyacrylamide and its copolymers with polyacrylic acid. The acrylamide monomer, making up the units of the polymer, is prepared by reacting natural gas and methane at high temperatures in the presence of controlled amounts of oxygen and ammonia to form hydrocyanic acid and acetylene, followed by catalysis with cuprous chloride. The acrylonitrile which results is then acid hydrolyzed to the acrylamide monomer using sulfuric acid. The acrylamide is then polymerized by catalysis.

Polyacrylamide itself is nonionic but on hydrolysis the macromolecule acquires carboxyl groups and assumes an anionic character. Increased negativity is usually accomplished by copolymerizing acrylamide with acrylic acid. The structure of nonionic polyacrylamide is as follows:

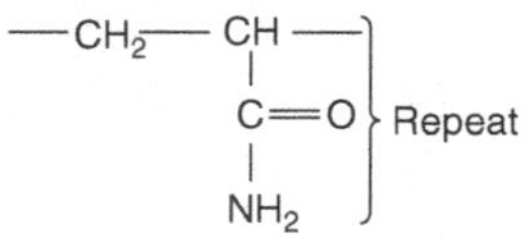

The structure of anionic hydrolyzed polyacrylamide is as follows:

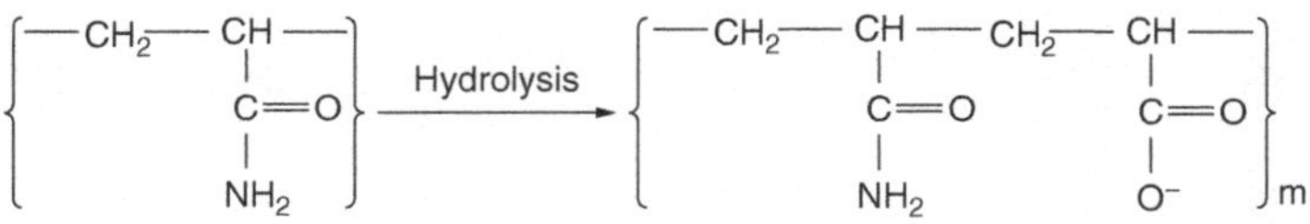

The structure of anionic polyacrylamide formed by copolymerization with sodium (or potassium, or hydrogen) acrylate is as follows:

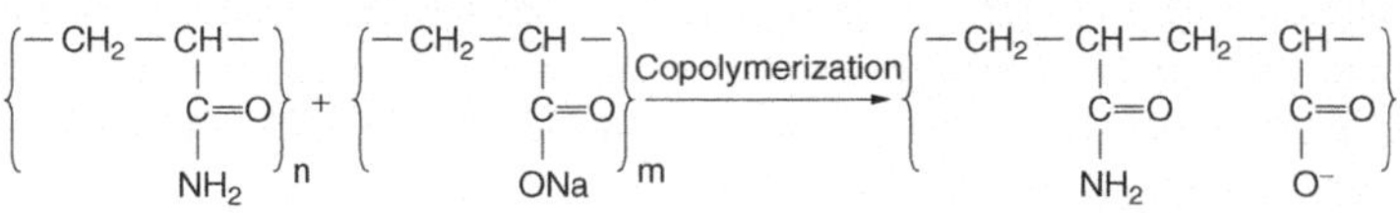

The structures of anionic polyacrylamide formed by hydrolysis, and by copolymerization with acrylate are seen to be identical. However, with copolymerization more control over molecular weight distribution and charge density is possible. The percentage of acrylate units copolymerized is referred to as the percent hydrolysis. The higher the degree of hydrolysis, the higher is the overall ionic charge. Typically, percent hydrolysis ranges from zero to 30%.

Cationic derivatives of polyacrylamide are available. For example, those produced by copolymerization of acrylamide and quaternary ammonium polyacrylamides, represented as follows:

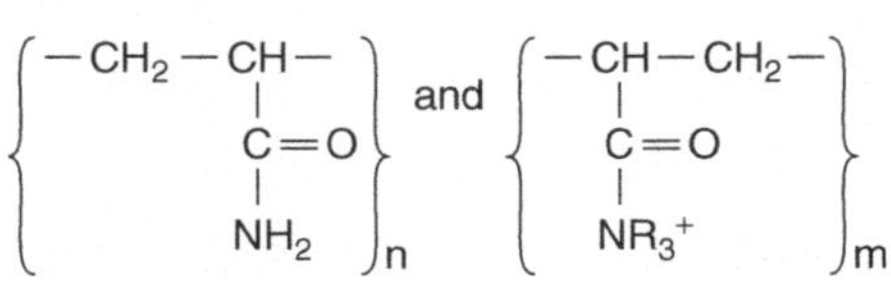

where R is usually a CH_4 or CH_3CH_2 derivative.

Cationic polyacrylamides are also prepared by the post reaction of polyacrylamide with formaldehyde and dimethylamine – known as *Mannich* polymers. Usually the end product of these polymers is liquid, although dried products are possible. Because of the high viscosity of Mannich polymers, these are usually supplied at 3 to 7% concentrations. Mannich polymers are widely used for biological sludge conditioning. One objection often expressed relates to the strong smell of formaldehyde, particularly in closed environments.

A common cationic quaternary ammonium compound is polydiallyldimethylammonium chloride (PDADMAC), which typically has the following structure:

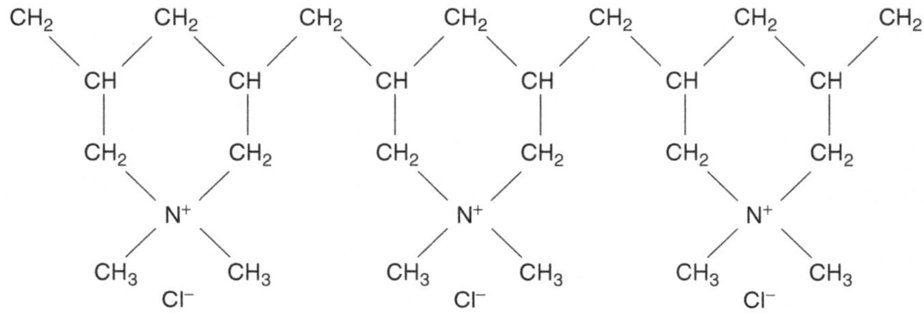

PDADMAC polymers are considered to have good chlorine resistance. Molecular weights of this polymer are typically 2 to 3×10^6 Daltons.

There are concerns with the formation of carcinogenic by products with these polymers (see later). PDADMAC polymers tend to have more unreacted monomer content than other polyelectrolyte products. As of 1990, there appeared to be little information available from EPA and others on the toxicity of PDADMAC and the monomer (Letterman & Pero, 1998, 1990).

Many cationics are based on polyethylenimine (PEI), which is usually of lower molecular weight than polyacrylamide. This is described as a highly branched polyamine produced by the acid catalyzed polymerization of ethylenimine. The polymer is composed of segments which have two carbons per nitrogen randomly distributed in the ratio of one primary nitrogen : two secondary nitrogens : one tertiary nitrogen, represented as follows:

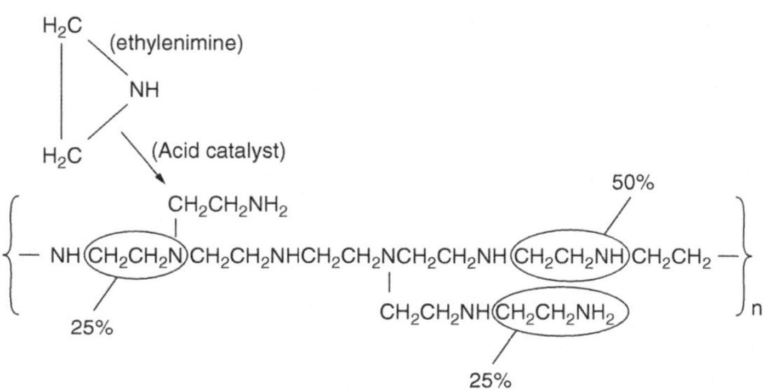

This distribution gives rise to what is believed to be an ellipsoid or spheroid shaped molecule composed of many branched segments with the tertiary amino nitrogens being the branching sites and the primary amino nitrogens being the terminal groups of each segment. Thus one segment proposed is as follows, although many others are possible:

$$-N-CH_2CH_2-NH-CH_2CH_2-NH-CH_2CH_2-NH_2$$

The primary and secondary amino nitrogens comprise reaction sites whereby the polymer may be conveniently modified. In water, the amino nitrogens react with H_2O to form positively charged nitrogens:

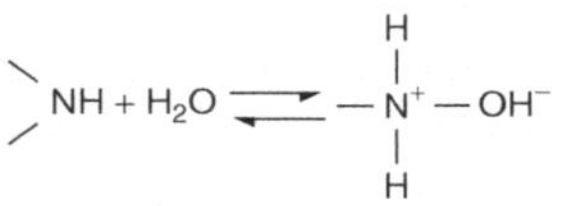

From the preceding reaction, it is seen that alkaline solutions are formed. Since the polymer is composed of many nitrogens per molecule, there will be many such positive sites; the addition of acid will increase the charge density whereas base will decrease it.

The ethylenimine monomer is particularly toxic, although PEI products typically contain less than 1 mg/kg residual monomer (Letterman & Pero, 1988).

Another common polyamine type of polyelectrolyte is epichlorohydrin-dimethylamine. This polymer tends to be of a linear nature, with molecular weights of the order 750,000 Daltons.

The polymeric structures and scheme of reactions shown above are intended to portray a general idea of the way in which polyelectrolytes are formed. There are other polymer formulations, such as polyethylene polyamine polymers; melamine/formaldehyde polymers; poly-dimethylaminoethyl polymers, and others. There are also many other reactions modifying the final product and producing polyelectrolytes with widely differing properties; with non-ionic, anionic, cationic or ampholytic characteristics, with different charge densities and distribution of charge, different molecular weights, and so on (see Letterman & Pero, 1988).

3.3.4.2 Charge density and molecular weight

The effect of charge density, or percentage hydrolysis, of polyelectrolytes is to influence the configuration in solution. For a given molecular weight, increasing charge density has the effect of increasing the viscosity of the polyelectrolyte solution (Katchalsky, 1953). Since viscosity increases with the length of the molecule, the implication is that with increasing charge density, polyelectrolyte chains are increasingly stretched by increasing electrostatic repulsion between charged units. At low charge densities, the polyelectrolyte assumes a tightly coiled, almost spherical configuration due to the predominance of Brownian movement. At medium charge density (10% to 20% hydrolysis) the configuration is that of randomly kinked or flexing coils. At high charge densities, the configuration is that of fully extended filamentous rods where the effects of electrostatic repulsion overcome those of molecular bombardment. It is appreciated that, for a given charge density, the longer the polyelectrolyte chain, the higher is the viscosity. Hence, it also follows that the higher the molecular weight of the polyelectrolyte, the higher is the viscosity.

Increasing the ionic strength of a polyelectrolyte solution decreases the range of influence of the charged sites on the polyelectrolyte chain (see Chapter 2) and an increasingly coiled configuration is achieved. The viscosity of the solution also decreases.

Charge density and molecular weight of a particular polyelectrolyte both influence the destabilization mechanism and resulting floc formation; the degree depending on the type of polyelectrolyte and the system to which it is applied. This will be discussed further in Chapter 5. As far as molecular weight is concerned, there does not appear to be an upper limit in terms of effective floc formation (Kitchener, 1972). However, there is a practical limit dictated by the difficulty of dissolving polymers with molecular weights in excess of 10^7. The grains of powder swell quickly, but considerable time is required to untangle the long chains. High shear stirring cannot be employed because of the breakdown of the expanding polyelectrolyte chains.

3.3.4.3 Forms of polymer

Synthetic polyelectrolytes are supplied as free flowing bead or granular powders, fine and dusty powders, or liquid types, including emulsions and solution polymers.

3.3.4.3.1 Dry polymers

Many polymers are supplied in dry form. The principal advantage of the dry form is that the products are approximately 80 to 95% active polymer, which economizes substantially on transport costs. Storage times are much longer than other forms, typically of the order 2 years if stored correctly. Some dry products comprise fine and dusty powders that can be a health and safety hazard depending on the handling methods employed. Although more expensive, bead types of dry polymer have a number of advantages. The dust problems are considerably reduced, and dissolution and preparation of the products is improved, because of the uniform size distribution of the particles (Kim, 1995).

Dry polymers are very hygroscopic and should be stored carefully in a cool, dry area until ready for use. If moisture penetrates the stored polymer, the effectiveness of the polymer can be reduced, and dissolution of the gels formed can be very difficult. Spilled polymer, particularly because of the hygroscopicity, is difficult to clean up. Even a thin film of dust can cause a severe slipping hazard when wetted. For this reason, using a hose for cleanup of spilled dry polymer is not advisable. If possible, spilled polymer should be removed with a vacuum or, if unavoidably wetted, some form of polymer adsorbent such as vermiculite or other proprietary preparation should be used. Polymer solutions can be broken down with a strong chlorine solution and hosed down. Aluminum sulfate solution has also been occasionally used for this purpose.

Dry polymers are usually supplied in 25; 500; 750 and 1,000 kg (55; 1,100; 1,650 and 2,200 lb) moisture resistant bags. The larger bags are mounted directly over volumetric feeders using a monorail hoist system.

3.3.4.3.2 Emulsion polymers

The process of drying polyacrylamide polymers is one of the most difficult and costly operations of the manufacturing process. Therefore, some manufacturers produce only emulsion types of these polymers. Emulsion polymers consist of micron-sized droplets of polymer and water suspended in mineral oil. Stabilizer surfactant additives are included in the emulsion preparation to maintain the polymer droplets stable within the hydrocarbon oil and to extend the shelf life of the products. Other surfactants are used to allow the polymer product to invert from a water-in-oil state, to an oil-in-water state during dilution, which makes the polymer easier to dissolve and prepare for use. A serious objection to these polymers is the adverse environmental effects of the surfactants. The mineral oil and surfactants add nothing to the action of the polymers during treatment, but are released into the treated water. The surfactants used include alkylphenolethoxylates (APEOs) which decompose to nonylphenol, a known endocrine disruptor which is mutagenic to aquatic organisms.

Emulsion polymers typically have active polymer contents of 25 to 60%. These polymer products tend to become stratified if stored for extended periods. Therefore, drum mixers or recirculation pumps are often used to re-homogenize the contents before transferring the contents to feeding and makeup equipment.

Liquid polymers (emulsions and solution polymers) can be delivered in 200 l (55 US gallon) drums or in larger self-contained containers that are returned to the supplier after use. These come in various sizes, although the 750 and 1,500 l (200 and 400 US gallon) containers appear to be the most common.

Liquid polymers can also be delivered in bulk and stored in on-site storage tanks. Large vent pipes are usually provided to relieve internal pressure and vacuum as the tanks are filled or drawn down. In the case of emulsion polymers, the vent pipe is often connected to a large dessicant vessel to avoid drawing in humid air. A separate large pallet valve is often provided on top of the tank to allow air to escape when the tank is being filled quickly and the tanker truck is being purged. The end of the overflow pipe from the tank is often immersed in several inches of mineral oil to avoid humid air being drawn in from the overflow pipe.

A new development in emulsion polymer manufacture has been to abandon the use of mineral oils and surfactants for a new class of water soluble emulsion products. The process essentially involves the dissolution of the monomer in an aqueous salt solution of ammonium sulfate before polymerization. A low molecular weight dispersant polymer is added to prevent aggregation of the polymer chains as they develop. Dissolution and preparation of these products for final use are easier because the polymers are produced as stable colloids in water.

3.3.4.3.3 Mannich polymers

These are a special kind of solution polymer. They consist of hydrolyzed polyacrylamide with very high molecular weights. Because of the very high viscosity, these polymers are usually supplied at 3 to 7% concentrations and contain less than 10% active polymer. Because of the high viscosity, Mannich polymers are usually prepared with a combined metering and dilution device with high mixing intensity.

Mannich polymers are manufactured by attaching aminomethylated tertiary amine to a polyacrylamide backbone using dimethylamine (DMA) and formaldehyde. One drawback with this polymer is that the Mannich reaction is reversible and DMA and formaldehyde can be released. A common objection with these polymers relates to the strong smell of formaldehyde, particularly in closed environments.

3.3.4.3.4 Solution polymers

These polymers typically contain 10 to 50% active polymer dissolved in water. Molecular weights tend to be low, from 5,000 to 200,000 and are cationic because of the presence of amine sites. The most widely used of these polymers are PDADMAC and epidimethylamine polymers. These polymers are readily dissolved in water simply by dilution. In-line combined metering and dilution devices are commonly used with these polymers.

3.3.4.4 Preparation of synthetic polymer solutions

Dry and emulsion polymer products require special dissolution procedures since there is the risk of forming considerable quantities of gelled (*fish-eye*) solids which do not subsequently dissolve.

A common technique for uniformly wetting solid polyelectrolyte particles is to use some form of eductor. As shown in Figure 3.7, the eductor comprises essentially a variable aperture venturi formed between the threaded upper tube and the lower delivery tube. Water entering via the side inlet flows through the venturi producing a vacuum through the powder feed tube, drawing polyelectrolyte into the water stream. The method of operation is to first gently screw the top tube down as far as it will go. Then, with water flowing through the device, slowly unscrew the tube again until maximum suction is obtained as felt by placing a finger over the top of the powder inlet. Normally, one to two turns only are required.

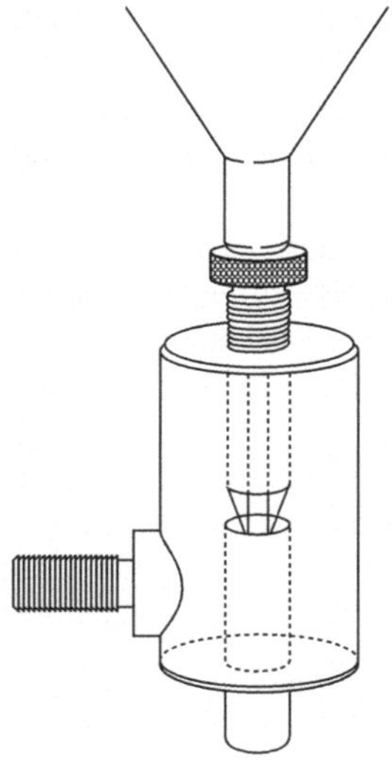

Figure 3.7 Eductor for evenly wetting polymer particles prior to dissolution (courtesy Allied Colloids).

When preparing polyelectrolyte stock solutions in batch mode, the following procedure is normally recommended (refer to Figure 3.8):

- Run water into the dissolving tank until it is approximately one-third full and covers the paddle stirrer, or air sparger. (Note that high speed stirrers are not recommended since breakup of the expanding polymer chains may occur).

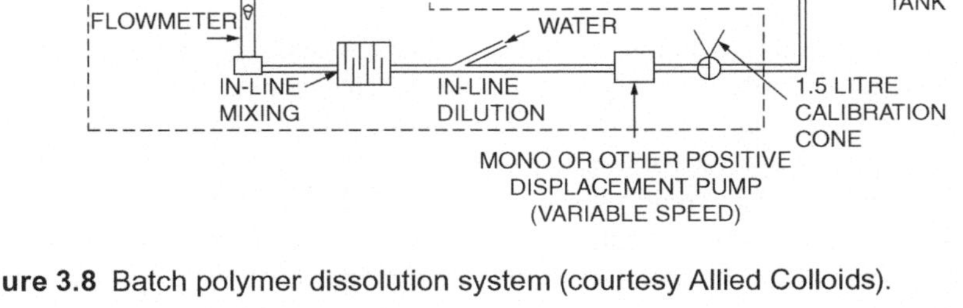

Figure 3.8 Batch polymer dissolution system (courtesy Allied Colloids).

- Commence stirring, turn on water to the eductor, insert dry funnel and commence feeding the desired quantity of polyelectrolyte through the funnel. (For emulsions, these are normally fed into the tank via a metering pump or a combined unit incorporating metering pump, in-line water feed and wetting device such as an in-line mixer).
- Continue adding water until a 0.5% solution is obtained (or value recommended by the manufacturer of the particular product).
- Continue slow stirring for a sufficient aging period to allow polymer coils to unwind. The period depends on the particular product but is usually of the order one hour.
- The prepared solution is passed to a stock solution holding tank.
- Dilution water is added downstream of the stock tank and blended with an in-line mixing device. The metering pump and in-line wetting device are often incorporated in a combined package unit.

Figures 3.9, 3.11 and 3.13 show continuous polyelectrolyte dissolution and feeding systems. The system shown in Figure 3.9 relies on a vibrating device (such as shown in Figure 3.10) to feed a continuous supply of powder to the eductor.

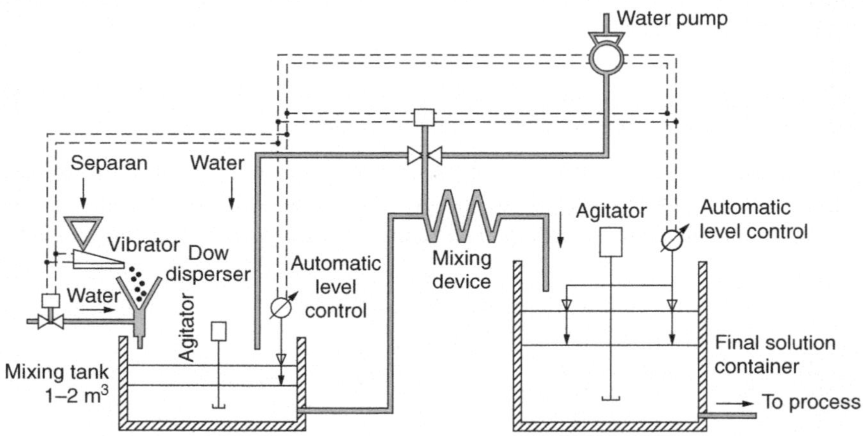

Figure 3.9 Continuous type polymer dissolution system (courtesy Dow Chemicals).

Figure 3.11 shows a continuous dissolution system specially designed for bead type solid polyelectrolytes. A dry chemical feeder (b) – for example, a screw type, as shown in Figure 3.12 – transports solid polyelectrolyte from the bulk container (j) to a hopper (d) and thence to a dry type eductor (e) through which is passed a high velocity air stream from a blower (c). Bead polyelectrolyte is thus passed to dispersers (a) within which an intense water spray pattern ensures uniform wetting of the solid particles. The hose connecting the dry eductor to the disperser is of an antistatic type.

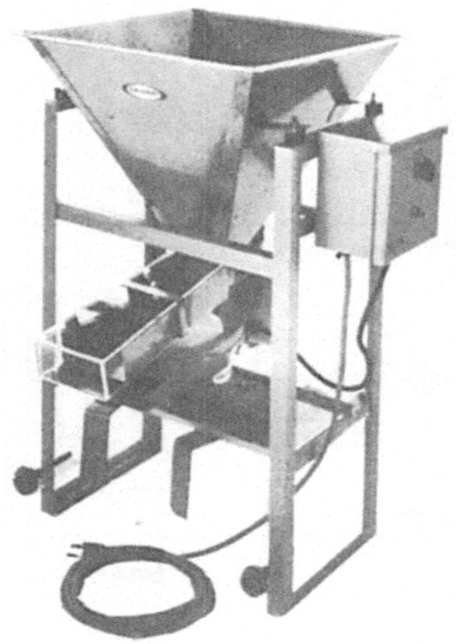

Figure 3.10 Example of volumetric dry polymer vibrator feeder (courtesy Calgon Corp.).

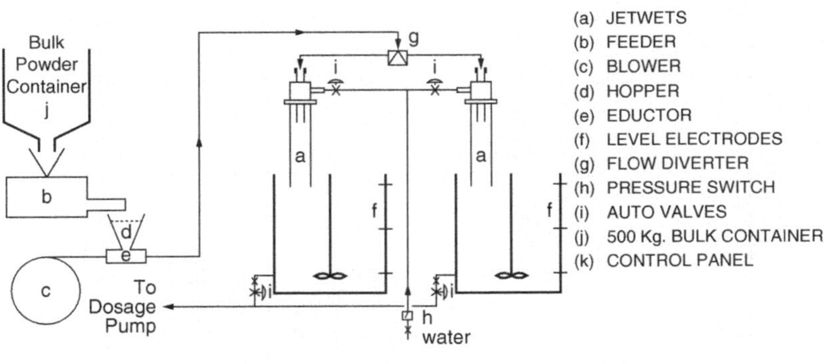

Figure 3.11 Example of bead type polymer continuous dissolution system (courtesy Allied Colloids).

In Figure 3.13, polyelectrolyte powder is fed into a disperser from a screw fed dry feeder. A small circular heater is fitted around the feeder outlet to prevent condensation and thus blockages. The disperser provides a high velocity water curtain directed towards the centre. The solid particles, thus wetted, are passed into the first of three in-series mixing/aging chambers to give a total retention time of approximately one hour. Diluting water is added downstream of the aging chambers.

With dry and emulsion type polymers, aging and proper activation is an important part of the preparation process. Some emulsion polymer manufacturers

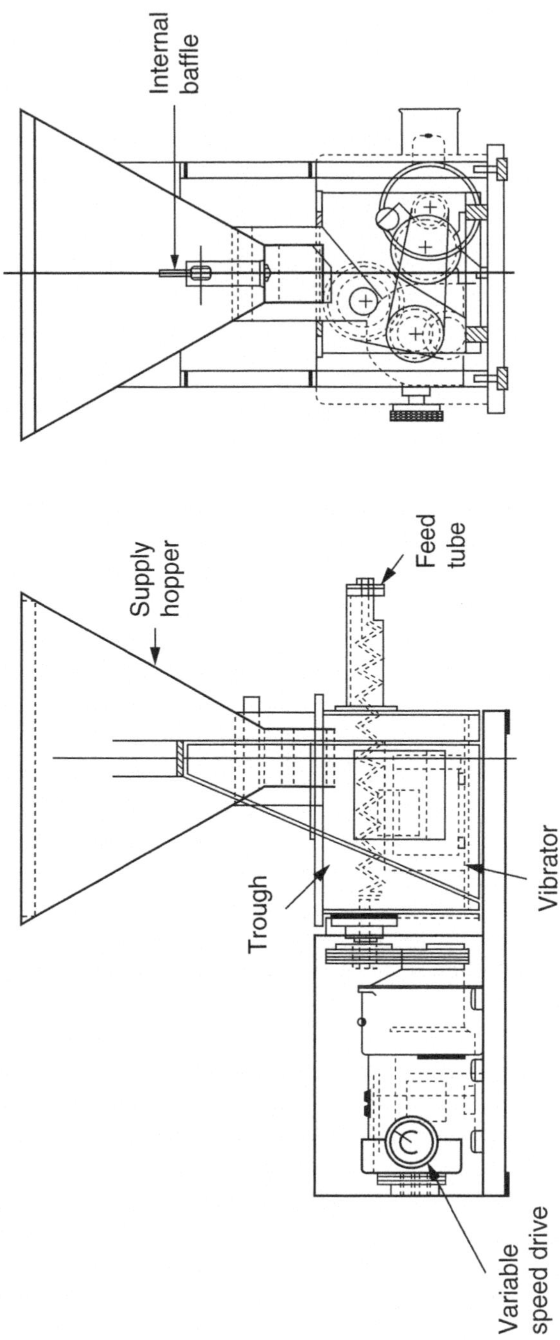

Figure 3.12 Example of screw type dry chemical metering feeder (courtesy Simon-Barron Ltd.).

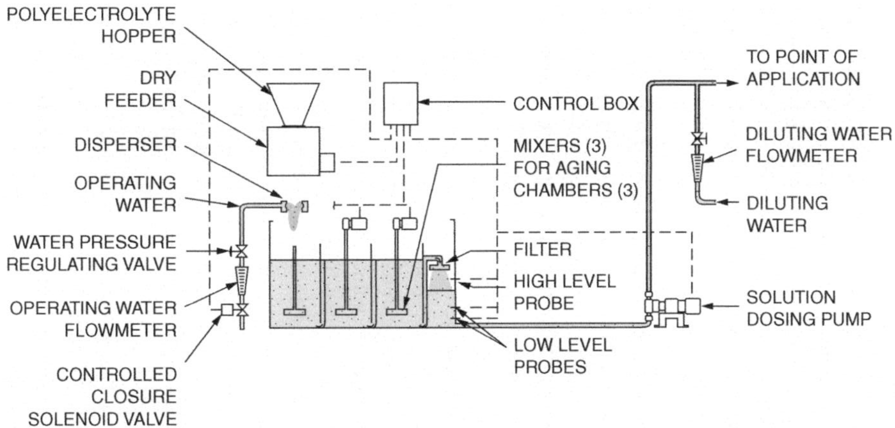

Figure 3.13 Example of continuous type dry polymer product dissolution system (courtesy Wallace Tiernan).

recommend in-line blenders and dosing systems without significant aging. In some cases this may be appropriate for small systems. Smaller users may wish to avoid the larger capital investment in a complete polymer preparation system. However, the penalty is to pay 10 to 20-percent more for polymer because of the lower efficiency of the polymer solution. For medium to high polymer users, it is normally appropriate to properly prepare the polymer before use. With these types of polymer, the appropriate aging time depends on the polymer, but one hour is not unusual.

Kim (1990) presented results investigating the factors that affect polymer activation. In general, the degree of activation achieved is higher with softer waters. As total hardness was increased from zero (distilled water) to 420 mg/l, the viscosity achieved (an indication of the degree of activation achieved) reduced by 87% for nonionic; 93% for cationic; and 95% for anionic polyacrylamide polymers. High electrolyte and hardness concentrations reduce the driving force for polymer swelling.

Temperature also retards the activation process, by reducing the diffusion rate of water into the polymer gel, particularly for ionic polyelectrolytes. The importance of aging time was illustrated by comparing viscosities at 8°C and 23°C. At zero time, the viscosity at 8°C was 80% the value at 23°C. However, after one hour of aging, the viscosities were almost equal (Kim, 1990).

Because of the risk of breaking polyelectrolyte chains both during mixing and during transfer from the stock tank to the point of application, care should be taken in designing feed lines, etc. Pumps should not be of the centrifugal type. Mono- or diaphragm types have been used extensively and give good service.

Final dilution of the polymer solution should be to a concentration of 0.05 to 0.1%, although this varies depending on the type of polymer. Dosing may be by means of a variable speed mono (or other positive displacement) pump with calibrated discharge. Flow meters (such as rotameters) may be used with polyelectrolyte solutions at concentrations less than 0.1%. An in-line 0.15 mm mesh filter should be installed to protect the flow meter from undissolved polyelectrolyte gel particles. Magnetic flow meters are often preferred for flow measurement of polymer solution.

Multipoint addition to the stream to be treated is often recommended by manufacturers. This could be merely a measure to overcome inefficient mixing (Packham, 1967) and local overdosing (Gregory, 1972), but it has been suggested (Akers, 1972) that loops formed on adsorption of polyelectrolyte chains, of importance for a bridging mechanism of destabilization, are retained for longer periods of time. However, there is no conclusive evidence on this point.

3.3.4.5 Toxicity of synthetic polyelectrolytes

Contaminants of synthetic polymers used in water and wastewater treatment generally arise from residual unreacted monomers (such as acrylamide, ethylenimine, diallydimethylammonium chloride and trimethylolmelamine), unreacted chemicals used to produce the monomer units (such as epichlorohydrin, dimethylamine, melamine and formaldehyde) and reaction by-products of the polymers in water.

With synthetic products, although there is no evidence that polymerized species are of high toxicity (Packham, 1967) the unpolymerized monomer species are. For example, acrylamide is extremely toxic producing severe neurotoxic effects. Packham (1967) reports on experiments carried out on rats, guinea pigs, rabbits, cats and monkeys, where it was demonstrated that ingestion of acrylamide had the following progressive effects: stiffness and weakness of hindquarters; loss of ability to control hindquarters; urinary retention; ataxia of the front legs; and inability to stand.

The 50% lethal dose, LD_{50}, for these animals varied from 100 to 250 mg/kg, depending on the animal (Letterman & Pero, 1990). When ingested, acrylamide is quickly distributed throughout the body and easily crosses the placenta in pregnant rats. Acrylamide is a cumulative neurotoxin. When the total acrylamide dose reaches 100 to 150 mg/kg body weight, over short- or long-term exposure, neurotoxic effects begin to appear in many species. Neurotoxic effects in humans exposed to acrylamide in drinking water has also been reported (Letterman & Pero, 1990).

The carcinogenicity of acrylamide, particularly affecting the thyroid, mammary and adrenal glands, as well as scrotum and oral cavity has been demonstrated in experimental animals.

For humans, it has been recommended that total absorption of acrylamide should not exceed 0.005 mg/kg · d. From this, for a polyacrylamide containing 1% monomer, the concentration in public water supplies that would be required for an 80 kg (175 lb) person drinking two litres (0.5 U.S. gal) of water per day, is 2 mg/l. This assumes that the acrylamide monomer is not absorbed by floc particles. However, many commercial polyacrylamides contain considerably less than 1% monomer (generally below 0.10%) and most manufacturers produce a wide range of special potable water grades where the monomer content is maintained at a low figure. Furthermore, in potable water treatment, the polyelectrolyte dosage is usually much less than 2 mg/l. For use in drinking water, the acrylamide content is generally mandated to be less than 0.05% by polymer weight (Dentel *et al.* 2000). Mallevialle *et al.* (1984) determined that the monomer content of a polyacrylamide based polymer used by them was at a low level and was not of concern, as long as an adequate quality control program was instituted at the water plant.

Polymers added to chlorinated waters often result in a reduction in the effectiveness of the polymer due to degradation by chlorine. Another effect is the formation of disinfection by-products. Chlorination of waters containing acrylamide produces chloroform ($CHCl_3$) and 2,3-dichloropropionic acid (Aizawa *et al.* 1990).

The rate of $CHCl_3$ formation is relatively slow and fractional yields of less than 0.01 mg $CHCl_3$/mg polyelectrolyte were obtained at pH 7. Yields increased at higher pH and holding times, increasing to 0.019 mg $CHCl_3$/mg polyelectrolyte at pH 9 and 3 days (Gehr & Soponkanaporn, 1990).

The formation of N-nitrosodimethylamine (NDMA) by chlorination of waters treated with polymers is a concern. NDMA is a human carcinogen and USEPA has recommended that levels in lakes and streams be less than 0.69 ng/L to protect drinking water sources and fish consumption.

Dimethylamine (DMA) is a NDMA precursor and is used as a starting reagent in the production of many polymers. Park *et al.* (2007) compared the residual DMA in several polymers investigated for their NDMA formation potential. A cationic polyacrylamide (PAM) product had 0.04–0.14 μg/mg active polymer; a PDADMAC product had 0.90–1.40 μg/mg active polymer; a polyamine product had 0.81–1.84 μg/mg active polymer; and a Mannich product had 5800–7800 μg/mg active polymer. The reason for the exceptionally high DMA residual is the reversible nature of the Mannich reaction and the release of DMA with time.

The NDMA formation from the same polymers tested by Park *et al.* was in the order Mannich polymer $\gg$ polyamine > PDADMAC > cationic PAM. The Mannich polymer had very high levels. In the tests, monochloramine yielded the highest NDMA compared with free chlorine, chlorine dioxide or nitrite although, with the Mannich polymer, all the oxidants yielded significant levels.

Wilczak *et al.* found significant NDMA formation when chloramines were dosed to waters containing PDADMAC polymers. The levels of NDMA formed increased with increasing PDADMAC concentrations and contact times. However, negligible levels of NDMA were formed using free chlorine in the ranges typically used in water plants. Also, lower NDMA levels were formed when free chlorine was contacted for 1 to 4 hours before chloramination. At two water plants tested, the recommendation was to reduce PDADMAC polymer dosages from 0.8–1.2 mg/l, to 0.3 mg/l, and maintain a free chlorine contact before ammonia addition and chloramine formation. This procedure reduced NDMA concentrations in the distribution system from over 10 ng/l, to approximately 2 ng/l (Wilczak *et al.* 2003).

Epichlorohydrin, together with acrylamide, is also of significant concern. The LD_{50} for rats, mice and guinea pigs is between 90 and 240 mg/kg. Death is caused by respiratory failure through depression of the central nervous system. Ingestion of epichlorohydrin leads to rapid and complete absorption and distribution to many tissues. Epichlorohydrin is both a potent carcinogen and mutagen.

Biesinger and Stokes (1986) investigated the effects of a number of nonionic, anionic and cationic polyelectrolytes on aquatic organisms. They found that nonionic and anionic polymers were not acutely toxic to *daphnids*, *fathead minnows*, *midges*, and *gammarids*. However, cationic polymers were very toxic. The toxicity of the 15 cationic polymers tested varied widely with the chemical structure. *Daphnids* were the most sensitive species. The 48-hour LC_{50} ranged from 0.09 mg/l to >100 mg/l for this species, depending on the polymer.

Environmentally, polymer molecules tend to be recalcitrant. Therefore, one expects a long term presence of these substances in biosolids that have been processed using polymers. However, no harmful effects of residual polymers in aerobic soil environments have been detected (Dentel *et al.* 2000).

It is advisable to check the likely toxicity of any chemical additives to water supplies with organizations such as the Department of the Environment, U.K., the Environmental Protection Agency, U.S.A., and the National Science Foundation (NSF), U.S.A. that publish lists of approved products, together with concentration limits.

3.4 REFERENCES

Aizawa T., Magara Y. and Musashi M. (1990). Problems with introducing synthetic polyelectrolyte coagulants into the water purification process. Water Supply, **8**, Jönköping, 27–35.

Akers R. J. (1972). Factors affecting the rate of the polyelectrolyte flocculation reaction. *Filtration and Separation*, 423–424.

Anon (1975). Aluminum sulphate in wastewater treatment. *Wat. and Sew. Wks.*, Ref. No., R189–R192.

Anon (1990). Carcinogen found in Canadian water plant. *World Water and Env. Eng.*, 10.

Basolo F. and Johnson R. (1964). Coordination Chemistry. W.A. Benjamin, Inc., N.Y.

Batchelor B. and Yang P.-d. (1982). Technical note: removing trihalomethane formation potential by coagulation with activated silica and cationic polymer. *Jour. AWWA*, 494–496.

Beck M. T. (1970). Chemistry of Complex Equilibria. Van Nostrand Reinhold Co., Ltd.

Benedek A. and Bancsi J. J. (1977). Comparative evaluation of commercial polyelectrolytes for flocculating alum precipitated domestic wastewater. *Prog. Wat. Tech.*, **9**, 33–42.

Biesinger K. E. and Stokes G. N. (1986). Effects of synthetic polyelectrolytes on selected aquatic organisms. *Jour. WPCF*, **58**(3), 207–213.

Bratby J. and Marais G. vR. (1977). Thickening of brown water sludges by dissolved air pressure flotation. *Water SA*, **3**(4), 202–212.

Brosset C. (1954). Studies on the hydrolysis of metal ions – XI. *The Aluminum ion. Acta Chem. Scand.*, **8**, 1917.

Brown R. and Wierenga J. (2005). Problems with chemical shipments. *AWWA Opflow*, 10–13.

Bulusu K. R. and Pathak B. N. (1974). Seeds of red sorrela. A new coagulant – boon to villages. *Indian J. Env. Health*, **16**(1) 63–67.

Campos J. R., Vieira M. B. S. and Campos L. V. (1984). Use of starch derived from arrowroot gelatinized with sodium hydroxide as flocculant in potable water treatment (in Portuguese). *Revista DAE*, **44**, 136, Brazil.

Campos J. R., Vieira M. B. S. and Villela L. M. O. (1984). Comparison of the efficiency of various natural sources of starch when used as flocculants in potable water treatment (in Portuguese). *Revista DAE*, **44**, 137, Brazil.

Cohen J. M., Rourke G. A. and Woodward R. L. (1958). Natural and synthetic polyelectrolytes as coagulant aids. *Jour. AWWA*, **50**, 463–478.

Dentel S. K., Chang L.-L., Raudenbush D. L., Junnier III R. W. and Abu-Orf M. M. (2000). Analysis and Fate of Polymers in Wastewater Treatment. Wat. Env. Res. Foundation, Alexandria, Virginia.

Dousma J. and DeBruyn P. L. (1978). Hydrolysis-precipitation studies of iron solutions II. Aging studies and the model for precipitation from Fe(III) nitrate solutions. *Jour. Coloid and Interface Sci.*, **64**(1), 154–170.

Fan M., Sung S., Brown R. C., Wheelock T. D. and Laabs F. C. (2002). Synthesis, characterization, and coagulation of polymeric ferric sulfate. *Jour. Env. Eng. ASCE*, 483–490.

Fan M., Brown R. C., Sung S. W., Huang C.-P., Ong S. K. and van Leeuwen J. H. (2003). Comparisons of polymeric and conventional coagulants in arsenic(V) removal. *Wat. Env. Res.*, **75**(4), 308–313.

Folkard G. and Sutherland J. (2002). Development of a naturally derived coagulant for water and wastewater treatment. *Wat. Sci. Tech.–Water Supply*, **2**(5–6), 89–94.

Gao B., Yue Q. and Miao J. (2002). Evaluation of polyaluminum ferric chloride (PAFC) as a composite coagulant for water and wastewater treatment. *Wat. Sci. Tech.*, **47**(1), 127–132.

Gehr R. and Soponkanaporn T. (1990). Assessing polyelectrolyte behavior by size-exclusion chromatography. *Jour. Env. Eng., ASCE*, **116**(2), 343–360.

Ghosh M. M., Cox C. D. and Prakash T. M. (1985). Polyelectrolyte selection for water treatment. *Jour. AWWA*, 67–73.

Gibson S. J. (1994). Enhanced coagulation with activated silica. AWWA Enhanced Coagulation Research Workshop, Charleston, SC, 4–6.

Gillberg L., Hansen B., Karlsson I., Enkel A. N. and Pålsson A. (2003). About Water Treatment. Kemira Kemwater, Sweden.

Grabow W., Slabbert J. L., Morgan W. S. G. and Jahn S. A. A. (1985). Toxicity and mutagenicity evaluation of water coagulated with moringa oleifera seed preparations using fish, protozoan, bacterial, coliphage, enzyme and ames salmonella assays. *Water SA*, **11**(1), 9–14.

Gray K. A., Yao C. and O'Melia C. R. (1995). Inorganic metal polymers: preparation and characterization. *Jour. AWWA*, 136–146.

Gregory J. (1972). Discussion to factors affecting the rate of the polyelectrolyte flocculation reaction. *Filtration and Separation*, 425.

Gregory J. and Dupont V. (2001). Properties of flocs produced by water treatment coagulants. *Wat. Sci. Tech.*, **44**(10), 231–236.

Grobbelaar N. J. H. and Ernst H. R. (1977). The use of ferric chloride as flocculating agent. *IMIESA*, April, 9 and 18.

Guibal E., Milot C. and Roussy J. (1999). Molybdate sorption by cross-linked chitosan beads: dynamic studies. *Wat. Env. Res. WEF*, **71**(1), 10–17.

He Y., Li F., Jiang J. and Wang H. (2011). Preparation and application of polyferric sulfate in drinking water treatment. *Proc. 12th International Conf. on Environmental Sci. and Technol., Rhodes, Greece*, 8–10.

Jahn S. A. A. (1988). Using Moringa seeds as coagulants in developing countries. *Jour. AWWA*, 43–50.

Jiang J., Graham N. and Harward C. (1993). Comparison of polyferric sulphate with other coagulants for the removal of algae and algae-derived organic matter. *Water Sci. Tech.*, **27**(11), 221–230.

Jiang J., Graham N. J. D. and Harward C. (1996). Coagulation of upland coloured water with polyferric sulphate compared to conventional coagulants. *J. Water SRT–Aqua*, **45**(3), 143–154.

Jiang J. (2001). Development of coagulation theory and new coagulants for water treatment: its past, current and future trend. *Wat. Sci. Tech.: Water Supply*, **1**(4), 57–64.

Jiang J. and Graham N. J. D. (2003). Development of optimal poly-alumino-iron sulphate coagulant. *Jour. Env. Eng., ASCE*, 699–708.

Jones C. (1990). Personal Communication, Mesa CAP WTP.

Katchalsky A. (1953). Polyelectrolytes. Endeavour, 90–94.

Kawamura S. (1991). Effectiveness of natural polyelectrolytes in water treatment. *Jour. AWWA*, 88–91.

Kim Y. (1990). Maximize polymer activation. *Water/Eng. & Management*, 28–31.

Kim Y. H. (1995). Coagulants and flocculants. Theory and Practice. Tall Oaks Publishing, Colorado.

Kirchmer C. J., Arboleda J. and Castro M. L. (1975). Natural Polymers and Their Application as Flocculant Aids (in Spanish). Centro Panamericano de Ingenieria Sanitaria y Ciencias del Ambiente, Lima, Peru.

Kitchener J. A. (1972). Principles of action of polymeric flocculants. *Brit. Polym. J.*, **4**, 217–229.

Leprince A., Fiessinger F. and Bottero J. Y. (1984). Polymerized iron chloride: an improved inorganic coagulant. *Jour. AWWA*, 93–97.

Letterman R. D. and Pero R. W. (1988). Polyelectrolyte Coagulants in Water Treatment–An Assessment of Research Needs. AWWA Research Foundation, Denver, Colorado.

Letterman R. D. and Pero R. W. (1990). Contaminants in polyelectrolytes used in water treatment. *Jour. AWWA*, 87–97.

Mallevialle J., Bruchet A. and Fiessinger F. (1984). How safe are organic polymers in water treatment? *Jour. AWWA*, 87–93.

Matijevic E. (1961). Detection of metal ion hydrolysis by coagulation. *III–Aluminum. J. Phys. Chem.*, **65**, 826.

Najm I. and Trussell R. R. (2001). NDMA formation in water and wastewater. *Jour. AWWA*, 92–99.

Narasiah K. S., Vogel A. and Kramadhati N. N. (2002). Coagulation of turbid waters using Moringa oleifera seeds from two distinct sources. *Wat. Sci. Tech.–Water Supply*, **2**(5–6), 83–88.

O'Brien J. H. and Novak J. T. (1977). Effects of pH and mixing on polymer conditioning of chemical sludge. *Jour. AWWA*, **69**, 600–605.

Ockershausen R. W. (1965). Use of liquid alum in water plants. *Jour. AWWA*, **57**, 309–313.

Orgel L. E. (1970). An Introduction to Transition Metal Chemistry: Ligand Field Theory. Methuen and Co., Ltd.

Packham R. F. (1967). Polyelectrolytes in water clarification. *Proc. Soc. Wat. Treat. Exam.*, **16**, 88–111.

Packham R. F. and Sheiham I. (1977). Developments in the theory of coagulation and flocculation. *Jnl. Inst. Water Eng. and Sci.*, **31**(2), 96–108.

Park S., Wei S., Huang C., Mizaikoff B. and Aral M. (2007). A study of the effects of polymers on potential N-nitrosodimethylamine (NDMA) formation at water and wastewater treatment plants. Report from Multimedia Environmental Simulations

Laboratory (MESL-03–07), School of Civil and Environmental Engineering, Georgia Institute of Technology, Atlanta, GA.

Patel N. S. and Shah A. N. (1978). Coagulation aids – a review. *Jour, Indian Wat. Works Assoc.*, **10**(4), 357–362.

Polasek P. and Mutl S. (2002). Cationic polymers in water treatment. Part 1: treatability of water with cationic polymers. *Water SA*, **28**(1), 69–82.

Robinson C. N. Jr. (1974). Polyelectrolytes as primary coagulants for potable water systems. *Jour. AWWA*, **66**, 252–257.

Ruehl K. (1999). Aluminum chlorohydrate: a tool to help meet stricter standards and improve operations. *Water World*, 14–17.

Sekine M., Takeshita A., Oda M., Imai T. and Higuchi T. (2006). On-site treatment of turbid river water using chitosan, a natural organic polymer coagulant. *Wat. Sci. Tech.*, **53**(2), 155–161.

Singley J. E. and Black A. P. (1967). Hydrolysis products of iron III. *Jour. AWWA*, **59**, 1549–1564.

Singley J. E. and Sullivan J. H. (1969). Reactions of metal ions in dilute solutions: recalculation of hydrolysis of iron III data. *Jour. AWWA*, **61**, 190–192.

Stumm W. and Morgan J. J. (1962). Chemical aspects of coagulation. *Jour. AWWA*, **54**, 971–991.

Stumm W. and O'Melia C. R. (1968). Stoichiometry of coagulation. *Jour. AWWA*, **60**, 514–539.

Sullivan J. H. and Singley J. E. (1968). Reactions of metal ions in dilute aqueous solution: hydrolysis of aluminum. *Jour. AWWA*, **60**, 1280–1287.

Tripathi P. N., Chaudhuri M. and Bokil S. D. (1976). Nirmali seed–a naturally occurring coagulant. *Indian J. Env. Health*, **18**(4), 272–281.

Vogelsang C., Andersen D. O., Hey A., Hàkonsen T., Jantsch T. G., Müller E. D., Pedersen M. A. and Vàrum K. M. (2004). Removal of humic substances by chitosan. *Wat. Sci. Tech.–Water Supply*, **4**(5–6), 121–129.

Wang J., Deevanhxay P., Hasegawa T., Ehara Y., Kurokawa M., Hashimoto K., Nishijima W. and Okada M. (2002). A pilot study of polysilicato-iron coagulant. *Wat. Sci. Tech.–Water Supply*, **12**(2), 107–113.

Wang D., Luan Z. and Tang H. (2003). Differences in coagulation efficiencies between PACl and PICl. *Jour. AWWA*, **95**(1), 79–86.

Wilczak A. J., Assadi-Rad A., Lai H. H., Hoover L. L., Smith J. F., Berger R., Rodigari F., Beland J. W., Lazzelle L. J., Kincannon E. G., Baker H. and Heaney C. T. (2003). Formation of NDMA in chloraminated water coagulated with DADMAC cationic polymer. *Jour. AWWA*, **95**(9), 94–106.

Worthington R. (1978). New alums for old. *Water Services*, **82**, 100–112, 143–144, 211–215.

4

Treatment with metal coagulants

4.1 INTRODUCTION

A mechanism of destabilization of hydrophobic colloids with indifferent electrolytes has already been discussed in Chapter 2. It was seen that by increasing the ionic strength of the solution, compression of the double layer occurs thereby reducing the range of interparticle repulsion. For a given suspension, as the ionic strength is incrementally increased, the transition from stability to destabilization occurs over a narrow range of electrolyte concentration. Furthermore, if the indifferent electrolyte is increased to excess, there is no effect on destabilization.

Since the effect of an indifferent electrolyte is merely to increase ionic strength and, thereby, to increase the effective concentration of counter-ions in the vicinity of the charged surfaces without the occurrence of adsorption, the electrolyte (or coagulant) concentration required for destabilization is independent of particle concentration. Furthermore, for the case of indifferent electrolytes, the efficacy of a coagulant is strongly dependent on the valency of the counter-ions such that for mono-, di-, and trivalent counter-ions, the concentration of coagulant required would be in the ratio 800:12:1, respectively.

However, in the majority of cases, the above phenomena may not occur simultaneously and, in some cases, not at all. For example, restabilization of a suspension often takes place if an excess of coagulant is applied; The particle concentration often has an effect on the concentration of coagulant required for destabilization (in some cases a stoichiometric relationship is evident) and the efficacy of multivalent coagulant species may be even more so than predicted by the $1/z^6$ Schulze-Hardy rule.

From the above, it is evident that further considerations beyond that of double layer repression by increasing ionic strength are of importance. As stated previously, the latter mechanism does not consider the possibility of adsorption at particle surfaces. As seen in Chapter 2, the adsorption of species in solution does occur at the particle surface and may augment the original charge carried by the particle or, in the case of adsorption of counter-ions, may negate or even cause charge reversal. Furthermore, as inferred above, adsorbable coagulant species are observed to destabilize hydrophobic colloids at much lower concentrations than is the case with species of low adsorptivity.

With the acknowledgement of the role of adsorption during destabilization with metal coagulants, two further mechanisms may be advanced: the first is described as *adsorption destabilization* which ultimately has a similar effect to that of the physical double layer mechanism already described. That is, through adsorption of charged coagulant species of opposite sign to the particle surface, the effective surface charge is reduced and, as a consequence, the extent of double layer repulsive interaction between adjacent particles is reduced.

The second mechanism considered, again as a result of adsorption of coagulant species at the particle – solution interface, is that described as a *bridging* mechanism. As seen in Section 3.1.2, metal coagulants have a pronounced tendency to polymerize during hydrolysis reactions. As the extent of hydrolysis increases, progressively higher polynuclear species form. On adsorption of such polymeric species to particles a coagulant bridge spanning between adjacent particles is formed thereby promoting destabilization.

It will be seen in Chapter 5 that the above mechanisms are not confined to metal coagulants. Similar considerations apply to destabilization by polyelectrolytes where both bridging and an electrostatic phenomenon are accepted as mechanisms of destabilization.

A further mechanism of destabilization by metal coagulants is that of precipitate enmeshment. Under appropriate conditions of coagulant concentration and pH, metal coagulants in aqueous solution form metal-hydroxide precipitates. Such species serve to enmesh particulate material thus effecting destabilization essentially by a sweep action. Here, apart from perhaps serving as nuclei for precipitation, the presence of particulate material would be incidental.

Precipitation mechanisms are also of importance during destabilization of hydrophilic colloids. Here, because of the extent of hydration, electrostatic effects are relatively unimportant. Coordination reactions occur between metal ions and certain functional groups on the particle surface which serve as ligands. Destabilization is visualized as being the result of metal ion – functional group – hydroxide precipitate formation. Furthermore, with the removal of certain anions from solution (for example, phosphorus removal in tertiary wastewater treatment) precipitation reactions occur between metal ions, anions and hydroxides.

The four mechanisms of destabilization put forward for metal coagulants will be discussed more fully in the following sections. A greater in-depth coverage of the bridging mechanism is reserved for Chapter 5, where destabilization mechanisms for polyelectrolytes are dealt with. To give a preliminary overall picture of the four mechanisms, Table 4.1 (adapted from Stumm & O'Melia, 1968) presents various pertinent characteristics. From these characteristics it is evident that destabilization with metal coagulants cannot be exclusively attributed to any one

Table 4.1 Characteristics of destabilization mechanisms with metal coagulants.

Parameter	Influence of Indicated Parameter According to Mechanism			
	Physical Double Layer	**Adsorption Destabilization**	**Bridging**	**Precipitation**
Electrostatic interactions	Predominant	Important	Subordinate	Subordinate
Chemical interactions and adsorption	Absent	Important	Predominant	May occur but not essential for removal
Zeta potential for optimum destabilization	Near zero	Not necessarily zero	Usually not zero	Not necessarily zero
Addition of excess coagulant	No detrimental effect	Restabilization usually accompanied by charge reversal; may be blurred by precipitation	Restabilization due to complete surface coverage	No detrimental effect
Fraction of surface coverage (θ) for optimum floc formation	Negligible	$0 < \theta < 1$ (See Chapter 5)	$0 < \theta < 1$ (See Chapter 5)	Unimportant
Relationship between optimum coagulant dosage and particle concentration	Optimum dosage virtually independent of colloid concentration	Stoichiometry possible but does not always occur	Stoichiometry between dosage and particle concentration	Optimum dosage virtually independent of colloid concentration
Physical properties of flocs produced	Dense, high shear strength but poor filterability in cake filtration	Flocs of widely varying shear strength and density	Flocs of 3-dimensional structure; low shear strength, but excellent filterability in cake filtration	Flocs of widely varying shear strength and density

particular mechanism; in a particular instance, destabilization may be contributed to by one or several mechanisms.

4.2 DESTABILIZATION OF HYDROPHOBIC COLLOIDS

4.2.1 Extent of hydrolysis and adsorption

It was seen in Chapter 3 that metal coagulants during the transition from the free aquo metal ion to the insoluble metal hydroxide precipitate, undergo a series of hydrolytic reactions. It is strongly evident that the hydrolysis products of metal coagulants in aqueous solution are adsorbed more readily than free aquo metal ions; the greater the degree of hydrolysis, the more extensive is adsorption. In fact, replacing at least only one coordinated water molecule of the hydrated metal ion by a hydroxyl ion enhances significantly the adsorptivity of the ion (Matijevic, 1973).

The reasons for the enhanced adsorptivity of hydrolyzed metal ions on hydrophobic colloid surfaces are not well understood. Evidently the presence of hydroxyl ions in the coordination sheath is to a large extent responsible but just why hydroxyl ions are able to promote counter-ion adsorption remains to be resolved. As seen in Chapter 2 the electrostatic contribution to counter-ion adsorptivity may be negligible. For this reason adsorption of hydrolyzed neutral species, such as $Al(OH)_3$, and co-ions does occur, the latter often evidenced by an increase in the original charge carried by the particles. Accepting that the presence of hydroxyl ions does play a part during adsorption, it is appreciated that, as stated before, the greater the degree of hydrolysis and polymerization, the more extensive is the adsorption. One can visualize that with extensive hydrolysis, successively higher numbers of hydroxide groups per compound can be adsorbed at the interface.

One may further postulate that the replacement of water molecules by hydroxyl ions in the coordination sphere of a metal ion imparts a certain degree of hydrophobicity to the complex (Stumm & O'Melia, 1968). Hydrophobic colloidal particles do not adsorb unhydrolyzed aquo metal ions for the same reason that they do not adsorb water molecules (Matijevic, 1973). Furthermore, with a dispersion destabilized with hydrolyzed metal coagulants, if the suspension is made sufficiently acidic, restabilization occurs as evidenced by reversion of particle surface charge back to the original value, determined by electrokinetic measurements. Such desorption of metal ion species is due to de-hydrolyzation and a return to the aquo metal ion state. Such a desorption process is slow when compared to the rapidity of adsorption of hydrolyzed species (Matijevic, 1973). This indicates a slow replacement of complexed hydroxyl ions by water molecules on acidification.

4.2.2 Effect of coagulant dosage

As seen in Table 4.1, the effect of coagulant dosage on destabilization depends very much on the destabilization mechanism operative. For example, in certain instances where the mechanism of electrical double layer repression by increasing ionic strength is the predominant mechanism, as may be assumed for synthetic rubber latex particles destabilized with sodium or barium ions (Gregory, 1977), then a destabilization response to coagulant concentration curve of the form shown in Figure 4.1 will probably be obtained. The suspension passes from stability to destabilization over a narrow range of coagulant concentration. The critical coagulant concentration (CCC) is identified as the inflection point on the curve.

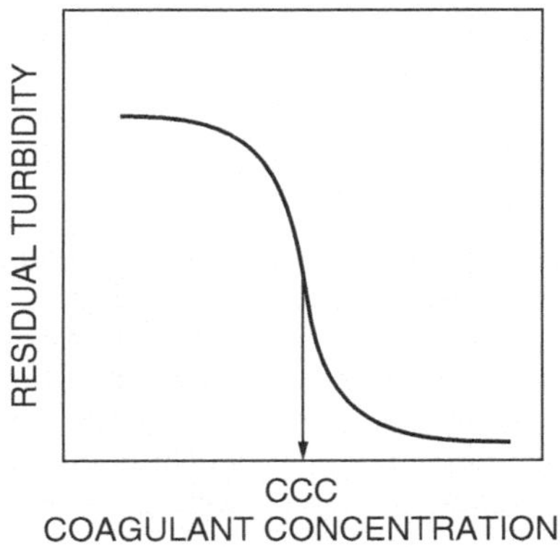

Figure 4.1 Destabilization characteristics where an electrical double layer repression mechanism is predominant. Increasing the coagulant (indifferent electrolyte) concentration beyond the CCC (critical coagulant concentration) has little effect.

In such instances, the CCC is independent of particle concentration since destabilization relies solely on the concentration of counter-ions in solution. Furthermore, increasing the coagulant concentration beyond the CCC has little effect: once the ionic strength is sufficient to compress the diffuse part of the double layer sufficiently for spontaneous destabilization to occur, no further advantage in increasing ionic strength is attained. Moreover, it is in such cases where the Schulze-Hardy $1/z^6$ rule is most likely to be applicable. Consequently, the valency of the counter-ions has a profound effect on destabilization: the higher the valency, the lower is the CCC for destabilization.

Often, from results obtained during destabilization of certain suspensions, more than one destabilization mechanism is evident, depending on the coagulant concentration applied, the concentration of particles in the suspension, and the pH.

Considering for the time being coagulant concentration only, Figure 4.2 (adapted from Stumm & O'Melia, 1968) shows results obtained for destabilization of a silica dispersion, with a surface concentration of $10 \, m^2/l$, using Fe(III) at a pH of 4.0. With increasing coagulant concentration, there are two regions where destabilization is obtained.

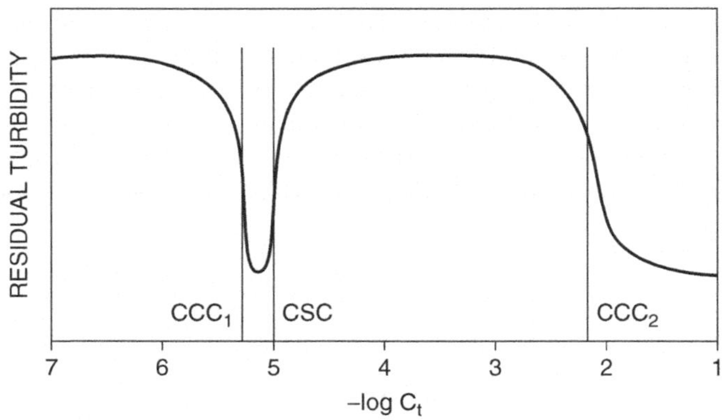

Figure 4.2 Destabilization characteristics where adsorption of coagulant species to colloidal particles is operative. CCC_1 and CSC signify the concentrations, C_t of coagulant necessary to destabilize and restabilize respectively the dispersion. A further critical coagulant concentration, CCC_2 indicates the occurrence of a double layer repression or enmeshment mechanism at higher coagulant dosages. The data was taken from Stumm and O'Melia (1968) for destabilization of a silica dispersion (surface concentration = 10 m^2/l) with Fe(III) at a pH of 4.0.

The first region, identified as CCC_1, occurs when sufficient metal-hydroxide coagulant species are present to become adsorbed and thereby destabilize the suspension. The mechanism here is one of either adsorption of charged metal hydroxide species, giving rise to reduction of charge or charge neutralization, or of adsorption of polymerized metal hydrolysis species giving rise to a bridging mechanism of destabilization. The first mechanism is probably predominant at low colloid surface concentrations, whereas with relatively high colloid concentrations, the second bridging mechanism may become predominant. A mechanism of destabilization predominantly by diffuse layer compression is discounted because if a higher coagulant dosage is initially applied, restabilization occurs. The initial concentration of coagulant required for restabilization is identified on Figure 4.2 as the CSC (critical stabilization concentration). Restabilization could occur either because of an excess of counter-ion adsorption giving rise to charge reversal, or because of excess adsorption of polymeric species occupying too many adsorption sites thereby preventing bridge formation between adjacent particles.

By increasing the initial applied coagulant concentration sufficiently, a further destabilization region is evident, the critical coagulant concentration for this region being identified as CCC_2 on Figure 4.2. Again, there are two possible mechanisms, the predominance of each depending on the colloid concentration. At relatively high colloid concentrations, the predominant mechanism is probably due to ionic strength effects. That is, the presence of charged counter-ions in solution gives rise to double layer repression. Alternatively, perhaps predominantly at lower colloid concentrations, metal hydroxide precipitates are formed. Destabilization in this case arises from a sweep mechanism of colloid enmeshment by precipitates.

The above discussion suggests that besides coagulant concentration, there are at least two other factors influencing the destabilization characteristics of a given suspension of hydrophobic colloids. These are the concentration of colloids in the suspension and the pH existing during destabilization. The following sections deal with each in turn.

4.2.3 Effect of colloid concentration

The preceding section discussed the effect of coagulant dosage on the destabilization of hydrophobic colloids. This section extends the discussion to include the effects of colloid concentration. Figure 4.3 (adapted from Stumm & O'Melia, 1968) demonstrates the relationship between coagulant dosage and colloid concentration, the latter in terms of surface area concentration. (Note that the following discussion may be taken as applicable for a pH of approximately 4.0 to 5.0). The shaded area indicates zones where destabilization has occurred. The relationships shown in Figure 4.3 are for one particular value of pH.

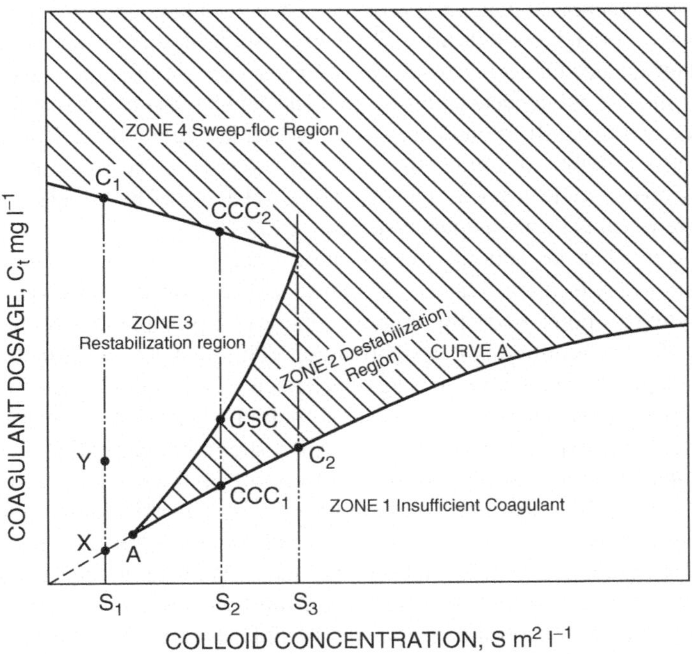

Figure 4.3 Zones of destabilization and restabilization at a given pH value as related to colloid concentration, S and coagulant dosage, C_t. (Adapted from Stumm & O'Melia, 1968).

Note that there will be several such diagrams, each at a particular pH value. Four zones are identified as follows:

- **Zone 1** indicates that insufficient coagulant has been applied to the colloidal suspension and that destabilization does not take place.
- **Zone 2** refers to the region in which destabilization has taken place.
- **Zone 3** is that region where destabilization and then, restabilization has taken place, due to excessive coagulant addition.
- **Zone 4** is the region where the coagulant dosage is high enough for over-saturation, and precipitation of metal hydroxide species occurs.

Figure 4.3 is explained as follows: At very low concentrations of colloids (S_1, say) there is very little contact opportunity for colloid – colloid interactions. When a quantity of metal coagulant such as X is introduced, even though the extrapolated stoichiometric coagulant dosage may be applied, uneven adsorption occurs and, therefore, a certain number of particles will remain stable, some will suffer charge reversal, some will be destabilized, and so on. The problem here, therefore, is visualized as being one of inadequate dispersion.

Suppose an increased coagulant dosage, Y, had been applied. The situation now exists where, although colloids are still greatly dispersed, there is an ample supply of coagulant species for each colloid. However, the concentration of coagulant with respect to the colloid concentration is relatively high so that excessive adsorption takes place and restabilization occurs as described in Section 4.1.2. The only way in which it is possible to remove the colloidal solids is to increase the coagulant concentration to a value of at least that given by C_1. Here metal hydroxide precipitates form and removal is by a sweep mechanism.

It is seen that at colloid concentrations such as S_1, increasing colloid concentration results in a decreased coagulant concentration, until at a concentration corresponding to point A on the diagram, there is a precipitous decrease in the coagulant dosage required for destabilization. This may be explained in terms of contact opportunity between coagulant species and colloids. As the colloid concentration progressively increases, adsorption of transition hydrolysis products occurs to a certain extent thereby forming nucleic sites for precipitation. In this way, less coagulant is progressively required to form precipitates. At a certain colloid concentration, such as that corresponding to A, contact opportunity is sufficient for hydrolysis species to become adsorbed and effect destabilization by charge effects or bridging, before precipitation takes place. With kaolin clay suspensions, point A probably corresponds to a suspended solids concentration of the order 50 to 100 mg/l, based on kaolinite (Kim *et al.* 1965; Packham, 1963, 1965).

Referring to Figure 4.3 it is seen that waters of low solids concentration may be more efficiently destabilized by augmenting the total surface area to a value such as that given by S_2, for example. This is carried out in some cases of water treatment practice where bentonite or activated silica may be added.

At colloid concentrations such as S_2, it is seen that the critical coagulant concentration for destabilization is much less than for S_1. However, destabilization occurs over a relatively narrow range of coagulant dosage. Beyond a critical dosage (CSC) restabilization of the suspension occurs. Significant colloid removal for coagulant dosages greater than CSC only occurs when a quantity of coagulant sufficient for precipitation is added. As discussed in the preceding section, the mechanism for destabilization for coagulant dosages between the CCC and CSC is one dependent on adsorption. Because of the predominance of an adsorption type mechanism in this region of colloid concentration, one would expect a dependence of the critical coagulant concentration on colloid concentration. This is seen to be so by a consideration of the linearity of curve A in this region. Such stoichiometry probably occurs when suspended solids concentrations are greater than that represented by point A, that is, of the order 50 to 100 mg/l (based on kaolinite).

At high colloid concentrations (S_3 and higher) it is seen that restabilization does not occur. This may be explained by considering chronologically the action of metal hydrolysis species in solution: The first effect must be compression of the diffuse part of the double layer, rather than extensive adsorption. As stated earlier, adsorption is more favourable with the higher metal-hydroxide species. Hence, before extensive adsorption takes place with its eventual related charge reversal, colloids, being in closer proximity than the preceding cases considered, will be destabilized and will have already formed flocs. Hence the region corresponding to colloid concentrations in excess of S_3 does not imply a sweep floc region, but rather a region of destabilization brought about by a double layer repression mechanism. This is supported by the shape of line A which is horizontally asymptotic with increasing colloid concentration. For a physical double layer mechanism, colloid concentration has no effect on the coagulant dosage for destabilization. It should be noted, however, that if the coagulant dosage is very high, then metal-hydroxide precipitates will form, but there would appear to be little advantage from the point of view of destabilization efficiency.

Summarizing the above discussion:

- For low colloid concentrations, less than that corresponding to point A, the destabilization mechanism is one governed by precipitate formation.
- For medium concentrations, between A and S_3, the destabilization mechanism is governed by adsorption and charge reversal may occur. Here, a model of the form proposed by Stern (see Chapter 2) would adequately describe the phenomena.
- For high colloid concentrations (greater than S_3) a double layer mechanism, such as that described by the Gouy–Chapman model, is probably dominant.
- For medium and high colloid concentrations, destabilization may be brought about by high coagulant dosages where metal-hydroxide precipitates are induced to form.

4.2.4 Effect of pH

4.2.4.1 General

The preceding discussions were directed towards the combined influence of colloid concentration and coagulant dosage on the characteristics of destabilization of hydrophobic colloids. A further factor, of paramount importance in destabilization with hydrolyzing metal coagulants, is the pH. As seen in Chapter 3, the predominance of a particular hydrolysis species during destabilization is very largely dependent on the pH value. For a particular colloidal suspension, it is logical to consider that there exists a particular hydrolysis species most effective for destabilization, whether by virtue of the charge carried, or the adsorptivity of the species. To quote Black and Chen (1965):

> The adjustment of pH to a range where the most effective hydrolysis species of the coagulant is formed is shown to be very essential in producing optimum coagulation.

Figure 4.4 shows the interrelationship between coagulant dosage, solids concentration and pH for the particular case of a silica dispersion using Fe(III) as coagulant.

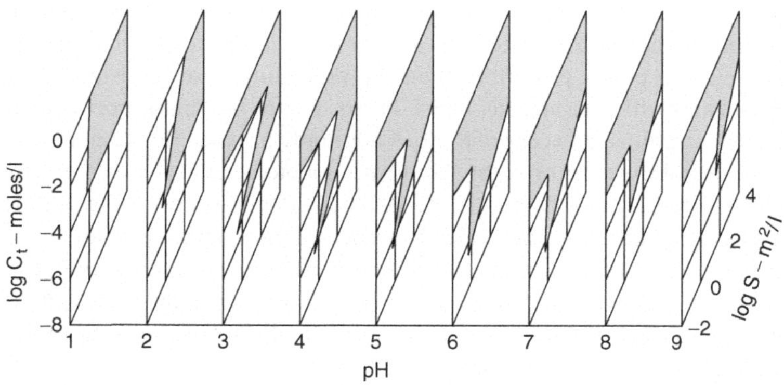

Figure 4.4 Zones of destabilization and restabilization as related to pH, colloid concentration, S and coagulant dosage, C_t. (From Stumm & O'Melia, 1968).

At each pH value, curves of a similar form to that of Figure 4.3 are obtained. It is instructive to study each pH value in turn:

4.2.4.2 pH 1.0

From Figure 4.4 it is seen that at low colloid surface concentrations, appreciable destabilization does not occur. Furthermore, it is seen that the coagulant concentration for destabilization is independent of colloid concentration and restabilization does not occur. This may be explained by a consideration of the probable hydrolysis species existing at this pH value. From Figure 3.5(a) it is seen that hydrated aquo metal ions are predominant. Therefore, it follows from the above observations that the mechanism of destabilization is almost totally due to double layer repression by ionic strength considerations. The absence of destabilization at low colloid concentrations can be explained in terms of kinetic effects where contact opportunities are too low for appreciable perikinetic flocculation in the time allowed for destabilization.

4.2.4.3 pH 2.0

Here, it is seen that similar characteristics to that at pH 1.0 are evident, with one important exception: at low to medium colloid concentrations, restabilization occurs. Referring to Figure 3.5(a) for, say, a total Fe concentration of 10^{-3} M, it is apparent that hydrolysis products begin to have significance at pH 2.0. As stated previously, hydrolyzed metal coagulant species have a pronounced tendency for adsorption. Therefore, at pH 2.0 destabilization is significantly influenced by adsorption such that excessive dosages lead to surface charge reversal and restabilization. At high colloid concentrations, destabilization is due to double layer repression by positively charged transition coagulant species.

4.2.4.4 pH 3.0 to 5.0

From the destabilization curves for pH 3.0 to 5.0 it is seen that at a given pH the coagulant dosage–colloid concentration relationship is as discussed previously. To briefly recapitulate: for colloid surface concentrations greater than approximately 1.0 m^2/L (for the colloid system considered here) the destabilization mechanism is one of adsorption, as evidenced by the dependence of coagulant dosage for destabilization on colloid concentration (lower boundary of the destabilization zone) and the occurrence of restabilization. The upper destabilization zone is one of metal-hydroxide precipitate enmeshment. At higher colloid concentrations the mechanism is increasingly one of double layer repression.

From Figure 4.4, it is seen that the slope of the critical coagulant dosage–colloid concentration boundary increases with increasing pH. This indicates an increasing dependence of the destabilization mechanism on adsorption. Referring to Figure 3.5(a) it is seen that within the range of critical coagulant dosage, of around 10^{-5} M, seven percent of coagulant species are hydrolyzed at pH 3.0. At pH 5.0, full hydrolysis is apparent.

Restabilization at each pH value is also explained by increasing hydrolysis. For example, at a pH of 3.0 the percentage hydrolysis of coagulant species at the critical dosage for destabilization is approximately 7%. Increasing the dosage to 10^{-4} M (approximately corresponding to the CSC) the percentage hydrolysis is 60%. Hence, increasing dosage in each case gives rise to increased hydrolysis and consequently increased adsorption. At low pH values (pH 3.0) the predominant species are positively charged; hence, restabilization is due primarily to charge reversal. At higher pH, at around 5.0, the predominant species are negative (for example, $Fe(OH)_4^-$) and one may postulate a mechanism of bridging with excessive surface site coverage resulting in restabilization.

4.2.4.5 pH 6.0 to 9.0

For pH values of 6.0 and higher, it is seen that restabilization does not occur. This may be explained by a destabilization mechanism of bridging by negative hydrolysis species with a continuous transition to precipitate enmeshment as the dosage is incrementally increased. The progressively increasing dosage for destabilization with increasing pH supports the predominance of negative hydrolysis species since with increasing negativity statistically less contact opportunity is afforded between coagulant species and colloidal particles.

Evidence for the increasing predominance of negative hydrolysis species would be provided if electrophoretic mobilities on coagulant addition were to become increasingly negative with increasing pH. This is seen to be so from Figure 4.5 which shows (a) ferric sulfate added to destabilize a montmorillonite suspension; (b) a Fullers earth suspension and (c) a kaolinite suspension. In general, initially (at low pH values) the mobility values are seen to increase to positive values (which follows from the discussions above where positive hydrolysis species were postulated); at higher pH values the mobilities are seen to increase in negativity.

4.2.4.6 Further considerations

Rubin and Blocksidge (1979) concluded from studies on the destabilization of montmorillonite sols with alum that in the pH-concentration regions common during water treatment, the precipitation of aluminum hydroxide species controls destabilization reactions. They also suggested the following treatment options to overcome the restabilization regions shown in Figures 4.3 and 4.4:

(1) increase the coagulant dosage;
(2) decrease the coagulant dose (settling will suffer, but since turbidity is likely low, flocs will be retained by filters);
(3) raise the pH;
(4) add coagulant aids (polymers).

Black and Chen (1967) found that with aluminum sulfate applied to kaolinite suspensions, destabilization at pH 3.0 was due to electrical double layer depression by unhydrolyzed $Al(H_2O)_6^{3+}$. At pH 5.0, destabilization was due to adsorption of positively charged polynuclear aluminum hydrolysis products with restabilization occurring after a narrow range of coagulant dosage due to charge reversal. The optimum dosage for destabilization was 5 mg/l which corresponds to an Al(III) concentration of 1.5×10^{-5} M. Referring to the work of Singley *et al.* 1966 (see Chapter 3) it is seen that this is the order of concentration where complex polynuclear hydrolysis products form, such as those shown in Figure 3.4(b). With such highly charged strongly adsorptive species the occurrence of restabilization is strongly suggested.

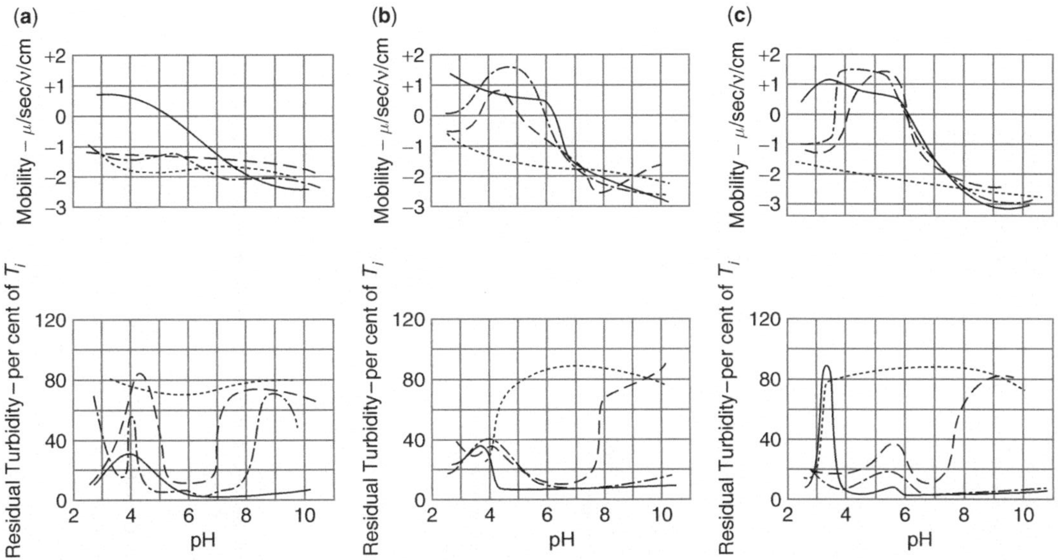

Comparison of the Effects of pH and Mobility on Coagulation of Montmorillonite Suspensions With Ferric Sulfate

Comparison of the Effects of pH and Mobility on the Coagulation of Fuller's Earth Suspensions With Ferric Sulfate

Comparison of the Effect of pH and Mobility on the Coagulation of Kaolinite Suspensions With Ferric Sulfate

Figure 4.5 Evidence for the increasing predominance of negative hydrolysis species with increasing pH. At higher pH values mobilities increase in negativity. T_i represents initial turbidity. The dotted curves are for no ferric sulfate added; the dashed curve for 3 ppm ferric sulfate; the dot-dash curves for 5 ppm; the solid curves for 50 ppm (from Black & Walters, 1964).

At pH 8.0 the destabilization mechanism with increasing coagulant dosage is a gradual transition from adsorption of positively charged hydrolysis species to precipitate enmeshment with $Al(OH)_3$ species.

Dentel and Gossett (1987, 1988) suggested the following destabilization mechanisms which modify but are still somewhat consistent with the explanations presented earlier. Destabilization is seen as involving positively charged metal hydroxide precipitates that attach at discrete locations to the surfaces of the original suspension particles, rather than as uniform coated layers, as schematically depicted in Figure 4.6.

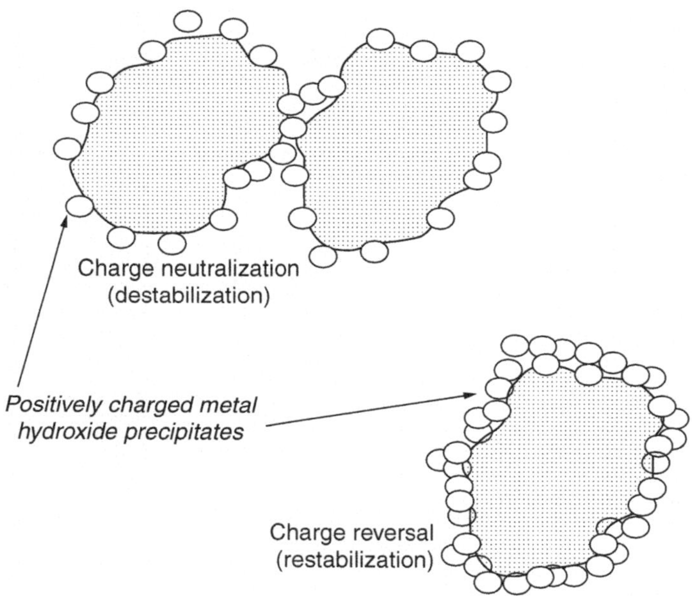

Figure 4.6 Schematic of possible metal hydroxide-particle interaction (adapted from Dentel & Gossett, 1987 and 1988).

Attachment is envisaged as arising from either precipitation of metal hydroxide in solution, followed by attachment to particles; or by an initial adsorption of metal hydroxide species to the particles, which then serve as nucleation sites for further precipitation. Since the precipitate has a positive charge, then charge neutralization and destabilization would occur – as well as eventual restabilization.

Therefore, precipitation is seen as preceding charge neutralization, and the effects of metal hydroxide precipitation are not limited to a sweep-floc action.

Licsko (2004) argued for the same dominant mechanism. Although water soluble polynuclear metal complex species are formed (such as $Al_8(OH)_{20}{}^{4+}$

or $Al_{13}(OH)_{32}^{7+}$) they are relatively short lived, and are not able to significantly connect with colloidal particles and effect destabilization before positively charged metal hydroxide precipitates are formed.

Shin and O'Melia (2004) confirmed the destabilization model described earlier for the destabilization of silica particles (see Figures 4.3 and 4.4). However, for the removal of natural organic matter, there was a strong stoichiometry between coagulant dosage and the DOC concentration. The aluminum species bound with the NOM were hypothesized as moderately neutralized polynuclear species, and/or aluminum hydroxide precipitates.

Amirtharajah and co-workers developed a series of coagulation domain diagrams in which different conditions for treatment are identified together with the predominant coagulant species present (Amirtharajah & Mills, 1982; Amirtharajah, 1990; Johnson & Amirtharajah, 1982). Figures 4.7 and 4.8 present the coagulation domain diagrams for alum and iron.

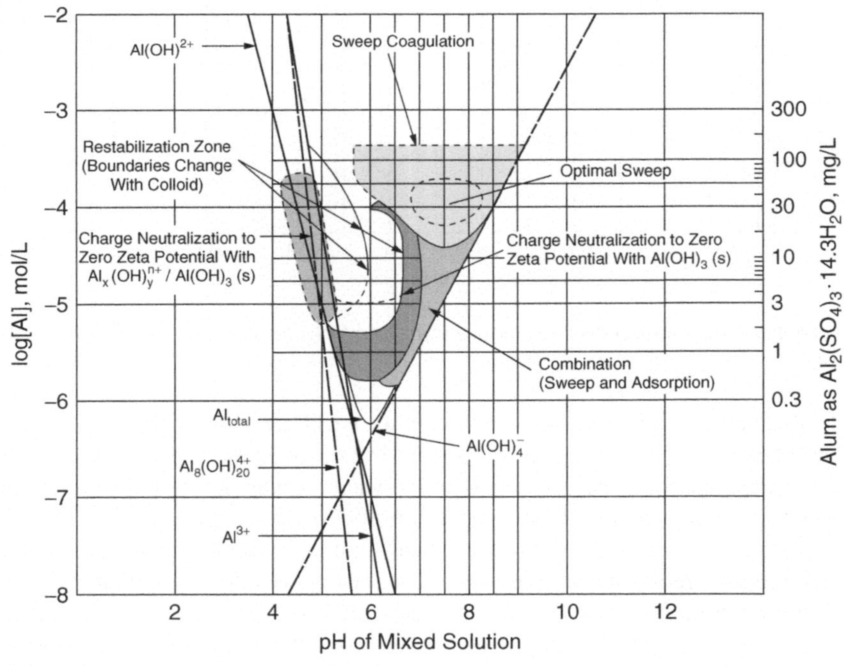

Figure 4.7 Coagulation domain diagram for alum (after Amirtharajah & Mills, 1982).

Figure 4.7 suggests that the coagulation conditions for conventional treatment would be in the region of $Al(OH)_3$ precipitation and optimum sweep floc formation. Conditions for direct filtration or flotation would likely still be in a region of $Al(OH)_3$ formation, but the mechanism would be influenced by charge neutralization with

$Al(OH)_3$ species. Color and TOC removal would likely be in a region of charge neutralization with $Al(OH)_3$ species and/or highly charged polynuclear species. Figure 4.8 also suggests that a sweep floc mechanism would be dominant at pH values above 6 for ferric coagulants.

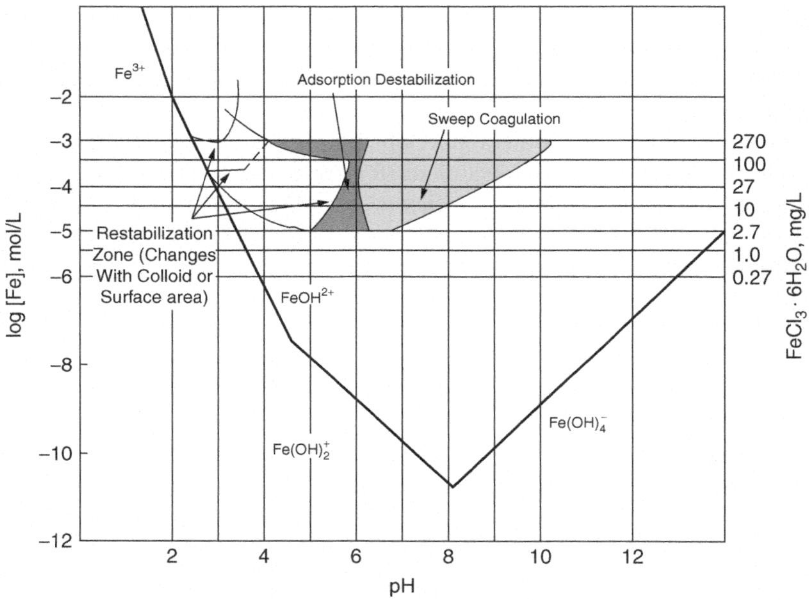

Figure 4.8 Coagulation domain diagram for iron III (after Johnson & Amirtharajah, 1982).

In this and preceding sections, although a logical account has been attempted of the phenomena occurring during destabilization of hydrophobic colloids with metal coagulants, there are certain instances where explanations are not so readily forthcoming. For example, Black and Chen (1965) during investigations on the destabilization of various clay suspensions with aluminum sulfate, found that a region of high turbidity removal was consistently obtained at pH values of 10.0 and higher.

The predominance of negatively charged hydrolysis species at such pH values (see Figures 3.5a and b) especially at lower coagulant dosages would suggest an adsorptive bridging type mechanism. However, this was not reflected in the electrophoretic mobility values obtained by Black and Chen which remained relatively constant with increasing pH at values greater than about pH 8.0.

Mackrle (1962) has suggested the influence of hydrous-oxide crystals at higher pH values of the form $nAl_2O_3 \cdot H_2O$ and $nFe_2O_3 \cdot H_2O$. A possible explanation for the results given above may be a combined effect of, for example, negatively charged $Al(OH)^{4-}$ and positively charged Al_2O_3 species.

4.3 DESTABILIZATION OF HYDROPHILIC COLLOIDS

Hydrophilic colloids, because of hydrated layers surrounding them, are not destabilized by such considerations as depression of the electrical double layer. However, hydrophilic colloids do carry a surface charge although the effective range of influence may be confined in some cases to within the hydrated layer.

The surface charge carried by hydrophilic colloids is often due to the ionization of ionogenic groups. Such groups may be carboxylic, aliphatic or aromatic hydroxyl, sulphato, phosphato, and aminogroups. Stumm and Morgan (1962) by means of alkalimetric titration curves, demonstrated that metal ions form complexes readily with ligands such as phosphates, pyrophosphates, salicylates (which contain a carboxyl and an aromatic hydroxyl group) and oxalates. Such ligands are representative of the functional groups of many hydrophilic colloids.

The above substances, as ligands, do not have as great an affinity with metal ions as do OH^- ions. However, statistical considerations play a significant role in selective coordination and the preference of, say, phosphates over OH^- will depend on the ratio [phosphate]/[OH^-]. Phosphates will, therefore, tend to be preferentially coordinated at lower pH values. This reasoning also applies to other ligands. Furthermore, to form an electroneutral precipitate, less OH^- ions will be required in the coordination sphere of the metal-ion complex so that the pH for precipitation will be lower in the presence of such ligands than if OH^- were the only ligand present.

Stumm and Morgan considered the destabilization of hydrophilic colloids carrying functional groups such as those described above, to be principally due to precipitation of the metal-ligand complexes. This is substantiated by two facts:

(1) It is known that destabilization of hydrophilic colloids does take place at acidic pH values, which tends to indicate preferential coordinative reactions with metal ions and functional groups at the colloid surfaces;
(2) It is further known that to destabilize a dispersion of hydrophilic colloids requires higher coagulant dosages than their hydrophobic counterparts. This suggests that the system will be oversaturated with respect to the metal ion–ligand complexes in precipitation.

4.4 REMOVAL OF NATURAL ORGANIC MATTER

4.4.1 Organic color

4.4.1.1 Nature of organic color

Organic color is usually associated with soft surface waters containing little or no alkalinity. It arises from the aqueous extraction of living woody substances, the solution of degradation products in decaying wood, the solution of soil organic matter, or a combination of these. Christman and Ghassemi (1966) suggested that the manifestation of color is an intermediate step in the transformation of organic matter from living or decaying woody tissue to the soil organic complex.

Black and Christman (1963), by filtration and electrophoretic measurements, concluded that organic color causing compounds were true hydrophilic colloids possessing a negative surface charge. They further suggest that color is due to a light scattering and fluorescence effect rather than to molecular absorption. However, Shapiro (1957) suggested that color compounds were straight chain substituted fatty acids in true solution and that colloidal properties exhibited by these compounds are due to complex formation with iron if present in solution. Packham (1964) suggested that of the humic fractions present in organic colored waters, the humic acid fraction is of colloidal size whereas the fulvic fraction is in true solution. He pointed out that a proportion of the molecules are large enough to exhibit colloidal properties. Furthermore, in the presence of iron, coordination reactions may occur giving rise to colloidal characteristics.

The molecules, some large enough to exhibit colloidal properties, are hydrophilic in nature – with the fulvic fraction exhibiting the most hydrophilic properties, and the humic fraction the least – with some authors referring to the humic acid fraction as "hydrophobic". Humic substances in general are acidic, heterodisperse, polymeric, colored, chemically complex macromolecules.

4.4.1.2 Classification of organic color

Humic substances include a wide variety of compounds with similar constitution and properties. Oden (1919) classified humic substances in terms of four fractions, as shown in Table 4.2 (Oden's original terminology has since been modified).

Table 4.2 Oden's classification of humic substances.

Group Name	Alternative Names	Definition
Humus coal	Humin, ulmin	Insoluble in water and sodium hydroxide
Fulvic acid	Crenic and apocrenic acids	Soluble in sodium hydroxide and mineral acid
Hymatomelanic acid	Hematomelanic, ulmic acids	Soluble in sodium hydroxide and alcohol, insoluble in mineral acid
Humic acid	–	Soluble in sodium hydroxide, insoluble in mineral acid and alcohol

In natural waters, the predominant fractions present are fulvic, hymatomelanic and humic acids. The proportions of these fractions are dependent on the particular water. For example, Black and Christman (1963) determined a fairly consistent distribution between these species for ten colored water samples, with average percentages of 87%; 12%; and 1%, respectively. Packham (1964), for seven water samples taken throughout the U.K. determined an average distribution of 76%; 17%; and 7%, respectively. Common to both investigations was the predominance of the fulvic acid fraction.

However, a significant difference between the two investigations was the color contribution of each fraction. Black and Christman found that on a unit weight basis, the fulvic acid fraction exhibited a slightly greater capacity for color production than did the hymatomelanic acid fraction, with the humic acid fraction producing the least color. Packham found the opposite: the order of increasing color intensity was fulvic < hymatomelanic < humic. This finding appeared to be compatible with the order of increasing molecular complexity, with humic acid possessing the largest molecular weight.

Humic substances derived from a number of sources are often quite different from one water to another. Although chemical and spectral properties may be similar, other differences due to physical properties, molecular weight, particle size and so on, may give rise to different requirements for treatment. Packham (1964) suggested that differences in properties of humic substances of different waters are due principally to differences in the proportions of the various fractions, – each fraction having different properties.

Humic acids possess the highest molecular weights of the three groups, with values of up to 50,000. Hymatomelanic acids are regarded as intermediates between fulvic and humic acids with molecular weights of the order 800. Fulvic acids have the lowest molecular weights in the groups. Some workers suggest that fulvic acids represent a more complex mixture of substances than humic acid (Hall & Packham, 1965) whereas others suggest that fulvic acids are essentially small sized humic acid molecules (Rook, 1977). Black and Christman (1963) found from solubility relationships, chemical spot tests and infrared absorption spectra, indications that fulvic acids are aromatic polyhydroxy methoxy carboxylic acids. These workers suggest that ionization of carboxyl and aromatic hydroxyl groups are responsible for the negative charge of color compounds.

The reactivity of fulvic acids is largely attributed to carboxyl, phenolic, carbonyl and methoxyl functional groups. The carboxyl and phenolic groups provide most of the negative charge carried by the molecules.

It is likely that, despite the general structure of fulvic acids indicated above, no two molecules of color compounds are exactly identical; an indication of this is the variety of structures proposed by several workers (Rook, 1977).

4.4.1.3 Disadvantages of organic color in water supplies

The presence of organic color in water supplies is considered undesirable for a number of reasons:

- Consumers prefer a clear colorless supply on aesthetic grounds;
- Humic substances may impart taste to the water, although not necessarily unpleasant;
- The presence of organic color increases the disinfection (chlorine, ozone, etc) demand;
- Humic substances have a nutrient effect on microorganisms;

- Many industries require a low color level, for example, the pulp and paper industry for the production of high brightness bleached paper require a maximum concentration of 5 units or less (Black & Willems, 1961);
- Humic substances may foul ion exchange resins;
- The presence of color interferes with water analyses, particularly those based on colorimetry;
- High color may reduce the productivity of a water by limiting transmission of light for photosynthesis;
- Humic substances combine with iron, manganese and lead by peptization or chelation, forming humus metal complexes (Tryland, 1976);
- Humic substances act as a vehicle of transport for toxic substances and micropollutants, including heavy metals and organic pollutants;
- Besides increasing chlorine demand as mentioned above, humic substances react with chlorine to form potentially carcinogenic haloforms as by-products. This aspect of color removal will be discussed in more detail in subsequent sections dealing with enhanced coagulation.

The presence of organic material also imparts increased stability to inorganic colloids, probably by a coating of the colloids by the organic material, effectively acting as dispersion agents. The higher surface charge carried by natural organic material is considered to be the prime reason for the enhanced stability imparted to the inorganic colloids. Therefore, in the presence of natural organic matter, turbidity removal can be more difficult. In a clay suspension, the alum dosage required for turbidity removal was increased by over five times, when 3 mg/l fulvic acid (measured as TOC) was added (Yao & Yan, 1989).

The above undesirable properties notwithstanding, there is no evidence that humic substances in their native state are directly harmful to higher organisms. In fact, although no systematic studies appear to have been carried out, a certain therapeutic value has been claimed with regard to rheumatic fever and certain skin defects (Committee Report, 1967).

The apparent nutrient effect of humic substances may in some cases be advantageous. The stimulating effect of humics on algal growth has been demonstrated by several workers (Committee Report, 1967) and may be due to trace metal stabilization by the humics.

4.4.1.4 Measurement of color

Because of the complexity of humic substances, it has been necessary to adopt an arbitrary standard for the measurement of color. The standard accepted is based on platinum-cobalt solutions which produce a yellowish-brown color similar to that of humic substances. Often a number of color standards are prepared in Nessler tubes and colored samples compared visually. However, this method has the disadvantage of subjectivity. At the low color values of many natural waters, visual comparison is almost impossible; less than 3 Pt-Co units are virtually unnoticeable. An alternative is to calibrate spectrophotometric readings with standard platinum-cobalt standards and take sample color readings directly with the spectrophotometer. This is typically carried out in the UV range of approximately 254 nm. Adapted spectrophotometric instruments are available that give direct Pt-Co readings (Hach, 1977). Vik *et al.* (1985) found reasonable consistency between color and UV absorption at 254 nm, for three waters:

- *Småputten lake*: Color = 216 (UV absorption, cm^{-1}) – 6.4 ($r^2 = 0.97$)
- *Tjernsmotjern lake*: Color = 208 (UV absorption, cm^{-1}) – 9.2 ($r^2 = 0.93$)
- *Hellerudmyra lake*: Color = 244 (UV absorption, cm^{-1}) – 1.6 ($r^2 = 0.93$).

The presence of turbidity in water samples may give rise to an enhanced *apparent* color reading. For a *true* color measurement, it is recommended that samples be passed through a 0.45 μ membrane filter prior to color measurement (Hach, 1977). Removal of true color causing compounds by filtration is probably negligible.

A difficulty in having a consistent standard of color measurement is that the color intensity of waters varies with pH. Black and Christman (1963) found a concurrent pH effect on both the particle size and number of color producing particles: increasing pH causes both a decrease in particle size and an increase in particle numbers. Narkis and Rebhun (1977) refer to humic and fulvic acids as anionic polyelectrolytes, because of the presence of carboxylate and phenolate groups. The degree of ionization is dependent on pH. Assuming color to be due primarily to light scattering and fluorescence, raising the pH of a given water should increase the color intensity. This is shown in Figure 4.9.

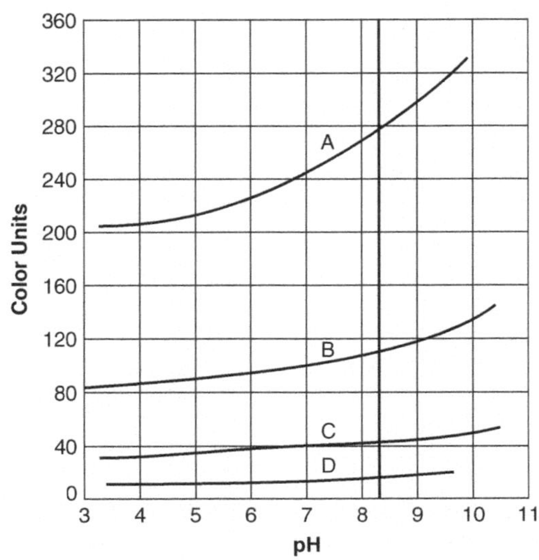

Figure 4.9 Effect of pH on color of four representative waters (from Singley *et al.* 1966).

It is further evident that this indicator effect is more prominent with higher initial color values. For color values less than 50 Pt-Co units, the indicator effect is negligible.

Singley *et al.* (1966) suggested reducing all color measurements to a standard value of pH. A value of 8.3 was chosen because this represents the final pH of most treated waters. For convenience, they presented the nomograph reproduced in Figure 4.10.

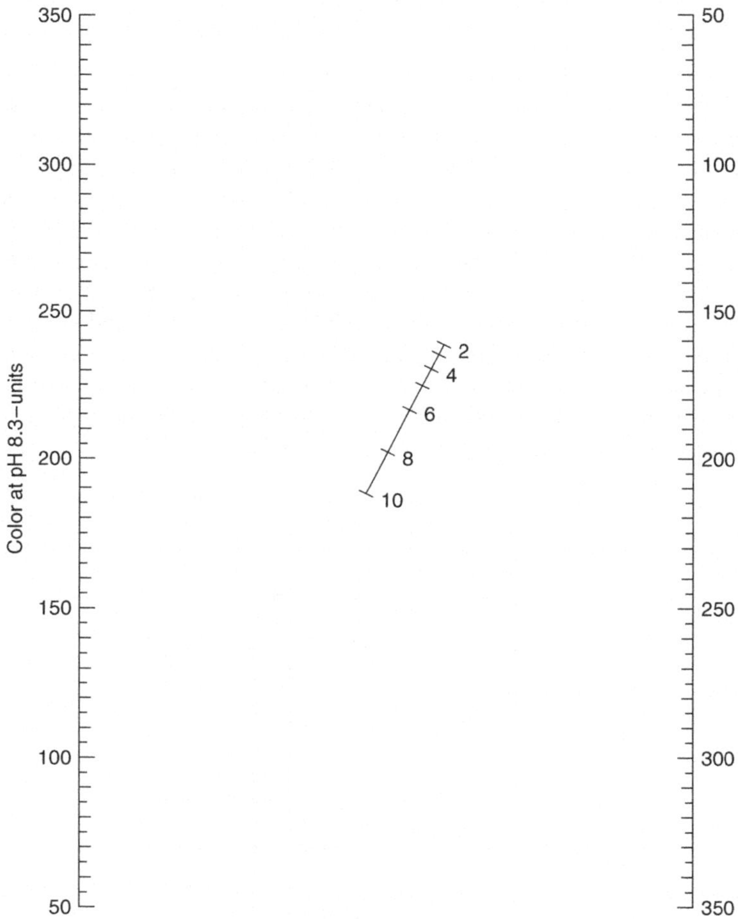

Figure 4.10 Color conversion nomograph for natural colored waters. The scale at the center is for pH at which color was read. To use the nomograph, (i) read color at any pH; (ii) place straight-edge through color read on right-hand scale and pH at which color was read on center scale, and (iii) read corrected color on left-hand side (from Singley *et al.* 1966).

A reading of color for a given water, at any particular pH (from 2 to 10) is readily converted to a standard reading at pH 8.3 using the nomograph. Color values less than 50 units were not included since, as stated above, the indicator effect is negligible at low values. Singley *et al.* suggested that use of the nomograph is more convenient than adjusting the pH of color samples, particularly in view of the fact that most colored waters are low buffered, which results in difficulties of adjusting pH to an exact value.

4.4.1.5 Destabilization of organic color with metal coagulants

The mechanism of destabilization of humic substances with metal coagulants is conceptually assumed to be one of precipitation between metal ion and functional groups on hydrophilic color particles, forming multi-ligand complexes together with OH^- and H_2O species. Such a mechanism explains the stoichiometric relationship between metal coagulant dosage and color intensity, and the lower pH for destabilization when compared with that for turbidity removal. Hall and Packham (1965) suggest such a chemical interaction involving metal hydrolysis species such as $Al_n(OH)_{2,5n}$ and probably carboxyl groups on the color particles. Such an interaction would further explain the near zero zeta potentials at optimum destabilization (Black & Willems, 1961).

Possible mechanisms contributing to color removal include:

- Charge neutralization-precipitation, consisting of the reaction between soluble polynuclear metal coagulant species and humic substances. This reaction produces a stoichiometry between the metal added and the color removed. Restabilization is also evident at higher coagulant dosages.
- Simultaneous precipitation consisting of charge neutralization-precipitation reactions, and reactions with metal hydroxide precipitates, occurring simultaneously.
- Adsorption of humic substances to metal hydroxide surfaces by van der Waals interactions, hydrogen bonding, hydrophobic bonding, ligand exchange, anion exchange, or dipole interactions.

Angbo *et al.* (1990) showed that metal hydroxide surfaces provide better adsorption of humic substances when compared with activated carbon or activated alumina. The efficiency of adsorption increases with increasing velocity gradient, where the smaller flocs formed provide a higher total specific surface area.

Table 4.3 summarizes possible mechanisms for the removal of humic compounds (after Randtke, 1988). The Table refers to potentially colloidal or dissolved humic substances. This reflects the fact that humic acids, although they may not be true colloids, are large enough that they can be referred to as macromolecules that behave in many respects as colloidal material, while at the same time exhibiting characteristics of soluble material. The largest fraction of humic substances in water is generally fulvic acid, which is considered to be a true solution.

Table 4.3 Possible mechanisms for the removal of natural organic matter.

Conditions	Removal Mechanisms Assuming Humus Species are Colloidal	Removal Mechanisms Assuming Humus Species are in True Solution	Comments
Metal salts at low pH (5 to 6)	Charge neutralization by positively charged hydrolysis products	Precipitation of aluminum or iron humate	Relatively low dosages may exist where hydroxides may not precipitate; dosages proportional to humic concentrations
Metal salts at high pH (7 to 8)	Enmeshment in a precipitate; adsorption and bridging; or charge neutralization with higher dosages required because of competition with hydroxide or less positively charged hydrolysis species	Precipitation of Al or Fe humate impeded by precipitation of hydroxides; or co-precipitation, involving adsorption of humic species onto metal hydroxide particles	A higher dosage of coagulant is required than at low pH; dosage increases as humics and pH increase; metal hydroxides are precipitated.
Polymers	Charge neutralization or adsorption and bridging	Precipitation of insoluble polymer-humate compexes	Dosage proportional to humus concentrations
Lime softening (pH 9.5 to 10.5)	Enmeshment in precipitating calcite, perhaps with heterogeneous nucleation on humic particles	Precipitation of calcium humate, adsorption of humics onto calcite nuclei and crystals	Removal may be poor unless another coagulant is added
Lime softening (pH > 11)	As above, aided by charge neutralization by positively charged magnesium hydroxide complexes and nuclei	As above, but with adsorption onto positively charged magnesium hydroxide particles the primary mechanism	Magnesium hydroxide also precipitated; excellent removal possible

The stoichiometry between raw water color and optimum coagulant dosage has been reported by several workers for both aluminum and ferric salts (Black *et al.* 1963; Hall & Packham, 1965). In general, the optimum coagulant concentration for fulvic acid is higher than for humic acid at the same concentration indicating that two waters, even though possessing similar color intensities, may require different coagulant dosages. Hall and Packham found that although both fulvic and humic fractions were removed by coagulation, fulvic acid behaves as a more complex mixture, a fraction of which is not readily removed by coagulation. This has important significance when the issue of the removal of disinfection by-product precursors by enhanced coagulation is discussed.

Ferric salts, in general, have been found to yield slightly lower color residuals after destabilization. However, the optimum dosage for lowest residual color in each case appears to be the same if concentrations are expressed in terms of the metal ion. The mechanism of destabilization for the different coagulants, therefore, is evidently similar.

Although the optimum dosage–color relationship with ferric and aluminum salts is essentially the same, the optimum pH values for these coagulants are different. With ferric salts, the optimum pH is typically within the range 3.7 to 4.2 and with aluminum sulfate within the range 5.0 to 5.5 (Black *et al.* 1963; Hall & Packham, 1965; Mangravite *et al.* 1975). The differences may be due to the greater affinity of ferric ions for OH^- (Hall & Packham, 1965). From previous discussions, it is seen that for functional groups to serve as ligands in the metal complex, lower pH values are required thereby limiting the statistical opportunity for OH^- ions to be complexed. Furthermore, the lower residual humic concentrations attained with iron indicates that the humic-iron bond is stronger than the humic–aluminum bond (Hall & Packham, 1965).

For a given colored water, Black *et al.* (1963) showed that, at the optimum coagulant dosage (specifically ferric sulfate in their experiments) there is a relationship between initial raw water color and optimum pH, such that for color values ranging from 50 to 450, the optimum pH varied from 3.8 to 3.45. With the coagulant dosage–color stoichiometric relationship mentioned previously, it is seen that control of treatment of a given colored water is greatly facilitated. Optimum dosage and pH changes in color intensity of the water may be determined from pre-calibrated color-dosage-pH curves, with all color measurements carried out or reduced to the standard pH of 8.3. However, such a procedure would still require confirmatory routine jar test experiments.

The presence of turbidity in a colored raw water may complicate treatment, since the mechanisms of removal are different. Usually, the presence of clay has little effect on color removal (that is, if optimum conditions for destabilization of such a mixed water are assessed in terms of color only). However, the presence of color has an effect very similar to that of the presence of anions such as phosphates on clay turbidity. That is, the optimum pH for destabilization of clay dispersions is progressively shifted to the acid side with increasing phosphate, or color concentration.

Hundt and O'Melia (1988) found that for a given fulvic acid removal, the addition of calcium reduced the required aluminum dosage, as well as extending the pH range for required removals. In general, divalent cations tend to broaden the range for effective removal and lower the required coagulant dosage. The mechanism appears to be by complexing with organic functional groups that otherwise would need metal hydrolysis complexes to complete the destabilization reactions.

In general, when adjusting pH for optimal coagulation, the pH adjusting chemical is added first, before the metal coagulant. However, this may not always be the case. Vik *et al.* (1985) investigated the best sequence of chemicals for TOC and color removal from lake waters and found no significant differences in results between adding sodium hydroxide first for pH adjustment, immediately followed by alum; and adding alum first, immediately followed by sodium hydroxide, at pH values less than 7.

Randtke (1988) listed general guidelines for the sequence of chemical addition:

- Acid should be added before metal coagulants to enhance formation of positively charged polynuclear hydrolysis species;
- Base should be added after metal coagulants (however, see Vik *et al.* above);
- Clay or PAC for adsorption should be added before coagulants to allow adsorption to occur before coagulation;
- There will always be exceptions to the above so jar test evaluations are essential to optimize a given application.

Figures 4.11 through 4.13 show the color removal domains obtained from three studies treating (a) humic acid solutions; (b) fulvic acid solutions; and (c) natural colored waters.

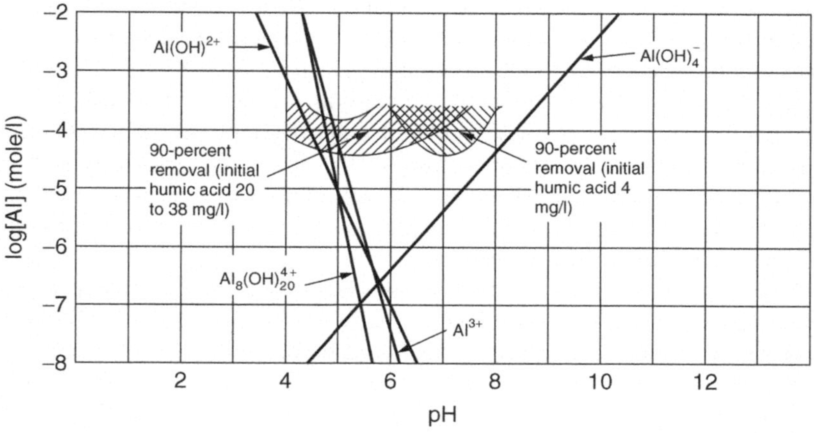

Figure 4.11 Humic acid removal domains (after Edwards & Amirtharajah, 1985).

Figure 4.11 shows the 90-percent color removal domains for varying concentrations of humic acid using aluminum sulfate (Edwards & Amirtharajah, 1985). At a low humic acid concentration of 4 mg/l (representing 100 color units) 90-percent color removal occurred within the pH range 6 to 7, with alum dosages greater than approximately 12 mg/l (log[Al in Mole/l] = –4.4). This is the region of aluminum hydroxide precipitation with a likely removal mechanism of adsorption to the hydroxide floc surfaces. At higher humic acid concentrations, of 20 to 38 mg/l (450 to 900 color units), the band of color removal spreads to the acid side. The area of 90-percent removal occurred in the pH range 4.0 to 7.5. Optimal removal occurred within the narrower band of pH 4.5 to 5.5 and alum dosage 15 to 30 mg/l (log[Al in Mole/l] = –4.3 to –4.0].

Figure 4.12 shows 80-percent removal domains obtained for fulvic acid solutions (Hundt & O'Melia, 1988).

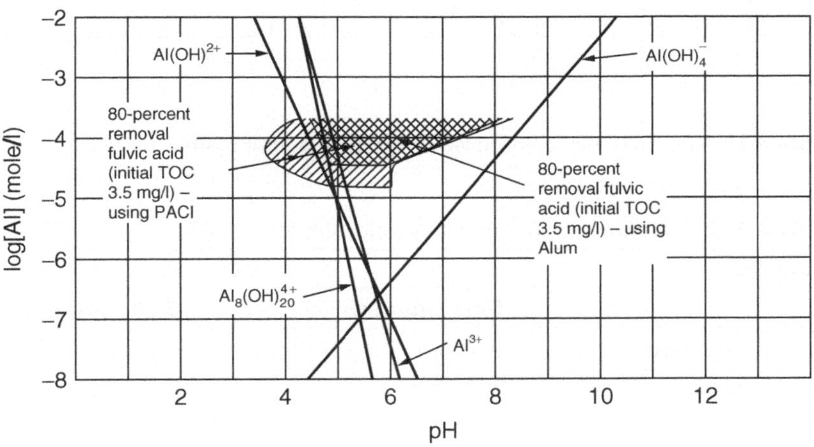

Figure 4.12 Fulvic acid removal domains (after Hundt & O'Melia, 1988).

In Figure 4.12, the fulvic acid concentration expressed as TOC was 3.5 mg/l. Results for aluminum sulfate and polyaluminum chloride (PACl) are presented. With alum, the greatest removals were achieved between pH 5.0 and 5.5 and down to an alum concentration of 11.4 mg/l (log[Al in Mole/l] = –4.42). The majority of the region of 80-percent removal was obtained where aluminum hydroxide is expected to precipitate, although a part of the region does correspond with a charge neutralization-precipitation mechanism.

The use of PACl also produced a region of 80-percent removal that overlapped and extended the region to lower Al dosages and towards the acid side. The region begins at a lower pH of 3.75 where aluminum polynuclear species are present. The region also extends to where $Al(OH)_3$ is expected, at higher pH.

Figure 4.13 shows the 90-percent removal domain for color removal from two different waters in Norway (Vik *et al.* 1985). Table 4.4 summarizes general characteristics of the two lake waters.

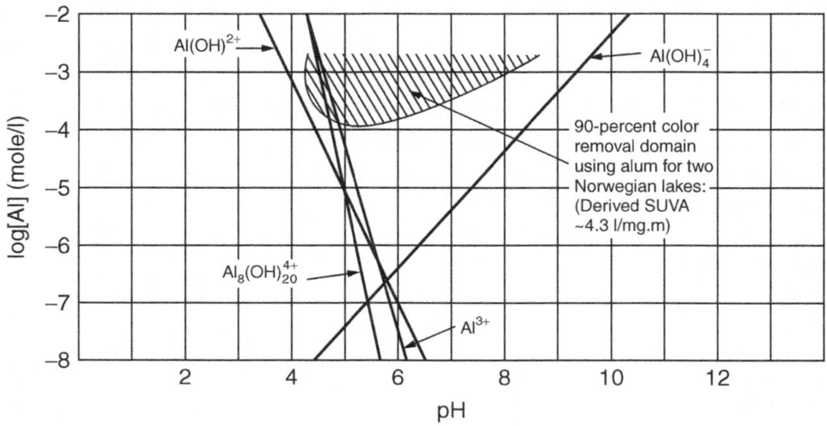

Figure 4.13 Natural humic colored water removal domain (after Vik *et al.* 1985).

Table 4.4 General water characteristics (after Vik *et al.* 1985).

Parameter	Lake	
	Tjernsmotjern	Hellerudmyra
Color (mg Pt-Co/l)	110–140	105–110
UV254 (cm-1)	0.53–0.65	0.50–0.52
TOC (mg/l)	12.2–15.6	10.8–12.2
Turbidity (NTU)	0.90	0.90
pH	6.0–6.5	4.3–4.9

In all cases a pH of around 5.5 was optimal for color removal. The minimum alum dosage for 90-percent color removal was approximately 3.5 mgAl/l (log[Al in Mole/l] = –3.9).

In general, the influence of pH results from the competition between metal hydrolysis products and hydrogen ions for organic ligands; and between hydroxyl ions and organic anions for metal hydrolysis products. As pH increases, natural organic compounds become more negatively charged and the metal hydrolysis species become less positively charged, resulting in less adsorption propensity and an increased coagulant demand (Randtke, 1988).

Vik *et al.* showed that, in terms of TOC removal, the lake Småputten water contained a high proportion of low molecular weight species and was more difficult to treat. This is generally the case, where higher molecular weight humic acid species are easier to remove than the low molecular weight fulvic acid species. Randtke (1988) presented a compilation of reported experiences, summarized as follows:

- Higher molecular weight (MW) natural organic matter (NOM) fractions more readily adsorbed to activated alumina (ozone increased adsorption);
- Higher MW NOM fractions more strongly adsorbed to aluminum oxides;
- Higher MW humic acid fractions better removed by alum coagulation. (Ozone decreased the removal of dissolved organic compounds, DOC);
- Preferential removal of higher MW NOM fractions and disinfection by-product (DBP) precursors by alum coagulation;
- Poor removal of low MW DBP precursors by coagulation with alum, iron, polymer or lime;
- Lowest MW humic material exerted the highest coagulant demand and was least removable with alum coagulation;
- Preferential removal of higher MW, hydrophobic, neutral NOM fractions with alum, iron, polymer or lime coagulation. Poor removal of low MW NOM fractions;
- After coagulation, low MW, neutral NOM compounds are dominant in the treated water;
- Adsorption of humic compounds to granular activated carbon (GAC) better with the low MW fraction;
- Adsorption of highest MW humic fractions least adsorbable to GAC;
- Adsorption of humic compounds with GAC better after coagulation;
- With conventional treatment, direct filtration or softening:
    - Higher MW fractions better removed;
    - MW < 500 not effectively removed;
    - More hydrophobic fractions better removed;
    - NOM fraction with highest carboxylic acidity difficult to remove;
    - Higher MW fractions generally yield highest DBP.

4.4.2 Enhanced coagulation

Humic compounds are part of a general class referred to as natural organic matter (NOM). After chemical disinfection, these compounds produce potentially carcinogenic disinfection by-products (DBPs). In drinking water, the regulated DBPs include trihalomethane species (THMs), haloacetic acid species (HAAs), bromate and chlorite. Since coagulation is so effective at removing a wide variety of organic contaminants either directly, or by removing the particles to which other contaminants are adsorbed, the United States Environmental Protection Agency (USEPA) introduced *Enhanced Coagulation* as an accepted treatment process to meet the Stage 2 Disinfection By-Products Rule (DBPR) finalized in January

2006. This rule requires locational (in the distribution system) running annual average concentrations of total trihalomethane species (TTHM) and haloacetic acid species (HAA5) of less than 80 and 60 μg/l, respectively, as well as other DBP concentrations.

The goal of the enhanced coagulation technique is to provide the removal of natural organic matter (NOM), measured as total organic carbon (TOC). The advantages of promoting such a treatment technique are that not only are the regulated DBPs addressed, but also DBPs and other organic compounds not yet regulated are also removed by enhanced coagulation. Enhanced coagulation is an elaboration of the long practiced techniques of organic color removal by coagulation. Therefore, this section on enhanced coagulation should be considered an extension of the preceding section.

Enhanced coagulation requires the removal of NOM material, while still achieving good turbidity removals. The means by which these dual objectives can be met is by (a) selecting the best coagulant type; (b) applying the best coagulant dosage; (c) adjusting the pH to a value where best (or adequate) coagulation conditions are achieved.

Different coagulant types such as aluminum or iron based coagulants, function with different solubilities and coordination chemistries under operating conditions. Higher coagulant dosages provide more metal for either floc or complex formation. Lower operating pH values affect the metal complexes formed, and reduce the charge density of humic and fulvic acids, making them more hydrophobic (and adsorbable).

The USEPA recognized that the effectiveness of enhanced coagulation to remove TOC is largely dependent on natural water characteristics, and the respective fractions of humic and fulvic acids, and other constituents. The higher molecular weight fraction of NOM is generally easier to remove, and should be required to be removed since this fraction largely contributes to DBP formation. The lower MW fraction is required to be removed to a lesser extent since it is more difficult to remove, and tends to be more recalcitrant and contributes to a lesser extent to DBP formation.

Recognizing this, the USEPA developed a two-step standard: Step 1 defines percentage removals of TOC that are dependent on alkalinity and initial TOC levels (Table 4.5). It allows for the more difficult TOC removal expected in waters with higher alkalinity and lower TOC.

Source water TOC levels generally range from 1 to 20 mg/l. A survey of 65 waters in the United States found a median level of approximately 3.5 mg/l (Symons *et al.* 1975). Some swamp and bog waters may have values as high as 300 mg/l.

The TOC concentration is a collective measure of organics in water and provides no information on the composition and distribution of the organic constituents. Because of unique raw water characteristics and the particular mix of organic constituents, some plants may be unable to meet the requirements in Table 4.5. Therefore, Step 2 allows for jar testing (or pilot or full-scale testing) to establish what the removal requirements should be. After determining the TOC removals that the water plant should achieve, the plant is free to implement any combination of chemicals and coagulation conditions to achieve that removal, within maximum concentration limits established by the National Science Foundation (NSF) – based largely on the impurities contained in the chemicals. (For example, the NSF limit for sulfuric acid addition, to minimize the effects of impurities on treated waters, is 50 mg/l).

Table 4.5 Step 1 TOC removal percentages (EPA, 1999).

Source Water TOC (mg/l)[1]	Source Water Alkalinity (mg/l as $CaCO_3$)		
	0 to 60	**>60 to 120**	**>120[2]**
≤2.0		Not required	
>2.0 to 4.0	35%	25%	15%
>4.0 to 8.0	45%	35%	25%
>8.0	50%	40%	30%

[1]Calculated quarterly as running annual average.
[2]Plants practicing precipitative softening must meet the TOC removal requirements in this column.

Under the Step 2 procedure, 10 mg/l increments of alum are added (or a molar equivalent amount of ferric salt – where 10 mg/l alum = 0.03367 mMAl/l = 0.03367 mMFe/l = 5.5 mg/l $FeCl_3$, *etc.*). The coagulant increments are added without acid addition for pH adjustment, and incremental TOC removals determined. The range of coagulant added (or the number of increments) must be up to the point that at least certain target pH values are reached, shown in Table 4.6. The final pH at the end of the series of tests would be at or below the target values. The target pH values are dependent on raw water alkalinity to allow for the fact that higher coagulant dosages are required to reduce pH to the optimum required for NOM removal. Waters with alkalinities less than 60 mg/l $CaCO_3$ are required to maintain the pH at 5.3 to 5.7 using alkaline chemicals (preferably the same chemical as used in the full scale plant for pH adjustment) if the pH is driven below these values before the PODR is reached (see below).

Table 4.6 Target pH for Step 2 requirements (EPA, 1999).

Raw Water Alkalinity (mg/l $CaCO_3$)	Target pH
0 to 60	5.5
>60 to 120	6.3
>120 to 240	7.0
>240	7.5

The objective of the Step 2 procedure is to identify a point of diminishing returns (PODR) where a coagulant increment (10 mg/l for alum) results in a TOC removal increment of less than 0.3 mg/l. This is best identified graphically, as shown in Figure 4.14.

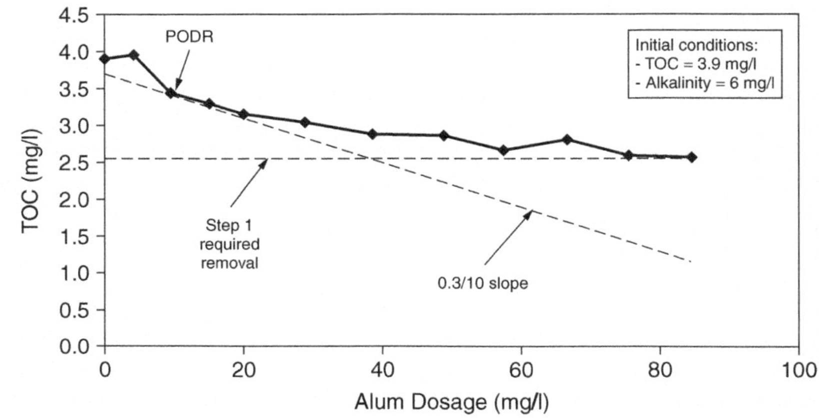

Figure 4.14 Alum jar test profile illustrating the PODR and Step 1 criteria.

In Figure 4.14, the Step 1 criterion is not met at a dosage of over 80 mg/l. However, the Step 2 criterion achieves the point of diminishing returns removal at a dosage of 9.5 mg/l alum.

To further recognize waters that have a low potential for forming DBPs, the USEPA allowed any one of the following alternative criteria that allow exemptions to the above criteria (all values calculated quarterly as running annual averages):

- The raw source water contains less than 2.0 mg/l TOC;
- The raw source water specific ultraviolet absorption (SUVA) is less than or equal to 2.0 l/mg · m;
- Treated water contains less than 2.0 mg/l TOC;
- Treated water SUVA (with no oxidant addition) is less than or equal to 2.0 l/mg · m;
- Raw source water TOC is less than 4.0 mg/l *and* alkalinity greater than 60 mg/l $CaCO_3$ *and* TTHM less than 40 μg/l *and* HAA5 less than 30 μg/l;
- For systems using only free chlorine for disinfection and distribution system residuals, TTHM less than 40 μg/l *and* HAA5 less than 30 μg/l.

The specific ultraviolet absorption (SUVA) is related to the dissolved organic carbon (DOC) concentration by the following relationship:

$$SUVA = UV_{254}/DOC \ (l/mg \cdot m)$$

where

UV_{254} is the ultraviolet absorbance measured at a wavelength of 253.7 nm, after filtration through 0.45 μm filters (m^{-1})
DOC is the dissolved organic carbon measured after filtration through 0.45 μm filters (mg/l)

SUVA varies with the type of DOC present and is highest when long-chain humic acids are present (Edzwald, 1994):

- *SUVA of 4 to 5 l/mg·m*: The DOC of these waters is composed largely of aquatic humics high in hydrophobic material, high in aromatic chemical structure, and with a high negative charge.
- *SUVA less than 3*: The DOC of these waters is composed largely of non-humics high in hydrophilic material, low in aromatic chemical structure, and with a low negative charge.

4.4.2.1 Effectiveness of metal coagulants

Figure 4.15 shows the influence of SUVA on the alum dosage required to remove DOC. Data was derived from White *et al.* (1997) for 25 water sources from different regions in the United States. In general, for waters with SUVA values of 4 to 5 l/mg·m, alum dosages vary from approximately 0.6 to 2 mg/l Al per mg/l DOC removed. For waters with SUVA values less than 3 l/mg·m, alum dosages vary from 2 to 13 mg/l Al per mg/l DOC removed.

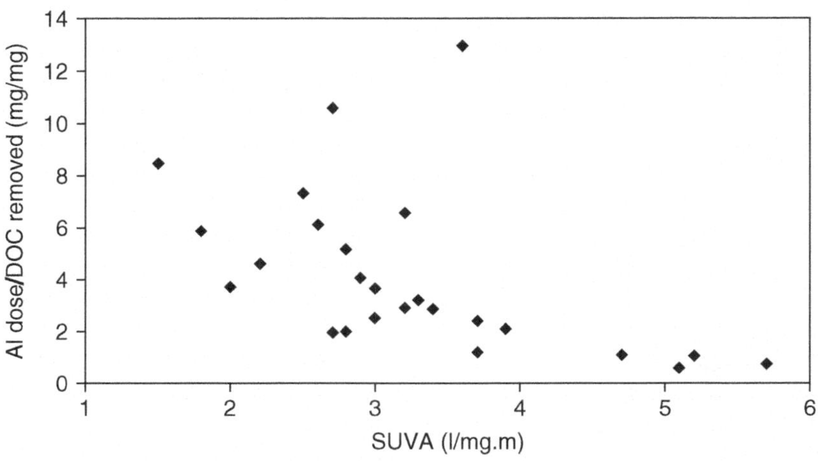

Figure 4.15 Influence of SUVA on alum (as Al) requirements for DOC removal.

Ruehl (1999) reported on numerous experiments carried out with enhanced coagulation on a number of waters. In general, for comparable treated water quality, the ratio of active metal (mgAl/mgFe) required with alum and ferric sulfate varied from 1:3 to 1:4.5. This relationship is higher than the equivalent dosages indicated by EPA for the Step 2 procedure, where equal molar dosages are required. In this case a molar ratio of 1.0 Al/Fe is equal to 1:2.1 in terms of mgAl/mgFe.

White *et al.* (1997) reported on enhanced coagulation tests carried out on 31 natural waters around the United States. Of the 31 waters, 17 would be subject to the PODR treatment requirement as the controlling criterion. However, most of the waters were able to achieve the required TOC removals.

Those waters that met the Step 1 TOC removal requirements were mostly low in alkalinity and moderate to high in TOC. Waters with low alkalinity (<60 mg/l $CaCO_3$) and higher TOC (>4 mg/l) are generally easier to remove TOC. Also, waters with high humic contents (and high SUVA values) are also easier to remove TOC.

The waters with the lowest raw water TOCs were generally the most difficult in terms of meeting Step 1 TOC removal requirements. Those waters for which the PODR was the controlling criterion, were mostly low in TOC, high in alkalinity, or both.

For many waters, the addition of threshold metal dosages is required before appreciable TOC or turbidity removals occur. The threshold metal dosage is largely dependent on the initial DOC concentration content of the water.

In none of the waters tested by White *et al.* did enhanced coagulation interfere with turbidity removal.

However, this may not be true in all cases. Waters that are coagulated under charge neutralization conditions are the least likely to have coincident conditions for TOC removal and particle destabilization. This makes it especially important to conduct a comprehensive jar testing campaign to determine optimum coagulation conditions to meet overall water quality goals.

Organics removal and enhanced coagulation are effective with traditional coagulants like aluminum sulfate, ferric chloride and ferric sulfate, as well as formulations like polyaluminum chloride (PACl) and acid alum. PACl coagulants are polymer-like aggregates described by their basicity and aluminum content. The basicity describes the amount of hydroxide in the PACl formulation. The basicity ranges from low (10-percent), to moderate (50-percent), to high (more than 75-percent). Higher basicity products tend to have a higher positive charge.

Acid alum formulations are aluminum sulfate with 1 to 15-percent free sulfuric acid. Their effectiveness with TOC removal applications is due to the enhanced depression of pH.

From extensive tests reported by Lind (1996) low basicity PACl and acid alum generally provided the best results for TOC removals. Acid alum reduced TOC at 0.1 to 0.5 higher pH units than alum, and 1.4 to 2.5 higher pH than iron compounds. PACl combined good TOC removal with lower residuals generation and alkalinity consumption at many sites. It also proved effective at a lower dose and active metal content than alum or iron salts.

Table 4.7 presents the average TOC removals and metal dosages required, averaged at 46 treatment plants (after Lind, 1996).

Table 4.7 Overall averages of TOC removals at 46 treatment plant locations.

Coagulant	TOC Removal (%)	TOC Removed per Metal Dose (mg/mg Metal)
Alum	32.3	0.71
Ferric sulfate	12.7	0.34
Ferric chloride	12.7	0.34
Acid alum	42.7	1.14
Low basicity PACl	38.4	1.14
Medium basicity PACl	30.0	0.88
High basicity PACl	26.5	0.94
PACS	28.1	0.82

TOC or NOM reductions depend on the type and dosage of coagulant, the pH, temperature, raw water quality and NOM characteristics. In general, the optimum pH for ferric salts is in the range 3.7 to 4.2, and for aluminum sulfate in the range 5.0 to 5.5. Jiang *et al.* (1996) showed that in terms of mass metal dosages, aluminum sulfate was superior to ferric sulfate or polyferric sulfate in the narrow pH range of 4.5 to 5.5, as shown in Figure 4.16. The mass ratios for alum, polyferric sulfate and ferric sulfate were 0.67, 0.82 and 1.18 mg/mg C removed, respectively. However, in terms of molar ratios, the polyferric sulfate product was superior to the other two coagulants, with a molar ratio of 0.015 M Fe/g C removed, compared with alum at 0.025 M Al/g C removed. The polyferric sulfate product was also less restrictive in terms of operating pH, compared with the conventional coagulants.

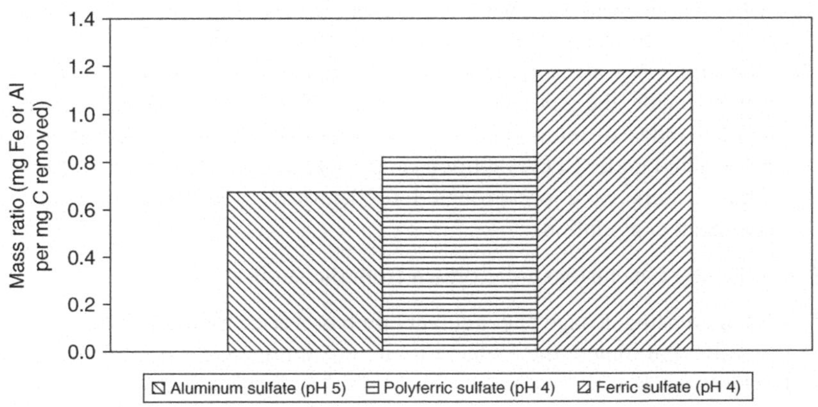

Figure 4.16 Mass ratio (mg Fe or Al per mg C removed) for 75-percent DOC removal of 6.5 mgC/l, at best coagulation pH for each coagulant (data from Jiang *et al.* 1996).

4.4.2.2 Mechanisms of NOM removal with metal coagulants

As discussed in the previous section on color removal, metal coagulants remove NOM by two general mechanisms: (1) adsorption onto metal hydroxide [$Me(OH)_3$] floc and (2) formation of insoluble complexes (aluminum or iron humates or fulvates) in a manner analogous to charge neutralization.

The first mechanism is dominant generally at higher coagulant dosages and higher pH, in the range 6 to 8. Under these conditions, the surface charge of the floc affects the ability of the coagulant to electrostatically adsorb anionic humic and fulvic acids.

The second mechanism is influenced by the acidity of the NOM representing ligand sites for metal complexation, and is relatively more dominant at lower dosages and pH conditions (Krasner & Amy, 1995). This mechanism could result from specific chemical reactions between positively charged species and the negatively charged humic molecules, or it could arise from shielding of the negatively charged NOM sites, causing a contraction and collapse of the molecule, thus leading to precipitation (Edwards & Amirtharajah, 1985).

Krasner *et al.* (1994) investigated an experimental protocol to characterize the NOM in waters and thereby predict their treatability by coagulation. Their procedure was to

(1) Use a 1,000 dalton molecular weight (1 K) ultrafiltration membrane to determine what fraction of the water was lower versus higher in MW
(2) Use a polymeric adsorbent (XAD-8 resin column) to separate and quantify the humic (hydrophobic) versus non-humic (hydrophilic) content of the NOM. These absorbents are basically macroporous strongly basic anion exchangers with a high affinity for organic acids. They are "macroporous" with an open structure enabling the large humic molecules to be adsorbed, then removed when regenerated with sodium chloride.
(3) DOC and UVA measurements were also made of the bulk and fractionated waters, with the specific absorbance, SUVA (UVA/DOC) providing another index of humic character. In addition, selected bulk waters were chlorinated under SDS conditions and analyzed for THMs and HAAs.
(4) Some waters were pre-ozonated to determine the effect of ozone on enhanced coagulation.

The following conclusions were drawn from the studies:

- With increasing dosage of coagulant (alum or ferric chloride), DOC removals generally paralleled the reduction in the humic and larger MW (>1 K) fractions of DOC.
- For most of the water samples, the residual DOC remaining after enhanced coagulation was primarily made up of low-MW and non-humic material. These fractions represent that part of the DOC that is recalcitrant to removal by coagulation.

- Acid addition prior to coagulation in high-alkalinity waters resulted in a higher removal of DOC, demonstrating that the humic material was better removed under more acidic pH conditions.
- Reductions in UVA, as a result of either removal by coagulation or destruction by ozonation, typically corresponded to reductions in humic DOC. Waters with low specific absorbance (UVA/DOC) were also low in humic matter.
- The removal of UVA provided an indication of the reduction in SDS DBP levels. However, in some waters, the non-humic fraction was still a source of DBP precursors.
- Ozonation, especially at high ozone-to-DOC ratios, transformed high MW and humic NOM into low MW and non-humic NOM, respectively. As a result of preozonation with a high ozone dose, the coagulation efficiency was reduced, as there was now a higher percentage of NOM that was less amenable to coagulation.

Ma *et al.* (2001) found that pre-oxidation with potassium permanganate assisted in the destabilization of surface waters with high organic contents. The mechanism was thought to be adsorption of NOM to manganese dioxide, forming larger and denser flocs. Removal of NOM and inorganic turbidity was enhanced. Care is needed to avoid overdosing $KMnO_4$ and increasing treated water manganese levels.

In some cases, to improve the removal of lower molecular weight organics and supplement the removals attainable with metal coagulants, powdered activated carbon (PAC) could be added to supplement enhanced coagulation. Pizzi and Rodgers (1997) found that a combination of an alum-polymer blend coagulant at 25 mg/l with PAC at 10 mg/l was optimal to achieve a 39-percent TOC reduction, and minimize sludge production. In their studies the raw water TOC was 2.4 mg/l and turbidity 2.5 NTU.

Najm *et al.* (1998) also investigated this concept using Colorado River water from Lake Mead in Nevada. This water had a low humic content and low SUVA (1.43 l/mg · m). Fifty-six percent of the TOC was non-humic and 46-percent of the TOC had molecular weights less than 1,000. Table 4.8 summarizes their results.

Since the raw water SUVA was less than 2 l/mg · m, the enhanced coagulation requirements did not strictly apply to this water. However, in terms of reducing distribution system disinfection by-products, additional TOC removal was important. Based on these investigations, the least cost combination of ferric chloride and PAC was determined to be 65 mg/l $FeCl_3$ and 23 mg/l PAC. The evaluation criteria were based on meeting 80-percent of the Stage 2 D/DBP MCL requirements for THMs and HAAs, as well as minimizing sludge handling costs.

4.4.2.3 Predictive models for enhanced coagulation

From a nationwide database of 127 waters in the United States, Edwards (1997) and Tseng and Edwards (1999) developed a model to predict dissolved organic carbon (DOC) removal during enhanced coagulation. The model used coagulant dosage, coagulation pH, raw water UV_{254}, and raw water DOC as inputs. The model assumed that all DOC can be divided into two fractions: DOC that strongly complexes to hydroxide surfaces, and DOC that does not. These fractions are referred to as sorbing and nonsorbing DOC.

Table 4.8 Results of enhanced coagulation with and without PAC (data from Najm *et al.* 1998).

Ferric Chloride Dosage (mg/l $FeCl_3$)	PAC Dosage (mg/l)	TOC (mg/l)	TOC Removal (%)
0	0	2.8	0
0	10	2.6	7
0	40	2.1	25
0	80	1.7	39
30	0	2.3	18
30	10	2.0	28
30	40	1.5	46
30	80	1.1	61
60	0	1.8	36
60	10	1.6	43
60	40	1.2	57
60	80	0.8	71
100	0	1.5	46
100	10	1.3	53
100	40	1.0	64
100	80	0.6	79

Raw water turbidity = 0.4 NTU; TOC = 2.8 mg/l; SUVA = 1.43 l/mg · m; flocculation time = 30 minutes.

The model calculates the relative fractions of sorbing and nonsorbing DOC using an empirical relationship based on the value of SUVA:

$$\text{Fraction nonsorbing DOC} = K_1(\text{SUVA})_{\text{raw water}} + K_2$$

where K_1 and K_2 are empirical fitting constants.

The concentration of sorbable DOC is then given by:

$$\text{Sorbable DOC} = (1 - \text{fraction of nonsorbing DOC}) \times \text{DOC}_{\text{raw water}} \quad (\text{mg/l})$$

The equilibrium between sorbable DOC and the metal hydroxide surface is described by a Langmuir expression:

$$\frac{x}{M} = \frac{a \cdot b \cdot [C]_{eq}}{1 + b[C]_{eq}}$$

where:

x = concentration of DOC removed (mg/l)
M = metal coagulant added (mM/l)
a = maximum DOC sorption/mM coagulant (mg DOC/mM Al or Fe)
b = sorption constant for sorbable DOC to hydroxide surface (l/mg DOC)
$[C]_{eq}$ = sorbable DOC (mg/l) in solution at equilibrium.

The affinity of DOC for hydroxide surfaces was considered to be strongly influenced by pH, as described by the following expression:

$$a = x_3\text{pH}^3 + x_2\text{pH}^2 + x_1\text{pH}$$

where x_1 through x_3 are empirical fitting constants.

Consolidation of the above equations provides the following expression that can be solved for $[C]_{eq}$, given the coagulation pH, raw water UV_{254}, raw water DOC, and the six empirical constants K_1, K_2, x_1, x_2, x_3 and b:

$$\frac{(1 - \text{SUVA} \cdot K_1 - K_2)\text{DOC}_{\text{initial}} - [C]_{eq}}{M} = \frac{(x_3\text{pH}^3 + x_2\text{pH}^2 + x_1\text{pH})b[C]_{eq}}{1 + b[C]_{eq}}$$

The model prediction for DOC remaining after coagulation is then given by:

$$\text{DOC}_{\text{remaining}} = C_{eq} + \text{nonsorbable DOC (mg/l)}$$

Overall TOC removal is calculated assuming that nonsorbing DOC is not removed, sorbing DOC is removed according to the model, and a third fractional component, particulate organic carbon (POC), is completely removed through enhanced coagulation.

For raw water DOC less than 10 mg/l, a general model for alum and ferric coagulants, with the following empirical constants, had a correlation coefficient R^2 of 0.936:

$K_1 = -0.053$
$K_2 = 0.54$
$x_3 = 6.44$
$x_2 = -99.2$
$x_1 = 387$
$b = 0.107$

There were site specific differences in the model results that produced higher correlation coefficients of up to 0.993. In general, alum outperformed ferric coagulants particularly at lower coagulant dosages.

In terms of calibrating the model, the fraction of nonsorbing DOC can be determined by applying 2.5 mM/l Al or Fe at pH 5 to 6. The remaining DOC can be assumed to be essentially non-adsorbing.

The Langmuir adsorption constant b can be determined by applying a range of coagulant dosages at a constant pH, and applying linear fit algorithms following similar procedures to those used for activated carbon.

Edwards (1997) suggested that calibrating to either the nonsorbable fraction, or the sorption constant b, is sufficient for very good predictive capability of the model.

Kastl *et al.* (2004) pointed out that the model of Edwards appeared to provide credible results between the common pH range of 5 to 7 for DOC removal by metal coagulants, but was unreliable outside that range. By extending Edwards' model, Kastl *et al.* replaced the maximum sorption capacity of the floc, *a* by a term describing the effect of pH on the dissociation of humic acid.

Kastl *et al.* divided the sorbable DOC fraction into humic acid and nonpolar fractions. The "humic acid" fraction comprised humic and fulvic acids. The nonpolar fraction is adsorbed irrespective of pH. Only the associated form of humic acid is adsorbed – this fraction is dependent on pH. By lowering pH, the charge density of humic and fulvic acid groups is lowered as they are converted to their nonionized or associated form.

Therefore, the model of Kastl *et al.* included the following three groups:

(1) Non-polar, which adsorb independent of pH
(2) Polar humic acids, which do adsorb dependent on pH
(3) Non-adsorbing

Edwards had found that the maximum sorption capacity, *a* was relatively insensitive to pH. Kastl *et al.* therefore assumed that this parameter did not vary with pH. Like Edwards, the model of Kastl *et al.* assumed that adsorption followed a Langmuir isotherm and that equilibrium is reached faster than the development of removable flocs. Using similar nomenclature as presented above for Edwards' model, the model presented by Kastl *et al.* is as follows:

$$\frac{\mathrm{DOC}_{\mathrm{nonpolar},0} - \mathrm{DOC}_{\mathrm{nonpolar},l}}{M} = \frac{a \cdot b \cdot \mathrm{DOC}_{\mathrm{nonpolar},l}}{1 + b[HA]_l + \mathrm{DOC}_{\mathrm{nonpolar},l}}$$

where:

subscript 0 refers to initial conditions
subscript l refers to the treated water
M = metal coagulant added (mM/l)
a = maximum DOC sorption/mM coagulant (mg DOC/mM Al or Fe)
b = sorption coefficient (l/mg DOC)
$[HA]_l$ is the sorbable humic acid fraction (mg/l) in solution at equilibrium

Kastl *et al.* described how the data from two sets of jar tests are sufficient to accurately determine the five parameters required to activate the model:

- One jar is set at pH ~8.5 with high metal coagulant dose (say 2.5 mM/l Al or Fe)
- Five jars are set at lower pH (~5.5) with doses varying from low to high
- Six jars are set at pH 7 with doses varying from low to very high.

Figure 4.17 conceptually represents the DOC fractions remaining in solution during the jar test model calibrations.

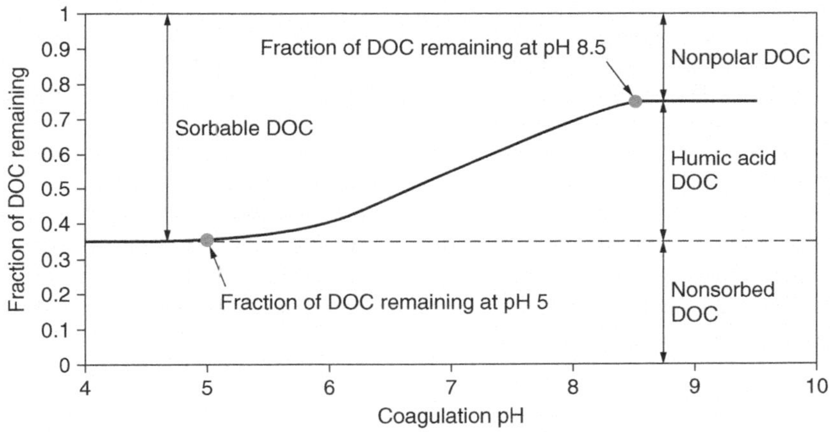

Figure 4.17 Conceptual diagram of DOC fractions remaining at high coagulant dosages (adapted from Kastl *et al.* 2004).

4.4.2.4 Importance of jar tests

Notwithstanding the feasibility of predicting TOC removals by models describing enhanced coagulation, there is still the necessity to treat water to an overall quality goal that includes not only TOC and disinfection by-products, but also particulates and pathogens. It is important to conduct comprehensive jar tests to evaluate the best combination of chemicals, pH and dosages to reduce not only TOC and DOC, but also to minimize turbidity. Such tests should also be conducted seasonally, to account for the variable nature of NOM in raw water sources. Floc characteristics have been demonstrated to change seasonally when treating NOM laden waters (Jarvis *et al.* 2004; Jefferson *et al.* 2004). Flocs formed in winter tended to be smaller and more fragile, had lower growth rates and settled slower. Possible reasons were related to higher fulvic acid fractions and lower SUVA values found in winter.

Cornwell (1994) showed the importance of conducting a series of pH and coagulant dosage tests to develop TOC and turbidity contour plots (see Chapter 8). This procedure facilitates the best treatment regime for optimizing overall treated water quality, and identifies the best criteria for controlling the coagulation process.

Gregory (1998) pointed out that particularly for low alkalinity and high TOC waters, the pH level at which NOM removal is optimized is also a region which could cause particle restabilization. This emphasizes the importance of identifying the coagulation conditions required for optimized overall water quality.

Carlson *et al.* (2000) provided an example with two basic alternatives to meet enhanced coagulation requirements. The first used alum alone at a dosage of 90 mg/l. The second also used alum, but with a reduced dosage of 45 mg/l and pH adjustment with sulfuric acid, at a dosage of 22 mg/l. The first alternative using alum alone represented a cost of US$18/Ml ($69/MG), including sludge handling costs. The second, using sulfuric acid and alum, had a corresponding cost of US$13/Ml ($50/MG). This again emphasizes the importance of evaluating a range of treatment options to meet overall treatment requirements.

When carrying out a series of jar tests to determine best treatment conditions, it is convenient to assess performance based on a surrogate parameter for TOC. When a reduced band of treatment options have been identified, more precise TOC measurements can be conducted to assess actual treatment performance. Probably the most convenient surrogate parameter for many waters is to use UV absorbance at 254 nm. This parameter has a correlation with true color arising from humic substances, as exemplified in Figure 4.18.

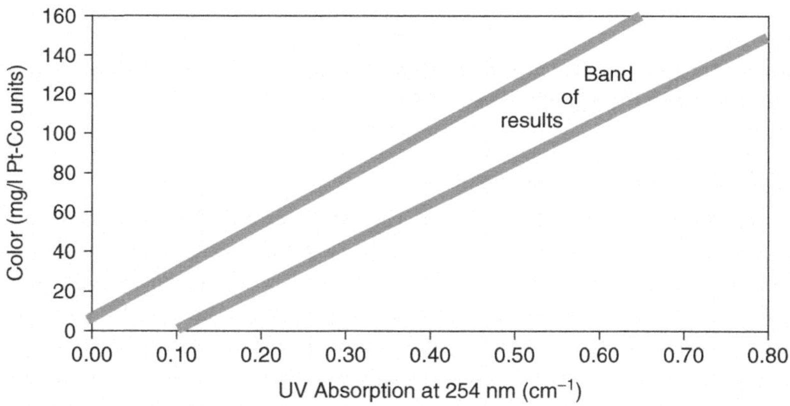

Figure 4.18 Correlation between UV absorption and color (data derived from Vik *et al.* 1985).

Data for this Figure were derived from Vik *et al.* (1985), who measured UV absorbance, color and TOC for a large number of raw and treated samples from several Norwegian lakes.

There is also often a relationship between true color and TOC. The relationship is not constant but varies depending on the raw water characteristics, as shown by data derived from various workers in Figure 4.19. However, during preliminary screening tests, measurement of UV absorption, which is related to color, can be a useful surrogate for TOC.

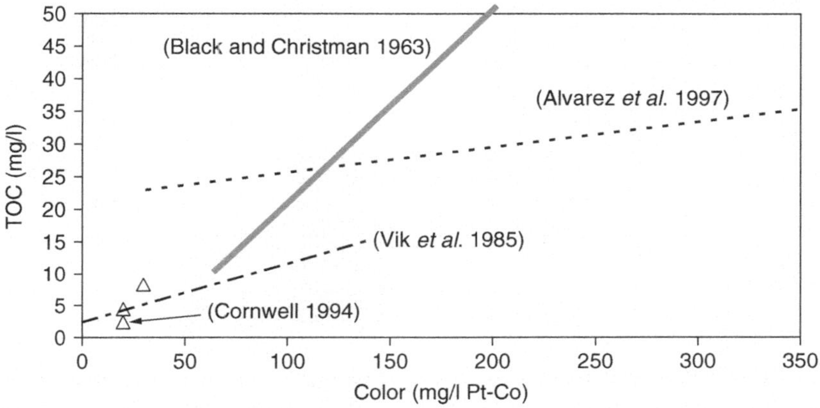

Figure 4.19 Examples of relationships between color and TOC (data derived from references shown).

4.4.2.5 Impacts of enhanced coagulation

Impacts of enhanced coagulation practices on water treatment plants include the following:

- Increased sludge generation due principally to the higher coagulant dosages. For example, Freese *et al.* (2001) found that depending on the raw water, the optimal coagulant dosages for NOM removal were 1.5 to 7 times higher than the optimal dosages for turbidity removal.
- The higher hydroxide content and the higher TOC content of the sludges produced generally contribute to poorer dewatering characteristics.
- Because of the higher organic content contained in the sludge, specific resistance values tend to be higher, and dewatered cake concentrations lower.
- Sludges derived from waters with higher molecular weight NOM species to which enhanced coagulation is applied tend to be more difficult to dewater than sludges from lower molecular weight NOM waters.

This latter point was demonstrated by Kelkar and Schafran (1994) where the specific resistance to filtration of different molecular weight fractions was measured after coagulation. Ultrafiltration was used to fractionate the NOM of different waters into five molecular weight ranges (>10,000; 3,000 to 10,000; 1,000 to 3,000; 500 to 1,000; less than 500). Figure 4.20 shows that the higher MW fractions had higher specific resistance values and were more difficult to dewater. The reason is that higher MW NOM is more readily removed by enhanced coagulation and, therefore, sludges from higher MW NOM waters would have a larger amount of NOM organic matter in the sludge matrix.

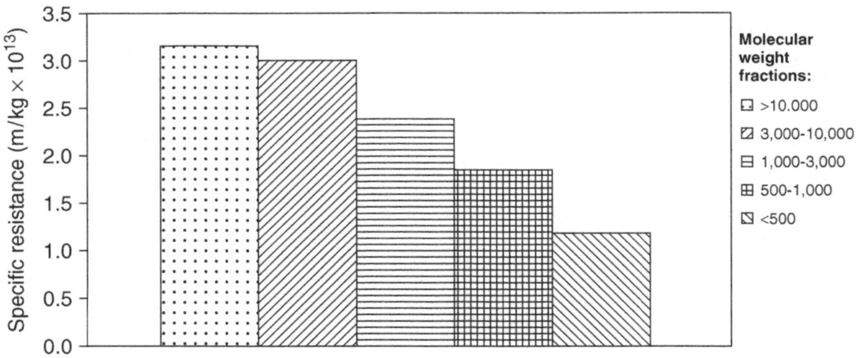

Figure 4.20 Effect of NOM molecular weight on sludge dewatering characteristics. In all cases alum dosage = 35 mg/l; TOC = 5 mg/l; coagulation pH = 5.6 (adapted from Kelkar & Schafran, 1994).

Enhanced coagulation in some cases also produces less-dense and more fragile flocs. This can increase floc carryover from clarifiers in conventional treatment plants, which in turn can shorten filter runs, lead to premature breakthrough, or increase filter backwash requirements.

A significant impact of enhanced coagulation relates to the integrity of treatment plant infrastructure. Enhanced coagulation generally operates at lower pH values which can be corrosive to concrete and mechanical equipment. A survey of 275 coagulation utilities in the United States showed that of those plants already practicing enhanced coagulation, approximately 15-percent had seen impacts such as increased concrete degradation; failure of metal pipes, pumps or structures; and negative effects on water quality of metal corrosion (McNeill & Edwards, 2003).

The issues with enhanced coagulation contributing to the negative impacts include decreased NOM; lower pH; increased chloride, sulfate, calcium or sodium; more iron or aluminum solids; higher residual aluminum levels; higher CO2 levels; and decreased alkalinity.

The following *Leaching Corrosion Index* was suggested as an indicator of the potential aggressiveness of a water to concrete (Alexander *et al.* 1994):

$$\mathrm{LCI} = 0.333\left[200(9.5-\mathrm{pH})-2{,}000(\Delta\mathrm{pH}) + 5.5(200-Ca)\right]$$

where

ΔpH = (actual water pH) – (saturation pH of $CaCO_3$)
Ca = calcium (mg/l)

The following flow regime corrections are recommended:

- *Turbulent flow*: multiply by 1.75
- *Stagnant*: multiply by 0.5.

The following temperature correction factor is also applied:

[1 + 0.05(T – 20)]

The following guides to LCI values were suggested:

- *LCI < 350*: protective coatings not required;
- *LCI 350–750*: higher grade cement required;
- *LCI > 750*: protective coating to the concrete recommended.

The LCI model is a simplification because of all the issues influencing aggressiveness mentioned above. Calibration of the model to water treatment plants is required for a more reliable index (McNeill & Edwards, 2003).

4.5 ALGAE REMOVAL AND HARVESTING

Algae production and harvesting for beneficial uses has been of interest for many years. McGarry *et al.* (1970, 1971) pointed out the considerable potential for protein production from high-rate algae ponds, with potential protein production rates per unit area approximately 85-times that of soybean fields. There is also considerable interest in large-scale algae production for the algae-to-biofuels industry.

One of the challenges with large scale algae production relates to harvesting, which could account for up to 30% the total cost of algae production (Oh *et al.* 2001). During the logarithmic phase of algal growth, the negative surface charge of the cells is relatively high and the algal cells remain stable and dispersed. The addition of metal or other coagulants is required to coagulate the algal dispersions for subsequent removal by some means. During longer residence times as the growth rate reduces, the negative surface charge reduces and the cells start to clump and settle, resulting in a process called autoflocculation. This is enhanced by the natural excretion of extracellular algogenic organic matter (AOM) that resembles organic polymers. If metal or organic coagulants are used to enhance the autoflocculation process, the AOM exerts most of the coagulant demand during coagulation operations. In general, coagulant dosages are lower during the stationary growth phase, compared with log-growth conditions.

Apart from harvesting for beneficial uses, algae removal is important for a number of applications, including:

- Upgrading of oxidation pond effluents, where the algae represent a considerable suspended solids and nutrient load that may impact receiving bodies of water;
- Pretreatment of eutrophied bodies of water used for drinking water sources which create difficulties in drinking water treatment such as coagulation,

sedimentation and filtration. The following are all motives for algae removal in potable water plants:

- algae and the associated algogenic organic material (AOM) are precursors for disinfection byproducts;
- there is the potential for toxin releases by cyanobacteria, such as *Microcystis*;
- high algae concentrations can release taste and odor compounds to the water;
- coagulation and separation of algae at the initial stages of drinking water treatment are often preferred to pre-oxidation to avoid cell lysis and the release of toxins or taste and odor compounds that could persist to the treated water.

This section describes algae removal and harvesting using metal coagulants alone, or enhanced with polymers. Chapter 5 continues the discussion using polymers as primary coagulants, without metal coagulants.

Bernhardt and Clasen (1991) demonstrated the importance of charge neutralization when using alum applied to *Scenedesmus obliquus* in a reservoir water. Zero electrophoretic mobility with alum addition coincided with maximum filter runs. Excess alum dosages caused charge reversal and restabilization, and a reduction in filter run times. These researchers suggested a stoichiometric, linear relationship between the optimum dosage of coagulant and the cell density in the suspension. However, Henderson *et al.* (2008, 2010) showed that the relationship is more complex and that treatment is influenced by algal morphology, motility, surface charge, cell density, and the concentration and composition of the AOM. It appears that the AOM controls the surface charge more than the algal cell surface itself (Edzwald & Wingler, 1990; Bernhardt & Clasen, 1991; Henderson *et al.* 2008). Coagulant demand is influenced strongly by protein complexation of the AOM before charge neutralization can take effect.

Complex formation between metal coagulants and proteins derived from AOM sequesters the coagulants, requiring much higher dosages for effective sweep-floc coagulation. Iron and aluminum ions form complexes with OH and COOH groups on the protein matter thereby hindering the formation of the metal-hydroxo complexes effective in coagulation (Pivokonsky *et al.* 2006). The protein-metal complexes are soluble or colloidal and can pass through filtration during water treatment. Bernhardt *et al.*,1986 (in Pivokonsky *et al.* 2006) showed that the sequestration effect can be reduced by the addition of calcium ions.

Surface area has been identified as a useful preliminary indicator of optimum coagulant dosage. Henderson *et al.* (2008) found a relationship between algal cell surface area and optimum coagulant dosage. They tabulated cell dimensions and calculated surface areas for a number of algae species. Their relationship can be expressed as follows:

$$\text{Coagulant dosage (ng Al/cell)} = 0.00010 \times [\text{Cell surface area } (\mu m^2/\text{cell})]^{1.1364}$$

However, they also pointed out that the actual coagulant demand depends on other factors, including the AOM composition and concentration that would impact surface charge, complex the metal coagulant, or sterically interfere with coagulation. They further suggest that a relationship between coagulant dosage and charge density is stronger than that of surface area, particularly when a relationship is sought across multiple species, rather than just one species. Their relationship for optimum coagulant dosage related to charge density measured by a back-titration method, can be expressed as follows:

$$\text{Coagulant dosage (ng Al/cell)} = 42.27 \times [\text{Charge/cell (neq/cell)}]^{0.813}$$

The above expression was derived using the algal species *Chlorella vulgaris*, *Microcystis aeruginosa*, *Asterionella formosa* and *Melosira sp.* The authors suggest that for a given system, an effective control strategy for coagulation demand of algal laden waters would be by charge measurements using zeta potential.

Tests showed that at high algae concentrations (200 NTU suspensions) *Spirulina sp.* required much higher dosages of a high basicity polyaluminum chloride coagulant compared with *Chlorella sp.*, attributed to the higher negative surface charge of *Spirulina sp.* (Cheng *et al.* 2010). Other tests with *Anabaena flos-aquae* showed that optimum coagulation conditions using ferric sulfate were predicted by the zero point of charge using streaming current measurements (Briley & Knappe, 2002). Jar test evaluations determined that a pH of 6.0 was optimal for coagulation of this algal species and that the optimal coagulant dosage at this pH corresponded with the zero point of charge detected by streaming current. Best performance for algae removal and NOM removal corresponded with these conditions when 0.1 mg/L anionic polymer was added to improve settling. Streaming current detection was considered to be an adequate control strategy for algae removal for a given system, after optimum conditions are defined by jar test evaluations.

Figures 4.21 through 4.23 show results of algae removal from oxidation pond effluents (principally *Chlorella vulgaris*) using ferric chloride, ferric sulfate and aluminum sulfate. Two sets of experiments were conducted: first, jar tests with settling were used to determine optimum dosages and pH for each coagulant. Second, continuous pilot flotation tests were conducted to confirm the jar test results and ascertain whether optimum conditions for coagulation were different when bubble-particle attachment was induced.

Figures 4.21 and 4.22 show that the optimum ferric chloride dosage was 50 mg/L as Fe and the optimum pH 5.7, determined in the jar test. Tests conducted with the pilot flotation test unit confirmed the optimum dosage of 50 mg/L as Fe, but indicated that the optimum pH had a more distinct optimum value. It appears that bubble-particle attachment to the algal flocs was significantly influenced by pH. The magnitude of the zeta potential difference between bubbles and coagulated algal particles at different pH was likely a contributing factor, as suggested by Kwak *et al.* (2006). At optimum conditions, the results of the pilot flotation unit and the jar tests were not substantially different when using ferric chloride.

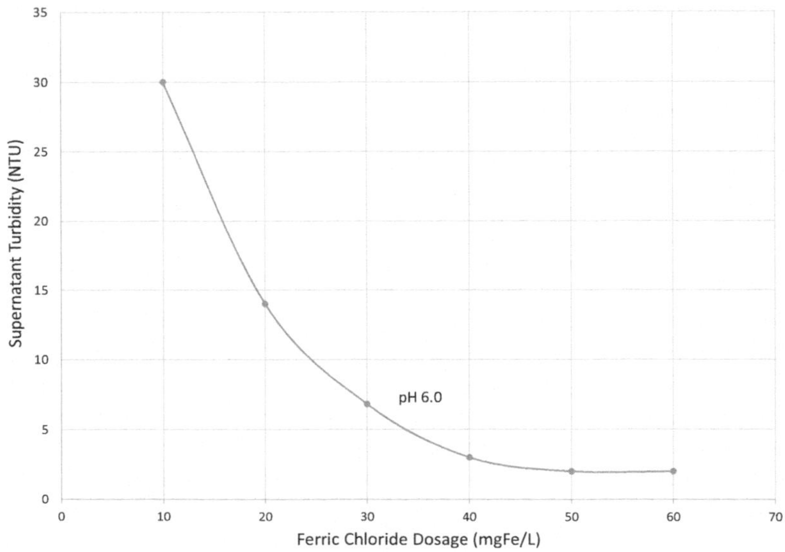

Figure 4.21 Results of jar test using ferric chloride at pH 6.0 on oxidation pond removal of algae (Bratby *et al.* 1974).

When ferric sulfate was used, although jar test results were similar to those using ferric chloride, the pilot flotation unit results were slightly worse, with a minimum turbidity of 11 NTU, compared with 8 NTU when using ferric chloride. It appeared that the sulfate anions could be impairing bubble-particle attachment. When aluminum sulfate was used in both the jar test and the pilot flotation tests, the optimum dosage was 25 mg/L as Al at an optimum pH of approximately 5.5. This optimized dosage and pH using alum was similar to subsequent findings elsewhere (Sandbank & Shelef, 1987). However, comparing these results on algae removal from oxidation ponds treating predominantly domestic wastewater, with ponds treating piggery wastewater using spiked algal populations in the laboratory (*Chlorella*, *Scenedesmus obliquus*, *Chlorococcum*), the generally higher coagulant dosages required in the latter case emphasize the importance of the organic content in the liquid phase and its associated coagulant demand (de Godos *et al.* 2011).

In the jar test using alum the supernatant was very clear with a turbidity of 2 to 3 NTUs (Figure 4.23). However the pilot flotation tests using alum produced a turbidity of approximately 17 NTU which did suggest that the sulfate anions may cause some interference with bubble-particle attachment, possibly by collecting more readily at the air-water interface and augmenting the already negative charge carried by the bubbles.

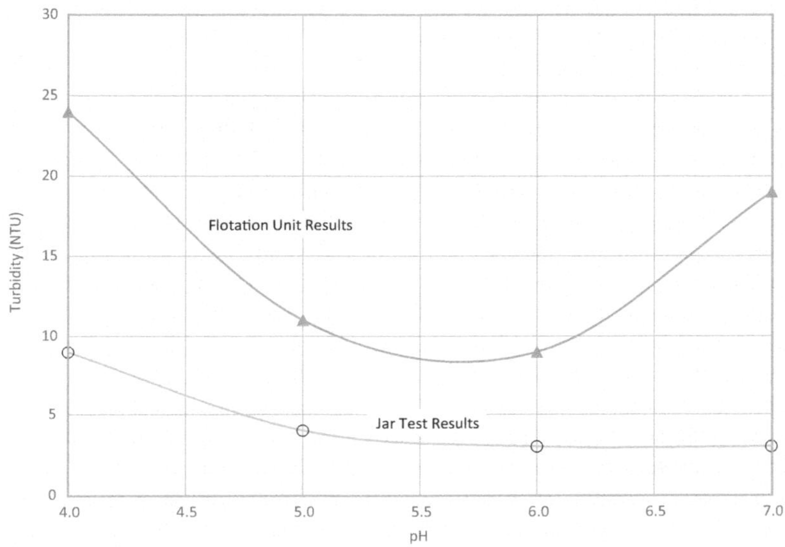

Figure 4.22 Determination of optimum pH using ferric chloride at optimum dosage (50 mg/L as Fe) in jar test and flotation unit (Bratby *et al.* 1974).

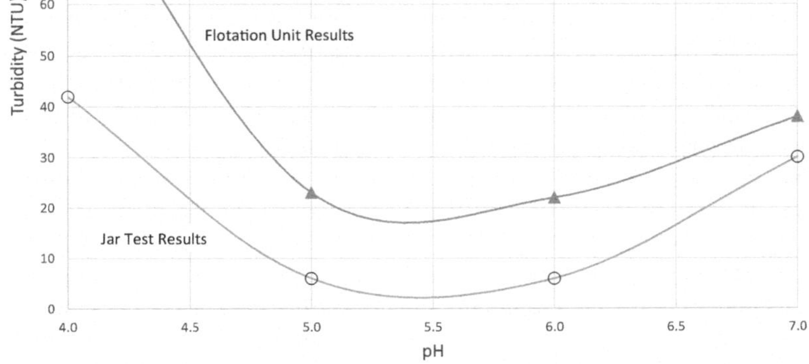

Figure 4.23 Determination of optimum pH using aluminum sulfate at optimum dosage (25 mg/L as Al) in jar test and flotation unit (Bratby *et al.* 1974).

Depending on the intended use of the harvested algae, the type and quantity of coagulant used could contaminate the final product. Despite the effectiveness of the common metal coagulants like aluminum sulfate, ferric chloride and ferric sulfate, there could be concerns with the resultant quality of the removed algae because of the high metal content. For example, one beneficial use of the separated algal sludge is to produce methane biogas in anaerobic digesters. Algae sludge coagulated with alum is successfully anaerobically digested (Wiley *et al.* 2011; Sukias & Craggs, 2011) although there is evidence of 13% to 40% reduction in methane production per volatile solids added when the algal sludge contains 16 g/m^3 to 130 g/m^3 aluminum (Sukias & Craggs, 2011). Suggested causes for the anaerobic inhibition include binding of substrates, immobilization of phosphates, and direct toxicity by the metal, particularly aluminum.

Besides the common metal coagulants, pre-hydrolyzed coagulants such as polyaluminum chlorides, polyferric sulfate and coagulant blends have been effective for algae coagulation. Comparative tests were conducted on suspensions of a blue-green species, *Anabaena flos-aquae* and a diatom, *Asterionella formosa* using polyferric sulfate, ferric sulfate, alum and polyaluminum chloride (Jiang *et al.* 1993). For *Anabaena*, the order of effectiveness was polyferric sulfate, alum, ferric sulfate then the polyaluminum chloride coagulant. For *Asterionella*, the order of effectiveness was polyferric sulfate, alum, PACl, then ferric sulfate.

The effect of increasing concentrations of humic substances in the water on algae removal was demonstrated with *Asterionella*: the effect was to reduce the total cell removal as the concentrations of humic materials increased. However, the maximum removal attained remained the same, although for given algae removals, the concentration of coagulant had to be increased as the humic concentrations increased. The effect of humic substances on coagulation performance strongly suggested the influence of charge interactions. Using colloid titrations it was shown that higher coagulant dosages are required for overall charge neutralization at higher humic concentrations.

The order of effectiveness of the coagulants in terms of algae removal was also influenced by the rates of floc formation and settling, presumably influenced by resultant floc densities. Tests of supernatant total particle counts with settling time, using *Asterionella* showed remaining particle counts in the order from lowest to highest: polyferric sulfate, alum, ferric sulfate then PACl (Jiang *et al.* 1993).

Coagulation tests were applied to a eutrophic drinking water source with a suite of algal species (3.52×10^4 cells/mL: the cyanobacteria *Microcystis*, *Merismopedia*, *Gloecapsa*, *Spirulina* and *Oscillatoria* comprised 45% of the total; the chlorophyte *Oocystis*, *Chlorella* and *Scenedesmus* comprised 20%; the bacillariophyta *Melosira*, *Cyclotella* and *Fragilaria* comprised 25%; and other algae made up the remaining 10%). Comparing alum and an alum-polymer blend comprising alum plus polydiallyldimethylammonium chloride (PDADMAC), the alum-polymer coagulant out-performed alum alone with higher algae removals at lower dosages expressed as aluminum. The alum-PDADMAC blends varied the polymer content from 0.05 to 0.20 g PDADMAC per g Al_2O_3. In general, algae removal performance increased as the polymer content increased (Zhao *et al.* 2010).

Formal flocculation has been found to be unnecessary for good performance using dissolved air flotation after adding metal coagulant at the rapid mix stage for algae removal from oxidation ponds; an in-line 14-second reaction time after coagulant addition was sufficient for efficient flotation and greater flocculation times than this provided no discernible improvement in flotation performance (Bratby *et al.* 1974). In this regard, reference should also be made to the section on CEPT later in this Chapter.

Edzwald and Wingler (1990), treating significantly lower algae concentrations from reservoir waters also found that extensive flocculation times were not necessary before flotation. In their case they found minimal improvement beyond 5-minutes of flocculation at temperatures as low as 4°C.

A common algae pretreatment strategy prior to drinking water treatment is pre-oxidation using oxidants such as chlorine, potassium permanganate, ozone or potassium ferrate. The benefit in many cases is to improve algae removal by coagulation, particularly for motile species, by the following mechanisms (Henderson *et al.* 2008):

- Change in external cells such as detachment of spinal appendages of some species, such as *Scenedesmus* and *Chlorococum*. There may also be damage and perforation to the outer cell and cell wall, although this is usually at higher oxidant dosages.
- The motion of species such as *Rhodomonas minuta*, *Cryptomonas* sp., *Euglena* sp., *Navicula* sp., and *Nitzschia* sp. is impeded.
- Excretion of extracellular polymeric chitin fibrils that aid coagulation.
- Degradation of extracellular AOM reduces the interference with coagulation.

Despite these benefits, there are disadvantages to pre-oxidation, including the formation of trihalomethanes when using chlorine or chlorine dioxide. When using ozone, the chlorophyll in the algae cells is oxidized and causes the release of intracellular organic matter (Ma *et al.* 2006). The dissolved organic carbon (DOC) can increase substantially thereby increasing the disinfection byproduct (DBP) formation potential. Over-dosing the oxidant can cause algal cell lysis which releases toxins and/or taste and odor compounds to the water.

The oxidant dose should be such that it improves algal coagulation but avoids cell lysis. This is often difficult since the dosage of a particular oxidant that causes cell lysis varies with the algae species. The disadvantages with pre-oxidation could be mitigated in treatment plants that include post-ozonation and GAC filters, where the increased DOC formed by pre-oxidation can be removed (Henderson *et al.* 2008).

The effectiveness of pre-oxidation on algae coagulation depends on the species. For example, pre-oxidation with potassium permanganate has had beneficial results with the coagulation of the green algae species *Chodatella* sp. and the cyanobacteria *Microcystis aeruginosa*. Pre-ozonation had an even greater beneficial effect on alum coagulation when applied to *Chodatella* sp. (Huang *et al.* 2009).

The beneficial effects of ozone and potassium permanganate on *Chodatella* sp. cultures are due to different mechanisms. Ozone causes cell stress that releases extracellular polymeric material and reduces the zeta potential of the cell. The released extracellular polymeric material also serves as a flocculant aid (Chen & Yeh, 2006; Bernhardt & Clasen, 1991). Potassium permanganate also causes secretion of extracellular material and the interaction of this material with MnO_2 within the algae floc improves algae removal. Pre-oxidation with potassium permanganate was more beneficial compared with ozone. The presence of calcium generally improved the effects of pre-oxidation with *Chodatella* (Chen & Yeh, 2006). These conclusions were also confirmed by pilot tests applied to a eutrophic drinking water lake source with green algae (*Chodatella* sp.), diatoms and blue-green algae as the dominant algal species. Pre-oxidation with potassium permanganate before alum coagulation was more beneficial compared with ozone. The high hardness of this water and the calcium content contributed to the beneficial effects (Chen *et al.* 2009).

Potassium permanganate was shown to have a marginal beneficial effect on the alum coagulation of the diatom *Navicula* sp. However, pre-ozonation had a marked deleterious effect on coagulation with this species, speculated to be due to the breakdown of extracellular polymeric material by the ozone, thereby reducing the beneficial effect of this material on coagulation. The conclusion was that ozone should be avoided when diatoms are the dominant species (Huang *et al.* 2009).

Potassium ferrate was applied for pre-oxidation of eutrophied lake water containing the green algae species *Chlorella*, *Spirogyra*, *Chlorococoum* and *Scenedesmus*. Algae removals with alum coagulation increased with increasing K_2FeO_4 dosages up to 5 mg/L. Part of the beneficial effect of K_2FeO_4 pre-oxidation arises from ferric hydroxide formation that assists in coagulation. For a given performance the subsequent coagulant dosage could be reduced (Ma *et al.* 2002).

The effects of ozone on algae cells and the release of extracellular polymeric material provides a removal strategy by flotation using ozone bubbles, without added coagulants. Tests with *Chlorella vulgaris*, targeting lipid extraction for biofuel production, showed that ozone dosages of 0.005 to 0.03 mg/mg biomass were effective for flotation. The collection mechanism was assumed to be due to the release of intracellular proteins that acted as surfactants which promoted ozone bubble-algae cell attachment. For this reason, oxygen flotation without ozone did not achieve separation (Cheng *et al.* 2010). Further tests with *Scenedesmus obliquus* FSP-3, also targeting lipid production, showed effective flotation and separation using ozone bubbles and confirmed that air flotation did not have the same effect. In this case, required ozone dosages were higher, from 0.2 to 0.5 mg/mg biomass. The higher dosages required for *Scenedesmus obliquus* FSP-3, compared with *Chlorella vulgaris*, were attributed to higher concentrations of humic-like substances excreted by the algae during the stationary growth phase, which exerted a considerable ozone demand (Cheng *et al.* 2011).

4.6 PATHOGEN REMOVAL

4.6.1 Removal of *Giardia* and *Cryptosporidium*

The U.S. EPA surface water treatment rule uses water treatment technology to control *Giardia* in drinking-water by requiring 99.9-percent (3-log) *Giardia* removal or inactivation. Adequately designed and operated water treatment plants, with coagulation, flocculation, sedimentation and filtration are assigned a 2.5-log removal credit, leaving only 0.5-log inactivation to be achieved by disinfection.

Cryptosporidium is also a protozoan parasite but is of particular concern in drinking water because of its resistance to disinfectants like chlorine. It has been associated with several waterborne disease outbreaks. Existing drinking water regulations required at least 99-percent (2-log) removal of *Cryptosporidium*. However the long term 2 enhanced surface water treatment rule (LT2ESWTR) requires the identification of filtered water systems in one of four risk bins, based on monitoring results. For those systems classified in higher risk bins, these must provide 90 to 99.7-percent (1.0 to 2.5-log) additional reduction of *Cryptosporidium*.

Horn *et al.* (1988) demonstrated that a dual stage filtration process was able to remove greater than 2-log Giardia cysts from cold waters with low turbidity (<1 NTU) and temperatures less than 8°C (46°F). Corresponding total coliform and turbidity removals were greater than 98 and 90-percent, respectively.

Edzwald *et al.* (2000) showed that by using dissolved air flotation (DAF) for clarification, average log removals of *Giardia* and *Cryptosporidium* were 2.4 and 2.1, respectively. Corresponding removals using a lamella clarifier were 1.1 and 1.0, respectively. It was shown that optimum coagulation conditions are governed by turbidity and NOM removal requirements, rather than by pathogen removals. Optimum conditions for the former correspond with those for pathogen removal. Overall *Giardia* and *Cryptosporidium* removals, including the filtration step were approximately 5-log.

Cryptosporidium oocyst surfaces are believed to consist of polysaccharide layers. The negative charge carried by the oocysts is believed to arise from carboxylic acid groups in surface proteins. Xagoraraki and Harrington (2004) suggested that removal of *Cryptosporidium* in their studies using alum coagulation was by a sweep floc mechanism. Zeta potential measurements suggested that removal did not appear to be by a charge neutralization mechanism at lower DOC concentrations. At higher DOC, it was suggested that the mechanism was mediated by a NOM-assisted bridging between aluminum hydroxide and oocyst particles. Similarly to the findings by Edzwald *et al.*, optimal coagulation conditions for oocyst removal coincided with optimum conditions for NOM removal.

4.6.2 Virus removal

The percentage of cases of virus infection which are directly attributable to drinking water is not conclusively known, although it had been suggested that before 1974, approximately 30-percent of infected cases could be attributable to the water route (McDermott, 1974).

The survival time of viruses in water varies widely and is dependent on the specific virus, temperature and general quality of the water. Under certain conditions and with certain virus types, survival times may be up to 200 days in river water (Cookson, 1974).

It has often been reported that chlorine is not wholly effective in virus inactivation (Shelton & Drewry, 1973; York & Drewry, 1974). A possible reason is that in cases where positive virus counts were obtained in treated waters, insufficient disinfection contact time was allowed. For example, some virus types (Polio 2) require a chlorine contact time of 40 minutes for 99.99% inactivation in the presence of 0.5 mg/l free chlorine (Cookson, 1974). The average contact time (for twenty human viruses) is of the order 15 minutes with the least resistant virus type requiring 2.7 minutes and the most resistant, 40 minutes.

Several workers have reported significant virus removals using metal coagulants and polyelectrolytes (Chaudhuri & Engelbrecht, 1970; Manwaring *et al.* 1971; Shelton & Drewry, 1973; York & Drewry, 1974; Rao *et al.* 1988; Nasser *et al.* 1995). With metal coagulants, removals of up to 99.9% have been reported for both aluminum and ferric salts. Various polyelectrolytes (cationic) have effected removals of greater than 99% but have the disadvantage that if other material is present in the form of color, turbidity, and COD, removal of such material is poor. Using metal coagulants and poyelectrolytes conjointly has the advantage that better floc characteristics are produced. If a variety of substances are present in water, it is possible that the use of both metal coagulants and polyelectrolytes will effect a higher overall removal. However, this very much depends on the conditions pertaining for each case. Some workers report that when using polyelectrolytes as flocculant aids, floc formation improves but does not improve virus removals beyond those achieved using metal coagulants alone.

Viruses are essentially DNA (deoxyribonucleic acid) or RNA (ribonucleic acid) units contained within a protein coat. The destabilization mechanism involves coordination reactions between metal coagulant species and carboxyl groups of the virus coat protein (Chaudhuri & Engelbrecht, 1970). From the similarity of the destabilization mechanisms for organic color and viruses, one would expect similar conditions of pH at optimum removal. From reported results this is apparently so. For example the optimum pH for virus removal with aluminum sulfate has been found to be in the region of 5.0. Percentage virus removals with aluminum sulfate were in the range 97.7 to 99.8%. Turbidity removals were in the range 98.3 to 99.3%. Using a cationic polyelectrolyte as flocculant aid, virus and turbidity removals were increased to 99.9 and 98.5% respectively (Chaudhuri & Engelbrecht, 1970).

Manwaring *et al.* (1971) applied ferric chloride to remove a bacterial virus (bacteriophage MS2 against E.Coli). With an initial virus concentration of 3.9×10^5 pfu (plaque forming units)/ml, removals of up to 99.7% at a ferric chloride dosage of 60 mg/l and pH 5.0 were achieved.

It should be noted that applying metal coagulants or polyelectrolytes does not fully inactivate viruses. Therefore, a potential health hazard exists with the ultimate disposal of water treatment plant sludges. Furthermore, complete virus removal by destabilization with metal coagulants has not been reported. For a safe drinking water, disinfection of the water before distribution is required, using chlorine, ozone, chlorine dioxide, or ultraviolet (UV) radiation.

However, there is some inactivation that accompanies virus removal by coagulation. Matsushita *et al.* (2004) reported that the infectious virus concentration only recovered partially after re-dissolution of the aluminum hydroxide precipitates. This phenomenon was interpreted as a virucidal activity of the aluminum. PACl coagulants appeared to have a higher virucidal activity compared with alum. The presence of NOM in waters appeared to inhibit the virucidal activity of the aluminum.

4.7 EFFECT OF ANIONS

4.7.1 General

In earlier sections it was mentioned that certain anions tend to displace hydroxyl ions from the coordination sphere of metal ions and that this displacement preferentially occurs at lower pH values. The effect of a particular anionic species on the destabilization characteristics of metal ions applied to a given suspension depends on a number of factors. In general, the tendency of an anion to alter the behavior of hydrolyzed metal coagulants is related to its electronegativity and its tendency to react with the metal ion or positively charged sites on the metal hydroxide precipitate. Marion and Thomas (1946) put forward the following general rules:

- If the anion is a strong coordinator with aluminum and not readily replaced by hydroxyl ions, the pH of optimum destabilization will drop sharply with increasing anion concentration.
- If the anion is a strong coordinator with aluminum but can be displaced by hydroxide ion, the pH of optimum precipitation increases with a very basic anion, and decreases with a weakly basic anion.
- If the anion is only a very weak coordinator with aluminum, it exerts only a slight effect on optimum precipitation, generally in the direction of lower pH values.

The above explains why nitrates and perchlorates have very little tendency to coordinate with metal ions and, therefore, do not have a notable influence on destabilization with metal coagulants. Chlorides, probably having a weak coordination capacity with metal ions, shift the optimum pH slightly to the acid side if the concentration is high enough. That is, if the $[Cl^-]/[OH^-]$ ratio is high enough, hydroxyl ions are displaced from the coordination sphere of metal ion complexes by chloride ions.

The relative significance of different anions in altering the pH range of effective coagulation, based on molar concentrations, is as follows: phosphate > silicate > sulfate > fluoride > bicarbonate > chloride > nitrate (Letterman & Venderbrook, 1983). One effect of this is that, in general, aluminum sulfate is a more effective coagulant than aluminum nitrate.

The effect of bicarbonate on coagulation is not only to increase the alkalinity and buffer capacity of the water, but also to alter the physical-chemical characteristics of the metal-hydroxide precipitates. Increasing the bicarbonate ion concentration tends to increase the rate of flocculation reactions (Letterman *et al.* 1979). For a given metal coagulant concentration, increasing bicarbonate concentration also tends to reduce the final turbidity of the treated water. Alternatively, increasing bicarbonate ion concentrations tend to reduce the metal coagulant concentration required to achieve a given treated water turbidity.

The sulfate anion has a moderate effect on the metal hydroxide system. It does not appear to form a basic metal-sulfate precipitate and take the place of hydroxide in the precipitate matrix, as does phosphate. With sulfate anions the effect is to *extend* the pH zone of destabilization towards the acid side whereas with phosphate anions, the effect is to *shift* the optimum pH zone to the acid side. These two cases will be discussed in greater detail in the following sections. Furthermore, the special case of phosphorus removal during tertiary wastewater treatment will be discussed in a separate section.

4.7.2 Effect of sulfate

Sulfate anions are relatively strongly coordinated with metal ions and are weakly basic. Their effect, therefore, is to broaden the pH zone of optimum destabilization to the acid side (Hanna & Rubin, 1970). This is apparent in Figure 4.21 where, with increasing SO_4^{2-} concentration, the pH zone for optimum destabilization is increasingly spread further towards the acid side. (Note that in Figure 4.21 the efficiency of destabilization is expressed in terms of the time for floc formation). With increasing sulfate concentration, progressively greater opportunity is afforded for mixed ligand complexes to form. This is even more pronounced at lower pH values where the opportunity for SO_4^{2-} to be complexed is further increased.

Pommerenk and Schafran (2005) showed how adsorption of sulfate to aluminum hydroxide complexes increases with decreasing pH. Sulfates reduce the zeta potential of the aluminum hydroxide precipitates but does not cause charge reversal, which helps to explain the results in Figure 4.21.

Matijevic (1973) reported the probable predominance of tetrapositively charged mixed ligand complexes of the form $[Al_8(OH)_{10}(SO_4)_5]^{4+}$ at pH values in the region

4.5 to 7.0 for aluminum sulfate applied to destabilization of silver halide sols. At lower pH values, relatively ineffective species of the form $Al(SO_4)^+$ are increasingly predominant. Matijevic pointed out that at low pH in the presence of sulfate, the residual hydrated Al^{3+} ion uncomplexed with SO_4^{2-} is the effective species during destabilization. Hence, since the $Al(SO_4)^+$ complex has no significant adsorptive capacity, the higher the sulfate concentration and the lower the pH, the less effective is destabilization.

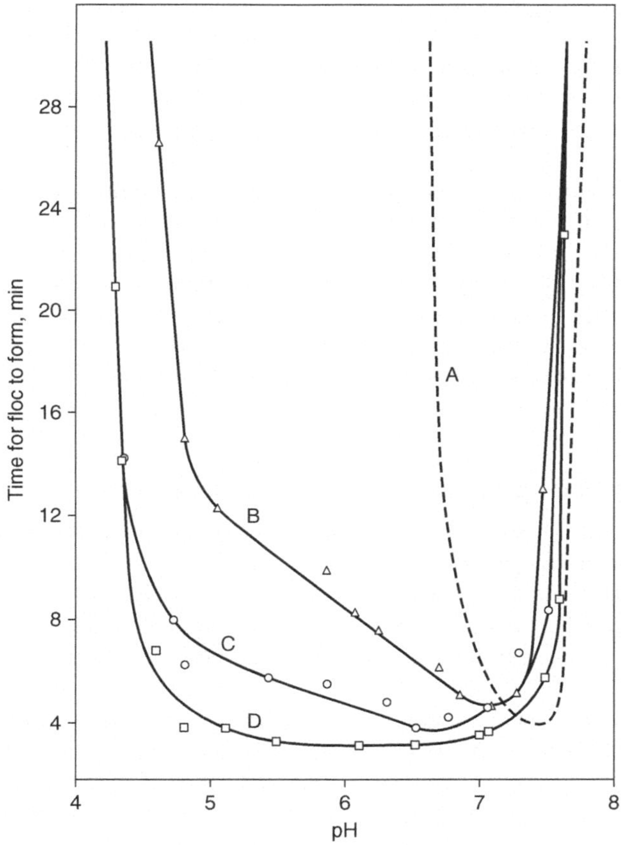

Figure 4.24 Effect of sulfate ion on time of formation of alum floc. Broadening of the pH zone of coagulation by sulfate ion. Curve A, 35 mg/l alum and sodium hydroxide; Curve B, addition of 25 mg/l SO_4; Curve C, addition of 50 mg/l SO_4; Curve D, addition of 125 mg/l SO_4 (from Black *et al.* 1933).

One can summarize the effects of sulfate anions on destabilization with metal coagulants as follows: at pH values greater than approximately 7.0, added sulfate has very little effect since minimal displacement of OH^- occurs. As the pH is lowered, SO_4^{2-} progressively displaces OH^- forming complexes with an [Al]:[OH] ratio of 1:1.25 (Matijevic, 1973) compared to complexes with an [Al]:[OH] ratio of 1:2.5 (Matijevic *et al.* 1961).

However, the presence of SO_4^{2-} in the coordination sheath of the metal ion complex renders the resultant complex with similar charge characteristics to the Al-OH complex at optimum pH in the absence of SO_4^{2-}. If the pH is excessively lowered $Al(SO4)^+$ species predominate and destabilization is minimal. From this it is appreciated that the effect of adding sulfate to a hydrolyzed metal ion system does not shift the optimum pH but merely extends it to the acid side to an extent depending on the sulfate concentration present. Within the (spread) optimum pH range there is a gradual transition from Al-OH to $A1\text{-}SO_4\text{-}OH$ species as the pH is lowered, each species retaining similar charge characteristics and destabilization efficacy.

The pH required for maximum precipitation of aluminum compounds is lower in the presence of sulfate anions. De Hek *et al.* (1978) attributed this to a catalytic reaction of SO_4^{2-} on precipitate formation. Although the sulfate ion enters into complexing reactions with aluminum, on precipitate formation the hydroxyl ion penetrates the inner coordination sphere. Hence, on growth of the precipitate, sulfate anions are progressively pushed outwards serving to initiate further precipitate growth by complex formation with aluminum and hydroxyl ions.

4.7.3 Effect of phosphate

In the preceding section it was seen that the effect of sulfate anions is to *extend* the optimum pH range to the acid side to an extent dependent on the sulfate concentration present. With phosphate anions the effect is somewhat different. As seen in Figure 4.22 (Packham, 1963), increasing phosphate concentrations *shift* the optimum pH to the acid side, to a value dependent on the phosphate concentration present.

In the presence of phosphate the principal ligands at pH values of 7.0 and above are OH^-, HPO_4^{2-}, and PO_4^{3-}. Phosphate species, through their displacement of OH^- ions, render the complex more negative than if hydroxyl ions were the only ligands present; such phosphate species are strong coordinators with metal ions. At lower pH values phosphate species are changed to singly charged $H_2PO_4^-$ and the availability of OH^- is reduced. Hence, to retain the same overall charge characteristics as at the optimum pH value when no phosphate is present, the optimum pH in the presence of phosphate is lower.

From the above argument it is appreciated that, as the phosphate concentration is increased, because of a greater statistical opportunity for incorporation of phosphate species in the coordination sheath, the lower will be the optimum pH for destabilization.

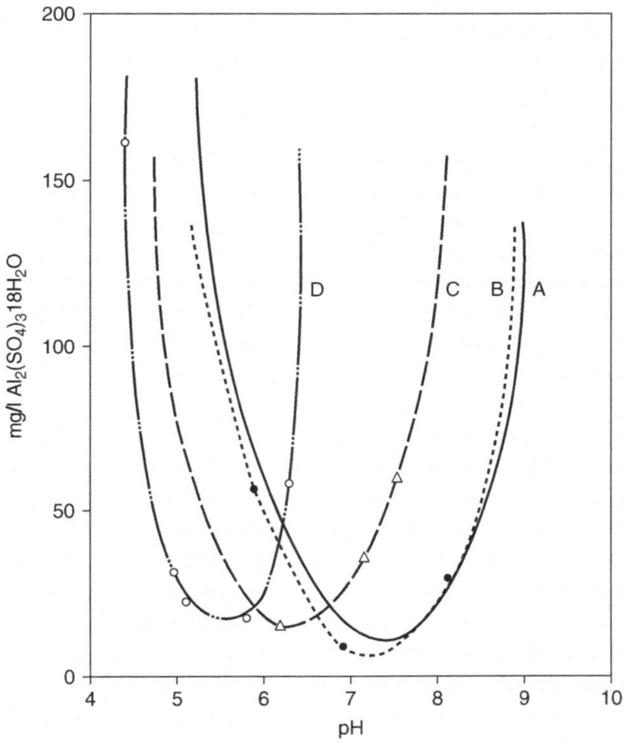

Figure 4.25 Effect of orthophosphate on pH zone of coagulation. Points on the curves represent the alum dosage required to reduce 50 mg/l of clay turbidity to one-half its original value. Curve A, alum alone, no PO_4; Curve B, 4.8 mg/l PO_4 added; Curve C, 48 mg/l PO_4 added; Curve D, 475 mg/l PO_4 added. Presence of orthophosphate displaces the pH zone of coagulation towards the acid side (after Packham, 1963).

4.8 CHEMICAL PHOSPHORUS REMOVAL IN WASTEWATER TREATMENT

4.8.1 General

Eutrophication is a process that results from the accumulation of nutrients in lakes or other bodies of water. In freshwater lakes and rivers, phosphorus is often the growth limiting nutrient. In clean, pristine waters the concentrations of phosphorus are usually very low. However, phosphorus is used extensively in fertilizer and other chemicals, and can be found in higher concentrations in areas of human activity. Domestic and industrial wastewater discharges are important sources of phosphorus to surface waters. Organic phosphates are contributed to sewage by body waste and food residues. Orthophosphate and polyphosphates can be contributed by detergents. Orthophosphates are the most stable form of phosphates and are used directly by plant cells. Various forms of phosphorus will convert to orthophosphate during wastewater treatment.

Increasing recognition in the 1970s and 1980s of the impact of phosphorus on eutrophication led to a movement to reduce phosphorus loads to the environment, particularly in the USA, Japan and some European Union member states. The two main areas of action were to reduce the amount of sodium tripolyphosphate used in detergent builders, and switch to alternative builders (such as zeolites, polycarboxylic acids, citrates, and nitrilotriacetic acid); and to improve wastewater treatment to include a higher degree of phosphorus removal. Measures to move from sodium tripolyphosphate to other detergent builders in domestic laundry detergents in most EU member states were initiated by 1990.

A ban on the use of phosphate based detergents can achieve a phosphorus load reduction of up to 40-percent entering surface waters (EC, 2002). This would not be sufficient on its own to result in substantial improvements in eutrophication (Lee & Jones, 1986). However, by combining a reduction in the sodium tripolyphosphate content of detergents, together with improved wastewater treatment, eutrophication related water quality would be significantly improved. Hartig *et al.* (1982) determined that by limiting the phosphorus content in detergents in Michigan to no more than 0.5-percent by weight, average cost savings in coagulant costs of $425 per Ml/d ($1,610 per mgd) were achieved for phosphorus removal.

In Switzerland, the phosphate content of detergents was progressively reduced by 20% in 1981, a further 10% in 1983, and by a complete ban in 1986. The phosphate ban resulted in reductions of over 40-percent in the phosphorus concentrations and loads to wastewater plants, as well as significant shifts in phosphorus species. Maurer and Boller (1999) present a comparison between phosphorus loads before and after the ban. In 1980, typical phosphorus loads were 1,720 gP/capita · year. In 1994, loads had reduced to 960 gP/capita · year. Table 4.9 shows the percentage phosphorus fractions in raw and settled sewage before and after the phosphorus ban.

Table 4.9 Percentage phosphorus fractions in raw and settled sewage before (1980) and after (1994) the phosphate ban in Switzerland (after Maurer & Boller, 1999).

Phosphorus Fractions	1980		1994	
	Raw Wastewater	Primary Effluent	Raw Wastewater	Primary Effluent
Total	100%	100%	100%	100%
Particulate	20%	13%	36%	27%
Dissolved	80%	87%	64%	73%
Organic	5%	4%	9%	11%
Polyphosphates	52%	58%	13%	13%
Orthophosphate	23%	25%	42%	49%

In areas where phosphate builders are allowed in detergents, phosphorus concentrations in wastewater discharges typically vary from 4 to 9 mg/l. By controlling the use of phosphates in household detergents, phosphorus concentrations in wastewaters have been reduced by 14 to 41-percent, to 3 to 7 mg/l (Hartig *et al.* 1982). Most of the remaining phosphorus is contributed by human wastes as a result of metabolic breakdown of proteins, and some remaining detergent uses. Although in some areas phosphate builders in household detergents are controlled, others (such as dishwashing formulations and industrial detergents) are not. Phosphorus contributions from human wastes vary from particulate forms, contained in faeces, to soluble orthophosphates contained in urine.

With or without phosphate detergents, the phosphorus concentrations in wastewater discharges are often high enough to be a major cause of eutrophication. In many cases, wastewater treatment employs the activated sludge process incorporating nitrification-denitrification and/or biological phosphorus removal. With biological phosphorus removal, a high percentage of phosphorus is incorporated in the biological mass and is removed from the system via sludge wastage. The effluent typically contains 0.5 to 1.0 mg/l phosphorus or less. In some cases this is adequate to maintain an oligotrophic state in the receiving body of water. In other cases, higher degrees of phosphorus removal are required.

In such cases, metal coagulants are added to precipitate and increase the removals of phosphorus from the effluent stream. In other cases, where biological phosphorus removal is not used, metal coagulants are the only means by which phosphorus is removed from the wastewater stream.

Germany, in 1989, enacted the requirement that all wastewater treatment plants treating more than 20,000 population equivalents should reduce effluent phosphorus to less than 2 mg/l, and those larger than 100,000 to less than 1 mg/l (Schaefer, 1990). It was estimated that over 1,000 wastewater treatment plants needed to implement chemical phosphorus removal. Prior to implementation, approximately 4-percent of all wastewater treatment plants in Germany practiced chemical precipitation, in contrast to approximately 90-percent in Scandinavian countries.

South Africa introduced an effluent phosphate standard that restricted effluent phosphorus discharges to 1 mg/l as PO_4-P (Government Gazette, 1984). It was estimated in 1983 that up to 1.3×10^6 m^3 (345 MG) of wastewater would have to be treated chemically to meet this standard (Wiechers, 1986). Since that time a number of biological nutrient removal wastewater treatment plant were constructed or modified from existing plants. However, in many cases, biological phosphorus removal is supplemented with chemical precipitation to comply with the phosphate standard (de Haas *et al.* 2000–2001).

Figure 4.26 summarizes the mechanisms involved with phosphorus precipitation and removal from water and wastewater streams (adapted from Maurer & Boller, 1999). These mechanisms involve chemical precipitation of metal-phosphate-hydroxo complexes; selective adsorption of dissolved phosphorus species onto precipitate surfaces; flocculation and co-precipitation of finely dispersed precipitate and other colloidal matter. These mechanisms are not independent but likely occur simultaneously when metal coagulants are used to remove phosphorus.

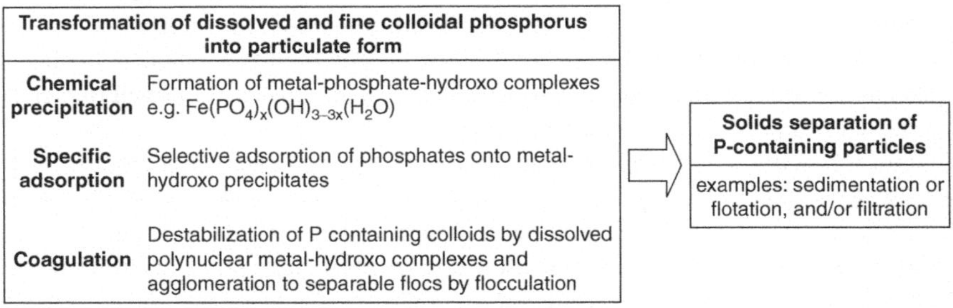

Figure 4.26 Overview of chemical precipitation and removal processes (after Maurer & Boller, 1999).

Phosphorus precipitates formed by metal addition are separated from the effluent stream either prior to discharge or within the receiving body of water. In the latter case phosphorus is fixed by precipitate formation and is incorporated in the sludge zone of, say, the lake receiving the discharge. Here, it is important that the coagulant species incorporated in the precipitate are not reduced under anaerobic conditions. For this reason, where coagulant addition to the effluent immediately prior to discharge is practiced, aluminum sulfate is often preferred to, say, ferric sulfate. In the latter instance, the eventual reduction of Fe(III) species to Fe(II) under anaerobic conditions in the sludge zone could give rise to re-dissolution and release of phosphorus back into the water body.

Coagulation for phosphorus removal has been applied to non-mechanical plants. For example, coagulants applied to stabilization pond effluents (Shindala & Stewart, 1971; Bratby, 1978) in conjunction with flotation or other processes effect algal, COD, nitrogen as well as phosphorus reduced loads to receiving waters. On the smaller scale, Brandes (1977) describes an automatic alum dosing system for households served by septic tanks – of particular interest in areas where infiltration to a body of water is likely and eutrophication of concern.

In mechanical wastewater treatment plants, chemical addition is typically at one or more of several locations within the wastewater treatment plant, as shown schematically in Figure 4.27.

In summary, the alternative dosage points are as follows:

- At primary clarifiers to remove much of the chemical solids before the biological process;

- Within secondary treatment (termed simultaneous precipitation) either within the activated sludge process, or at the secondary clarifiers (for both trickling filter and activated sludge systems);
- Tertiary treatment, with add-on processes after the main wastewater treatment processes. These could include one or more of the following: some form of clarifier, dissolved-air flotation, filtration, micro- or ultrafiltration.

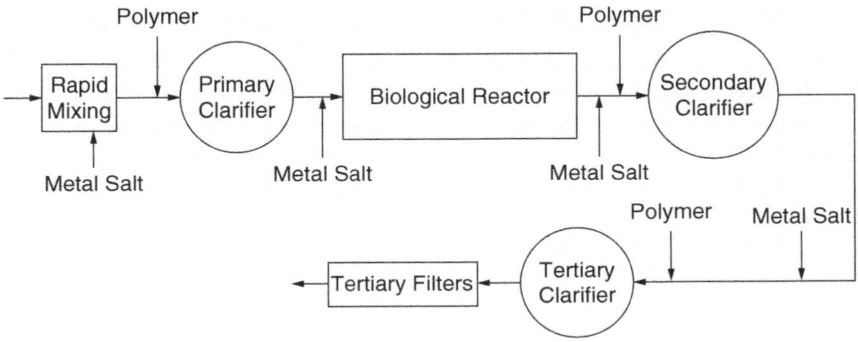

Figure 4.27 Possible chemical dosage points for phosphorus precipitation.

Adding coagulant chemicals at the primary clarifiers has the advantage that much of the chemical solids produced are removed before liquid stream biological processes. However, sludge handling process will still be required to process the increased sludge load.

A disadvantage of adding chemicals at the primary clarifiers for phosphorus precipitation is that a larger proportion of polyphosphates are present, before conversion to orthophosphates during biological treatment, thereby exerting a higher chemical demand.

Depending on the final effluent phosphorus concentration to be achieved, and the required coagulant dosage, adding metal coagulants to the biological process may give rise to pH inhibition problems and/or interfere with subsequent biological sludge conditioning by aerobic or anaerobic digestion. For example, each mg/l alum added potentially decreases the alkalinity by approximately 0.5 mg/l as $CaCO_3$. Depending on the initial alkalinity of the raw wastewater, lime may be required. The alkalinity provided by 0.39 mg/l lime is approximately equivalent to the alkalinity consumed by 1 mg/l alum added.

Metal coagulants also increase the amount of inert solids generated during wastewater treatment. For example, every mg/l aluminum sulfate added generates approximately 0.35 mg/l inert solids.

One particular issue with coagulant addition to biological systems is whether or not simultaneous phosphorus precipitation inhibits biological phosphorus removal.

Problems of coagulant addition to biological systems may be avoided by incorporating a tertiary treatment facility which may include some form of clarification and filtration. This approach will likely achieve lower effluent phosphorus concentrations. In some cases there will be benefits of doing both simultaneous phosphorus precipitation and tertiary treatment, to achieve low final phosphorus concentrations and reduce the total amount of metal precipitant added.

4.8.2 Mechanisms of chemical phosphorus removal

In general, phosphorus exists in soluble or particulate forms. The objective of phosphorus precipitation is to convert soluble phosphorus species to an insoluble form to facilitate removal. An operational definition of soluble phosphorus is that passing through 0.45 micron filters. In general, soluble phosphorus could include orthophosphates, polyphosphates, pyrophosphates and organic phosphates. Orthophosphates exist in a number of ionic forms, distributed according to the pH of the solution, as shown in Figure 4.28.

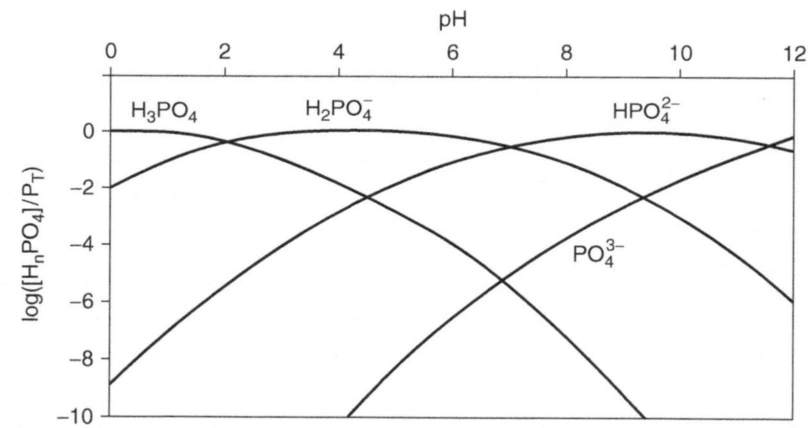

Figure 4.28 Distribution of phosphorus species with pH.

The predominant dissolved orthophosphate species over the pH range 5 to 9 are $H_2PO_4^-$ and HPO_4^{2-}. Predominant pyrophosphate species at neutral pH are $H_2P_2O_7^{2-}$ and $HP_2O_7^{3-}$. Polyphosphates can be important components of detergents – unless they are controlled. Depending on the product they may contain 2 to 7 phosphorus atoms. One example is $Na_5P_3O_{10}$.

Orthophosphates, and condensed forms such as pyrophosphates, triphosphates and higher polyphosphate anions are known to form complexes, chelates and insoluble salts with a number of metal ions. The actual form of the reaction between the metal ion and phosphorus depends on a number of factors, including relative concentrations of the metal and the phosphates, the pH, the presence of other ligands such as sulfates, carbonates, fluoride, and organic species (Stumm & Morgan, 1970, 1996).

Chemical precipitation removes orthophosphate and particulate phosphorus species the easiest. Polyphosphates and organic phosphorus forms may take part in some precipitation or adsorption reactions, but not as readily as the other species. During biological wastewater treatment, bacterial enzymes convert much of the condensed phosphate species to orthophosphate species. This is one reason why it is often found to be more efficient in terms of coagulant usage, to precipitate phosphorus after or during biological treatment, compared with upstream in primary clarifiers, where a larger proportion of the phosphorus is in the form of organic or condensed phosphates.

An understanding of different phosphorus species present is important to understand the potential for phosphorus removal to very low levels. The phosphorus species are identified in terms of the methods of analysis, as follows (refer to *Standard Methods*, 2005):

(a) *Total phosphorus (TP)*. In this case, samples are immediately preserved to pH 2 or less using sulfuric acid, without a filtration step. The acidified samples are digested to convert all the phosphorus forms to dissolved orthophosphate and measured by colorimetry. Digestion of the organic and condensed phosphates occurs and any precipitated phosphorus is dissolved.

(b) *Soluble total phosphorus (STP)*. Samples are immediately filtered through 0.45 micron filters and preserved to pH 2, following the procedures for total P.

(c) *Total reactive phosphorus (TRP)*. Samples are analyzed by direct colorimetry. This is principally a measure of the orthophosphate content of the sample, although part of the condensed species is included in the test results because a small amount is unavoidably hydrolyzed during the analytical procedure.

(d) *Soluble reactive phosphorus (SRP)*. Samples are immediately filtered through 0.45 micron filters and analyzed by colorimetry, as for (c).

(e) *Total acid hydrolysable phosphorus (TAHP)*. pH of the unfiltered sample is adjusted with acid until the phenolphthalein red color is discharged, then boiled gently for at least 90-minutes. Samples are then analyzed by colorimetry to measure condensed and polyphosphates plus any chemical and adsorbed P.

(f) *Soluble acid hydrolysable phosphorus (SAHP)*. Samples are immediately filtered through 0.45 micron filters then processed as for (e).

From the preceding analyses, the following phosphorus species can be identified (*Standard Methods* 2005; Neethling *et al.* 2007; Gu *et al.* 2011):

- Total phosphorus (TP) = (a)
- Soluble total phosphorus (STP) = (b)

- Total reactive phosphorus (TRP) = (c)
- Soluble reactive phosphorus (SRP) = (d)
- Soluble non-reactive phosphorus (SNRP) = (b)–(d)
- Particulate phosphorus = (a)–(b)
- Total organic phosphorus = (a)–(e)–(c)
- Soluble organic phosphorus (SOP, or DOP) = (b)–(d)–(f)

Metal coagulants complex and remove soluble reactive phosphorus the most readily. Particulate forms of phosphorus are also removed, principally by flocculation and by the solid-liquid separation process employed. Polyphosphates and condensed phosphate forms, expressed as soluble acid hydrolysable phosphorus are removed less readily compared with soluble reactive phosphorus; higher metal coagulant dosages are typically required and removal is probably by surface complexation reactions. Soluble organic phosphorus (SOP or DOP) is the most difficult species to remove and, like DON (discussed elsewhere in this Chapter) requires high coagulant dosages to effect some removal. The mechanism probably involves surface complexation and adsorption to metal coagulant complexes and precipitates.

There exists a stoichiometric relationship between the mass of metal coagulant added and the mass of phosphorus removed, down to certain phosphorus concentrations. Below such phosphorus concentrations, the relationship is no longer stoichiometric and higher coagulant dosages are required for successively lower phosphorus residuals. Understanding of the mechanisms involved in phosphorus removal generally evolved as the need for lower effluent phosphorus concentrations progressed. The following presents the development of phosphorus removal by metal coagulants from essentially stoichiometric relationships to the realization that phenomena other than those described by chemical equilibria were involved.

Figure 4.29 (after Ferguson & King, 1977) shows that for the oxidation pond effluent tested, the stoichiometric relationship was 1.75 moles Al added per mole phosphorus removed. Below approximately 3 mg/l P (0.1 mM/l), progressively inordinate amounts of alum were needed to achieve lower effluent phosphorus concentrations. The lowest effluent phosphorus concentration achieved was approximately 0.0065 mg/l (2×10^{-4} mM/l) at alum concentrations up to 700 mg/l.

Figure 4.30 shows that for another oxidation pond effluent, the molar ratio was 4.3 moles Al added per mole P removed. In this case, progressively inordinate amounts of alum were needed to achieve effluent phosphorus concentrations below approximately 0.3 mg/l P (0.01 mM/l). The lowest effluent phosphorus concentration achieved was approximately 0.045 mg/l (0.00145 mM/l) at alum concentrations up to 300 mg/l.

The stoichiometric Al_{added}:$P_{removed}$ ratio has been reported as varying from 1.4 to 4.3 (for example, Baillod *et al.* 1977; Brandes, 1977; Bratby, 1978; Ferguson & King, 1977). Assuming that all the reported values correspond to optimum pH values in each case, it is likely that the differences in reported Al:P ratios are due to different wastewater constituents and coagulant demands, and differing species of phosphorus. A wastewater with a predominance of orthophosphates will behave differently to a wastewater with high proportions of condensed and organic phosphorus forms.

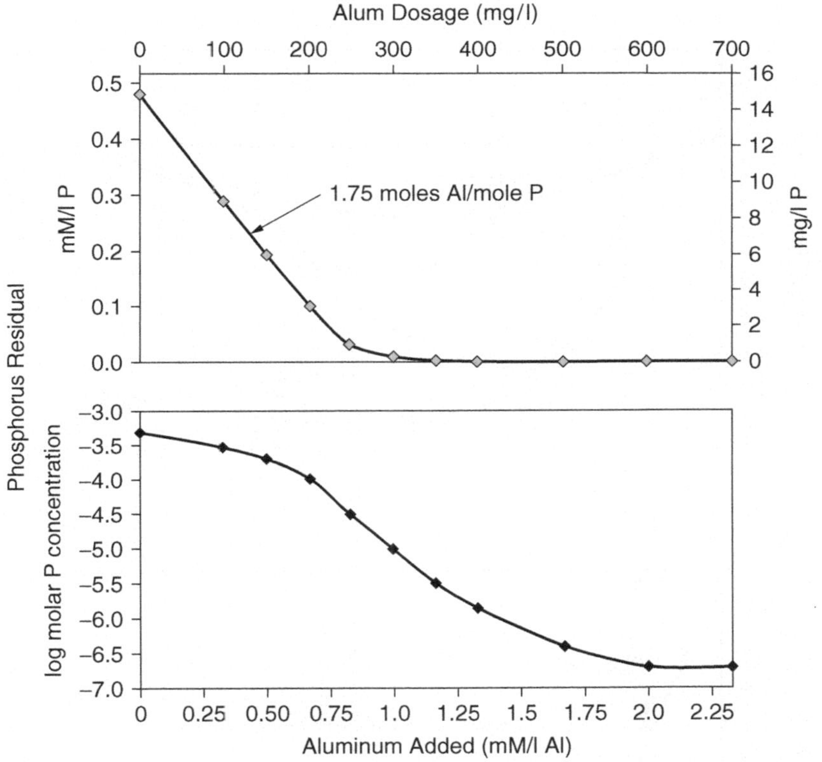

Figure 4.29 Phosphorus removal from oxidation pond effluent (after Ferguson & King, 1977).

Other factors that also may affect the stoichiometric ratios found in practice include the following:

- Acid or base for pH correction should be added before the metal coagulant is added, never after, since precipitation reactions are not reversible.
- Mixing at the point of metal addition should be as efficient as possible. The intensity should be high and the mode of rapid mixing should be plug flow.
- Chemical reactions between the metal and phosphate or polyphosphate ligands are rapid, complete in less than 1 second.
- Competing reactions with the hydroxide ligand leading to other multinuclear hydrolysis products can be avoided by efficient rapid mixing.

- Differences in performance could also be influenced by different methods of solid liquid separation, such as sedimentation, flotation, filtration, etc., which may require different conditions of coagulant addition to form settleable or filterable flocs that incorporate the aluminum-phosphorus precipitates.

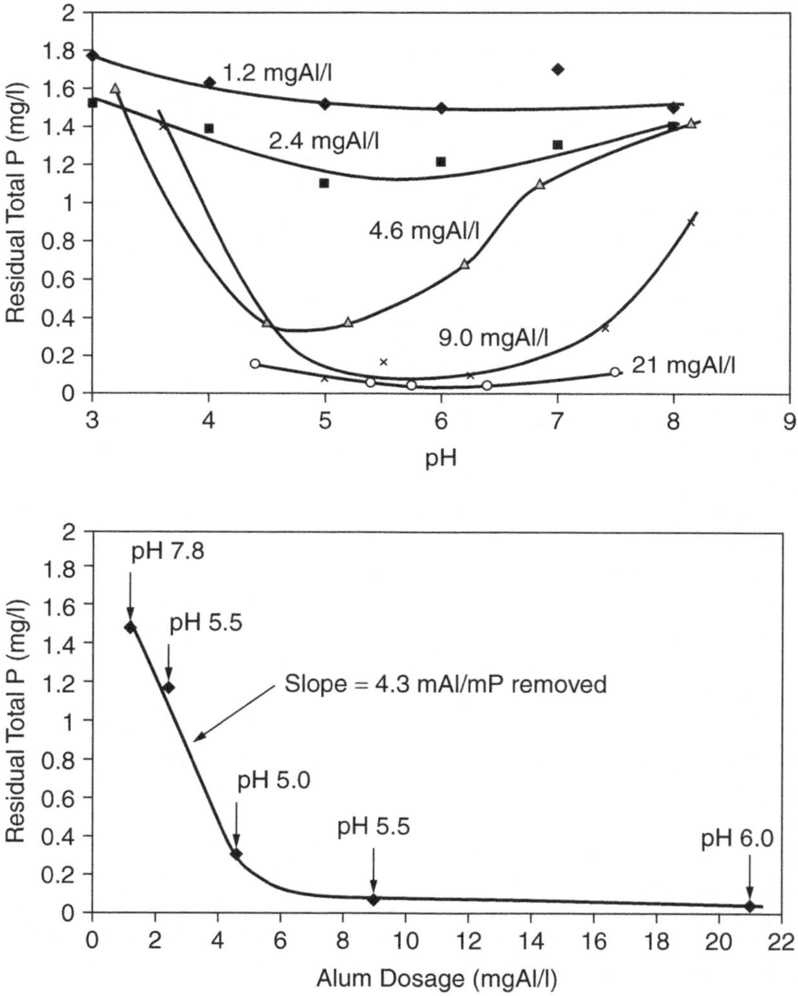

Figure 4.30 Optimum pH and molar ratio of Al/P removed for oxidation pond effluent (after Bratby, 1978).

Figure 4.30 shows that within the stoichiometric range of metal to phosphorus ratios, the pH range for optimized phosphorus removal becomes progressively tighter. However, after a certain metal to phosphorus ratio, the optimal pH range widens, but only in the direction of higher pH. In Figure 4.30, this effect is seen after an alum dosage of 64.9 mg/l, or a molar Al/P ratio of approximately 3.0. Results presented by Ferguson and King (1977), with molar Al/P ratios less than 3.0, appear to be consistent with this observation. Takacs *et al.* (2004, 2005) also found a wide optimum pH spread, from approximately 6.0 to 7.0 for very low residual phosphorus concentrations and Molar Al/P ratios above 3.0.

The widening optimal pH range, towards higher pH, at higher metal to phosphorus ratios can be explained in terms of competition between coordinating ligands: At a given pH value, at any particular time there will be a certain amount of phosphorus complexed with metal coagulant; the optimum pH represents that pH where the largest fraction of complexes are formed with phosphorus. If the metal coagulant dosage is higher than the stoichiometric requirements then there will be an increasing formation of mixed metal-hydroxide-phosphate complexes. The acceptability of higher pH values for optimal phosphorus removal suggests a shift in the predominant removal mechanisms, and an increasing influence of adsorption to the hydroxide complexes.

The optimum pH for phosphorus removal using aluminum coagulants is normally within the range 5.5 to 6.0 (Bratby, 1978; Ferguson & King, 1977; Shindala & Stewart, 1971). Again, differences in reported values are probably due to differences in wastewater constituents.

Without pH correction, increasing metal coagulant dosages will decrease the pH of the treated wastewater. Therefore, progressively higher dosages would initially reduce the residual phosphorus concentrations, until a minimum value is reached. Beyond this value, with increasing dosages and reducing pH, residual phosphorus concentrations could progressively increase. This is shown in Figure 4.31 (Bratby, 1982), where alum added to raw wastewater overflows treated with dissolved-air flotation without pH correction, effected a minimum phosphorus concentration of approximately 0.2 mg/l, at a pH of approximately 5.5. Higher dosages reduced pH further, and effluent phosphorus concentrations progressively increased. However, effluent suspended solids and COD removals did not increase within the alum dosage and pH ranges studied. The increase in residual phosphorus at lower than optimum pH values is due to the formation of soluble complexes, such as $AlH_2PO_4^{2+}$.

Various empirical models have been used to predict metal coagulant dosages required for phosphorus precipitation. These have generally been applicable to higher final effluent phosphorus concentrations, but are seriously inadequate for the very low effluent concentrations that have become necessary in recent years.

For example, the following expression was adapted from Shindala and Stewart (1971) from work carried out with stabilization pond effluents:

$$C = \frac{P_i}{0.06(P_e) + 0.043} - 8.0$$

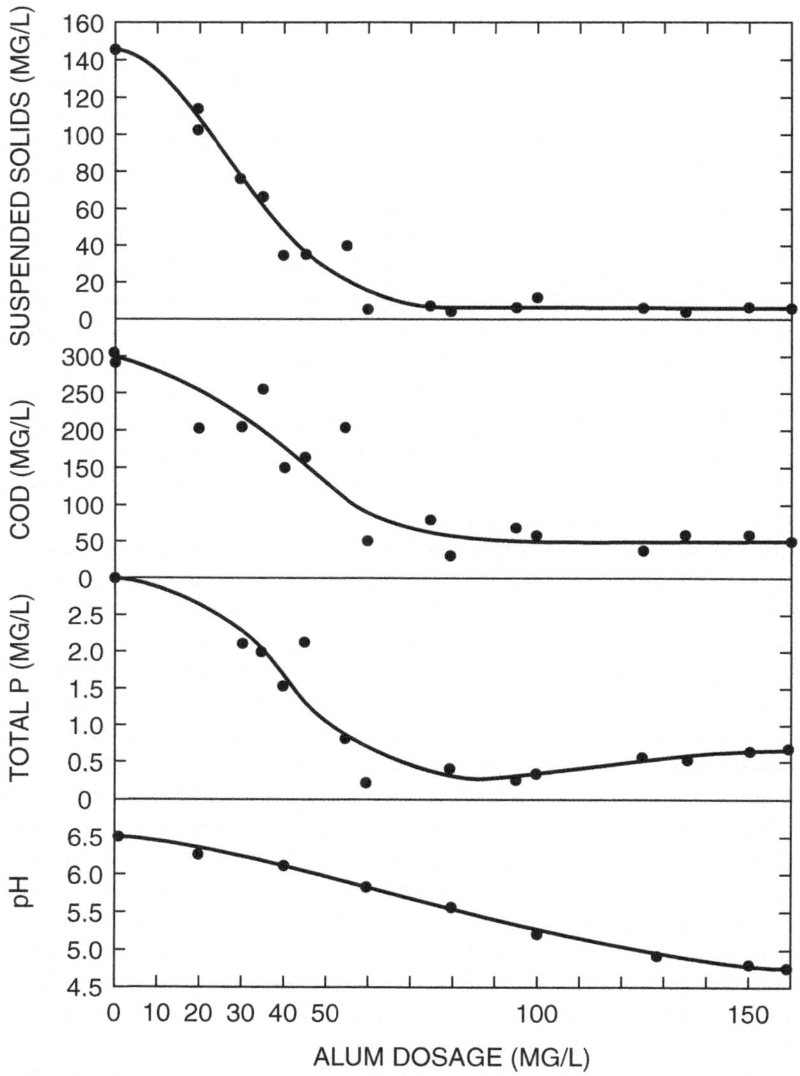

Figure 4.31 Effect of increasing alum dosage on pH and effluent phosphorus (after Bratby, 1982).

where

C = alum dosage, mg/l
P_i = initial total PO_4-P concentration, mg/l
P_e = final (effluent) total PO_4-P concentration, mg/l

Shindala and Stewart clearly state that the expression is valid for effluent phosphates in the range 1 to 5 mg/l PO_4 (or 0.3 to 1.6 mg/l PO_4-P). Figure 4.36 and Table 4.12 present a number of chemical phosphorus removal experiences. The accuracy of Shindala and Stewart's empirical model is tested using three separate examples from these experiences:

- The alum dosage required to reduce influent phosphorus of 5.17 mg/l to 0.97 mg/l, by adding alum at the head end of a contact stabilization plant, was 83.5 mg/l. The expression derived from Shindala and Stewart predicts 43 mg/l.
- The alum dosage required to reduce oxidation pond effluent from 2 mg/l to 0.05 mg/l, was 125 mg/l. The expression derived from Shindala and Stewart predicts 33 mg/l.
- The alum dosage required to reduce secondary effluent phosphorus from 2.7 mg/l to 0.024 mg/l, was 225 mg/l. The expression derived from Shindala and Stewart predicts 53 mg/l.

From these examples, it is evident that use of the Shindala and Stewart empirical expression would grossly underestimate the required alum dosage in each case.

Another empirical expression was developed by Fettig *et al.* 1990 (presented in Jiang & Graham, 1998):

$$P_e = 32.89 - 5.7(\mathrm{Al}/\mathrm{P}_i) - 9.77\mathrm{pH} + 1.65(\mathrm{Al}/\mathrm{P}_i)^2 + 0.989\mathrm{pH}^2 - 0.463\mathrm{pH}(\mathrm{Al}/\mathrm{P}_i)$$

where

P_e = final (effluent) total PO_4-P concentration, mg/l
(Al/P_i) = molar ratio of aluminum dosage to initial PO_4-P concentration

The same examples as above are used to test the accuracy of this expression:

- An alum dosage of 83.5 mg/l added at the head end of a contact stabilization plant, reduced an influent phosphorus concentration of 5.17 mg/l to 0.97 mg/l. The expression derived from Fettig *et al.* predicts an effluent phosphorus concentration of 4.9 mg/l.
- An alum dosage of 125 mg/l added to oxidation pond effluent reduced the phosphorus concentration from 2 mg/l to 0.05 mg/l. The expression derived from Fettig *et al.* predicts an effluent phosphorus concentration of 11.7 mg/l.
- An alum dosage of 225 mg/l added to secondary effluent reduced the phosphorus concentration from 2.7 mg/l to 0.024 mg/l. The expression derived from Fettig *et al.* predicts an effluent phosphorus concentration of 63 mg/l.

Clearly, the expression derived by Fettig *et al.* breaks down severely when applied to low final phosphorus concentrations.

Although the above examples of empirical expressions were probably useful to describe the reactions occurring within a certain set of experiments, they do illustrate the necessity of establishing predictions of phosphorus precipitation reactions on a more fundamental basis.

Ferguson and King (1977) attempted to derive such a model. Their model describes the precipitation of orthophosphate using alum. The model describes steady state reactions only, although this was not considered a significant limitation since the chemical reactions are virtually instantaneous. The solid-liquid separation step is also not accounted for – complete removal of solid phase precipitates is assumed.

The model also ignores any adsorption mechanism of phosphorus removal – only a precipitation model described by equilibrium chemistry is assumed.

Table 4.10 presents the chemical species and related stability constants assumed in the model.

Table 4.10 Chemical reactions and stability constants (from Ferguson & King, 1977).

Reaction	Stability Constant	log K
$H_2PO_4^- = HPO_4^{2-} + H^+$	K_2	–7.2
$HPO_4^{2-} = PO_4^{3-} + H^+$	K_3	–12.2
$H_2CO_3 = HCO_3^- + H^+$	K_1	–6.3
$HCO_3^- = CO_3^{2-} + H^+$	K_2	–10.3
$H_2O = H^+ + OH^-$	K_W	–14.0
$Al^{3+} + H_2O = AlOH^{2+} + H^+$	$^*K_1$	5.0
$AlOH^{2+} + H_2O = Al(OH)_2^+ + H^+$	$^*K_2$	3.7
$Al(OH)_{2+} + H_2O = Al(OH)_3^0(aq) + H^+$	$^*K_3$	6.5
$Al(OH)_3^0(aq) + H_2O = Al(OH)_4^- + H^+$	$^*K_4$	8.1
$Al^{3+} + H_2PO_4^- = AlH_2PO_4^{2+}$	K	6.0
$Al_{1.4}PO_4(OH)_{1.2}(s) = 1.4Al^{3+} + PO_4^{3-} + 1.2OH^-$	K_{SO}	–34.0
$Al(OH)_3^0(s) + 3H^+ = Al^{3+} + 3H_2O$	$^*K_{SO}$	10.3

The precipitate $Al_{1.4}PO_4(OH)_{1.2}$, together with its empirically derived solubility product, is a compound suggested by Ferguson and King to provide a better fit between measured and final phosphate concentrations modeled using equilibrium constants.

The model requires a prediction of the final pH of the water based on chemical dosages and the initial water conditions of pH and alkalinity. The model also describes three different conditions of aluminum-phosphate precipitation in terms of "Zones":

- *Zone 1* is the stoichiometric zone. Here, insufficient aluminum is added to precipitate all of the phosphate. Reactions in this zone are described by a stoichiometric ratio of 1.4 moles aluminum/mole phosphorus removed.

- *Zone 2* represents aluminum dosages somewhat higher, but relatively close to the stoichiometric requirement. The extent of phosphorus removal is calculated from the equilibrium between aluminum phosphate solid precipitates and the solution. No residual aluminum is available to produce aluminum hydroxide precipitates.
- *Zone 3* is where excess aluminum has been dosed and both aluminum phosphate and aluminum hydroxide precipitate.

Figure 4.32 demonstrates how aluminum sulfate dosages to achieve given final phosphorus concentrations are obtained together with acid or base requirements for pH adjustment. Figure 4.32, valid for an initial pH of 7.5, accommodates two conditions. The first represents stoichiometric removal where the dosage required for a 1 mg/l phosphorus residual is determined. The second represents the conditions for a phosphorus residual of 0.03 mg/l P, corresponding to an overdose of 60 mg/l aluminum sulfate beyond that required for stoichiometric removal.

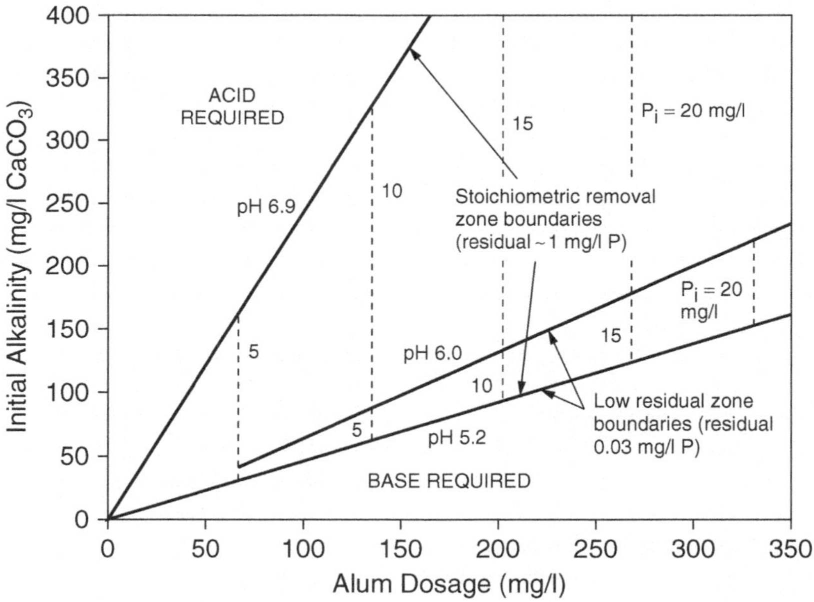

Figure 4.32 Relationship between alum dosage and initial alkalinity. Zone 1: stoichiometric removal and 1 mg/l P residual; Zone 3: lowest residual 0.03 mg/l P residual (after Ferguson & King, 1977).

Ferguson and King emphasized that such a model should not be considered a substitute for jar test or pilot plant evaluations of a particular wastewater application. However, the intention was to provide a guide for an experimental program and for preliminary chemical cost evaluations. For example, referring to Figure 4.32, for an initial phosphorus concentration of 10 mg/l, no acid or base is required for pH adjustment with initial alkalinities from 1.3 to 6.4 meq/l, for removal in the stoichiometric zone (to 1 mg/l phosphorus residual). The alum dosage required is 130 mg/l. However, 2.3 meq/l acid (5.0 to 2.7 meq/l at 200 mg/l alum) would be required to reach the pH 6 boundary for the lowest residual zone. An initial alkalinity of 1 meq/l would require 0.9 meq/l base for maximum removal with 200 mg/l alum.

To achieve a phosphorus residual of 1 mg/l, the model predicts that the molar Al/P ratio should be approximately 1.45. Figure 4.36 shows that this value would be close to the lowest boundary of results obtained in practice. To achieve the lowest phosphorus residual of 0.03 mg/l, with an initial phosphorus concentration of 5 mg/l, the model predicts a molar Al/P ratio of approximately 2.7. From Figure 4.36 this is barely below the lowest boundary.

Therefore, although the model of Ferguson and King uses a rational basis for predicting phosphorus precipitation with alum, the results do underestimate the alum dosages required, particularly for very low phosphorus concentrations.

A number of other authors have derived alternative empirical compounds and solubility products. De Haas *et al.* (2000–2001) accepted the general form of Luedecke *et al.* (1989): $Me_rPO_4(OH)_{3r-3}$ for both ferric and aluminum. However, they acknowledged the possibility of more complex compounds involving unknown cations, with the theoretical formulae $Al_{1.33}(X^{2+})PO_4(OH)_3$ and $Fe_{1.66}(X^{2+})PO_4(OH)_4$. X is postulated to be Mg^{2+} or Ca^{2+} or some other divalent cation.

Under very low phosphorus concentrations, under conditions of limiting P, the average formula for the ferric compounds, to match the observed alkalinity losses per mole Fe added was found to lie within the range $Fe_3(X^{2+})PO_4(OH)_8$ to $Fe_{2.33}(X^{2+})_{2.5}PO_4(OH)_9$.

Takacs *et al.* (2004) and Jones and Dold (2005) accepted the generalized forms of Gates *et al.* (1990): $Me_rH_2PO_4(OH)_{3r-1}$, with r = 0.8 for aluminum, and r = 1.6 for ferric. Like the model of Ferguson and King (1977), by incorporating weak acid/base chemistry and pH with the solubility and dissociation equations, close correspondence with full scale plant results were obtained. However, Takacs *et al.* pointed out that the accuracy of the model predictions relied heavily on calibration to plant data. As exemplified earlier with other models, application of a phosphorus precipitation model to other plants can result in inaccuracies.

In general, the limitation of equilibrium models that only take precipitation reactions into account is that they tend to underestimate the phosphorus removal achievable in practice. These models are unable to describe the marked adsorptive capacity of freshly precipitated iron or aluminum sludges for phosphates. Maurer and Boller (1999) pointed out that the recirculation of iron or alum precipitates contained in return sludge or the backwash from tertiary filters, can achieve reductions in dissolved phosphates of up to 30-percent – which indicates that adsorption is an important component of the overall phosphorus removal mechanism.

Luedecke *et al.* (1989) postulated a combined phosphorus precipitation and adsorption model. They suggested a general form for the precipitates of $Me_rPO_4(OH)_{3r-3}$. Adsorption of phosphate ions was assumed to occur on both $Me_rPO_4(OH)_{3r-3}$ and FeOOH precipitates. They proposed a first order adsorption equilibrium equation of the form:

$$[P]_{\text{adsorbed}} = k_a \cdot X_a \cdot \frac{[PO_4^{3-}]_{\text{residual}}}{[OH^-]^3}$$

where

$[P]_{adsorbed}$ = concentration of adsorbed phosphorus (mgP/l)
$[PO_4^{3-}]_{residual}$ = molar concentration of residual phosphate in solution after precipitation equilibrium is reached
$[OH^-]$ = molar concentration of OH^-
k_a = adsorption coefficient, experimentally determined to be 1.9 to 3.4×10^{-12} mol^2/l^2 at pH 7.2
X_a = concentration of the precipitated solids (mgTSS/l)

Maurer and Boller (1999) pointed out that although the combined precipitation and adsorption model of Luedecke *et al.* does provide a reasonable representation of the complex reactions occurring, results do not accurately predict residual phosphorus at high Me/P ratios. The models also were not able to adequately explain the superior performance of two-stage phosphorus addition, principally because the models did not include a rate step, but rather were based on equilibrium conditions. The rate at which metal-hydroxo precipitates grow from micro- to macro- flocs is a factor in phosphorus removal which was also not accounted for in the models.

Maurer and Boller (1990) proposed a combined precipitation and adsorption model similar to Luedecke *et al.*, but which included the variable adsorptive capacity of the precipitate-flocs. This latter was expressed as the alteration of the specific surface during precipitation and floc growth. They postulated that to achieve very low residual phosphorus concentrations, larger metal doses are required, which increases the rate of precipitate and floc formation, as well as increasing preferential binding of OH^- groups into the precipitates. The combination of a faster development of larger flocs and increased binding of hydroxyl groups, results in less overall surface area and less opportunity to bind and adsorb phosphates. In other words, adsorption of phosphates becomes less efficient as the metal dosage increases.

To describe this mechanism, Maurer and Boller included a variable and dosage dependent specific surface component to the adsorption model of Luedecke *et al.* The specific surface area of the adsorptive floc surfaces is decreased with increasing metal dosages according to the relationship:

$$S_{\text{eff}} = S_{\max} - k_{\text{Me}} \cdot [\text{Me}]$$

where

S_{eff} = effective specific surface for phosphate adsorption (m^2/g dry solids)
S_{max} = maximum possible specific surface at very low Me dosages (m^2/g dry solids)
k_{Me} = proportionality factor (m^4/g^2)
[Me] = Me dosage (mg Me/l)

For the proposed model to match experimental results, using Fe(III) salts at pH 6.8, Maurer and Boller assigned values to S_{max} and k_{Me} of 1,400 m^2/g and 23 m^4/g^2, respectively.

As with the other models, Maurer and Boller indicated that a generally valid model is difficult to develop. For this reason, empirical "calibration" information has to be assumed for reaction constants, solubility products, adsorption coefficients, specific surface for phosphorus adsorption to the precipitate-flocs, and so on. The difficulty lies with the unique nature of each wastewater or effluent in terms of dissolved and particulate wastewater constituents and, more importantly on the particular distribution of phosphorus species.

An added difficulty is that it is still unclear whether the mechanism of phosphorus removal beyond precipitation is indeed adsorption. As pointed out by He *et al.* (1996), phosphate adsorption is difficult to distinguish from other mechanisms such as ion exchange, or surface complexation of the phosphate and the metal hydroxide colloid.

An adsorption or surface complexation mechanism, similar in concept to that proposed by Maurer and Boller, has been tested and adopted more recently to explain phosphorus removal by metal coagulants to achieve very low residual phosphorus concentrations. The overall mechanism of phosphorus removal to very low residuals is seen as a complex hybrid process including a stoichiometric metal-phosphate precipitation phase, co-precipitation of metal-hydroxide-phosphorus complexes, surface complexation of phosphorus species to metal-hydroxide precipitates, and adsorption of phosphorus species on solid floc surfaces (Maher *et al.* 2015; Smith *et al.* 2008; Takács *et al.* 2011).

Smith *et al.* (2008) describe a model which involves precipitation of hydrous metal oxide together with co-precipitation of phosphate species and surface complexation of phosphorus species to the flocs. The co-precipitation mechanism involves diffusion of phosphate within the hydrous metal oxide flocs and sharing of oxygen atoms by covalent bonding.

A parameter was introduced that is proportional to the surface active reaction sites, known as the active site factor (ASF). In the model, ASF is the only parameter that requires adjustment to fit experimental data. ASF is a function of mixing intensity (*G* value) and the age of the floc, expressed as a solids retention time, SRT. Expressions describing the diffusion limitations within the floc are also included in the model.

Experiments demonstrated an initial fast rate of hydrous metal oxide-phosphorus complexation (termed co-precipitation) followed by a much slower rate of surface complexation or adsorption. The sorptive capacity of freshly formed hydrous metal oxides reduces quickly through surface consolidation processes, which suggests the importance of efficient rapid mixing at the point of metal coagulant addition. As hydrous metal oxide complexes start to form it is important that phosphate ions are available for co-precipitation reactions. This only occurs if efficient initial mixing occurs (see also Chapter 6). Higher G values have been found to result in lower residual phosphorus concentrations (Takács *et al.* 2011). In the model calibrations of Smith *et al.* (2008) ASF parameter values were higher at higher mixing intensities.

Fresh metal hydroxide sludge recycled back just upstream of the point of metal coagulant addition has been shown to be effective in phosphorus removal, reducing the required coagulant dosage to achieve low residual phosphorus concentrations (Takács *et al.* 2011; Maher *et al.* 2015). Initially, metal hydroxide precipitates are amorphous with a high porosity and total surface area for surface complexation sites. As the sludge ages, however, the surface contracts and the capacity for surface complexation reduces due to diffusion limitations (Smith, 2008). A similar effect is found when polymers are used as flocculant aids. Sludge flocs with polymer addition tend to be denser and more compact with fewer surface complexation sites for phosphorus uptake (Maher *et al.* 2015).

4.8.2.1 Simultaneous phosphorus precipitation

Figure 4.27 shows that metal coagulants could be added to activated sludge reactors for phosphorus precipitation, either indirectly to the influent of the secondary clarifiers, in which case precipitated solids are recycled back to the reactors with the return sludge, or directly to the reactors. When the biological process employs biological phosphorus removal, metal is added to supplement the biological process to achieve lower phosphorus residuals. This is termed simultaneous phosphorus precipitation.

Reported benefits of chemical addition to biological processes have been improved stability, lower effluent suspended solids, BOD and COD, and lower sludge volume index (Healey *et al.* 1989; Metsch *et al.* 1985).

In some cases, an increase of the volatile solids content of the sludge has been noted. One possible reason is that a larger quantity of colloidal organic material is enmeshed by the metal hydroxide precipitates, preventing appreciable biological degradation and contributing to the overall volatile solids fraction (de Haas *et al.* 2000–2001).

Several studies have shown that simultaneous precipitation, even with very low phosphorus residuals, does not interfere with nitrification. Healey *et al.* (1989) found that with alum dosage rates up to 250 mg/l, aluminum accumulation within the biological matrix did not inhibit nitrification.

Experiments carried out using ferric chloride added to activated sludge reactors, to achieve very low residual orthophosphate levels of less than 10 μgP/l, found no inhibition of nitrification or denitrification (Murthy *et al.* 2005). Despite the very low levels of residual phosphorus, the phosphorus bound in the chemical precipitates in the activated sludge matrix, was still available to the organisms because of the rapid kinetics of dissolution/dissociation of the precipitate.

This rapid dissociation and equilibration of the chemically removed phosphorus apparently supplied sufficient orthophosphate to the relatively slower growing anoxic/heterotrophic and aerobic/autotrophic biomass. The organisms appeared to be able to sequester the residual orthophosphate bound in the precipitates. As long as sufficient chemically removed phosphorus was present in the sludge, it did not appear to matter for metabolism and bioavailability whether the phosphorus was in the form of orthophosphate or chemically bound phosphorus.

In the case of simultaneous precipitation, in biological reactors that are achieving biological phosphorus removal, the objective is to supplement the degree of phosphorus removal achievable biologically, by precipitating phosphorus chemically within the same biological reactor.

There is ample evidence that simultaneous chemical precipitation, particularly when achieving low residuals, is partially inhibitory to the biological phosphorus removal process (de Haas *et al.* 1993; de Haas *et al.* 2000–2001; Lötter 1991).

Biological phosphorus removal relies on a sequence of phosphorus release under anaerobic conditions, and phosphorus uptake under aerobic conditions, in sequentially compartmentalized reactors. These reactions are mediated by a specific group of organisms that are environmentally advantaged under the sequential anaerobic and aerobic conditions.

With simultaneous phosphorus precipitation, some of the phosphates can migrate between biological storage and chemically precipitated/adsorbed solids in the mixed liquor. With high metal dosages and low phosphorus residuals, metal hydroxides accumulate in the mixed liquor, phosphorus concentrations become limiting, and phosphorus release under anaerobic conditions may give rise to a transient increase in chemical phosphorus precipitates in the anaerobic zone, followed by a reversal under aerobic (or anoxic) conditions when the equilibrium shifts due to phosphorus uptake by the biological mechanism (de Haas, 2000–2001).

In most cases where simultaneous precipitation is practiced, the objective is to achieve residual phosphorus levels much less than 0.5 mg/l. Therefore, these systems are essentially phosphorus limited. Under limiting phosphorus conditions due to high simultaneous phosphorus precipitation, the precipitation mechanism does limit the extent of biological phosphorus removal by removing a large part of the soluble phosphates from the system. The proportion of phosphorus removed by chemical precipitation is more significant when there is less phosphorus in the system.

Some circumstantial evidence exists that system pH and alkalinity play a role in the degree of inhibition of the biological phosphorus removal mechanism by simultaneous phosphorus precipitation.

When phosphorus is not limiting, the degree of interference on biological phosphorus removal is still detectable, but not severe. At high effluent and, therefore, high bulk liquid phosphorus concentrations, the fact that there is still some inhibition suggests that the phosphorus concentration in the bulk liquid may not reflect the localized orthophosphate concentration in close proximity to the sludge flocs with simultaneous precipitation (de Haas, 1993).

Extensive tests carried out by de Haas *et al.* (2000–2001) showed that when using aluminum sulfate, the degree of inhibition of the biological phosphorus removal mechanism was 11-percent for a low alum dose of 5 mgAl/l applied to the aerobic zone. When the same low alum dosage was applied to the anaerobic zone, the degree of inhibition was 16-percent. When a high alum dosage of 9 mgAl/l was applied, the degree of inhibition was 23 to 24-percent for either the aerobic or anaerobic zone.

Similar effects were observed with ferric chloride addition. When phosphorus was not limiting, the degree of inhibition was 3 to 20-percent at dosages of 10 to 20 mgFe/l. Under conditions of phosphorus limitation, the degree of inhibition to biological phosphorus removal was found to be 32-percent.

Simultaneous phosphorus precipitation exhibits a "persistence effect", where continued phosphorus removal occurs for several days after cessation of metal dosing. This effect may be due to the accumulation of metal hydroxide precipitates in the sludge mass, leading to phosphate removal through an adsorption or ion exchange effect between hydroxide and phosphate ions (de Haas *et al.* 2000–2001).

If the practice of simultaneous precipitation is interrupted, and a return to only biological phosphorus removal is attempted, this may not be possible without experiencing at least a transient increase in residual phosphorus concentrations, due to prior inhibition when chemicals were added. An acclimation period is often necessary for the biological phosphorus removal organisms to re-establish normal functions. The persistence effect described above may not be sufficient to cover the drop in performance during the recovery period (de Haas *et al.* 2000–2001). Therefore, the operation of combined simultaneous precipitation and biological phosphorus removal systems may be difficult to control, as demonstrated by Lötter (1991).

Maurer and Boller (1999) suggest that to minimize the inhibitory effects of simultaneous precipitation, the metal coagulant should be added as late as possible to the biological phosphorus removal treatment train – for example to the mixed liquor channel before the secondary clarifiers.

However, for consistently low residual phosphorus concentrations, tertiary chemical precipitation systems following BEPR plants may be the most efficient alternative. The advantage would be that the efficiencies of both the biological and chemical phosphorus removal stages can be optimized.

4.8.2.2 Sequential phosphorus precipitation

Figure 4.33 presents the general relationship between effluent residual phosphorus concentration and the ratio of metal added/P removed required to achieve the residuals. This data is also plotted on logarithmic plots in Figure 4.36 (see also Table 4.12). The data represents a wide range of applications and modes of solid-liquid separation.

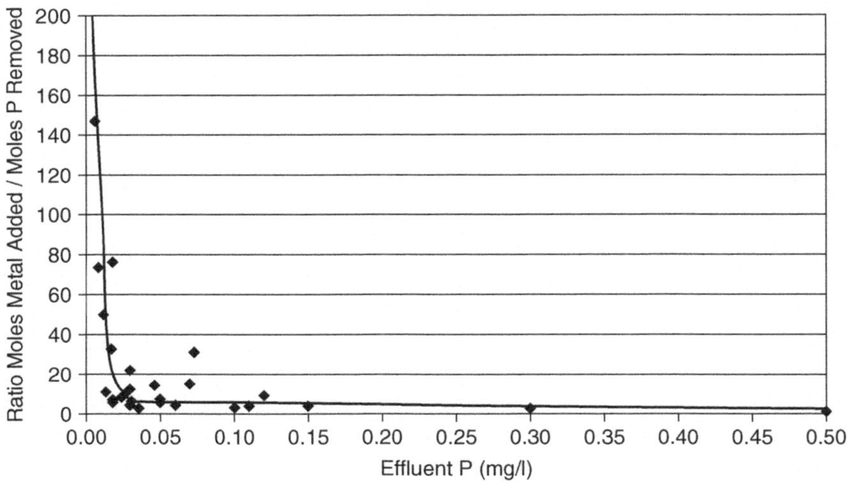

Figure 4.33 Residual phosphorus and metal/P ratios.

The benefits of sequential chemical addition for coagulation operations in general have been demonstrated on numerous occasions. In the case of phosphorus precipitation, Gros and Morgeli (1980) demonstrated the chemical savings possible with two-stage treatment. The first stage would achieve a large fraction of the overall removal of phosphorus at relatively low metal-to-phosphorus ratios. Then, to achieve very low phosphorus residuals, the high metal/P ratio required would be applied to a lower starting phosphorus concentration.

For example, with an initial phosphorus concentration of 4 mg/l, the molar metal/P ratio required for a phosphorus residual of 20 μg/l would be approximately 15. Therefore, using aluminum sulfate, the alum dosage would be approximately 575 mg/l.

An alternative would be two-stage treatment with, for example, the first stage in the biological reactors using simultaneous precipitation, and the second stage in tertiary treatment.

The first stage could reduce the phosphorus residual to 0.5 mg/l, using an alum dosage of approximately 75 mg/l (MAl/MP = 2.2). The second stage would reduce phosphorus from 0.5 mg/l to 20 μg/l. The required MAl/MP would be 15 as before, but now the alum dosage required is 72 mg/l. Therefore, the total alum dosage for two-stage treatment would be (75 + 72) = 147 mg/l – compared with 575 mg/l with single stage treatment.

The advantages of sequential phosphorus precipitation are more pronounced when the required phosphorus residual is at very low levels, below 0.5 mg/l. Maurer and Boller (1999) carried out pilot trials whereby they were able to optimize the distribution of metal addition in the first and second stages. The first stage was to an activated sludge system (simultaneous precipitation) and the second stage to tertiary filters downstream of the secondary clarifiers. Figure 4.34 presents their results.

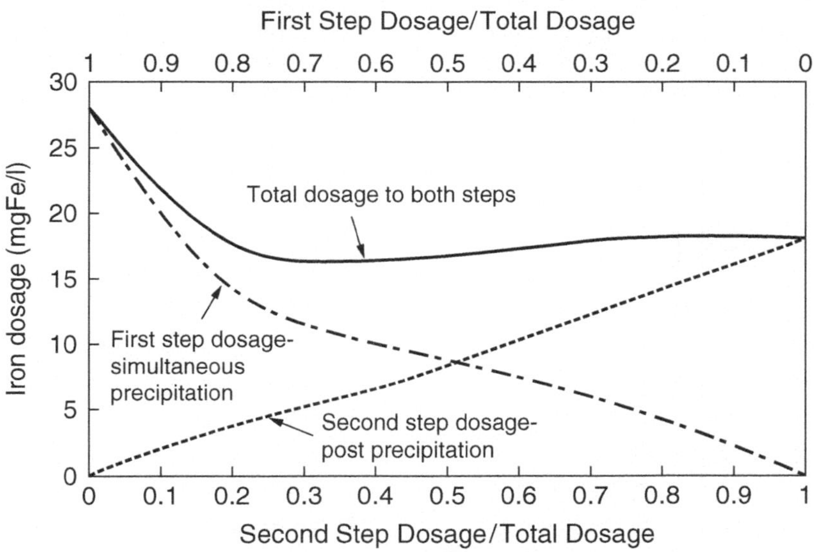

Figure 4.34 Relative dosages to first step simultaneous precipitation and second step direct filtration to achieve 0.1 mgP/l (after Maurer & Boller, 1999).

The data reflects the conditions to attain a final soluble phosphorus residual of 0.1 mg/l at all times. Primary effluent soluble phosphorus concentrations were approximately 4 mg/l. The results show a minimum total ferric dosage when the percentages of the total dosage to first and second stages were 60% and 40%, respectively. However, in their experiments, Maurer and Boller identified maximum practicable dosages of 5 mgFe/l and 3 mgAl/l to deep-bed filters, operating in direct filtration mode, to avoid premature head loss, breakthrough and excessive backwash frequencies.

4.8.3 Applications of chemical phosphorus removal

The following case studies present examples of simultaneous phosphorus precipitation as well as tertiary treatment, where the degree of phosphorus removal depends not only on the coagulant precipitant added, but also on the mode of solid-liquid separation employed. This is particularly important for those cases where very low final phosphorus concentrations are achieved.

Brandes (1977) found from studies on alum addition to septic tanks treating domestic wastewater, phosphorus contents of effluent suspended solids of approximately 0.032 mg/l P per mg/l TSS.

However, Figure 4.35 shows the phosphorus content of secondary effluent suspended solids for a range of different coagulant dosages (data from Levesque, 2001). The data shows that, for very low phosphorus residuals, and high metal coagulant dosages, the phosphorus content of effluent suspended solids is significantly reduced. The reason is that at very high metal dosages, a larger proportion of the precipitates formed are metal hydroxides.

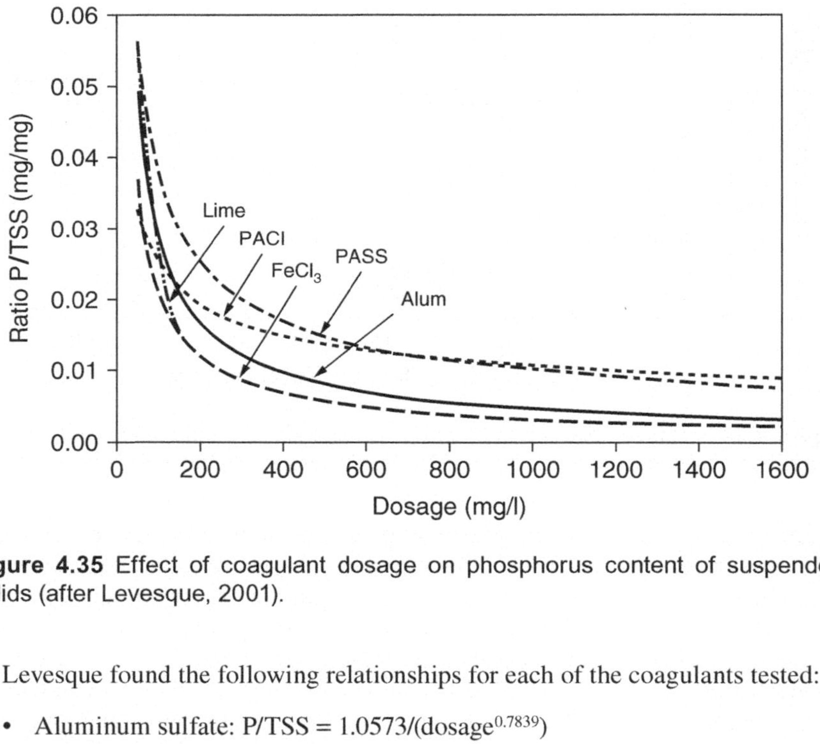

Figure 4.35 Effect of coagulant dosage on phosphorus content of suspended solids (after Levesque, 2001).

Levesque found the following relationships for each of the coagulants tested:

- Aluminum sulfate: P/TSS = $1.0573/(\text{dosage}^{0.7839})$
- Ferric chloride: P/TSS = $0.8945/(\text{dosage}^{0.814})$
- Polyaluminum chloride (PACl): P/TSS = $0.1411/(\text{dosage}^{0.3739})$
- Polyaluminum silicate sulfate (PASS): P/TSS = $0.5435/(\text{dosage}^{0.5795})$
- Lime: P/TSS = $5.2056/(\text{dosage}^{1.1655})$

(where P/TSS is the mass ratio in mg/mg, and dosage is the coagulant dosage in mg/l in each case)

Table 4.11 shows the influence of total suspended solids (TSS) on final effluent total phosphorus concentrations, for three different soluble phosphorus concentrations. These results illustrate that to achieve very low levels of effluent phosphorus using chemical precipitation, the mode of solid-liquid separation employed is very important.

Table 4.11 Effect of effluent TSS on total P.

Soluble P mg/l	Alum Dosage[1] mg/l	mg P/ mg TSS	TSS mg/l	Total P mg/l
0.01	400	0.01	0	0.01
			1	0.02
			2	0.03
			5	0.06
			10	0.11
0.02	150	0.02	0	0.02
			1	0.04
			2	0.06
			5	0.12
			10	0.22
0.05	70	0.04	0	0.05
			1	0.09
			2	0.13
			5	0.25
			10	0.45

[1]Assumes P before alum addition = 0.75 mg/l.

The mode of solid-liquid separation is important in terms not only of the final phosphorus residual achieved, but also in terms of the capacity to handle the additional solids load generated at high metal dosages. For example, direct filtration may be adequate for low effluent phosphorus residuals, but may not be suitable for high metal dosages. Boller (1984) suggested that direct filtration applied to tertiary phosphorus precipitation should be limited to an application rate of 10 to 15 mg/l suspended solids from metal addition, at filtration rates of approximately 10 m/h (4 gpm/ft^2), to maintain filter runs of 15 to 30 hours. This is equivalent to maximum dosages of 4.0 to 5.0 mg Fe/l, or 2.0 to 2.5 mg Al/l to the filter inlet. Polymer filter aids of 0.1 to 0.3 mg/l were considered essential to prevent solids breakthrough.

Table 4.12 summarizes the case studies evaluated in the following paragraphs. The information is presented generally in the order of decreasing final effluent phosphorus concentrations. Figure 4.36 presents a synthesis of all the data.

(1) Baillod *et al.* (1977) reported that 85 mg/l alum dosed to the head of the Portage Lake WWTP contact stabilization process reduced the influent total

phosphorus concentration from 5.2 mg/l to less than 1 mg/l. This dosage represents approximately 20 mg alum/mg P removed, or 2.1 M Al/M P removed. The results showed effluent TSS concentrations of approximately 11 mg/l. This effluent solids concentration represented approximately 0.15 mg/l P.

(2) Eklund *et al.* (1991) reported on the results obtained at six wastewater treatment plants using alum or ferric salts for post precipitation, or two-point precipitation to the secondary clarifiers, and to tertiary clarifiers and/or filters. An approximate linear relationship was found between the ratio of metal/influent phosphorus and the effluent total phosphorus concentration. At mole Al or Fe/mole influent P ratios of 0.85, 1.40 and 2.0, effluent total phosphorus concentrations were 0.5, 0.3 and 0.1 mg/l.

(3) Mueller *et al.* (1999) reported on full scale testing done at the Rock Creek Advanced Wastewater Treatment Plant, owned by the Unified Sewerage Agency of Washington County, Oregon. The objective was to reach a summer season plant effluent total phosphorus limit of 0.07 mg/l. Results showed that direct filtration using granular media filters was able to meet an effluent phosphorus of 0.15 mg/l with less than 10 mg/l alum dosage. However, the existing *Claricone* tertiary clarifiers followed by granular media filters are able to meet lower than 0.07 mg/l (typically 0.05 mg/l) so blending direct filtered effluent with effluent from the *Claricone* and filter train would achieve the treatment goal. No attempt was made in the study to optimize alum dosages, so better results might have been obtained in the filters at higher alum dosages.

(4) The City of Breckenridge in Colorado, evaluated both the *ActiFlo* and *DensaDeg* ballasted sedimentation systems for tertiary treatment to meet future effluent phosphorus concentrations. The evaluation included a pilot study using the *ActiFlo* system to treat secondary effluent in parallel with a lamella settling clarifier at the existing plant. With an alum dose of 50 mg/l, rise rates of 16 – 32 gpm/ft^2, and a polymer dose of 0.2 mg/l, the *ActiFlo* system reduced the total phosphorus from 1.2–1.7 mg/l to below 0.11 mg/l. Based on a competitive bid process, the *DensaDeg* process was ultimately selected for implementation at this plant. Operations personnel indicated that *Dynasand* continuous backwash filters downstream of the *Densadeg* process achieve effluent total phosphorus concentrations of less than 0.01 mg/l.

(5) Bratby (1978) applied alum to oxidation pond effluent to remove algae and phosphorus using dissolved-air flotation. Initial phosphorus concentrations were close to 2 mg/l. Jar tests using alum found a stoichiometric relationship between alum and phosphorus removed, down to a total phosphorus concentration of 0.3 mg/l. The lowest phosphorus concentration achieved was approximately 0.04 mg/l at an alum dosage of almost 300 mg/l. The optimum pH for each alum concentration ranged from 5.0 to 6.0.

(6) The 300 Ml/d (80 mgd) Metropolitan Syracuse WWTP, NY piloted technologies to meet low effluent total phosphorus limits (Hook & Ott, 2001). The first phase of the pilot study showed that high-rate flocculated and ballasted settling technologies, such as Actiflo and DensaDeg were able to meet a proposed limit of 0.12 mg/l total phophorus, treating the effluent from secondary treatment (without biological phosphorus removal). The second phase of the study investigated various filter technologies downstream of the *Actiflo* or *DensaDeg* processes to achieve an effluent below 0.02 mg/l total phosphorus. The results of the study are summarized below:

Technology	Process Variables		Media	Average Total P	
	Chemical Dose	Hydraulic Rate		Influent mg/l	Effluent mg/l
Dynasand	20 mg/l $FeCl_3$	9.1 m/h	1.2 mm	0.135	0.017
Hydroclear	20 mg/l PAC	6.1 m/h	0.45 mm	0.124	0.03
Tetra	10 mg/l $FeCl_3$	12.2 m/h	5 mm	0.135	0.073
ABW[1]	20 mg/l $FeCl_3$		0.6/0.4 mm[2]		
Supersand	30/10 mg/l $FeCl_3$[3]	9.8/4.4 m/h[3]	1.2/0.9 mm[3]	1.09[4]	0.024
Memcor	2.4 mg/l ACH	0.73 $m^3/m^2.d$	0.2 mm	0.11	0.014

[1]Operational problems due to clogging. Tests abandoned.
[2]0.6 mm anthracite; 0.4 mm sand.
[3]First and second stages.
[4]Blend tank TP concentration.

Operational issues experienced with the pilot equipment were minor and were attributed to the small scale of the units. No significant issues were experienced with the *Dynasand* or *dual-Supersand* alternatives. The *dual-Supersand* alternative was the only alternative tested without pre-treatment in the *Actiflo* system. It received secondary effluent with an average total phosphorus concentration of 1.09 mg/l. It did not meet the 0.02 mg/l limit, but did produce an average effluent total phosphorus below 0.03 mg/l. Further testing with a single stage *Supersand* filter downstream of the *Actiflo* process did achieve an average effluent total phosphorus of 0.012 mg/l at a $FeCl_3$ dosage of 20 mg/l. This dosage represents approximately 260 mg $FeCl_3$/mg P removed, or 50 M Fe/M P removed.

(7) Hunt *et al.* (2000) reported on the Henderson, NV AWT plant, which added alum after biological phosphorus removal to achieve less than 0.1 mg/l P. Pilot tests determined that 20 mg/l alum could be added in direct filtration mode, although tertiary clarifiers were added in the full scale plant for operational flexibility. The pilot tests indicated that inclined plate gravity

settlers and continuous-backwash upflow filters (*Dynasand*) in series were able to achieve treatment goals.

(8) Gupta *et al.* (1998) describe full-scale tests on tertiary phosphorus removal using a new coagulant – GCR20, a combination $AlCl_3$-$FeCl_3$ blend, in the ratio 79% $AlCl_3$ and 21% $FeCl_3$. The plant is the Lower Potomac Pollution Control Plant in Fairfax, Virginia. It used secondary treatment activated sludge followed by tertiary clarifiers and a combination of multimedia and monomedia gravity filters. Traditionally the plant used ferric chloride to precipitate phosphorus. However, the new chemical represented significant cost savings. Secondary effluent total phosphorus concentrations averaging 1.07 mg/l were reduced to 0.18 mg/l in the tertiary clarifiers and 0.06 mg/l in the filters. The coagulant dosage was 20 mg/l. This dosage represents a molar ratio of 4.13(Al + Fe)/P.

(9) Hartland Township in Michigan (1999) conducted pilot plant tests for tertiary treatment of an extended aeration oxidation ditch secondary treatment WWTP. The pilot system tested the use of the *DualSand* process to remove phosphorus to less than 0.03 mg/l. Ferric chloride was added to the influent of the oxidation ditches to reduce secondary effluent phosphorus to less than 1.5 mg/l. Sodium hypochlorite was added to the feed of the *DualSand* process to avoid biofilm growth. The coagulant used in the *DualSand* process was polyaluminum-silicate-sulfate (PASS) at a dosage of 4 gpd. The aluminum content of the PASS was 71,200 mgAl/l. Since the *DualSand* process was run at 50,000 gpd, the aluminum dosage appeared to be 5.7 mgAl/l. Backwash from the first filter was recycled to the headworks of the plant. Backwash from the second filter was recycled back to the inlet of the first filter. When the secondary effluent phosphorus concentration ranged from 1.5 to 2.3 mg/l, the *DualSand* effluent concentration ranged from 0.06 to 0.14 mg/l. With additional ferric chloride to secondary treatment, the feed phosphorus concentration to the *DualSand* process reduced to approximately 1 mg/l, and the effluent phosphorus concentration was 0.01 to 0.027 mg/l.

Stress testing showed that the *DualSand* process could reliably meet the 0.03 mg/l limit if filtration rates per filter were maintained below 11 m/h (4.5 gpm/ft^2).

(10) Arapahoe County, Colorado operated a 9 Ml/d (2.4 mgd) Membrane Bioreactor (MBR) wastewater treatment plant that discharged to the Cherry Creek basin. The discharge permit limit allowed a phosphorus load allocation of 183 kg/year (402 lb/year), which translated to an effluent limit of 0.055 mg/l. Build-out capacity was planned for 20 Ml/d (5.2 mgd). Therefore, the permit limit was expected to reduce further to 0.025 mg/l. Effluent concentrations as low as this had been attained so it was expected that with adequate controls, the lower limit should be attained on a consistent basis. Ferric chloride was added to the screened wastewater just

before entry to the anoxic zone in the bioreactor. The dosage used was approximately 100 mg/l to achieve less than 0.05 mg/l total phosphorus in the effluent. Operations personnel had tested sodium aluminate which appeared to be beneficial to meet an effluent limit of approximately 0.02 mg/l in the future.

(11) The Town of Parker South WWTP in Colorado operated a 7.6 Ml/d (2 mgd) Bardenpho biological nutrient removal process which typically produced a secondary effluent with total phosphorus concentrations ranging from 0.25 to 1.75 mg/l on a monthly average basis. Secondary effluent was further treated in a *Microfloc* process with an upflow media contactor followed by downflow mixed media granular filters. Alum was added upstream of the *Microfloc* process, at dosages typically ranging from 55 to 90 mg/l, to achieve less than 0.05 mg/l phosphorus. The discharge permit limit for phosphorus was 0.2 mg/l on a monthly average basis, and 0.5 mg/l on a daily maximum basis. However, because of a fixed waste load allocation for discharge to the Cherry Creek basin, this limit was subsequently reduced to 0.05 mg/l. Operations personnel had demonstrated that an effluent limit of 0.05 mg/l could be met on a monthly average basis; monthly average effluent total phosphorus concentrations as low as 0.02 mg/l had been achieved.

(12) The Durham Advanced Wastewater Treatment Facility in Tigard, Oregon typically achieved effluent phosphorus concentrations of less than 0.03 mg/l. The permit limit was 0.07 mg/l, based on a 30-day median value. When the plant was modified from a non-bio P activated sludge system to an enhanced biological phosphorus removal activated sludge system (bio-P), the alum dosage was reduced from approximately 170 mg/l to 50 mg/l. During this phase approximately 20 mg/l was added to the secondary clarifiers to supplement bio-P, and approximately 30 mg/l was added to the filters for polishing. Later phases optimized the bio-P process so that if secondary effluent was 0.5 to 1.0 mg/l phosphorus then only 30 mg/l alum needed to be added to the tertiary system. If secondary effluent after bio P was less than 0.2 mg/l P, then the alum dose was typically 20 mg/l alum to meet the 0.07 mg/l phosphorus limit. In these cases no alum was added to the biological process. Tertiary treatment comprised square clarifiers, arranged as two in-series for each of three trains. The first clarifier had a flocculating zone and achieved some solids removal; the second had solids removal and was equipped with launders. Mixed-media filters followed the clarifiers. (Baur *et al.* 1999; Benisch *et al.* 2001).

(13) The Village of Stamford, New York City Department of Environmental Protection (NYCDEP, 2000) conducted pilot plant tests for tertiary treatment of an extended aeration secondary treatment WWTP. The pilot tests involved side-by-side tests comparing the performances of *Memcor* microfiltration (MF) membranes and the *DualSand* process. The objective

was to assess the equivalency of the technologies to determine whether the *DualSand* process should be classified as equivalent to MF technology for tertiary phosphorus removal.

The Stamford WWTP routinely added ferrous sulfate to the aeration tanks for phosphorus precipitation down to approximately 1 mg/l. The pilot tertiary treatment trains were to reduce phosphorus further, to meet a future limit of 0.2 mg/l. Both pilot trains were capable of meeting levels much lower than 0.2 mg/l.

Secondary effluent phosphorus concentrations, fed to the pilot units, ranged from 0.5 mg/l to 4.0 mg/l, depending on whether ferrous sulfate was added to secondary treatment. For both trains, tertiary effluent phosphorus concentrations ranged from 0.013 mg/l to 0.06 mg/l. The MF train used either polyaluminum chloride (PAC) or polyaluminum-silicate-sulfate (PASS) as precipitants. The *DualSand* train used PASS throughout, at a dosage of 4.12 mg Al/l.

Interestingly, the *DualSand* process generally outperformed the MF train, with an average effluent phosphorus less than 0.05 mg/l throughout. Chlorine was added upstream of the DualSand process, to a 0.5–1 mg/l residual, to avoid biofilm growth. This was not possible in the MF train (because of the polypropylene membranes) and the MF train did exhibit blinding problems.

For the DualSand process, with feed phosphorus concentrations ranging from 0.24 to 0.77 mg/l, and averaging 0.5 mg/l, effluent phosphorus ranged from less than 0.01 mg/l to 0.06 mg/l, averaging 0.013 mg/l.

Solids load stress tests were conducted on both trains, where up to 270 mg/l TSS was passed to the units. The *DualSand* process exhibited no significant blinding problems. The effluent turbidity in the *DualSand* effluent rose to 0.078 NTU, which was only 0.018 NTU above the previous average levels. The MF train did exhibit some blinding during the stress tests, despite the self cleaning 500 micron strainer upstream of the MF train.

For a pilot flow of 36 gpm to each train, the MF train drew 80 amperes, whereas the *DualSand* train drew 10 amperes. Water production for the MF train was 90% of the feed, whereas for the *DualSand* train it was 95%. Downtime for the MF train was 20%, but for the *DualSand* train it was 0.1%.

Based on the results of these pilot trials, NYCDEP approved an equivalency rating for the *DualSand* process.

(14) The Town of Concord WWTP operated a trickling filter secondary treatment wastewater plant, with no appreciable nutrient removal. Pilot tests carried out at the plant showed that the *CoMag* process achieves less than 0.05 mg/l total P. The *CoMag* process primarily comprises an accelerated sedimentation process, using magnetite as a weighting and separation agent. Effluent from the sedimentation tanks passes through

magnet screens/filters that separate the remaining flocs from the liquid stream. With alum added directly upstream of the *CoMag* process, the molar ratio of aluminum (Al) added-to-P removed was approximately 2.0 for an effluent total phosphorus of 0.20 mg/l. The molar ratio increased to approximately 9.0 for an effluent total phosphorus of 0.025 mg/l.

If the alum dosage was split by adding 75 mg/l to the secondary treatment process at the secondary clarifiers, thereby reducing the phosphorus concentration to the *CoMag* process to less than 0.2 mg/l, the overall molar ratio reduced to approximately 5.5 for an effluent total phosphorus of 0.018 mg/l. The corresponding alum dosage to the CoMag process was approximately 110 mg/l. These results demonstrated that split dosage treatment resulted in lower overall dosages to meet a given phosphorus removal.

(15) Pilot plant trials carried out at the Ruhleben WWTP in Berlin investigated the use of microfiltration for secondary effluent phosphorus removal (Gnirss & Dittrich, 2000). The plant operated with biological phosphorus and nitrogen removal, with secondary effluent phosphorus concentrations to the pilot plant of approximately 0.15 mg/l. The coagulant used in the trials was ferric chloride sulfate ($FeClSO_4$). At iron dosages of 1.9 and 1.4 mg/l as Fe, microfiltration filtrate total phosphorus concentrations of 0.018 and 0.03 mg/l, respectively were achieved. The corresponding PO_4-P concentrations were 0.0025 and 0.0095 mg/l, respectively. Suspended solids concentrations were non-detectable.

(16) The South Florida Water Management District carried out extensive investigations to remove phosphorus from stormwater runoff entering the everglades. The goal for phosphorus discharge concentrations was 50 μg/l – with a possible final requirement of 10 μg/l (SFWMD, 2000).

First stage screening experiments evaluated the ability of conventional treatment to reduce phosphorus concentrations to less than 10 μg/l. The conventional pilot plant comprised rapid mixing, flocculation, clarifier with inclined plate settlers, and granular media filters. Two sites were tested: the first, the North site treated canal water collected from agricultural areas; the second, the South site treated effluent from constructed wetlands which already removed nutrients to a certain degree from agricultural and other runoff.

In both cases, treated effluent from the pilot plants had phosphorus concentrations significantly lower than 10 μg/l. At the North site, ferric chloride at a dosage of 40 mg Fe/l and 0.5 mg/l anionic polymer reduced the feed total phosphorus concentration from 164 μg/l to 6 μg/l. The clarifier overflow rate applied was 0.34 m/h (0.14 gpm/ft^2) and the filtration rate 12 m/h (4.9 gpm/ft^2). No pH adjustment was required. Initial pH and alkalinity of 6.8 and 129 mg/l $CaCO_3$ respectively were reduced to 6.0 and 38 mg/l $CaCO_3$ respectively. At the South site, alum was used at a dosage of 20 mg Al/l with 0.5 mg/l anionic polymer. The feed total phosphorus concentration was reduced from 22 μg/l to 6 μg/l. The clarifier overflow rate applied was 0.68 m/h (0.28 gpm/ft^2) and the filtration rate 12 m/h (4.9 gpm/ft^2). Initial pH and alkalinity of 7.1 and 220 mg/l $CaCO_3$ respectively were reduced to 6.4 and 114 mg/l $CaCO_3$ respectively.

Various alternative technologies were also tested. For example, the *Actiflo* process was capable of reducing feed total phosphorus concentrations of 16 μg/l to 5 μg/l using ferric chloride at a dosage of 33 mg/l. Sulfuric acid addition was required to maintain the pH at an optimum value of around 4.3. Similar tests using alum showed that the effluent phosphorus objectives could be met but with very high alum dosages. A higher total phosphorus feed concentration of 156 μg/l was reduced to 8 μg/l using approximately 56 mg/l ferric chloride. Again, the pH was reduced with sulfuric acid to 4.2. In all cases, 0.8 mg/l polymer was added to capture the precipitates in the flocs.

(17) Pilot plant trials carried out at the Coeur d'Alene WWTP in Idaho compared four technologies to reduce effluent total phosphorus levels to less than 50 μg/L and potentially less than 10 μg/L. Technologies investigated were Blue Water dual-stage Blue Pro™; Parkson DynaSand D2 two stage filters; US Filter Trident®; and Zenon ZeeWeed ultrafiltration (Benisch *et al.* 2007).

The Blue Water system comprised two Blue Pro™ moving bed filters in series. Ferric chloride was added to rapid mix reactors prior to each stage at an average dosage of 38 mg/L $FeCl_3$.

The Parkson D2 system comprised two DynaSand moving bed filters in series. Alum was added to a rapid mix reactor prior to the first stage filter at an average dosage of 40 mg/L alum. The first stage used a coarser sand grain to promote flocculation and solids retention. The second stage filter uses a smaller grain size for final polishing. Backwash water was conveyed to a lamella settler. Supernatant was returned upstream of the first stage filter.

The US Filter Trident® system comprised a tube clarifier, followed by a coarse floating media upflow clarifier used for flocculation and solids contact (Adsorption® clarifier), followed by a downflow mixed media filter. Alum was added upstream of the tube clarifier at an average dosage of 78 mg/L alum. Polymer was also added prior to the tube clarifier and prior to the adsorption clarifier.

The Zenon system comprised an aerated tank in which were immersed ZeeWeed™ 500 ultrafiltration membranes. Alum was added upstream of the membrane tank at an average dosage of 165 mg/L alum.

After eliminating results during equipment and operational problems of the pilots, all technologies were able to meet the 50 μg/L objective, but not the 10 μg/L objective. The secondary effluent phosphorus concentrations fed to the pilots was not reported. The Table summarizes the average results reported for each technology.

Phosphorus Species	Effluent Phosphorus (μg/L)			
	Blue Pro	D2	Trident	ZeeWeed
Total P	21.4	39.6	19.2	24.1
Soluble P	17.2	18.8	13.0	23.3
Particulate P	4.2	20.8	6.2	0.8
Soluble ortho-P	2.3	5.3	2.2	8.4

The authors pointed out that differences in overall performance may be due less to differences in technologies but more to mechanical and operational issues experienced during the pilots. A further factor is that chemical additions were quite different for each pilot.

(18) Pilot testing was carried out at the Lakeshore WPCP in Innisfil, Ontario to compare technologies to reduce effluent total phosphorus levels to less than 30 μg/L. Technologies investigated were Zenon ZeeWeed ultrafiltration; Blue Water dual-stage Blue Pro™; Veolia Actiflo® followed by multi-media filtration, and Parkson DynaSand D2 two stage filters (deBarbadillo *et al.* 2010).

The average secondary effluent TSS and turbidity feeding the pilot units was 6 mg/L and 2.5 NTU, respectively. Average secondary effluent total phosphorus to the pilot units during the testing period was 0.22 mg/L. Average results from each of the pilot units were as follows:

Phosphorus Species	Effluent Phosphorus (μg/L)			
	ZeeWeed	Blue Pro	Actiflo + Filter	Parkson D2
Total P	15	15	31	22
Soluble P	12	13	18	16
Ortho-P	8	8	7	8
Sol. acid hydrol. P[1]	7	6	9	8
Turbidity	0.5	0.5	0.6	0.7
Coagulant	93 mg/L as alum	32 mg/L as Fe	14 mg/L as Fe plus 1 mg/L polymer	113 mg/L ferric sulfate plus 92 mg/L PACXL6

[1]includes ortho-P.

(19) Tertiary treatment pilot tests were carried out at the Fond du Lac WRRC and the Waukesha WRRC, both in Wisconsin. The Fond du Lac WRRC is required to meet a future 6-month average total phosphorus limit of 40 μg/L and the Waukesha WRRC a future 6-month average limit of 75 μg/L total P.

Two technologies proved themselves capable of meeting the treatment goals. The Blue Water single-stage and dual-stage Blue Pro™ systems were tested at the Waukesha WRRC. The single-stage Blue Pro™ system was capable of meeting less than the 75 μg/L total P goal, and the dual-stage Blue Pro™ system was able to meet an average effluent of approximately 15 μg/L, well below the 30 μg/L target value assigned for the tests. Secondary effluent TP values during the tests varied from approximately 0.60 mg/L to 1.75 mg/L.

The CoMag® system was tested at the Fond du Lac WRRC. This system was able to meet effluent TP values less than the assigned 30 μg/L target value, using an average coagulant dosage of 24 mg/L as Fe plus 0.80 mg/L polymer. Secondary effluent TP values were approximately 1 mg/L. Effluent turbidities after CoMag® treatment were less than 0.30 NTU and UV transmittance was increased from a range of 65 to 70%, to more than 75% by coagulation and CoMag® treatment (Lee *et al.* 2015).

(20) The Town of Maynard in Massachusetts installed a tertiary treatment CoMag® system for phosphorus removal to meet an effluent phosphorus limit of 0.1 mg/L with the expectation that future limits would be less than 0.05 mg/L. During project startup and an extended process proving period, effluent daily composite total phosphorus concentrations from the tertiary system had a median value of 0.023 mg/L and were less than 0.04 mg/L at all times, even during stress testing with excessive solids carryover from secondary treatment. Effluent turbidities from the CoMag® system averaged 0.27 NTU with a median value of 0.18 NTU. Polyaluminum chloride was added to the secondary clarifier influent, and to the CoMag® influent. Dosages were 3.5 mg/L and 4.0 mg/L as aluminium, respectively. (Keaney *et al.* 2012).

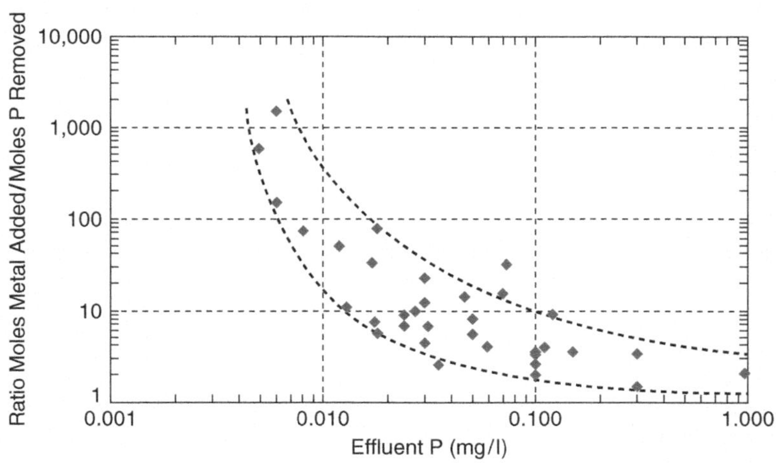

Figure 4.36 Synthesis of phosphorus removal results.

Table 4.12 Case studies of chemical phosphorus removal.

Case Study	Feed P mg/L	Final pH	Final Alkalinity mg/L CaCO3	Coagulant		Effluent P mg/L	Metal Added/P Removed	
				Type	Dosage mg/L		g Me/g P	M (Me)/M (P)
1	5.17	7.6		Alum	83.5	0.97	1.8	2.1
2	5.0			Alum	40	0.50	0.8	0.9
	5.0			Alum	65	0.30	1.2	1.5
	5.0			Alum	90	0.10	1.7	2.0
3	0.39	7.28	123	Alum	8	0.15	3.0	3.5
4	1.45			Alum	50	0.11	3.4	4.0
5	1.9	5.0		Alum	65	0.30	2.9	3.3
	1.9	5.5		Alum	125	0.050	4.8	5.5
6	0.65			$FeCl_3$	25	0.12	16.0	9.1
	0.65			$FeCl_3$	45	0.070	26.4	15.0
	0.65			$FeCl_3$	45	0.046	25.3	14.4
	0.135			$FeCl_3$	20	0.017	57.6	32.8
	0.124			PAC	20	0.030		
	0.135			$FeCl_3$	10	0.073	54.8	31.3
	1.09			$FeCl_3$	40	0.024	12.8	7.3
	0.09			$FeCl_3$	20	0.012	87.2	49.7
	0.11			ACH	2.4	0.014		

(*Continued*)

Table 4.12 (*Continued*) Case studies of chemical phosphorus removal.

Case Study	Feed P mg/L	Final pH	Final Alkalinity mg/L CaCO3	Coagulant		Effluent P mg/L	Metal Added/P Removed	
				Type	Dosage mg/L		g Me/g P	M (Me)/M (P)
7	2.5			Alum	75	0.10	2.8	3.3
	2.5			Alum	60	0.10	2.3	2.7
	0.6			Alum	15	0.10	2.7	3.2
	0.6			Alum	40	0.050	6.6	7.7
8	1.07			$AlCl_3$-$FeCl_3$ blend (Polymer added – no info.)	20	0.060	4.6	4.1
9	1.9			PASS		0.10	3.2	3.6
	1.0			PASS		0.019		
10	7.0	7.9	300	$FeCl_3$	90	0.035	4.4	2.5
	7.0	7.9	300	Sodium aluminate	70	0.026		
11	0.46	7.6	147	Alum	90	0.030	19.0	22.2
12	0.75			Alum	30	0.030	3.8	4.4
	0.20			Alum	20	0.030	10.7	12.5
13	0.45			PASS		0.013	9.4	10.7
14	2.0	5.3		Ferric sulfate + 2.3 mg/L polymer	123	0.027	17.1	9.5

	2.7	6.15		Alum + 1.7 mg/L polymer	225	0.024	7.7	8.9
	3.5	6.12		Alum + 2.5 mg/L poly. (75 mg/L alum added to secondary clarifier	180	0.018	4.7	5.5
	0.17	6.1			109	0.018	65.1	76.1
15	0.16			$FeClSO_4$	6.3	0.018	13.2	7.4
	0.15			$FeClSO_4$	4.7	0.031	11.8	6.6
16	0.164	6.0	38	$FeCl_3$	120	0.006	258	147
	0.022	6.4	114	Alum	220	0.006	1251	1461
	0.016	4.3	H_2SO_4 added	$FeCl_3$	33	0.005	1020	581
	0.156	4.2		$FeCl_3$	56	0.008	129	73
18	0.22			Alum	93	0.015	41	48
	0.22			$FeCl_3$	94	0.015	155	89
	0.22			$FeCl_3$ + 1.0 mg/L polymer	41	0.031	74	42
	0.22			Ferric sulfate + 92 mg/L PACXL6	113	0.022	159	89
20				PACl (dosages as Al to secondary clarifier influent and to CoMag)	3.5 mg/L + 4.0 mg/L	0.023		

4.8.3.1 Alternative methods of phosphorus precipitation

The effectiveness of lime for phosphorus precipitation is well established at pH values above 9. Major chemical reactions and stability constants are as follows:

- $CaHPO_4(s) = Ca^{2+} + HPO_4^{2-}$ (pH ~8.0) log K = −6.6
- $Ca_4H(PO_4)_3(s) = 4Ca^{2+} + 3PO_4^{3-} + H^+$ (pH ~9.0) log K = −46.9
- $Ca_{10}(PO_4)_6(OH)_2(s) = 10Ca^{2+} + 6PO_4^{3-} + 2OH^-$ (pH > 12) log K= −114
- $Ca_{10}(PO_4)_6(OH)_2(s) + 6H_2O = 4[Ca_2(HPO_4)(OH)_2] + 2Ca^{2+} + 2HPO_4^{2-}$ (pH > 12) log K = −17

Phosphorus precipitation is substantially complete when the lime dosage is sufficient to raise the pH to 9.5 to 10.0. However, separation of the pinpoint precipitates formed is difficult unless the pH is further raised to 11.0 to 11.5. A disadvantage of such a procedure is that the lime dosage necessary to raise the pH to 11.5 may be twice that necessary for phosphorus precipitation. A solution is to increase the pH with lime to 9.5 to 10, high enough for phosphorus precipitation, together with iron coagulant to aid flocculation of the precipitates formed (Cavagnaro *et al.* 1978). Lime dosages of around 350 mg/l, together with a ferric chloride dosage of 4.2 mg/l as Fe was found to be optimal to achieve a residual total phosphorus concentration of 0.2 mg/l.

Other sources of aluminum besides aluminum sulfate used for phosphorus removal have been powdered aluminum oxide and activated red mud. The application of powdered aluminum oxide for phosphorus removal was investigated by Huang (1977). The investigations utilized a commercial product comprising a fine powder (0.03 μm) of 90 percent γ-Al_2O_3, mixed in suspension with orthophosphate solutions. The results demonstrated that pH and the relative concentration of phosphate to alumina, P/Al_2O_3 (g/g) significantly influenced removal efficiencies. During the first hour of reaction, the phosphate was removed rapidly. From the second hour on, the reaction rate decreased and a linear relationship was observed. The maximum adsorption rate occurred at pH 4.5.

Huang suggested that powdered aluminum oxide could be used as a polishing agent for the removal of phosphate from secondary effluents, or from dilute aqueous solutions. The efficiency of phosphate removal generally increased with decreasing P/Al_2O_3 ratio. For a given quantity of Al_2O_3, the removal efficiency tended to be greater with more dilute influent concentrations. For example, a water containing 20 mg/l of phosphate (PO_4-P) and a neutral pH could be treated with 200 mg/l of Al_2O_3 to yield an effluent PO_4-P concentration of less than 3 mg/l or an 85 percent removal efficiency within less than 4 hours of contact time. At a P/Al_2O_3 ratio of 0.001, the residual PO_4-P concentration could be reduced to below 0.01 mg/l.

The rate of phosphate removal was found to be moderate. A large portion of the initial phosphate was eliminated within the first 1 to 2 hours of reaction time. The initial rate of phosphate adsorption was governed by pH and the P/Al_2O_3 ratio. The slow second phosphate removal reaction rate started after the second hour, and was prolonged for a long period of time.

Red mud is the waste product from the Bayer process for extraction of aluminum sulfate from bauxite. It is formed during the production of alumina when bauxite ore is subjected to caustic leaching. It is a brick red, high pH (10 to 12) sludge containing largely iron, aluminum, calcium, and silica. The disposal of red mud has been a problem for the aluminum industry. Activation is by removal of sodalite compounds with hydrochloric acid. Shiao and Akashi (1977) investigated activated red mud for phosphorus removal. The composition of the material used in their investigation was Al_2O_3 20.5 percent; Fe_2O_3 34.2 percent; SiO_2 17.2 percent; Na_2O 9.4 percent; and TiO_2 3.3 percent. The loss on ignition was 12.0 percent. The activated red mud was obtained by the treatment of batches of dried raw red mud with 5 to 30-percent by weight hydrochloric acid. They reported the long contact times required, of up to 2 hours for 72% phosphorus removal. Approximately 50-percent of the phosphate was removed in the first 10 minutes, and about 72-percent of it was removed within 120 minutes. Similarly for phosphate adsorption to alumina, the adsorption rate to activated red mud was slow. The phosphate removal capacity of activated red mud increased almost linearly with the increase of hydrochloric acid used for activation, up to 16-percent by weight. Beyond that point, the adsorption capacity of the activated red mud for phosphates did not improve.

Shannon and Verghese (1976) cited previous work on the potential for phosphorus removal using activated red mud solids where, at exceedingly high dosages, on the order of 5 g/l, red mud demonstrated reasonably good phosphorus removal from a typical domestic wastewater. At these excessively high dosages, however, removals were not consistent, red material remained in suspension causing an effluent turbidity problem, and sludge production was prohibitively large. Therefore, previously, it was concluded that the direct use of red mud as a phosphorus precipitant was not feasible.

Shannon and Verghese's work utilized alumized red mud, where red mud was slurried with sulfuric acid and the resulting product heat dried and crushed to a fine powder, with physical characteristics similar to commercial pulverized lime. They found that the alumized red mud was an efficient precipitant for phosphorus, resulting in average dissolved phosphorus levels of below 1 mg/l. They suggested that a practical dosage level for a 1 mg/l total phosphorus level was approximately 200 mg/l. The phosphorus removal capability lay between that of alum at a dose of 200 mg/l and lime at a dose of 250 mg/l.

One negative issue was that the alumized red mud contained significant concentrations of several heavy metals. The aluminum, iron, and calcium contents were high, giving the material its phosphorus removal properties. However, titanium, nickel, lead, chromium, copper, and mercury were also high.

Eggers *et al.* (1991) investigated the removal of phosphorus by crystallization of calcium phosphate in a granular form in a fluidized bed reactor. After pilot experiments, the first full-scale reactor was put into operation downstream of the biological sewage treatment plant at Westerbork, Drenthe in the Netherlands, in November 1988. The plant comprised a cascade CO_2 degassifier, a fluidized bed pellet reactor, and two pressure filters. The fluidized bed reactor had a height of 6 m (20 ft) and a diameter of 2.5 m (8 ft), for a design flow of 4.8 Ml/d (1.3 mgd). Lime solution was added to the reactor to maintain pH between 10.0 and 10.5. For removal of suspended calcium phosphate particles, the effluent of the pellet reactor was fed to two parallel dual-media pressure filters. Phosphate concentration in the effluent of the filters was less than 0.5 mg/l. When pellets grew to a certain size due to the crystallization process in the reactor, these were removed from the bottom of the reactor. The removed pellets were replaced by sand to maintain the total amount of grains in the reactor more or less constant. The pressure filters were backwashed daily. Backwash water was recycled to the reactor.

Basic operating conditions were as follows:

- pH cascade degassifier: 4.9 to 5.2
- pH reactor: 10.0 to 10.5
- Superficial velocity in reactor: 40 m/h (16 gpm/ft^2)
- Filtration rate in pressure filters: 20 m/h (8 gpm/ft^2)

Haruta *et al.* (1991) investigated the use of submerged corrugated iron contactors in activated sludge or biofilm reactors with agitation air for phosphorus precipitation. They found that the system operated best under sulfate reducing conditions. Therefore the degree of aeration was limited to avoid free oxygen in the tanks. The corrosion rate of the iron contactors was affected by the volumetric loading of BOD in the tanks where they were submerged. The BOD loading needed to be at least 0.3 kg/m$^3 \cdot$day (0.02 lb/ft$^3 \cdot$d) to make bacterial corrosion dominant about one month after the iron contactors were submerged. They estimated that the corrosion rate was about 1 mm or less in 30 years. From their results, the lowest residual total phosphorus concentration achieved appeared to be approximately 1 mg/l.

Ayoub *et al.* (2001) conducted laboratory experiments to determine the efficacy of iron aluminum hydroxy (oxide) coated filter media to remove low concentrations of phosphates. 600 to 700 μm (25 to 30 mesh) sand was soaked in a solution of 0.2 M ferric chloride and 0.4 M aluminum chloride for 30 minutes to coat the sand. The results indicated that the process achieved more than 90-percent P removals when the phosphorus was present in low concentrations of approximately 0.5 mg/l, when applied to distilled water and artificial groundwater. The process was less successful when applied to tap water and treated wastewater, with less than 70-percent removals. The success of the process relied on the absence of competing chemicals such as sulfates, nitrates, bicarbonates, chlorides, and organic matter.

The coated media achieved orthophosphate removal levels greater than 90%, with effluent concentrations less than 0.05 mg/L, from distilled water and artificial groundwater at flowrates as high as 27 m/h (11 gpm/ft^2).

James *et al.* (1992) investigated the feasibility of amending wetland peat and sand beds with oxidizing steel wool to precipitate and remove phosphorus from wastewater effluents. From the results, the authors suggested that steel wool offered a low-cost, efficient amendment for peat and sand beds designed for phosphorus removal. Phosphorus sorption on iron oxides in sand and peat required aerobic conditions to avoid the reduction of iron (III) and the release of soluble phosphorus to the effluent. Average phosphorus concentrations in the treated effluent were approximately 0.65 mg/l.

4.9 DISSOLVED ORGANIC NITROGEN (DON) REMOVAL IN WASTEWATER TREATMENT

4.9.1 General

As Utilities around the world are tasked with reducing wastewater effluent nutrient levels to increasingly lower limits, effluent nitrogen and phosphorus concentrations are at the forefront of treatment efforts. The total nitrogen concentration present in secondary or tertiary treated wastewater effluents comprises a number of major components: ammonia, nitrates, nitrites, and organic nitrogen (ON). The organic nitrogen component includes both particulate organic nitrogen (PON) and dissolved organic nitrogen (DON).

Figure 4.37 demonstrates a possible distribution of the various components in an effluent meeting a 3 mg/L total nitrogen effluent limit. Such a plant could be considered as providing excellent treatment in terms of total nitrogen removal. However, when effluent nitrogen limits are lower than approximately 3 mg/L, there is cause for particular concern and the options available to a Utility include reducing the organic nitrogen component even further.

Effluent DON concentrations reported at various secondary treatment plants vary widely. Figure 4.38 presents secondary effluent DON values measured at a number of treatment plants. The DON concentrations vary from 0.4 to 2.9 mg/L. From the results the 50th-percentile value is 1.12 mg/L, and the standard deviation 0.47 mg/L. The results presented employed a range of filters and analytical methods.

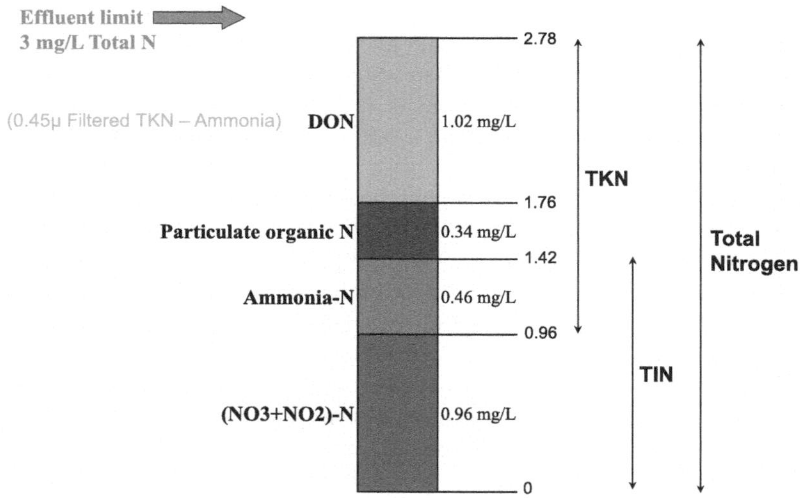

Figure 4.37 Relative concentrations of effluent total nitrogen components.

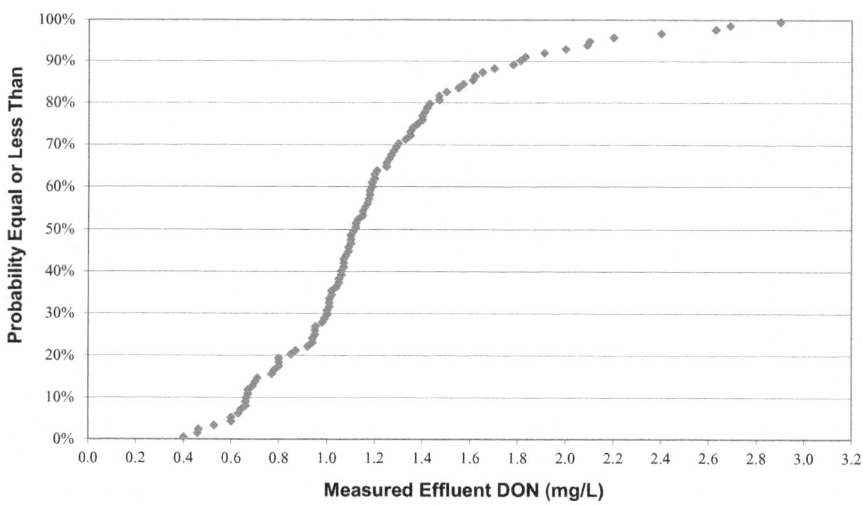

Figure 4.38 DON Concentrations at Diverse Locations. The 50th-percentile (median) value is 1.12 mg/L and the standard deviation 0.47 mg/L.

4.9.2 Characteristics of effluent DON

Effluent DON typically comprises two forms of organic nitrogen. The first is produced when proteins are metabolized and soluble microbial products are released by bacteria during biological wastewater treatment. The soluble microbial products result from the lysis and degradation of microorganisms. A large portion of the specific effluent DON compounds are unidentified although it is thought that these are a complex mixture of partially metabolized compounds. The second form is derived from humic substances contained in source waters that remain after drinking water treatment and are eventually discharged for wastewater treatment. This second fraction tends to be more recalcitrant than the free and combined amino acids, soluble microbial products, and other biomolecules produced during biological treatment.

The molecular weight distribution of DON varies from one source to the other. In general it appears that the low molecular weight fraction (less than 1.8 kDa) varies from approximately 60 to 70% of the effluent DON. The generally predominant smaller molecular weight fraction of wastewater effluents points to an added difficulty in DON removal. Higher molecular weights tend to be more readily removed.

DON varies within the treatment system, and generally increases during passage through the aerated portion of activated sludge reactors. DON can also be higher during plant startup and during plant upsets and corresponding molecular weight distributions can be higher during these conditions.

The DON component of effluent total nitrogen is typically defined as the fraction of the organic nitrogen that passes a 0.45-micrometer (0.45μ) filter. However, in reality the organic nitrogen measured in the filtrate from 0.45μ filters comprises both colloidal and dissolved (or soluble) organic nitrogen. The fraction of colloidal material comprising the DON is one of the factors that determine how much, and how easy it will be to reduce the effluent DON concentration.

Research at seven secondary treatment plants in the United States (Sattayatewa *et al.* 2009) found the following colloidal fractions of the effluent DON at each of the plants: 5.3%, 12.9%, 22.4%, 23.6%, 24%, 24%, and 62%. The colloidal fraction (referred to as colloidal organic nitrogen, CON) was determined in this case as the difference between the organic nitrogen measured in the filtrate from 1.2μ filters and that measured after coagulation with zinc sulfate at pH 10.5 and filtration through 0.45μ filters (following the procedures presented by Mamais *et al.* 1993). This is expressed as colloidal organic nitrogen, $CON = ON_{1.2\mu} - FFON_{0.45\mu}$.

Research at eight secondary treatment plants in Poland (Czerwionka *et al.* 2012) found the following colloidal fractions of the effluent DON at each of the plants: 36%, 37%, 37%, 47%, 48%, 52%, 52%, and 75%. In this case the colloidal fraction of the effluent DON was determined as the difference between the organic nitrogen measured in the filtrate from 0.45μ filters and that measured in the filtrate after passing further through 0.10μ filters. Therefore, in this case the colloidal organic nitrogen was expressed as $CON = ON_{0.45\mu} - ON_{0.10\mu}$.

Figure 4.39 shows the collective results from these two sets of research. In general, it appears that higher effluent DON concentrations tend towards higher colloidal fractions, although this relationship varies on a case-by-case basis.

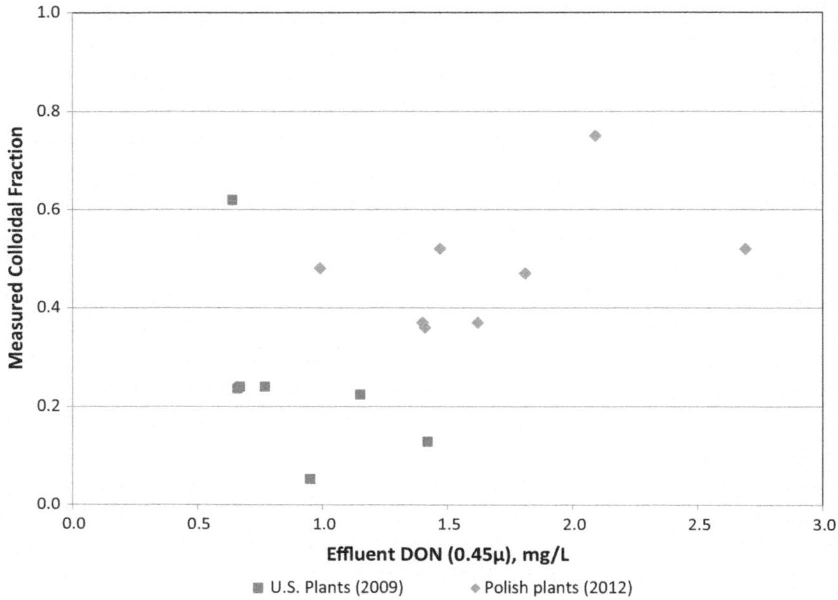

Figure 4.39 Colloidal fractions of the DON measured at 15 plants in the U.S. and Poland.

4.9.3 Impacts of effluent DON

It is not only the colloidal fraction of the effluent DON that determines the effectiveness of given treatment technologies. Another important factor is the hydrophobic or hydrophilic nature of the DON. Hydrophobic compounds are generally easier to remove, for example, by coagulation. Nitrogen bound to the functional groups of humic substances is an example of the hydrophobic portion of effluent DON. Table 4.13 shows the results of research at eight U.S. wastewater treatment facilities (Liu *et al.* 2012) where the hydrophilic and hydrophobic fractions of the effluent DON were determined by resin extraction. In this case the DON was separated from the effluent ON by filtration through 0.22μ filters. The hydrophobic fraction of the effluent DON varied from approximately 11% to 30% among the eight plants. Correspondingly, the hydrophilic portion varied from 70% to 89%.

The significance of this, apart from treatment considerations, is the bioavailability of the effluent DON to bacteria and algae in receiving waters. The hydrophilic portion of the effluent DON is the fraction that tends to stimulate algae growth, particularly when mediated by bacterial action in the receiving waters to convert the hydrophilic DON into lower molecular weight forms that are available to algae. The hydrophobic portion has little effect on algae growth. With an average hydrophilic fraction of approximately 78%, the results suggest that a relatively large fraction of the effluent DON is bioavailable in receiving waters, based on the 14-day algal bioassay tests employed in the research.

Table 4.13 Hydrophobic fractions of effluent DON at eight U.S. WWTPs.

WWTP	Effluent DON (0.22μ), mg/L	Hydrophobic Fraction of Effluent DON (%)
1	0.66	11.3
2	1.83	12.7
3	1.25	21.7
4	1.01	22.8
5	1.17	23.7
6	1.01	28.7
7	1.02	29.1
8	0.94	29.6

Bacteria and algae both uptake as well as release DON through metabolic activities during the growth phase. In the case of estuarine and marine discharges, photochemical reactions can affect the lability of organic material and readily convert “recalcitrant” compounds into reactive material. Photochemical reactions can release biologically available N and can convert DON to inorganic nutrients such as nitrite and ammonium. Many components of DON are bioavailable to microorganisms, including phytoplankton, cyanobacteria, and bacteria. For example, research has shown that within the residence times of the Chesapeake Bay, approximately 50% of the effluent DON was biodegradable within that timeframe. Recent research has shown that the high salinity in estuarine and marine environments contributes significantly to the abiotic release of ammonium from the effluent organic nitrogen. This is thought to be an important mechanism contributing to the bioavailability of effluent organic nitrogen in saline environments (Bronk *et al.* 2010).

Apart from the bioavailability of the effluent DON, there are other concerns. The identification of potent carcinogenic nitrogenous disinfection by-products, such as nitrosamines, nitromethanes and N-nitrosodimethylamine (NDMA) is a cause for concern over the discharge of DON in effluents from wastewater or agricultural sources. NDMA precursors (a component of DON) have been shown to be stable for at least 30 days during incubation with bacteria under aerobic conditions indicating that NDMA precursors are resistant to biodegradation in receiving streams, and therefore of concern for downstream drinking water treatment. Most of the NDMA precursors have been shown to be associated with the lower molecular weight fractions (less than 1 kDa) of the effluent DON. These concerns have led the European Community to establish a maximum admissible drinking water concentration for organic nitrogen of 1 mg/L Kjeldahl nitrogen, and some States in the United States to impose specific organic nitrogen limits in effluent discharge permits.

4.9.4 Measurement of DON

Generally, the measurement of DON involves subtracting the sum of the inorganic nitrogen species (ammonia, nitrates and nitrites) from the total nitrogen from filtered samples. However, since concentrations of the inorganic nitrogen species are generally considerably higher than the organic nitrogen, DON measurements are prone to significant errors.

Traditionally, DON has been determined by measuring the total Kjeldahl nitrogen (TKN) concentration of a filtered sample, using the macro-Kjeldahl method, and subtracting the corresponding ammonia-nitrogen concentration from it. Analytical methods typically follow Standard Methods procedures. To maintain reasonable accuracy the macro-Kjeldahl method has large sample requirements (typically 500 mL) and is subject to interference when nitrate concentrations in the sample exceed approximately 10 mgN/L. The method detection limit is relatively high at 0.1 to 0.2 mgN/L. Incomplete oxidation of some of the nitrogenous compounds is also an issue with the method.

An alternative approach preferred by many workers is to use the persulfate oxidation method whereby total nitrogen is oxidized to nitrates using persulfate digestion. Nitrates of the sample can then be determined by an accurate method such as ion chromatography. DON is determined by subtracting the ammonia, nitrate and nitrite concentrations from the total nitrogen value.

Filtration of the sample before DON determinations often employs 0.45μ membrane filters, although a range of filters have been used by various workers, including 1.2μ, 0.45μ, 0.22μ, 0.1μ, etc. The specific filter used affects the ultimate DON measurement since part of the measured DON is colloidal in nature. One suggestion has been to report DON in terms of the filter used, for example: "$DON_{0.45\mu}$".

4.9.5 Strategies for DON removal

4.9.5.1 Enhanced coagulation

Plants faced with the prospect of removing DON to meet tighter effluent nitrogen limits may look to coagulation as an initial approach. Enhanced coagulation is a technique designed to remove natural organic matter (NOM). In general, enhanced coagulation is optimized by (a) selecting the best coagulant type; (b) applying the best coagulant dosage; (c) adjusting the pH to a value where optimum (or adequate) coagulation conditions are achieved.

Different coagulant types such as aluminum or iron based coagulants, function with different solubilities and coordination chemistries under operating conditions. Higher coagulant dosages provide more metal for either floc or complex formation.

Lower operating pH values affect the metal complexes formed, and reduce the charge density of humic and fulvic acids, making them more hydrophobic (and adsorbable).

The effectiveness of enhanced coagulation depends largely on the water or wastewater characteristics and constituents. DON of biological origin, such as wastewater treatment effluents, tends to be of lower molecular weight and, therefore, more difficult to remove to a great extent by coagulation. Coagulation preferentially removes higher molecular weight fractions of DON, as well as of dissolved organic carbon (DOC).

Figure 4.40 demonstrates the importance of pH adjustment, particularly at the higher coagulant dosages typically required for DON removal. The results demonstrate the use of alum as coagulant, but similar considerations apply when using iron coagulants. By operating at the optimal pH, found by jar tests, the required coagulant dosage can be minimized. The impact of pH results from the competition between metal hydrolysis products and hydrogen ions for organic ligands; and between hydroxyl ions and organic anions for metal hydrolysis products. As pH increases, natural organic compounds become more negatively charged and the metal hydrolysis species become less positively charged, resulting in less adsorption propensity and an increased coagulant demand.

Figure 4.41 presents a compilation of DON removal results, using aluminum or iron based coagulants. The results have been rationalized in terms of dosage, expressed as mole metal added (as Al or Fe) per mole initial DON concentration, with corresponding DON removals. The spread of results reflects not only the wide variety of DON constituents in different wastewaters, but also the variety of coagulation conditions employed, many of which appeared to be without pH correction. The data is also influenced by different analytical methods employed for determining DON, and the variety of filters used for sample preparation.

The results suggest that two separate trends describe the data. The reason for the apparent separation is not clear, besides the reasons suggested above, but there does appear to be a group of results where the DON is more difficult to remove than the other.

An example of the use of Figure 4.41 is as follows:

- A plant can reliably meet a 3 mg/L total nitrogen limit but would be unable to meet a future limit of 2 mg/L total nitrogen on an average basis. Average values of nitrogen components are given in Figure 4.37.
- Effluent ammonia, nitrate and nitrite nitrogen are as low as expected with available technology. The particulate organic N and part of the residual DON would need to be removed to meet the new total nitrogen limit.
- Laboratory jar tests have indicated that all the particulate organic nitrogen and part of the DON can be removed by coagulation, together with an efficient solid-liquid separation process. The results obtained in the jar tests fall within the upper (grey) band of Figure 4.41.

- To meet an average effluent total nitrogen concentration of 2 mg/L, approximately 55% of the DON would need to be removed by coagulation and separation.
- From Figure 4.41, approximately 7.5 moles metal per mole initial DON is required.
- Given an initial DON of 1.02 mg/L, 0.546 mM/L of metal are required.
- Assuming aluminum sulfate, approximately 14.75 mg/L of aluminum are required.
- This is equivalent to 162 mg/L aluminum sulfate (assuming dry $Al_2(SO_4)_3.14H_2O$).

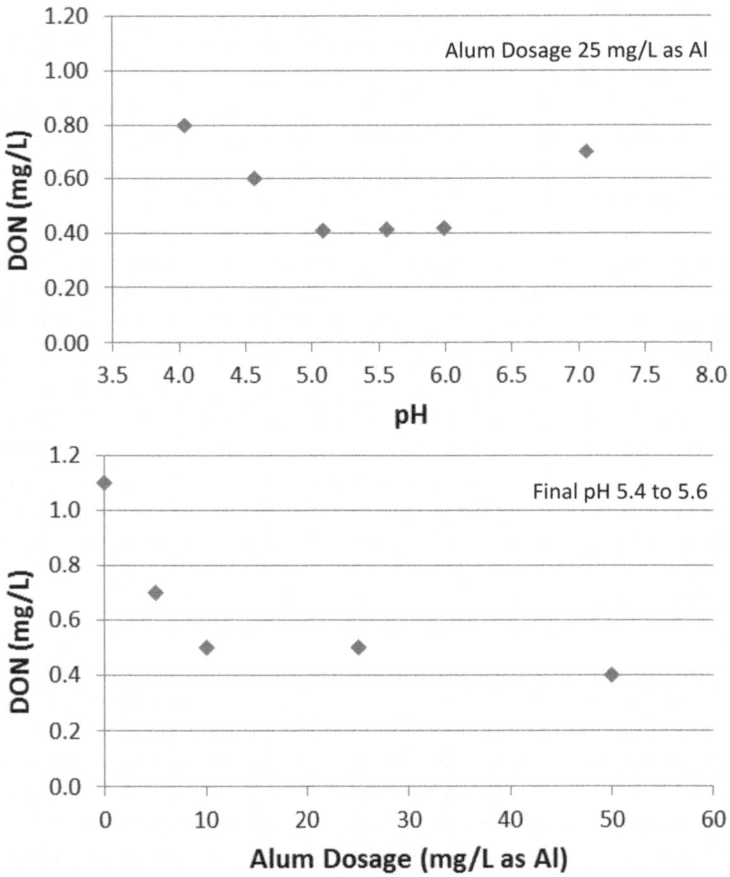

Figure 4.40 Effect of pH: Adjusting pH to an Optimal Value Maximizes DON Removal.

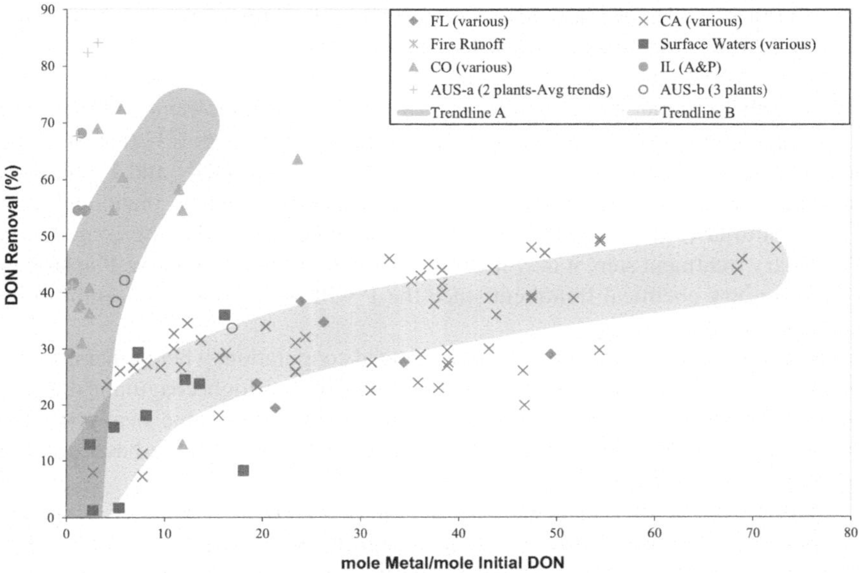

Figure 4.41 Compilation of Coagulant Dosages Required to Achieve DON Removals. The spread of results appears to follow two general trends.

4.9.5.2 Impacts of DON removal on phosphorus and other constituents

If low total phosphorus limits are in place, and a Utility is required to address low total P at the same time as a strategy directed towards reducing DON concentrations to also meet low total nitrogen limits, the question arises as to whether there will be interference between these two simultaneous objectives.

Referring to Figure 4.36 and continuing with the previous example:

- Together with the total nitrogen limit of 2 mg/L, the plant in the example must also meet a total phosphorus (P) limit of 0.10 mg/L.
- The current average effluent total P concentration is 1.0 mg/L.
- From Figure 4.36, the coagulant dosage required to achieve approximately 0.08 mg/L total P is approximately 6 mole metal per mole P removed.
- P removed = 0.92 mg/L (= 0.0297 mM P/L)
- Coagulant dosage = 0.178 mM Al/L (assuming the same coagulant as the original example).
- Assuming aluminum sulfate, approximately 4.8 mg/L of aluminum is required.
- This is equivalent to 53 mg/L aluminum sulfate (assuming dry $Al_2(SO_4)_3 \cdot 14H_2O$).

- In other words, the alum dosage required for DON removal is approximately three-times the dosage required for P removal.

From this example, it is clear that if a Utility is required to remove DON, and coagulation is the process to be adopted, then phosphorus removal could become incidental to the DON removal, depending on the relative DON and P removals required. This has significant implications at the planning stage for plant upgrades. In cases similar to this example, where DON is to be removed using coagulation in a tertiary treatment step, it may not make economic sense to remove P upstream of final tertiary chemical treatment since the P will, in any case, be removed by chemical precipitation.

There are other potential benefits of enhanced coagulation, where other effluent components will be removed to a certain extent by the high coagulant dosages required for DON removal. Examples include arsenic and some heavy metals. The actual removals would need to be evaluated on a case-by-case basis, but the potential exists and has been demonstrated.

Another benefit of enhanced coagulation observed at a number of plants that practice ultraviolet disinfection is a significant increase in effluent UV transmittance. One example is shown in Figure 4.42 where the UV transmittance increased from less than 60% to almost 75% after enhanced coagulation. This significantly increases the efficiency of UV disinfection.

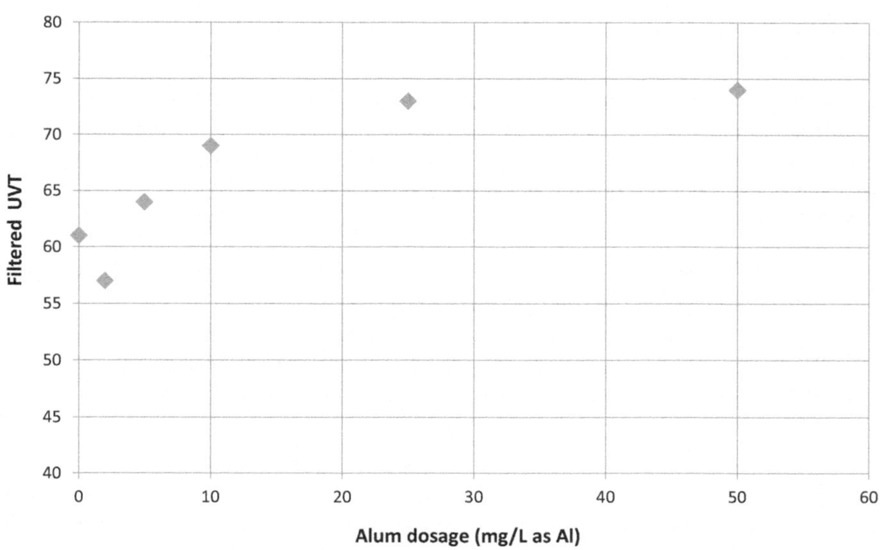

Figure 4.42 Results Show Significant Increase in UV Transmittance with Coagulant Addition.

4.9.5.3 *Solid-liquid separation technology*

Selection of the solid-liquid separation technology in the tertiary treatment process is an important component of the DON removal strategy using coagulation. The high coagulant dosages required will produce relatively high concentrations of inert chemical sludge. Some processes that have been successful in high phosphorus removal applications appear to have the potential for success also in DON removal applications using coagulation. These processes include ballasted sedimentation using microsand or recycled sludge, plus filtration; dissolved-air flotation plus filtration; and plate settlers plus filtration. Some processes that have targeted DON removal specifically include ballasted flocculation and sedimentation processes using magnetite (experiences to date have demonstrated that post-filtration is not required with this technology) and a dual filter process, where the first coarser stage stores much of the solids and the second stage polishes the effluent (experiences to date have demonstrated that a tertiary sedimentation process prior to the dual filter is not necessary).

4.9.5.4 *Other approaches for DON removal*

Depending on the characteristics of the effluent DON, it is possible that excessive coagulant dosages may be required to achieve the required DON removals. Alternatively, it could be the case that the required effluent TN limit is so low, or the magnitude of the required DON removal is so high, that the required removals are beyond the capabilities of coagulation. In this case it is possible that other treatment approaches may be appropriate. These treatment alternatives may include the following:

- Activated carbon. Studies have demonstrated removals of over 70% of effluent DON. Most of the effluent DON appears to be relatively non-polar, making this fraction particularly suitable for activated carbon adsorption.
- Reverse Osmosis (RO). Experiences at a number of locations have shown average ON removals of 68 to 94% through RO membranes with ultrafilter pretreatment. Some reported feed ON concentrations to the membrane systems appear to have been unfiltered prior to analyses, so actual DON removals may be lower than reported.
- Soil-Aquifer Treatment (SAT). Research has demonstrated efficient DON removals by SAT with greater than 90% removals. Corresponding dissolved organic carbon (DOC) removals ranged from approximately 50 to 75%.
- Ozonation. Approximately 14% DON destruction has been observed at absorbed dosages above 100 mg/L. Ozonation significantly increases the biodegradability of the remaining DON suggesting that biological active carbon filtration (BAC) using activated carbon beds may improve overall DON removal.
- Biological Active Carbon (BAC) – Followed by Enhanced Coagulation. Although this does not appear to have been applied specifically for DON

removal, there have been promising results with DOC removal. It appears that the BAC step biologically modifies the organic material making it more humic-like and more amenable to coagulation. For a given coagulant dosage the DOC was reduced by almost 50% below the concentration achieved without BAC.

- Membrane Bioreactor (MBR). Experiments have suggested that effluent DON from MBR systems are lower than from conventional BNR systems, in the approximate ratio 0.68:1. It is possible that the ultrafiltration membranes of the MBR system captured more of the colloidal portion of the DON.

4.10 WASTEWATER TREATMENT BY COAGULATION AND CHEMICALLY ENHANCED PRIMARY TREATMENT, CEPT

Physical-chemical treatment of wastewater was widely practiced until the advent of the trickling filter for biological treatment in the late 19th century. The early 1970s saw a partial revival of interest that has continued to the present day, particularly for treatment plants that are overloaded during peak flow events and where bypassed discharges of excess wastewater flows during storm events are not permitted.

In the United States, regulatory restrictions on combined sewer overflows (CSOs) has resulted in these wet weather flows being conveyed to wastewater treatment plants, thereby markedly increasing peak flows experienced at these plants. For example, if diversion of excess storm flows were allowed one might expect the hydraulic capacity provided in the wastewater treatment plant to be approximately two-times the average dry-weather flow. However, if the whole flow at all times is to be treated, the hydraulic capacity could be on the order of four or more times the average dry-weather flow which would considerably increase the capital costs of treatment. An alternative to treatment of the whole flow is to provide physical-chemical treatment of the excess wet-weather flow in parallel to biological treatment of the main design flow.

The addition of coagulant chemicals to primary clarifiers, or to other dedicated physical separation processes, is an effective way of reducing the load to downstream biological processes, or in some cases for direct discharge. This practice is generally referred to as chemically enhanced primary treatment, or CEPT. The practice has also been referred to as enhanced high rate treatment (EHRT), and others.

In general, the objectives and benefits of CEPT include the following:

- Reduced capital costs and land requirements because of higher overflow rates in clarifier units, and reduced strength of CEPT effluent to downstream biological treatment.
- Effective treatment for excess wastewater flows during storm events, avoiding high capital investments for full biological treatment, particularly with infrequent storm events.

- Chemical addition and precipitation reactions in CEPT can remove appreciable concentrations of toxic metals that otherwise might affect downstream biological processes.
- Phosphorus removal in CEPT can be very efficient when metal coagulants are used.
- Rapid start-up and shutdown of chemical addition, making this particularly appropriate for infrequent storm events.

Principal disadvantages that might preclude a wholly physical-chemical solution to wastewater treatment are the problems associated with the highly putrescible sludge produced, and the high operating costs of chemical addition. However, much of the current interests in physical-chemical treatment stem from its suitability for treatment under emergency measures; for seasonal applications, to avoid excess wastewater discharges during storm events; and for primary treatment before biological treatment, where the above disadvantages become of lesser impact. Moreover, physical-chemical treatment has a well established position in tertiary wastewater treatment.

Harleman and Murcott (2001) promoted the application of CEPT as an effective first step for pollution control in developing countries – particularly in large urban areas that have evolved with sewerage systems but without centralized wastewater treatment, and have limited financial resources for more complete, but capital intensive biological treatment options such as activated sludge systems. Such urban developments also may not have the areas available for appropriate technology options such as stabilization pond processes.

A part of Harleman and Murcott's thesis was that CEPT, if not complete treatment, is much better than no treatment. By implementing chemical treatment as an initial stage, biological polishing of some sort can be implemented later for soluble BOD removal and nitrogen conversion, if required, as funds become available.

4.10.1 Dependence of CEPT removals on wastewater characteristics

The efficiency of CEPT, in terms of TSS, BOD and/or COD removal, depends on wastewater characteristics. With CEPT, one can expect to remove particulate components, together with some portion of the colloidal components.

In warmer climates, with larger collection systems and relatively flat sewers, one would expect a higher degree of hydrolysis of particulate matter resulting in higher soluble fractions, and lower overall removals with CEPT. On the other hand, if the collection system is relatively small, the climate is cold, and wastewater is relatively fresh, there may be a higher proportion of particulate material, and CEPT removals could be higher. An example of this is in Norway, where typical removal efficiencies reported with CEPT are as follows (Ødegaard, 1992):

- Chemical oxygen demand (COD) 370 to 99 mg/l (73% removal)
- Biochemical oxygen demand (BOD) 140 to 27 mg/l (81% removal)

- Total organic carbon (TOC) 70 to 24 mg/l (65% removal)
- Total suspended solids (TSS) 190 to 17 mg/l (91% removal)
- Total phosphorus (TP) 4 to 0.25 mg/l (94% removal)
- Total nitrogen (TN) 37 to 27 mg/l (28% removal)

To achieve the results reported by Ødegaard, most plants used aluminum sulfate with typical dosages reported as approximately 16 mg/l as Al. Anionic polymer, when used, was typically at 0.5 mg/l.

Tebutt *et al.* (1975) and Wahlberg *et al.* (1998) pointed out that in primary sedimentation tanks, the maximum TSS removal efficiency possible corresponds with the fraction of TSS that is settleable. The objective of coagulant addition and CEPT is to convert some of the non-settleable material into settleable material thereby increasing overall removals. Narayanan *et al.* (2000) tested a procedure based on particle weight distributions. Sequential filters were applied to samples of primary influent wastewater samples: 8 micrometer (μ), 5 μ, 2.5 μ, and 0.7 to 1.0 μ. The weights retained on each filter allowed the distribution of the TSS concentrations in the four different size ranges to be determined: 1 to 2.5 μ, 2.5 to 5 μ, 5 to 8 μ and greater than 8 μ.

For example, a primary influent sample had the following percentages of TSS:

- 1 to 2.5 μ: 22.8%
- 2.5 to 5 μ: 15.7%
- 5 to 8 μ: 5.5%
- greater than 8 μ: 56.1%

There was a predominance of larger sized particles (greater than 8μ) but also a significant fraction of smallest sized ("non-settleable") particles.

After settling without CEPT in a 2-liter settleometer, the following percentages of TSS in the supernatant, relative to total primary influent TSS, were found:

- 1 to 2.5 μ: 20.1%
- 2.5 to 5 μ: 5.2%
- 5 to 8 μ: 4.5%
- greater than 8 μ: 4.1%

The overall TSS removal without CEPT was 66.1%. The results of the test showed that most of the largest sized particles were removed but removal of the smallest sized particles was minimal. This removal in the laboratory batch test compared closely with actual TSS removals in the full-scale primary clarifiers, suggesting that full-scale performance was not influenced to a large extent by hydraulic issues in the sedimentation tanks.

In general, by comparing results in the laboratory settleometer with full scale results, conclusions about the relative impacts of sedimentation tank characteristics and of wastewater characteristics and particle size distributions on plant performance can be made.

After CEPT of the primary influent sample, the following percentages of TSS in the supernatant, relative to total primary influent TSS, were found:

- 1 to 2.5 μ: 16.4%
- 2.5 to 5 μ: 1.4%
- 5 to 8 μ: 0.6%
- greater than 8 μ: 0.5%

The overall TSS removal with CEPT was 81%. From the results, the effect of CEPT was to increase the removal of all size ranges, with almost complete removal of the larger sizes. There was also significant removal of the smallest size range, demonstrating the conversion of some of the non-settleable material into settleable material.

4.10.2 Case studies of CEPT

The following paragraphs present an overview of CEPT experiences. The case studies concentrate on the use of metal coagulants, either alone or with polymer flocculants to accelerate floc separation. Chapter 5 includes examples of the use of polyelectrolytes instead of metal coagulants for CEPT applications.

- *Raw sewage jar tests.* Raw wastewater TSS 608 mg/l; BOD 383 mg/l; COD 940 mg/l. Using ferric chloride, an optimum dosage of 34 mg/l as Fe was found, producing the following percentage removals after sedimentation: TSS 96; BOD 63; COD 67. Similar removals were obtained with alum at 32 mg/l as Al (Cooper & Thomas, 1974).
- *CEPT full scale trials in Tampa, Florida.* Alum was added just before the influent screens, and polymer added to the inlet channel to the primary clarifiers. Raw wastewater characteristics were TSS 225 mg/l (180 to 320 mg/l) and BOD 270 mg/l (250 to 300 mg/l). Coagulants were alum and polymer. The alum dosage varied from 12 to 19 mg/l as Al, and the anionic polymer 0.6 to 0.7 mg/l. Removals obtained were: TSS 82 to 91% (primary effluent TSS 29 to 39 mg/l) and BOD 39 to 45% (primary effluent BOD 153 to 185 mg/l). As a comparison, TSS and BOD removals without alum were approximately 58% and 20%, respectively. These latter results reflect the high soluble BOD fraction of the wastewater of 50 to 60-percent (Wilson *et al.* 1975).
- *Jar tests of raw domestic wastewater.* Investigation of 32 commercial polyelectrolytes as flocculant aids, using alum as the primary coagulant. Alum dosages for different wastewaters varied from 10 to 16 mg/l as Al. The best polyelectrolyte was found to be an anionic hydrolyzed polyacrylamide with 10 to 20% hydrolysis and greater than 8×10^6 molecular weight. Typical dosages were 0.2 to 0.5 mg/l. TSS removals of up to 93-percent were achieved, with apparent overflow rates up to 1.3 m/h (Benedek & Bancsi, 1977).

- *Jar tests of raw domestic wastewater.* Characteristics were TSS 200 to 600 mg/l and BOD 250 to 350 mg/l. Using an alum dosage of 5.4 mg/l as Al, BOD removals were 52 to 62-percent. Without coagulation, removals were 30 to 35-percent (Thakur *et al.* 1977).
- *Jar tests of raw domestic wastewater.* Characteristics were BOD: 171 mg/l; COD: 460 mg/l; TOC: 165 mg/l; DOC[1]: 73 mg/l. At an optimum dosage of ferric chloride (62 mg/l as Fe) and pH 5.3, removals were TOC 55% and DOC 12%. At very high dosages, up to 100 mg/l as Fe at pH 7.7, increased soluble carbon removal was obtained – up to 39%. However, at the high pH, TOC removals reduced to 42%.

  Using alum, the optimum dosage was 32 mg/l as Al at pH 5.5. Removals were 59% TOC and 22% DOC (Leentvaar *et al.* 1978).[1]
- *Pilot plant tests of dissolved-air flotation applied to primary settled domestic wastewater.* Primary settled wastewater characteristics were TSS 150 mg/l; COD 300 mg/l; total P 3 mg/l; pH 6.5. Optimum performance for flotation separation was achieved at an alum dosage of 6.3 mg/l as Al. The final pH was 5.5. Removals were as follows: TSS 97%; COD 84%; total P 92%. No advantages were found by using polymer as flocculant aid. Alum alone was used. Lime was also investigated as a coagulant and precipitant. The cost for using lime was determined to be up to 12-times that for using alum, so was not considered further. Using alum without polymer, no formal flocculation was required. A residence time of 30-seconds in the transfer channel between alum addition and the flotation tank was sufficient. Overflow rates applied were up to 30 m/h (17,600 gpd/ft^2) without significant solids carry-over. For dewatering of the flotation thickened alum sludge, a filter press achieved 17-percent cake concentration without conditioning. A cake concentration of 22-percent was attained with polymer conditioning of the sludge. Using lime, cake concentrations of 40-percent were achieved (Bratby, 1982).
- *Full scale operation of physical-chemical treatment of raw domestic wastewater.* The 38 Ml/d (10 mgd) treatment plant included screening, grit removal, coagulation and flocculation, sedimentation, activated carbon adsorption – filtration, breakpoint chlorination. Influent TSS and BOD averaged 200 mg/l and 120 mg/l, respectively. Best overall results were achieved with ferric chloride at 5.2 mg/l as Fe, together with 0.15 mg/l anionic polymer. TSS and BOD removals through coagulation and sedimentation averaged 87% and 56% respectively (Lynch & Potter, 1981).
- *Pilot plant tests of CEPT before biological treatment.* Characteristics of the raw wastewater were: TSS 255 mg/l; BOD 273 mg/l; COD 650 mg/l; PO_4-P 14 mg/l. Removals achieved with alum were TSS 87%; BOD 52%; COD 58%; PO_4-P 93%. Final pH was 7.6. An important conclusion from this

[1] DOC is a functional description and is the organic carbon after filtration through 0.45 μm filters. Therefore, DOC contains both soluble and colloidal material.

work was that the degree of denitrification did not reduce with coagulation and CEPT, indicating that the soluble substrate is the important component for denitrification, and is still available after CEPT. The critical ratio of soluble BOD and total nitrates was identified as approximately 2.5 (Rebhun *et al.* 1985).

- *Full scale operation of CEPT upstream of activated sludge system.* The Hyperion WWTP had a capacity of 1,600 Ml/d (420 mgd). All of the wastewater treated received CEPT using ferric chloride and polymer. Dosages were 8.6 mg/l as Fe, and 0.3 mg/l polymer. Removals were TSS 80 to 85% and BOD 50 to 55%. Without CEPT, removals were TSS 65% and BOD 30%. CEPT increased the primary sludge production by about 45% and decreased secondary sludge by about 25%. The net change is an overall increase by 6% of the total sludge produced at the plant (Chaudhary *et al.* 1991).
- *Full scale application of CEPT upstream of biological treatment with trickling filters.* Flow to the primary sedimentation tanks averaged at about 0.4 Ml/d (0.11 mgd). Best results were obtained with alum at 19.5 mg/l as Al, followed by a cationic polyacrylamide polymer at 1 mg/l. A lag period of 5 minutes was found to be optimal, between alum addition and polymer addition. Raw wastewater TSS and BOD were approximately 350 mg/l and 380 mg/l, respectively. Removals were TSS 76% and BOD 59%. Removals without chemical addition were TSS 30% and BOD 9%. Part of the poor BOD removals without chemical addition was due to recycling of humus sludge to the primary clarifiers (Jones *et al.* 1991).
- *Full scale application of CEPT as an interim measure during construction of permanent secondary treatment at the Owls Head WWTP, New York.* Ferric chloride and polymer were used. Optimum dosages were ferric chloride 17 mg/l as Fe, and anionic polymer 1 mg/l, dosed for 6-hours per day during high flows, then reduced to 5 mg/l as Fe and 1 mg/l, respectively, for the remainder of the day. Primary clarifier overflow rates of up to 3.4 m/h (2,000 gpd/ft^2) were treated. Average monthly BOD and TSS removals were 36% and 55%, respectively. With plain sedimentation, without CEPT, corresponding removals were less than 20% and 40%, respectively.

  At periods of low flows, ferric chloride alone provided satisfactory treatment. At higher flows polymer was required to prevent floc carryover from the clarifier weirs. Alum performed as well as ferric chloride, although thickened solids deteriorated markedly – reducing from approximately 5.5% with ferric, to 2% with alum.

  The optimal location of the polymer feed was in close proximity to the primary tank influent channel.

  The 6-hour spike of ferric chloride dosage of 17 mg/l as Fe had to be reduced at times to allow the anaerobic digesters to recover from imminent failure. At these times the dosage was reduced to a constant feed of approximately 7 mg/l as Fe (Chack *et al.* 1994).

- *Full scale application of CEPT during high flow storm events to treat raw wastewater in existing primary clarifiers.* CEPT was tested at full scale at the King County South Plant in Seattle using a staged coagulation sequence comprising ferric chloride, followed by a polyaluminum chloride, followed by an anionic polymer. Dosages were ferric chloride 15 to 23 mg/l as Fe; polyaluminum chloride 8 to 15 mg/l; polymer 0.4 to 0.6 mg/l.

  Raw wastewater characteristics during the storm events were: TSS 112 to 326 mg/l; BOD 102 to 270 mg/l; COD 250 to 720 mg/l. Approximately 60-percent of the influent BOD was particulate. The test showed that the particulate BOD as well as a part of the colloidal BOD was removed.

  At a peak surface overflow rate of 6 m/h (3600 gpd/ft^2) removals were TSS 80 to 95%, and BOD 58 to 68%. Without chemicals, corresponding removals were TSS 50% and BOD 25%.

  At surface overflow rates as high as 8.8 m/h (5200 gpd/ft^2), removals obtained with the optimized chemical sequence were TSS 65 to 80% and BOD 40 to 50%.

  Mixing time between the ferric chloride and PACl addition points was approximately 20 seconds at peak flows. Flocculation time before polymer addition was approximately 6 minutes at peak flows. After polymer addition, flocculation time before entering the primary clarifiers was approximately 2 minutes at peak flows. Rapid mixing of the coagulant chemicals into the channels was achieved by using relatively simple linear sparger tubes but with compressed air introduced to increase the coagulant dispersion into the liquid flow.

  Prior tests using just ferric chloride and polymer showed inferior performance compared with those obtained with staged coagulation. With ferric chloride dosages of 3.5 to 10.5 mg/l as Fe and an anionic polymer dosage of 0.2 mg/l, removals were TSS 80 to 55% and BOD 65 to 40% at applied SOR of 2.2 to 4.8 m/h (1300 to 2800 gpd/ft^2). (Krugel *et al.* 2005; Melcer *et al.* 2005, 2011, 2014).
- *Full scale pilot testing of CEPT during wet weather events to treat raw wastewater in existing primary clarifiers.* CEPT was tested at full scale in two existing 180-ft primary clarifiers at the Southerly WWTP in Columbus, Ohio. Ferric chloride had been selected as the most cost-effective after jar testing compared ferric chloride, alum, and polyaluminum chloride. The dosage applied during the full-scale tests was 40 mg/L as ferric chloride. An unspecified polymer was added at the clarifier splitter box at 0.5 mg/L. Raw wastewater characteristics during the tests were: TSS 84 to 130 mg/l; CBOD 44 to 82 mg/l; Total phosphorus 1.0 to 1.9 mg/L. The average surface overflow rate during the tests appeared to be approximately 3.1 m/h (1860 gpd/ft^2). Average removals during the tests were TSS 78%, CBOD 64% and TP 78%. (Blake *et al.* 2015).
- *Jar, pilot and full-scale tests of raw domestic wastewater during storm events.* Jar testing was carried out at the Southerly Wastewater Treatment

Center (WTC) in the Greater Cleveland Metropolitan area, Ohio during storm events. A Consent Decree called for pilot demonstrations to meet the following treatment criteria over seven consecutive wet weather events: less than 40 mg/L TSS; E-coli 126 CFU/100 mL (daily geometric mean); E-coli 284 CFU/100 Ml (rolling geometric mean per 7 consecutive day period); 0.038 mg/L total residual chlorine. The TSS of the influent screened and degritted wastewater during the jar tests was 39 mg/L. Ferric chloride was optimized at 30 mg/L together with 1.0 to 2.0 mg/L of high MW anionic polymer. Similar results were obtained with 35 mg/L sodium aluminate plus 1.0 to 2.0 mg/L of the anionic polymer. At the optimum dosages both TSS and BOD removals of close to 80% were achieved.

Chlorine disinfection of raw sewage using sodium hypochlorite showed that 20 mg/L as chlorine and a contact time of 10 minutes was required to meet the disinfection requirements. A phenomenon was observed whereby overdosing at a chlorine dosage of 30 mg/L prior to coagulation resulted in wholesale flotation of the flocs during the jar tests.

Chlorine applied to settled supernatant was advantageous, compared with pre-chlorination of raw sewage. In this case a chlorine dosage of 5 mg/L and contact time 3 minutes applied to settled supernatant were sufficient to meet the disinfection requirements.

Subsequent full-scale trials at the Southerly WTC demonstrated successful operation at primary clarifier surface overflow rates (SOR) up to 6.8 m/h (4,000 gpd/ft^2) with influent TSS concentrations ranging from approximately 55 mg/L to 600 mg/L reduced to approximately 20 mg/L to 32 mg/L. Pilot scale trials at the Easterly plant and full-scale trials at the Westerly WTC also demonstrated that treatment requirements of less than 40 mg/L TSS were met by CEPT treatment. During a number of wet weather events the drop in alkalinity was such that ferric chloride became less effective. Polyaluminum chloride was found to be more effective in these cases. Alternative coagulants were included in the design of the facilities. (Melcer *et al.* 2012; Davis *et al.* 2014, 2015).

Figure 4.43 presents a compilation of various experiences with CEPT. The data includes jar test, pilot scale and full-scale results. The spread of results is likely influenced by the range of wastewater characteristics of the different wastewaters. The data includes the use of metal coagulants alone, metal coagulant plus polymer, and the use of dual metal coagulants plus polymer. The dual metal coagulants include ferric chloride followed by a polyaluminum chloride product (aluminum chlorohydrate). Discrete groupings of performances related to these three coagulation conditions are not evident in the results.

Part of the spread of results is also due to different coagulation, flocculation and settling conditions in each case. For example, Figure 4.44 shows the trend of TSS removals in one case at different SORs applied during CEPT in full-scale primary clarifiers. In general, TSS removals reduce at higher SORs, as expected.

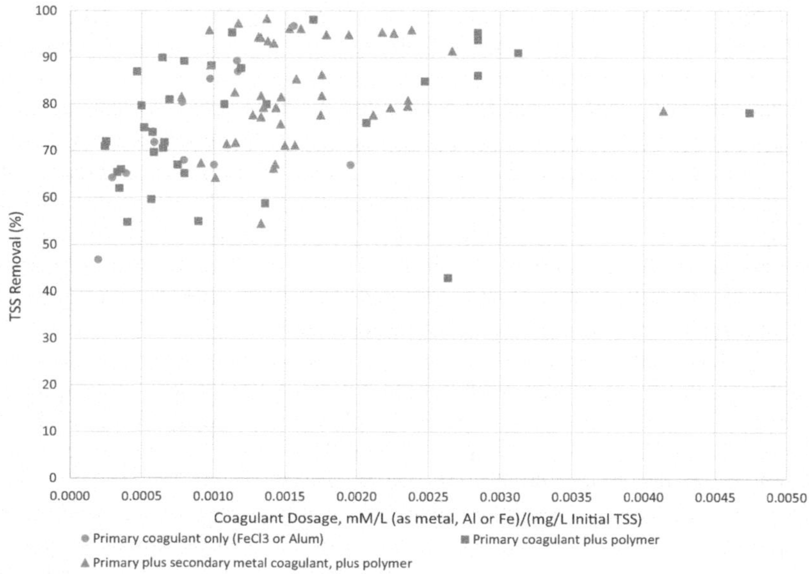

Figure 4.43 Synthesis of various CEPT results.

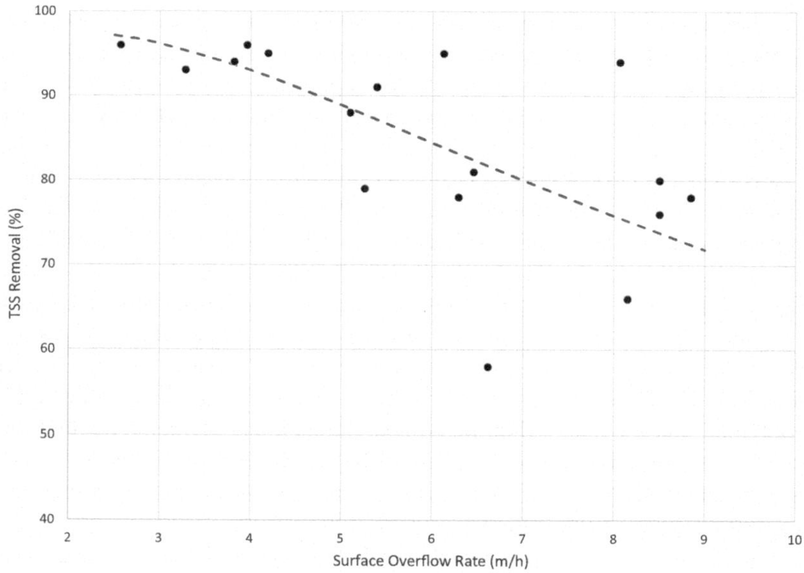

Figure 4.44 Effect of SOR on TSS removals during CEPT (Melcer, 2005, 2014; Fitzpatrick, 2012).

Also shown in Figure 4.44 is a relationship of the type suggested by Fitzpatrick (2012):

TSS removal (%) = (Maximum TSS removal) × $(1\text{-}e^{-\lambda/SOR})$

and (Maximum TSS removal) = $100(1\text{-}TSS_{non}/TSS_{inf})$;

TSS_{non} = influent non-settleable TSS concentration (mg/L);

TSS_{inf} = influent TSS concentration (mg/L);

(assume TSS_{non}/TSS_{inf} = 0.02 with CEPT, in this case)

λ = settling parameter (assume = 11.88 m/h or 7,000 gpd/ft^2)

SOR = surface overflow rate (m/h or gpd/ft^2)

4.10.3 Parameters for CEPT control

One of the issues with chemical treatment of wastewater, including CEPT, is dosage control of the coagulants. Leentvaar *et al.* (1979) investigated the dependence of coagulation on a number of raw wastewater parameters, to optimize TOC removal. These parameters included pH, alkalinity, conductivity, transmission, suspended solids, TOC, DOC, ortho-P, total P, and TKN. By using multiple regression analysis, the best control parameter was determined to be total phosphorus. The orthophosphate concentration of the raw wastewater was the second best parameter for control. For the particular wastewater used in their pilot plant experiments, the ferric chloride dosage was given by

Ferric chloride dosage = 24.9 + 3.11(ortho-P) mg/l as Fe

where ortho-P = orthophosphate concentration in raw wastewater, mg/l as P.

Leentvaar *et al.* postulated that the reason for the good correlation between orthophosphate, coagulant dosage, and TOC removal was because the metal-phosphorus complexes form the nucleus for the adsorptive and sweep coagulation reactions that occur. Related to this are the findings of Tenney and Stumm (1965) who found that metal coagulants first react with phosphate before destabilizing particulate matter.

Ødegaard (1992) reported that in Norway, typical control strategies were as follows:

- Flow proportioned (31% of plants)
- Flow proportioned with pH override (52% of plants)
- Flow proportioned with conductivity override (12% of plants)
- By schedule or empirical program (5% of plants)

4.10.4 Degree of flocculation required for CEPT

An important issue with the physical-chemical treatment of wastewater is the degree of flocculation required before solid-liquid separation. Bratby (1974) found that formal flocculation was unnecessary for good performance using dissolved air flotation after adding metal coagulant at the rapid mix stage for algae removal from oxidation ponds: an in-line 14-second reaction time after coagulant addition was sufficient for efficient flotation.

Similarly, when treating domestic wastewater with coagulation and flotation, a 30-second in-line residence time after metal coagulant addition was required for optimal performance. In this case, removals were TSS 97%, COD 84% and total phosphorus 92%, using alum without polymer, at a dosage of 6.3 mg/l as Al (Bratby, 1982).

In contrast to these experiences, Ødegaard (1992, 1995, 2001) advocated a high degree of flocculation. For sedimentation, flocculation in two to four stages was recommended, with total residence times of 20 to 30 minutes and tapered velocity gradients, G of 40 to 20 s^{-1}. For dissolved air flotation, the recommendation was for total flocculation residence times of 25 to 30 minutes, divided into at least two chambers. G values of 60 to 80 s^{-1}, without the need for tapering, were also recommended.

Krugel *et al.* and Melcer *et al.* (2005) achieved excellent results without extensive flocculation when treating peak flows with chemical addition before primary sedimentation tanks. Using a staged coagulation approach, ferric chloride was added in-line into an existing distribution channel, using an air-sparged diffuser. At peak flows, the residence time in the channel after ferric chloride addition was 20 seconds, at which point polyaluminum chloride was added, also with an in-line diffuser. The point of PACl addition was at the entrance to an aerated grit tank. At peak flow the residence time in the grit tank was 6 minutes. At the end of the grit tank, and at the entrance to the primary clarifiers, an anionic polymer was added in-line, using the same type of diffuser. The residence time in the channels between polymer addition and the primary inlets was 2 minutes at peak flow. At a primary clarifier surface overflow rate of over 6 m/h (3,600 gpd/ft^2) TSS and BOD removals up to 95% and 68% were achieved, respectively. At higher overflow rates of up to 8.8 m/h (5,200 gpd/ft^2), TSS and BOD removals up to 80% and 50%, respectively were achieved.

Newman *et al.* (2013) investigated the optimum coagulation time after ferric chloride addition and before polymer addition for CEPT applied to raw sewage in primary clarifiers at the Hartford Water Pollution Control Facility in Connecticut. They found that ferric chloride dosages required 1 to 2 minutes of reaction time before adding 0.5 to 1.5 mg/L of high molecular weight anionic polymer. The flocculation time after polymer addition was 2 minutes.

Davis *et al.* (2014) conducted pilot and full-scale tests at the Southerly, Easterly and Westerly wastewater treatment plants in the Greater Cleveland Metropolitan area. Under normal circumstances a reaction time after metal coagulant addition (ferric chloride) of approximately 90 seconds was sufficient before the addition of anionic polymer. Polymer was added almost immediately before entry to the primary clarifiers. However, during very cold periods when influent wastewater temperatures dropped to less than 6°C and coagulation performance reduced, increasing the metal coagulant reaction time before polymer addition, to approximately 150 seconds was sufficient to compensate for the detrimental temperature effect.

4.11 ACTIVATED SLUDGE BULKING AND FOAMING CONTROL AND ENHANCED BIOFLOCCULATION

There is an essential relationship between activated sludge biological flocs and metal cations. Activated sludge produces extracellular anionic biopolymers, composed principally of proteins and polysaccharides, which are attached to the activated sludge flocs, but also released into the surrounding liquid (Novak *et al.* 1977). The role of cations on activated sludge properties is explained by a divalent cation bridging theory whereby negatively charged sites on extracellular anionic biopolymers are bridged by divalent cations, helping to improve floc properties, settling and dewatering performance (Sobeck & Higgins, 2002).

The action of divalent cations in solution, such as Ca^{2+} and Mg^{2+}, is to overcome the repulsion between the anionic polymers and form inter-particle bridges between adjacent particles, as shown schematically in Figure 4.45. Monovalent cations, such as Na^{+} and K^{+}, are not capable of such charge reduction and, when in large enough concentrations, displace the divalent cations and reduce the inter-particle bridging effect.

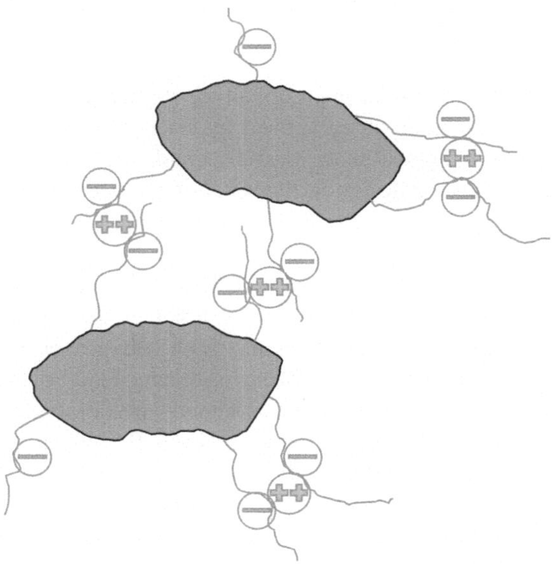

Figure 4.45 Schematic showing the bridging action of divalent cations such as Ca^{2+} and Mg^{2+} on negatively charged sites in extracellular biopolymers (after Sobeck & Higgins, 2002).

The ratio of monovalent to divalent cations (M/D) is a reliable indicator of potential activated sludge bioflocculation problems (Murthy & Novak, 1999, 2001). In general, a M/D ratio greater than 2.0 is expected to cause floc formation problems with higher sludge volume index (SVI) values and dewatering problems, requiring higher polymer dosages to maintain desired dewatered cake concentrations. A demonstrated solution is to add calcium or magnesium ions to the activated sludge to reduce the M/D ratio of the activated sludge mixed liquor.

Higher M/D ratios tend to release higher concentrations of extracellular anionic biopolymers into solution. Higher divalent cation concentrations tend to restrict the release of the biopolymers (Murthy & Novak, 2001). The concentration of these biopolymers in solution exerts a prior demand on cationic polymers used for thickening and dewatering, resulting in an overall increase in the polymer dosages required to achieve thickened and dewatered cake concentrations.

Extracellular material also exerts a demand on the action of trivalent metal coagulants. In this case a further demand is exerted by the solution phosphate concentration since the metal ion has a stronger affinity for phosphate than for the ionogenic groups on the extracellular material (Tenney & Stumm, 1965). Metal coagulants have been shown to be effective at overcoming difficulties caused by filamentous organisms. A bulking sludge caused by unidentified filamentous organisms was controlled by the addition of alum (Finger, 1973). It appears that the dosage of alum was approximately 50 mg/L for 3 days, controlled by maintaining the mixed liquor pH at 6.5. After alum addition the SVI reduced from 130 mL/g to 75 mL/g and filamentous organisms were no longer evident in the microscope.

A major wastewater treatment plant operating challenge arises from the onset of biological foaming, usually characterized by viscous, chocolate-colored stable foam at the surface of biological reactors and secondary clarifiers. The problem is particularly severe when the design of the reactors includes sub-surface passages and prevents the free passage of the floating foam from one compartment to the next, or to the clarifiers.

Narayanan *et al.* (2010) demonstrated that the foaming propensity of nocardioforms such as *Gordonia amarae* came less from floc-bound filaments but more from free-floating dispersed filaments. With sub-surface withdrawal from biological reactors, dispersed, free-floating filaments tend to proliferate. With surface withdrawal, floc-bound *G. amarae* filaments predominate (Blackall *et al.* 1991). By dosing either 0.5 mg/L cationic polymer (based on influent flow) or 2 g/kg MLSS.d of PACl, the total nocardioform filament counts did not reduce, but the dispersed filament counts did reduce by incorporating the dispersed filaments into the activated sludge flocs, converting the system to floc-bound filaments and reducing the ability to generate foam.

Mamais *et al.* (2007) tested various coagulants and found that the dosages of various coagulants required to reduce the total volume of foam produced during foam tests by 80% were as follows:

- Ferrous chloride: 32 g Fe^{2+}/kg MLSS
- Ferric chloride: 25 g Fe^{3+}/kg MLSS
- Alum: 30 g Al/kg MLSS
- Polyaluminum chloride: 6.6 to 11.5 g Al/kg MLSS (dosage depends on the specific product)
- Cationic polymer: 4.4 g/kg MLSS.

Polyaluminum chloride (PACl) was the most efficient coagulant. It produced more dense and compact flocs, and embedded *Microthrix parvicella* and *G. amarae* filaments within the flocs, thereby limiting their ability to assimilate colloidal substrate due to diffusional resistance. This general effect of PACl, on *M. parvicella* and *Nostocoida limicola*, was also reported by Linderoth *et al.* (2004).

Nitrifying and heterotrophic bacteria appear to be unaffected by PACl addition at the recommended dosages (Mamais *et al.* 2007; Rossetti *et al.* 2005). The mechanism by which PACl compounds affect *M. parvicella* is not fully understood although it appears that the compound inhibits the ability to assimilate substrates such as lipids by decreasing the activity of exoenzymes such as lipase, or by sequestering the lipid substrate by coagulation and entrapment in flocs (Rossetti *et al.* 2005).

Roels *et al.* (2002) reported on experiences at 9 full-scale wastewater plants in reducing high SVIs and foaming problems caused principally by *M. parvicella* using a specific PACl product. This product is described as a highly charged PACl compound, comprising 7.2% aluminum with a basicity of 26%. The experiences showed that treatment dosages should be 1.5 to 4.5 g Al/kg MLSS.d. Laboratory batch tests showed that there is a critical dosage between 150 and 250 μl/L at an MLSS concentration of 2,000 mg/L, or a dosage of 7 to 12 g Al/kg MLSS. At 250 μl/L the protozoa were seen to be less active and at 500 to 1,000 μl/L death of part of the biomass occurred. After 17 to 25 days, alterations in the surface of the filament cells were observed: cells became swollen and some lysis was observed with filaments broken up into smaller threads. A decrease in the amount of filaments was observed within 20 to 25 days. Although the PACl compound was very effective against *M. parvicella*, other filaments including nocardioforms and *N. limicola* were largely unaffected.

The PACl compound should be added to the return activated sludge (RAS) stream for approximately 3 weeks, or more. It is advisable to first remove accumulated scum/foam from the surface of the basins before applying the product. The objective is to reduce further foam causing propensity while avoiding excessive dosages or application times. Returning filaments by intra-plant streams such as thickener returns should also be avoided during treatment.

The effective dosage is dependent on the SRT of the system; the higher active fraction at lower SRTs requires higher dosages. From a number of experiences, a reasonable guide to the required PACl dosage appears to be:

Dosage = 30/SRT (g Al/kg MLSS.d) where SRT = solids retention time (days)

The calculated daily dose is based on the total mass of activated sludge solids in the basins. For example, if the MLSS concentration is 3,000 mg/L and the total basin volume is 7,000 m^3, the total mass of sludge is 21,000 kg.

If the total SRT is 10 days, an estimate of the required PACl dosage is 30/10 = 3 g Al/kg MLSS.d. Therefore, the required PACl dosage is 63 kg Al/d.

Assuming the Al content of the PACl is 7.2%, the required PACl dosage is 875 kg/day of product. By observing the system SVI response to the PACl dosage, the applied maintenance dosage can be increased or reduced depending on the SVI response.

4.12 INORGANICS REMOVAL

Metal coagulant addition can be effective for the removal of inorganics although, in some cases, removals can be incidental to other treatment objectives. Table 4.14 shows the removals achieved of several inorganic contaminants from activated sludge secondary effluent. Results are from pilot plant tertiary treatment directed towards phosphorus and dissolved organic nitrogen removal, using aluminum sulfate and the CoMag process with magnetite and polymer for enhanced flocculation and floc ballast. Other contaminants were monitored but are not presented since they were non-detectable in the samples. Removals of inorganic contaminants generally depend largely on the species and oxidation state in which the contaminants exist. Removals also depend heavily on the extent to which contaminants are adsorbed onto particulates, evidenced by the higher removals when influent samples were stressed by mixed liquor solids.

The following sections discuss the removal of a number of selected contaminants for which coagulation may be applied as a viable treatment option. These contaminants are arsenic, copper, fluoride, and manganese.

4.12.1 Arsenic removal

Arsenic is a commonly occurring metalloid element that enters waterways from both natural and anthropogenic sources. Arsenic occurs naturally in areas of high geothermal activity. However, smelting, petroleum refining, pesticides and herbicides manufacturing, and other industrial activities can introduce significant quantities of this toxic element to the environment.

The health effects of arsenic are well documented and it is regulated based on its impact to human health. Hyperpigmentation, skin and liver cancers, and circulatory disorders are attributed to higher levels of arsenic ingestion, particularly over extended periods. The World Health Organization in 1993 reduced the arsenic limit for drinking water from 50 μg/l to 10 μg/l. In the United States, since passage of the Safe Drinking Water act in 1976, the maximum allowable arsenic concentration in drinking water was 50 μg/l. In 2002 this limit was lowered to 10 μg/l. Some states may adopt arsenic limits below the Federal limit. For example, the State of New

Jersey in 2005 announced a plan to adopt a 5 μg/l limit and the State of California appeared to be considering an arsenic limit of approximately 4 μg/l.

Table 4.14 Metals removal from activated sludge secondary effluent.

Metal	Sample	Alum Dosage (mg/L as Al)				
		2.0	**10**	**15**	**20**	**15 mg/L + 165 mg/L MLSS**
Copper	Influent (μg/L)	10.7	9.7	5.9	7.2	79.8
	Effluent (μg/L)	7.0	6.9	6.9	3.9	6.6
	Removal (%)	**35**	**29**	**(17)**	**46**	**92**
Lead	Influent (μg/L)	0.94	0.81	0.60	0.62	4.7
	Effluent (μg/L)	0.60	0.57	0.51	0.51	0.51
	Removal (%)	**37**	**30**	**15**	**18**	**89**
Nickel	Influent (μg/L)	4.5	4.7	4.4	4.5	7.1
	Effluent (μg/L)	4.0	4.0	3.9	4.1	4.2
	Removal (%)	**11**	**15**	**12**	**11**	**41**
Selenium	Influent (μg/L)	17.6	17.5	16.5	17.0	34.7
	Effluent (μg/L)	14.6	13.9	12.9	13.3	14.2
	Removal (%)	**17**	**21**	**22**	**22**	**59**
Zinc	Influent (μg/L)	43.4	41.4	38.3	48.3	109
	Effluent (μg/L)	38.1	34.0	35.7	31.0	37.8
	Removal (%)	**12**	**18**	**7**	**36**	**65**

Arsenic is stable in several oxidation states, under different redox conditions in water. However, when present in groundwater, arsenic occurs mostly in the forms of arsenite, As(III) and arsenate, As(V). As(III) is usually the predominant form in many groundwaters since it is more likely to be found in oxygen free (anaerobic) conditions. As(V) is more common in aerobic waters.

The toxicity of arsenic varies greatly according to its oxidation state. For example, As(III) is far more toxic than As(V). Both As(III) and As(V) are more toxic than organic forms of arsenic, more commonly present as contaminants in surface waters. Both As(III) and As(V) are dissolved species in water. In most treatment technologies, As(V) is more readily removed than As(III). Therefore, if As(III) is the predominant form, oxidation may be required to convert As(III) to As(V) and increase overall arsenic removal efficiency. As(III) is oxidized to As(V) in the presence of oxygen, but in practice is effectively oxidized by chlorine, potassium permanganate and ozone. Chlorine dioxide is ineffective for arsenite oxidation.

There are various technologies used for removing arsenic from drinking water. These include adsorption on granular iron based media; adsorption on ion exchange resins; adsorption on activated alumina; coprecipitation in iron removal plants; coagulation with alum or ferric followed by conventional filtration; and coagulation with ferric followed by membrane filtration. Table 4.15 summarizes the general effectiveness of treatment technologies for arsenic removal.

Table 4.15 Effectiveness of treatment technologies for arsenic removal.

Treatment Technology	Removal Efficiency	
	As(III)	As(V)
Ion Exchange	–	+++
Activated alumina	+/++	+++
Porous media sorbents (iron oxide coated sand, greensand, etc.)	+/++	++/+++
Coagulation with alum	–	+++
Coagulation with iron salts	++	+++
Lime softening	+	+++
Membrane methods	–/+++	+++

Key: +++ (Consistently >90% removal).
++ (60 – 90% removals).
+ (30 – 60% removals).
– (<30% removal).

With coagulation for arsenic removal, iron based coagulants are generally more effective than aluminum coagulants. Iron coagulants added to water hydrolyze to form ferric hydroxide with a net positive charge. This net positive charge is a function of pH. As the pH decreases, the number of positively charged sites on the ferric hydroxide particles increases. Arsenate, As(V) is an anion and since it is negatively charged, it will adsorb to the positively charged ferric hydroxide particles by surface complexation. Arsenic removal is generally optimized at pH values of less than approximately 7.0 (Chwirka *et al.* 2004).

Fan *et al.* (2003) tested various aluminum and ferric based coagulants for the removal of arsenic(V) from water. Of the aluminum coagulants, the efficiencies of arsenic(V) removals were in the order polyaluminum chloride > polyaluminum sulfate > aluminum chloride > aluminum sulfate. Best results were obtained at a pH of 5.5. With an initial arsenic concentration of 50 μg/l and a combination of coagulation and sedimentation, As(V) removals ranged from approximately 41 to 89-percent at dosages of 0.8 to 1.9 mg/l as Al. By adding filtration after sedimentation, corresponding removals increased from 59 to 99-percent.

Using ferric coagulants under the same conditions, the efficiencies of arsenic(V) removals were in the order polyferric chloride > polyferric sulfate > ferric chloride > ferric sulfate. Best results were also obtained at a pH of 5.5. With an initial arsenic concentration of 50 μg/l and a combination of coagulation and sedimentation, As(V) removals ranged from approximately 44 to 98-percent at dosages of 1.7 to 3.8 mg/l as Fe. By adding filtration after sedimentation, corresponding removals increased from 70 to 99.6-percent.

The results confirmed that ferric coagulants are generally superior to aluminum coagulants for arsenic removals.

Smith *et al.* 2002 showed that coagulation followed by microfiltration provided the best overall removal efficiencies of the coagulation assisted systems tested. Microsand ballasted sedimentation (without filtration) also provided good results, similar to conventional sedimentation treatment, but generally not as efficient as microfiltration at lower iron dosages. Initial As concentrations of 45 μg/l were removed by over 99-percent by the addition of 8 mg/l Fe.

Chwirka *et al.* (2000) compared the overall costs of three alternative processes for arsenic removal from groundwater: ion exchange, activated alumina adsorption, and coagulation-microfiltration. All processes were able to reduce initial arsenic levels of approximately 40 μg/l to less than 2 μg/l. Although first implementation costs for ion exchange were the lowest of the three alternatives, facilities for brine handling and disposal were the highest. This caused the costs for this alternative to be the highest. The lowest overall cost was the coagulation-microfiltration alternative. The relative ratios of implementation costs for the three alternatives were 1.26 : 1.10 : 1.0, respectively. The relative ratio of operations and maintenance costs were 1.64 : 1.63 : 1.0, respectively.

Chwirka *et al.* (2004) tested the coagulation-membrane filtration process further on a groundwater with a total arsenic concentration of 25 μg/l. Of this, 10 μg/l was As(III) and 15 μg/l As(V). Treatment conditions were optimized with 1 mg/l chlorine to oxidize As(III) to As(V) followed by 10 mg/l $FeCl_3$ and 35 mg/l CO_2 to adjust pH to 6.8. A flux rate of 153 l/m^2.h (90 gfd) was maintained through the microfilters. Treated water arsenic concentrations were less than 2 μg/l.

The removal percentage of arsenic(V) by coagulation processes is largely insensitive to initial arsenic concentration. Fan *et al.* (2003) showed that for initial As(V) concentrations varying from 6 to 160 μg/l, the removal percentages of As(V) were constant, within 1 to 2-percent. This suggests that there may be merit in two-stage arsenic removal by coagulation, particularly with high initial arsenic concentrations.

Figure 4.46 presents a synthesis of arsenic removal results obtained by a number of workers. The approximate average trend through the results can be expressed as:

$$\text{M metal added/M As removed} = 17.0/(\text{effluent As, mg/L})^{0.3117}$$

The relatively large spread of results reflects differing degrees of oxidation, As(III) contents, and coagulation pH. However, the general trend of results does suggest that if a relatively high initial As concentration is to be reduced to low concentrations, the overall percentage removal could be achieved with lower coagulant dosages by two-stage treatment. This is somewhat analogous to the benefits of two-stage treatment for phosphorus removal discussed in an earlier section.

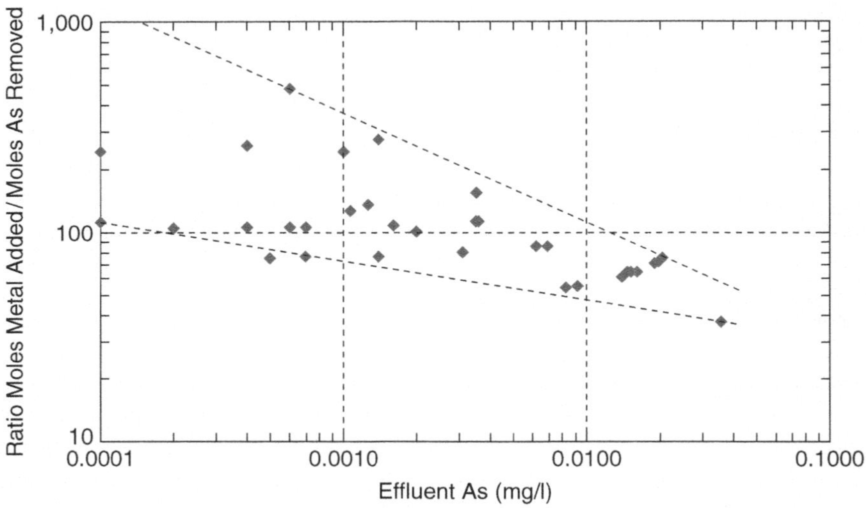

Figure 4.46 Synthesis of arsenic removal results.

For all methods of arsenic removal, consideration must be given to the disposal of residuals. Residuals containing arsenic may be considered a hazardous waste under RCRA (the Resource Conservation and Recovery Act) which means that, for disposal in a landfill, the leachate produced by the Toxicity Characteristic Leaching procedure (TCLP) must contain 5.0 mg/L of arsenic, or less. Blending of the arsenic residuals with other wastewater sludge may be one disposal option considered on a case-by-case basis.

4.12.2 Copper removal

Wastewater treatment plant (WWTP) effluents are often regulated for metals concentrations related to projected aquatic toxicity in receiving waters. The toxic effect of metals is complex and dependent on a number of factors, including the ratio of dissolved to total metal species, pH, hardness, TSS and concentrations of substances in the water that can bind to the metal and reduce the potential toxicity. Regulated discharge concentrations of metals are case specific. In the case of copper, examples of regulated limits at specific WWTPs include 8.5 μg/L chronic (average monthly) and 13.7 μg/L acute (maximum daily); 18 μg/L chronic and 27 μg/L acute; and an annual 95-percentile limit of 4.9 μg/L.

Swift *et al.* (2006) showed that for the City of Puyallup, WA, approximately 70% of the copper in wastewater could be attributed to corrosion of domestic plumbing materials. Other sources were minor in comparison. They compared drinking water sources with WWTP influent and effluent for a number of municipalities. For the City of Puyallup they found the following average values:

- First-draw tapwater (standing more than 6 hours): Cu = 444 μg/L
- Drinking water alkalinity: 55 mg/L as $CaCO_3$
- Drinking water pH: 7.1 to 7.2
- Domestic wastewater: Cu = 111 μg/L
- Wastewater treatment plant influent: Cu = 100 μg/L
- Wastewater treatment plant effluent: Cu = 9.46 μg/L

WWTP influent is diluted by infiltration flows and the considerable reductions within the WWTP itself are due to adsorption onto biological flocs. Ekster and Jenkins (2002) found that approximately 43% of influent total copper was removed in the primary clarifiers. All of this removal was due to particulate-bound copper. There was no removal of dissolved copper through the primaries. Secondary treatment (high rate activated sludge) removed approximately 93% of total copper, with 75% removal of dissolved copper and 94% of particulate-bound copper.

A copper profile through the City of Boulder biological nutrient removal activated sludge plant also found significant uptake of copper through biological treatment. Filtered influent to the aeration basins had a copper concentration of 44 μg/L and the secondary effluent had a concentration of 11 μg/L.

Cannon and Yeh (2001) attempted to enhance biological removal of copper, first by stimulating the growth of extracellular polysaccharide material by subjecting the activated sludge organisms to oxygen depleted anoxic conditions, then by adding ferric chloride to complex, sorb and enmesh both dissolved and colloidal copper into the activated sludge flocs. These experiments achieved removals of copper from 300 μg/L to less than 22 μg/L. Without iron or biosolids storage, supernatant copper levels were approximately 35 μg/L.

Despite the significant removals achievable in WWTPs, the overall removals are often insufficient to meet effluent permit limits. For this reason several workers have investigated copper removal using specific precipitants. Table 4.14 shows that metal coagulants, although effective to a degree may not provide sufficient removals to meet effluent limits in some cases. Sedlak *et al.* (2000) pointed out that metals, including copper are strongly complexed in wastewaters, including by chelating agents such as EDTA, thereby resisting further complexation by metal-hydroxides.

Gerhardt et al. (2002) compared the use of ferric chloride and two proprietary sulfur based organic polymers specifically formulated for metals precipitation. Ferric chloride did not provide sufficient copper removals. When ferric chloride was added to a headworks sample, residual dissolved copper reduced from approximately 36 μg/L to 27 μg/L. However, when adding the organic precipitant to a headworks sample, residual dissolved copper reduced from an initial concentration of 44 μg/L to approximately 9 μg/L. Full-scale trials added the organic precipitant to the secondary clarifiers and confirmed its effectiveness in meeting the effluent copper limit.

Tests carried out at the City of Hayward treatment plant also showed that a proprietary organic precipitant was effective at reducing dissolved copper concentrations from microfiltration backwash and reverse osmosis membrane concentrate streams (Walkowiak *et al.* 2002). However, poor flocculation of the precipitates occurred which reduced the overall copper removals. To remedy this a combination of chemicals were added, comprising first, a cationic polymer, followed by the organic metals precipitant, then followed by an anionic polymer to produce good settling in the clarifier.

Jar testing was carried out using mixed liquor from the solids contact aeration basins from the City of Boulder TF/SC WWTP (Bratby, 2006; Linenfelser *et al.* 2006). The nature of the TF/SC flocs is such that capture by flocculation of the precipitates formed by the proprietary sulfur based organic polymer was inadequate, although it was evident that dissolved copper was being precipitated. A combination of the organic metals precipitant with an aluminum chlorohydrate (PACl) coagulant, followed by a high molecular weight anionic polyacrylamide polymer provided good flocculation and settling and lowered the supernatant copper to the required levels.

4.12.3 Fluoride removal

Fluoride is considered an essential nutrient in human metabolism and is added to many drinking water supplies in small quantities, up to 1.5 mg/L, to strengthen tooth enamel and protect against dental caries. Excessive concentrations, however, are detrimental to human health. At 1.5 to 4 mg/L, dental fluorosis occurs with mottling and embrittlement of teeth and, at 4 to 10 mg/L or more, dental fluorosis progresses to skeletal fluorosis, with crippling bone changes. Natural waters most commonly have low fluoride concentrations between a few μg/L to 1 or 2 mg/L. The World Health Organization limit for fluoride is 1.5 mg/L. In some cases, natural waters contain excessive concentrations and fluoride removal is required.

Effective technologies for fluoride removal include adsorption using bone char activated carbon, reverse osmosis, nanofiltration, electrodialysis, and adsorption with variations of impregnated activated alumina using alum, manganese dioxide, magnesium oxide, iron oxide, and calcium (Mohapatra *et al.* 2009). Selection of an appropriate treatment process for fluoride removal is case specific and depends on water characteristics, such as the presence and concentrations of competing ions, and the raw water fluoride concentrations. In some cases, particularly those with relatively modest elevated fluoride concentrations, coagulation may be an appropriate treatment approach.

Fluoride is typically removed by aluminum coagulants by adsorption and co-precipitation with the metal-hydroxide. The extent of fluoride removal is affected by the coagulant concentration and the pH. Culp *et al.* (1958) and Cushing *et al.* (2000) showed that for the respective waters tested the optimum pH for fluoride removal using aluminum sulfate was approximately 7.0.

Sollo *et al.* (1978) determined in their tests that the optimum pH for fluoride removal using alum was in the range 6.2 to 6.4. They also found that the addition of calcium assisted the alum but resulted in only small improvements in fluoride removal. They determined that the use of coagulation was viable for waters containing less than 4 to 5 mg/L fluoride, where relatively small amounts of fluoride needed to be removed. At higher initial concentrations, the dosage of alum would be excessive, making other treatment approaches preferable. Cushing *et al.* also concluded that to reduce an initial fluoride concentration of 2.5 mg/L, coagulation treatment was a viable option.

Culp *et al.* applied an alum dosage of 225 mg/L to reduce an initial fluoride concentration of 3.6 mg/L to 1.5 mg/L. Recycling of alum sludge with or without the addition of fresh alum was ineffective and they concluded that the sludge did not have further capacity to adsorb fluoride.

Sollo *et al.* compared results with alum and sodium aluminate and showed that alum was only slightly more effective on an equivalent aluminum basis. Cushing *et al.* similarly concluded that the form of aluminum added (alum, sodium aluminate, or PACl) made little difference to fluoride removals when considered in terms of the mg/L aluminum added.

At the optimum pH of 7.0, Culp *et al.* found that the residual aluminum was a minimum and, therefore, not considered a treatment issue. This was confirmed in the tests of Cushing *et al.* who also found that optimal turbidity removal corresponded with the pH required for minimum aluminum residual.

Several workers pointed out the negative effects of fluoride on turbidity and organics removal, and residual aluminum (Feld *et al.* 2000; Cushing *et al.* 2000; Pommerenk & Schafran, 2002). Fluoride exerts a coagulant demand thereby reducing the effectiveness of coagulation on water treatment objectives. The stable complexes formed by fluoride and aluminum remain in solution and can increase residual aluminum in finished water. By optimizing pH and alum dosage for fluoride removal, improvements in finished water quality and residual aluminum concentrations are achieved. Cushing *et al.* pointed out the anomaly that in the presence of fluoride, high final residual aluminum levels can be reduced by increasing the alum dosage. Higher alum dosages (at the correct pH) form aluminum-hydroxide flocs that adsorb fluoride and reduce the soluble aluminum-fluoride complexes formed at lower alum dosages. It is the stable soluble aluminum-fluoride complexes that manifest as residual aluminum in the finished water.

Feld *et al.* (2000) and Pommerenk and Schafran (2002) showed that prefluoridation significantly impacts water treatment quality. Fluoride addition prior to or at the point of alum addition interferes with turbidity and NOM removal and results in increases in final TOC, residual aluminum levels, and filtered turbidity breakthrough. Filter run times are reduced because of poor floc formation and increased settled water turbidity. The final fluoride concentration is also reduced because of partial adsorption onto flocs. Minimal impact of fluoridation on finished water quality occurs when fluoride is added after filtration.

Figure 4.47 presents a synthesis of fluoride removal results by coagulation obtained by a number of workers. The relatively large spread of results probably reflects differing water qualities, such as the cations associated with the fluoride ions in the water. However, the general trend of results does indicate the inordinate increase in coagulant dosage required as the effluent fluoride concentrations are progressively lowered.

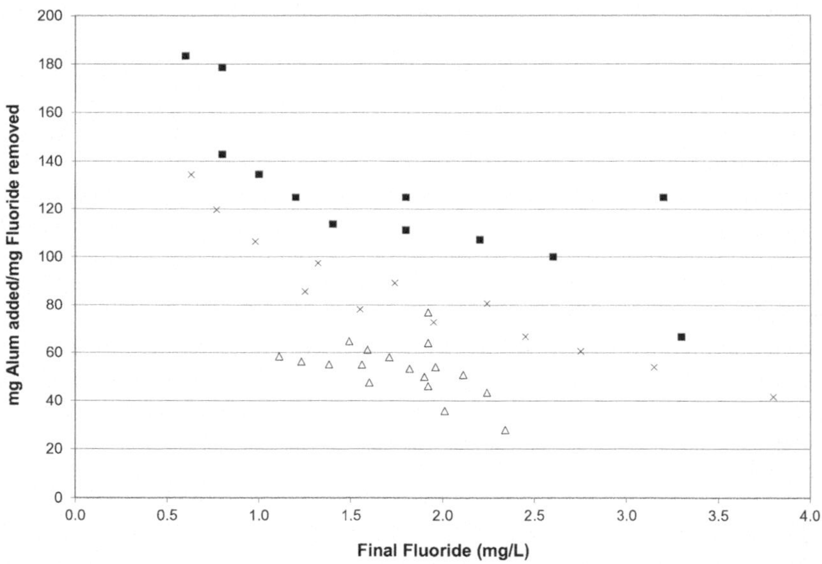

Figure 4.47 Synthesis of fluoride removal results using alum.

4.12.4 Manganese removal

Many groundwaters and some surface water supplies contain appreciable quantities of iron and manganese which require efficient removal prior to distribution. In the case of manganese, concentrations as low as 0.05 mg/L have given rise to customer complaints, making this contaminant particularly critical during water treatment. Unlike the case of iron removal alone, waters containing soluble manganese are not effectively treated by aeration, requiring strong oxidants for precipitation. The most stable oxidation state for manganese is Mn^{2+}. Manganese is typically removed by first oxidizing Mn^{2+} to Mn^{4+} and thence to manganese dioxide, MnO_2. The precipitates can then be removed using coagulating chemicals. The most commonly used oxidant is chlorine. However, chlorination typically results in slow precipitation of manganese dioxide unless the pH is maintained above 8.5 to achieve effective manganese removal.

For waters containing appreciable quantities of soluble manganese, a common sequence of pretreatment is prechlorination, followed by the elevation of pH (usually with lime), then the coagulation of suspended material and precipitates with ferric salts. Although the oxidation of soluble manganese is often efficient with this sequence of treatment, final water quality is often poor, with a relatively high final turbidity. The reason is that although high pH values are necessary for manganese oxidation with chlorine, optimal destabilization of the suspended material occurs at pH values lower than neutral.

Tests on two manganese-bearing waters from different catchments determined that the use of copper sulfate as catalyst for the oxidation reactions, determined elsewhere to reduce the pH required for oxidation (James, 1971), was found to be ineffective. Therefore, a pH of approximately 9.5 was required for satisfactory manganese oxidation using chlorine, as shown in Figure 4.48 (Bratby, 1988, 1990).

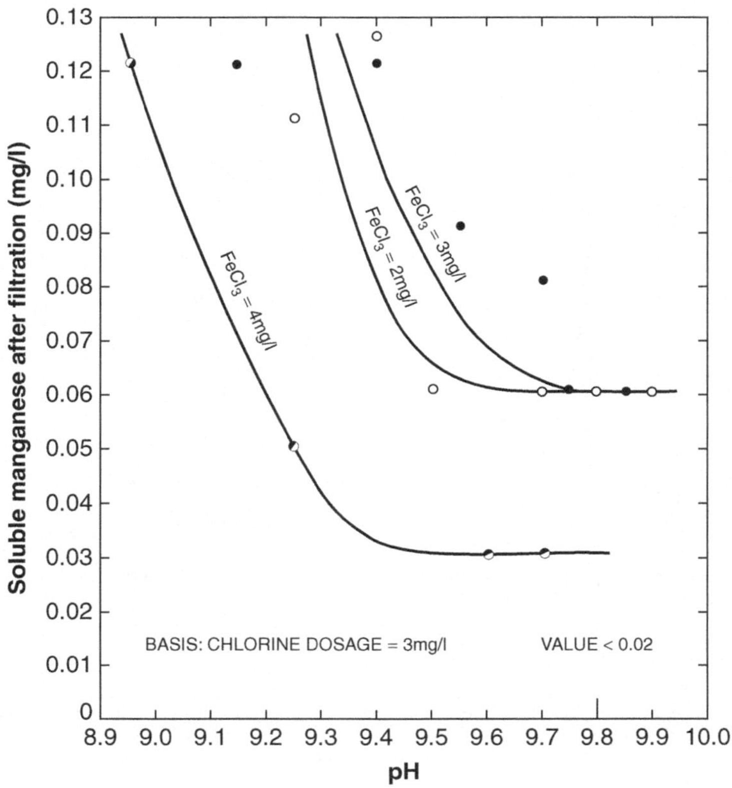

Figure 4.48 Optimization of pH for manganese oxidation (Bratby, 1990).

It was also demonstrated that when using lime for pH adjustment, a detrimental intermediate pH range of around 8.5 produced worst results in terms of filtered turbidity (see Figure 4.49), although this approximate pH is often cited in the literature as adequate for manganese oxidation. As expected, using ferric chloride, filtered turbidity results at lower pH were superior to those obtained at higher pH, as shown in Figure 4.50.

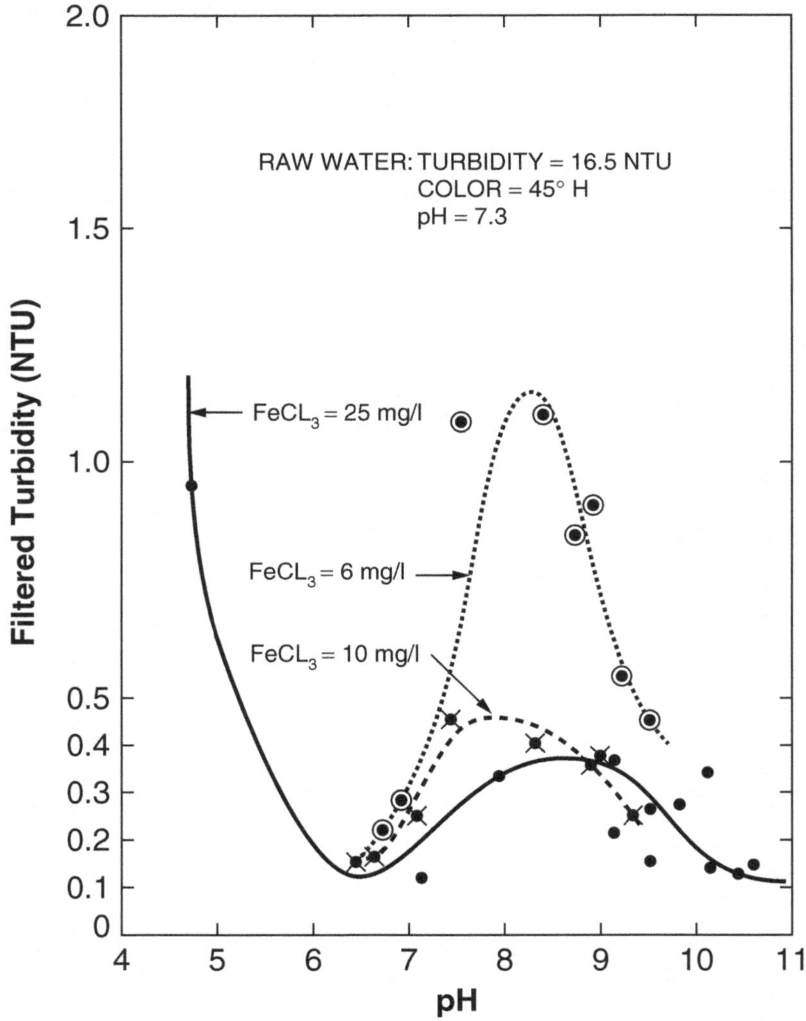

Figure 4.49 Effect of pH on filtered water quality (Bratby, 1990).

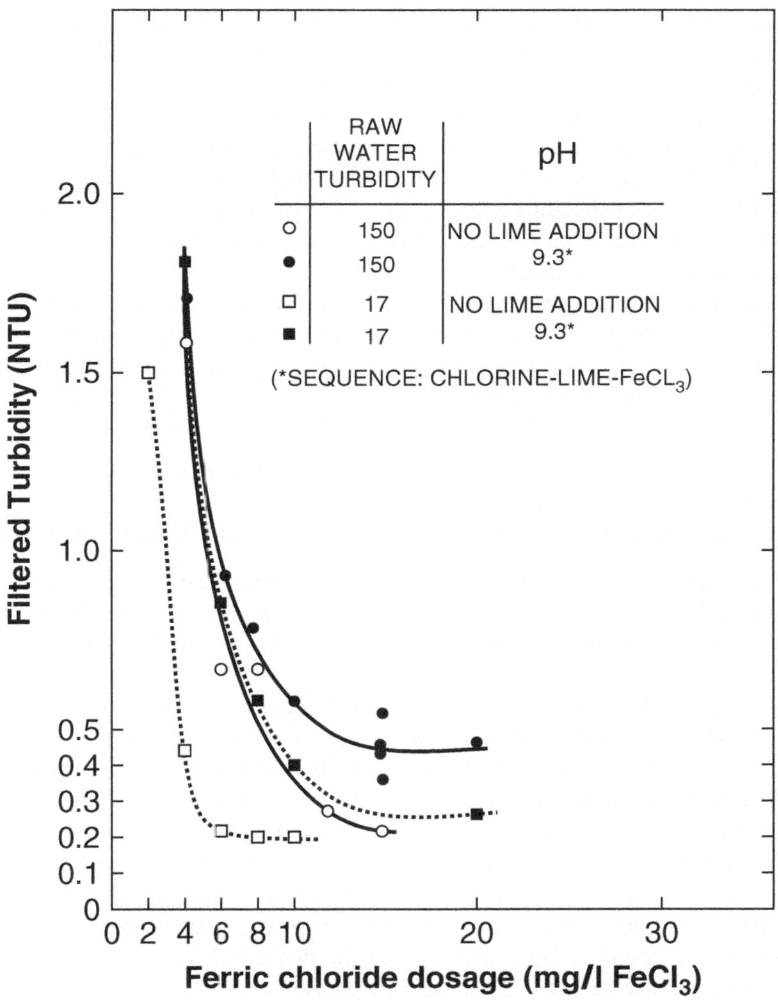

Figure 4.50 Filtered water qualities at low and high pH (Bratby, 1990).

The use of sodium hydroxide for pH adjustment did not result in the intermediate detrimental pH range – the higher the sodium hydroxide dosage and the higher the pH, the higher were filtered turbidity and color values.

It was postulated that at near neutral pH values, positively charged iron hydrolysis species are present with a high adsorption propensity and a high destabilization efficiency for the removal of turbidity causing material. By progressively increasing pH with sodium hydroxide, the positive charges on iron species are reduced and eventually reversed, thus progressively reducing the efficiency of destabilization.

When lime is added for pH adjustment, there is an additional effect from the calcium cations introduced. When the pH is raised with lime, an effect similar to that with sodium hydroxide is initially evident: increasing pH progressively reduces destabilization efficiency. However with increasing pH, manganese precipitates are increasingly formed. These precipitates, although normally negatively charged, possess a high propensity for cation adsorption (Jenkins *et al.* 1984; Posselt *et al.* 1968). By progressively adding lime, manganese precipitates adsorb calcium and thereby decrease the overall negative charge of the precipitates to a low constant value, beyond a certain calcium concentration.

Therefore, although relatively inefficient iron hydrolysis species are present at high pH values, the presence of sufficient calcium causes an overall destabilization whereby iron species and manganese precipitates are incorporated together with suspended turbidity causing material.

Conclusions drawn from the above investigations were as follows:

- For efficient overall destabilization of suspended material, low near-neutral pH values are preferred;
- For efficient manganese oxidation, high pH values of at least 9.5 are preferred.

To reconcile these conflicting requirements, tests were carried out where, after chlorine addition, ferric chloride was added to destabilize the suspended material at low, near neutral pH, and then pH was post-adjusted to 9.5 for manganese oxidation.

Figure 4.51 demonstrates that consistently superior results were obtained with the inverted sequence of (chlorine – ferric chloride – lime). Treated water turbidities obtained at the high pH value of approximately 9.5 approached the optimum turbidity values obtained at lower, near neutral pH.

Suspended material is destabilized more efficiently by adding ferric chloride before lime because of the predominance of positively charged iron hydrolysis species at lower pH values. When the pH is subsequently raised to precipitate manganese, the preformed microflocs probably serve as nucleation sites for the manganese precipitates. This suggests that the dosage of ferric chloride applied influences the efficiency of manganese precipitation.

Figure 4.52 shows that this supposition appears to be correct. The results present long term data where, in all cases the chlorine dosage was 3 mg/l and always sufficient for manganese oxidation, with final residuals of 0.5 to 1.0 mg/l. pH was maintained at approximately 9.5 throughout and the ferric chloride dosage was varied from zero to 8 mg/l. Values of soluble manganese less than the detection limit of the analytical method used, of 0.02 mg/l, are represented on Figure 4.52 by groups of vertical bars – the numbers in parentheses refer to the numbers of analyses represented in each group.

A clear trend was evident in the results: with increasing ferric chloride dosages, manganese precipitation was more complete. Interestingly, a similar beneficial nucleation effect was not obtained by recycling a portion of the filter washwater.

It is possible that the material in the filter washwater was formed at higher pH values with iron hydroxide precipitates containing negatively charged sites. Such species would not be as beneficial for manganese precipitation as the intermediate positively charged ferric species that are predominant in the results shown in Figure 4.52.

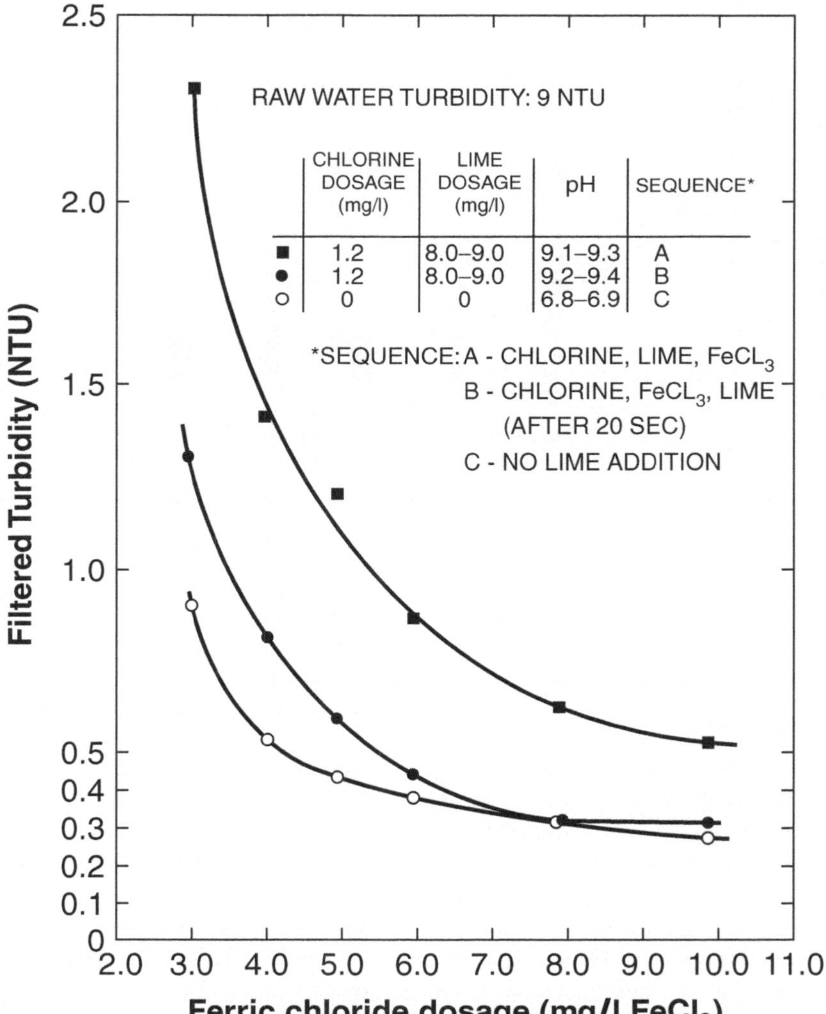

Figure 4.51 Effect of the sequence of chemical addition on filtered water quality (Bratby, 1990).

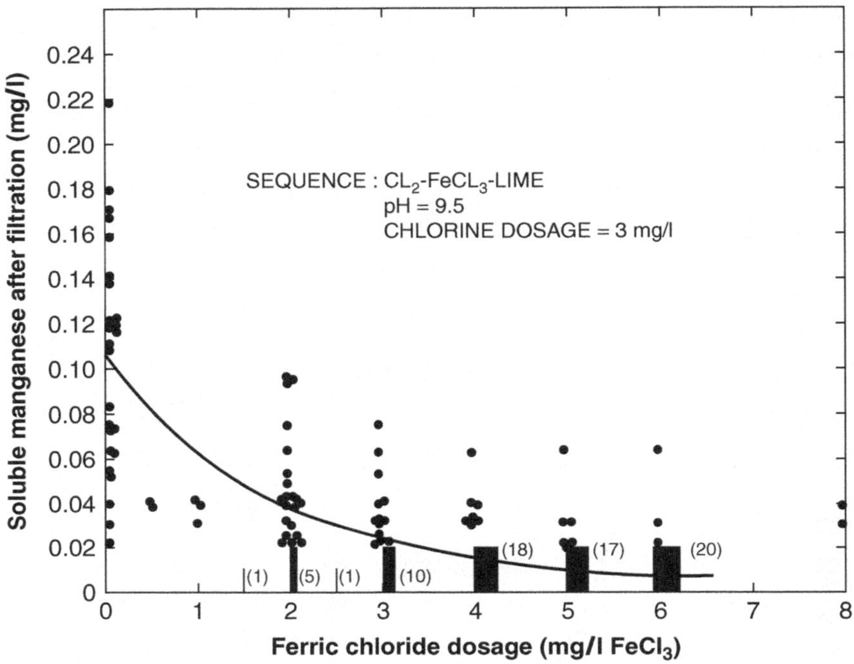

Figure 4.52 Effect of ferric chloride dosage on manganese precipitation (Bratby, 1990).

Figure 4.53 shows that there was an optimum interval between ferric chloride addition and lime addition for adjustment of pH to 9.5. A time lag of 15 to 30 seconds was optimal in terms of final filtered turbidity and observed floc formation.

In summary, conclusions from the studies described above were:

- A pH of 9.5 was optimal for manganese precipitation with chlorine. Because of low alkalinity and calcium concentrations, final conditioning of the treated waters in any case required final pH adjustment to 9.5;
- Copper sulfate, added with the objective of catalyzing the oxidation reactions, was ineffective with the two waters studied;
- Operation at the high pH values necessary for efficient manganese oxidation significantly worsened treated water turbidity. This effect was most pronounced within an intermediate pH range of around 8.5;
- By inverting the sequence of chemical addition, to (chlorine – ferric chloride – lime), treated water turbidities at high pH approached the optimum values obtained at lower pH. Lime should be added 15 to 30 seconds after ferric chloride addition;

- Apart from coagulating suspended material and precipitates, ferric chloride appeared to play a role in providing nucleation sites for manganese precipitation. Recycled filter washwater did not have the same effect.

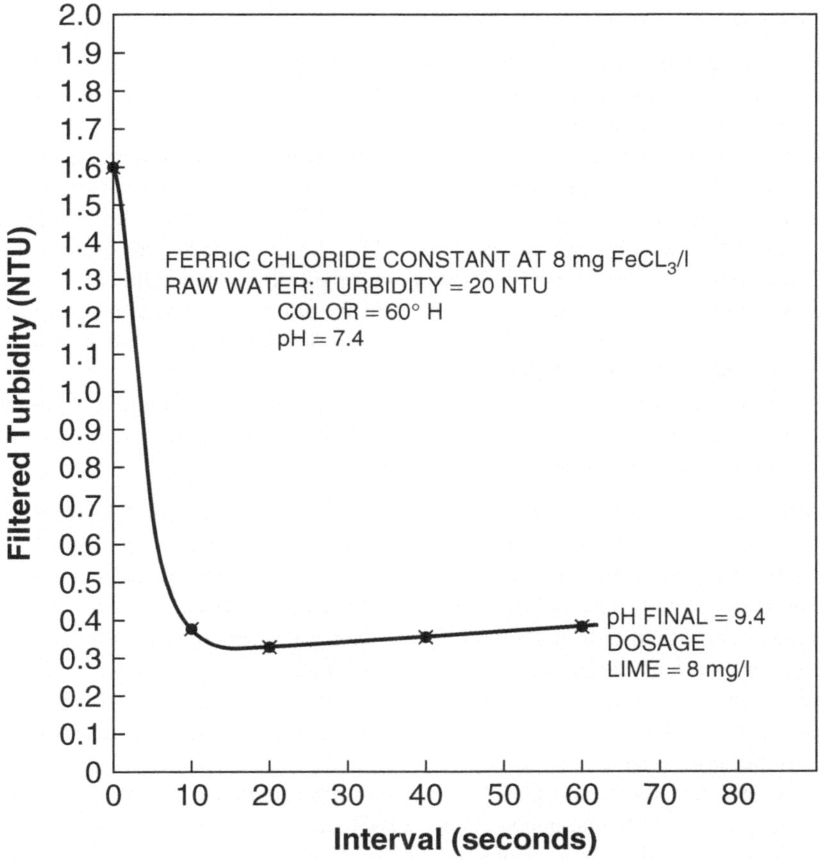

Figure 4.53 Optimization of the interval between ferric chloride and lime addition (Bratby, 1990).

4.13 STAGED COAGULATION AND SEQUENCING

Traditionally, the sequence of chemical addition for coagulation operations is to first add chemicals for pH correction (lime, sodium hydroxide, acid, etc.); then add the metal coagulant; then add the flocculant aid. Not all these chemicals are necessarily added of course, but the sequence logic is often as described. However, there are instances when other sequences are more effective, including inverting the sequence of metal coagulant and polymer addition, and the sequence of metal coagulant addition and pH adjustment.

Metal coagulants and cationic polymers tend to have complementary roles, and various sequences of addition of these two coagulants have been employed (from Ammary & Cleasby, 2004):

- Alum added first at 12 mg/l followed by 0.3 mg/l cationic polymer was found to be equivalent in performance to either 28 mg/l alum alone, or 2.3 mg/l cationic polymer alone.
- For coagulation of a silica suspension (turbidity 70 NTU), the sequence of addition of alum and cationic polymer had no effect on residual turbidity.
- For a natural lake water with turbidity 26 NTU, TOC 8 mg/l, and pH 6.5, alum followed by cationic polymer added 1 minute later gave best results, compared with polymer first then alum, or simultaneous addition.
- Cationic polymer added first, followed by alum was found to be optimal when treating a water with low turbidity, low alkalinity, and low humic content, with TOC < 5 mg/l.
- For a latex suspension, various sequence combinations of alum and cationic polymer were tested. Adding alum first followed by cationic polymer after 5 minutes provided lowest final turbidities.

Ammary and Cleasby (2004) tested the sequence of addition of ferric nitrate and cationic (poly-DADMAC) polymer applied to kaolin clay suspensions, with turbidities ranging from approximately 5 to 90 NTU. At higher turbidities, the use of dual coagulants produced best results. At the lowest turbidities, there was little advantage with the use of dual coagulants. When an advantage was demonstrated with the use of dual coagulants, compared with metal coagulant alone, or cationic polymer alone, adding the metal coagulant simultaneously with the polymer produced best results. Simultaneous addition provided higher floc growth rates with higher floc resistance to shear.

As shown previously in this Chapter, a common sequence of pretreatment for the removal of soluble manganese is prechlorination, followed by the elevation of pH, then the coagulation of suspended material and precipitates with ferric salts. Although the oxidation of soluble manganese is often efficient with this sequence of treatment, final water quality is often poor, with a relatively high final turbidity. The reason is that although high pH values are necessary for manganese oxidation with chlorine, optimal destabilization of the suspended material occurs at pH values lower than neutral.

Bratby (1988, 1990) showed that by inverting the normal sequence of chemical addition, destabilization of suspended material can occur more efficiently at lower pH values, with the subsequent oxidation of manganese at high pH values after a suitable interval. The optimized sequence of treatment was to first prechlorinate without pH adjustment, then add ferric chloride , then add lime 15 to 30 seconds after ferric chloride addition to adjust the pH to 9.5. This inverted sequence resulted in both efficient manganese precipitation and low final effluent turbidities. Apart from coagulating suspended material and precipitates, ferric chloride appeared to play a role in providing nucleation sites for manganese precipitation. Recycled filter washwater did not have the same effect.

In some cases, splitting the dosage of metal coagulant, or staged coagulation, has improved overall results. This has already been discussed in the case of phosphorus precipitation where staging the coagulant feed often results in a lower overall metal coagulant dosage. It was also suggested that the same benefit may be obtained in the case of arsenic removal.

For the treatment of waters with high humic contents, and high TOC, staged coagulation can also be advantageous. In the Cape peninsula and southern Cape coastal areas of South Africa, many of the catchment areas produce waters discolored by the presence of humic and fulvic acids. Water plants in the Table Mountain catchment area used the following sequence of chemicals for treatment of these highly colored waters: 10 mg/l sodium aluminate, followed by lime, followed by 50 mg/l aluminum sulfate. The dosage of lime was adjusted to achieve a pH after alum addition of 5.5. The raw water pH of these waters is typically between 4 and 5 (Bratby & Marais, 1977).

Carlson and Gregory (2000) also applied staged alum coagulation to waters with high TOC contents. Raw water conditions were turbidity 11.5 NTU and TOC 9.1 mg/l. Coagulation pH was maintained at approximately 6.5. They found marked benefits of staged coagulation: the settled water turbidity was consistently 25-percent less compared with one-stage alum addition. Average settled water turbidities were 1.6 NTU and 2.1 NTU for the two-stage and single-stage pilot trains, respectively.

Filtered water particle counts were also approximately 50-percent less with two-stage treatment. The one-stage train also exhibited particle breakthroughs after approximately 6 hours of filter run time. The two-stage train showed no evidence of particle breakthrough up to 10 hours of filter run time.

The mechanism proposed for improved two-stage treatment was that aluminum hydroxide colloids are formed relatively quickly, whereas the sorption reactions of the NOM compounds with the aluminum hydroxide colloids are relatively slow. Therefore, stabilization of the aluminum hydroxide colloids occurs, preventing further agglomeration of the small, poorly settleable and filterable flocs.

The advantage of two stage coagulation is that the second dosage of coagulant serves to further agglomerate the initial particles formed, producing more of a sweep floc regime, with improved settling and filtration characteristics. The mechanism proposed by Carlson and Gregory suggests that staged coagulation may not be as advantageous with low TOC waters.

With the objective of optimizing the removal of TOC, and disinfection byproduct formation from a water in the U.K., Fearing *et al.* (2004) fractionated and isolated the different fractions of DOC and determined the optimum coagulation conditions for each fraction. The objective was to understand and identify best coagulation conditions for each fraction, and the water as a whole. Table 4.16 summarizes their results.

Table 4.16 Optimum coagulation conditions for DOC fractions (after Fearing *et al.*).

Sample Source	DOC Fraction	Percent in Sample (%)	Optimum Dose Ferric Sulfate (mg/l as Fe)	Optimum pH
Raw water	Fulvic acid	55	12	4.8
Raw water	Humic acid	20	8	4.8
Raw water	Hydrophilic acid	16	8	4.2
Raw water	Hydrophilic nonacid	9	15	4.2
Treated water	Fulvic acid	36	8	4.2
Treated water	Humic acid	23	8	4.0
Treated water	Hydrophilic acid	25	8	4.0
Treated water	Hydrophilic nonacid	16	15	3.5

DOC removal at the treatment plant, practicing single stage coagulation using ferric sulfate at the background pH of approximately 6, was 79-percent. Using two-stage coagulation under laboratory conditions, with optimized dosage and pH conditions, the total DOC removal achieved was 80-percent. Although the improvement did not appear to be highly significant, it is not clear whether true side-by-side tests, including actual plant conditions would have identified a larger improvement.

Staged coagulation is also beneficial for chemical treatment of wastewater. Suspended solids removals up to 95-percent have been achieved in primary clarifiers at wastewater treatment plants at clarifier surface overflow rates as high as 10 m/h (6,000 gpd/ft^2), using a sequential, dual coagulant approach (Hanner *et al.* 2004). The sequence of chemicals has been ferric chloride, followed by polyaluminum chloride (PACl), followed by anionic polymer.

Melcer *et al.* and Krugel *et al.* (2005) followed this approach to optimize the performance of chemically enhanced primary treatment of wastewater during storm flow conditions at the South Plant in King County, Washington. Dosages were optimized at 60 to 65 mg/l ferric chloride, 12 to 15 mg/l PACl, and 0.4 to 0.6 mg/l anionic polymer. At primary clarifier overflow rates of 6 m/h (3,600 gpd/ft^2), associated removals were 80 to 95-percent TSS, and 58 to 68-percent BOD. At higher clarifier overflow rates, up to 8.8 m/h (5,200 gpd/ft^2), removals were 65 to 80-percent TSS and 40 to 50-percent BOD. Without chemicals, the corresponding removals were 50-percent TSS and 25-percent BOD.

Rapid mixing of the chemicals in the wastewater stream was considered efficient, using an atomizing injection system. This comprised an air operated eductor that created a turbulent environment after the coagulant-air mixture was passed through 75-mm (3-inch) long diffusers, perforated with 2-mm holes, and mounted at 90-degrees to the direction of channel flow.

Mixing time between the ferric chloride and PACl addition points was approximately 20 seconds at peak flows. Flocculation time before polymer addition was approximately 6 minutes at peak flows. After polymer addition, flocculation time before entering the primary clarifiers was approximately 2 minutes at peak flow.

4.14 EFFECTS OF PREOZONATION

For many years, anecdotal evidence has pointed to ancillary effects when using pre-oxidation before coagulation. In the 1980s these effects were given the name "microflocculation". Since the effects were particularly noticeable when ozone was used as pre-oxidant, they were also referred to as ozone induced particle destabilization. The following summarizes the effects of preozonation, based on the results of various workers:

- Increasing ozone dosage has led to increases in metal residuals for both iron and aluminum based coagulants (Edwards & Benjamin, 1992).
- TOC removal sometimes increased, and sometimes decreased, depending on the ozone and coagulant dosages (Edwards & Benjamin, 1992).
- Overall TOC removal slightly increased with increasing ozone dosage at very high and low coagulant dosages, but decreased at moderate coagulant dosages (Edwards & Benjamin, 1992).
- At constant pH, the dosage of metal coagulants required for optimal particulate removal in the presence of NOM increased with increasing ozone dosage (Edwards & Benjamin, 1992).
- Oxalate is a common ozonation by-product with NOM. It has an adverse effect on turbidity and TOC removals when alum is used as the only coagulant (Becker & O'Melia, 2001).
- When alum was used as coagulant, preozonation reduced removals of turbidity, TOC and DOC (Schneider & Tobiason, 2000).
- When polyaluminum chloride was used as coagulant, preozonation either increased or decreased turbidity, TOC and DOC removals (Schneider & Tobiason, 2000).
- The beneficial effects of microflocculation will not be observed when alum or ferric coagulants are used for moderate to high DOC waters (Becker & O'Melia, 2001).
- Microflocculation may occur using alum or ferric coagulants with low DOC waters (Becker & O'Melia, 2001).
- When organic (cationic) polymer coagulants are used alone, the polymer dosage for turbidity removal is reduced (Tobiason *et al.* 1995).
- When organic (cationic) polymer coagulants are used alone, although turbidity removals increase, NOM removals decrease (Tobiason *et al.* 1995).

- With organic (cationic) polymer coagulants used alone or with metal coagulants, ozonation decreased the quantity of NOM that could be precipitated with coagulant. However less coagulant was needed to precipitate these NOM molecules (Edwards & Benjamin, 1992; Becker & O'Melia, 2001).
- Calcium can reduce the adverse effects of preozonation at dosages as low as 100 mg/l as $CaCO_3$ (Becker & O'Melia, 2001).
- It is unlikely that preozonation will be adopted because of any microflocculation effects. The design of ozone facilities will be based on disinfection and/or oxidation kinetics (Schneider & Tobiason, 2000).
- Intermediate ozonation, upstream of filtration (after coagulation) significantly reduced filtered water particle count – up to one order of magnitude (Becker & O'Melia, 2001).

The reasons for the preozonation phenomena noted above include the following (Edwards & Benjamin, 1992; Krasner *et al.* 1994; Schneider & Tobiason, 2000; Becker & O'Melia, 2001):

- Ozonation converts NOM into smaller compounds (such as oxalic acid) that adversely affect coagulation; ozonation, especially at high ozone-to-DOC ratios, transforms high molecular weight and humic NOM into low MW, non-humic, more hydrophilic, less adsorbable molecules. Therefore, coagulation efficiency can be reduced because of the formation of a higher percentage of NOM that is less amenable to coagulation.
- Ozonation of NOM leads to an increase in the negative charge of the organic molecules that sorb to metal hydroxide surfaces. This increase in negative charge can increase the overall coagulant demand.
- The detrimental effect of preozonation on NOM removal with cationic polymers is explained by a decrease in NOM size and an increase in the degree of hydrophilicity, inhibiting direct precipitation of the NOM by polymers.

Ozone has been beneficial in the case of physical-chemical treatment of the effluent from anaerobic domestic wastewater treatment. Preozonation in this case allowed a reduction in the ferric chloride dosage from 65 mg/l without ozone, to 45 mg/l with 6 mg/l ozone. If the ferric chloride dosage was maintained at 65 mg/l, ozone at 6 mg/l increased removal efficiencies through dissolved air flotation, as follows: COD 74 to 80%; BOD 63 to 79%; Total P 90 to 93%; TKN 28 to 21% (reduction); and turbidity 90 to 97%. In all cases pH was controlled to 5.5 (Marchioretto & Reali, 2001).

4.15 EFFECTS OF TEMPERATURE

Coagulation reactions and metal coagulant chemistry are strongly affected by temperature. For example, with decreasing water temperature, the minimum solubility of aluminum hydroxide species shift to higher pH, and the "optimum" operating pH also shifts to higher pH.

The effect of temperature appears to be more pronounced when destabilization reactions rely on an enmeshment mechanism (Morris & Knocke, 1984). For an adsorption type mechanism, temperature effects were less severe. The effects of temperature on the coagulation of low turbidity waters were particularly severe.

Low temperatures (1°C) did not inhibit the rate of metal hydroxide precipitation. The deleterious effects of temperature were related to changes in floc characteristics. At low temperatures, flocs were smaller thus inhibiting an enmeshment mechanism of particle removal, especially important when turbidities are low.

Low temperatures affected both ferric chloride and alum, but affected ferric chloride to a lesser extent. Therefore, different coagulants are influenced differently by temperature. Figure 4.54 illustrates the detrimental effect of over-dosing alum at low temperatures. This effect was not apparent with ferric chloride.

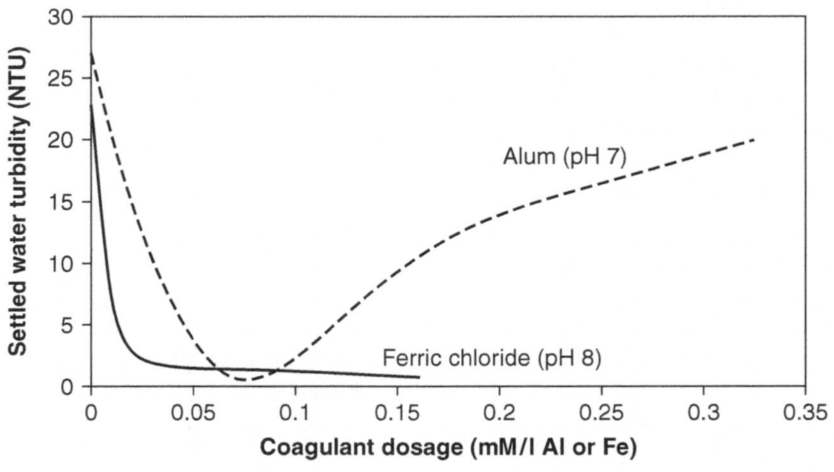

Figure 4.54 Performance of ferric chloride and aluminum sulfate at 1°C (adapted from Morris & Knocke, 1984).

Haarhoff and Cleasby (1988) also showed that on an equimolar basis, ferric chloride removed turbidity more efficiently than alum at low temperatures (approximately 3°C) and low turbidities (less than 2 NTU). However, equivalent results in terms of filter headloss and filtered turbidity were obtained when the molar Fe-to-Al dosage ratio was 3.0:5.6.

Hanson and Cleasby (1990) explained the marked deterioration of alum coagulation at low temperatures on the shift in optimum pH for sweep coagulation at low temperatures. They suggested that if the pH was adjusted based on the equilibrium chemistry of water, and maintained at a constant pOH, then equivalent coagulation performance would be maintained. This was based on the relationship

$$pH + pOH = pK_W$$

where $pK_W = 0.01706 \cdot T + \frac{4470.99}{T} - 6.0875$

and T = temperature in °K = 273.15 + °C

Therefore, if the optimum pH at 20°C is 7.5, at 2°C the pH should be 8.2. The idea of a constant pOH is consistent with the fact that species of importance during metal coagulant hydrolysis are hydroxide precipitates.

Hanson and Cleasby also concluded the following from their work:

- Alum flocs were significantly weaker than ferric chloride flocs under all conditions.
- Iron flocs at 20°C and 5°C had similar strengths if a constant pOH was maintained.
- Iron flocs formed at 5°C were significantly weaker than flocs formed at 20°C if a constant pH was maintained.
- Alum flocs formed at 5°C and constant pOH were considerably stronger than flocs formed at 5°C and constant pH.
- However, even with constant pOH, alum flocs formed at 20°C were stronger than flocs formed at 5°C.

The increase in optimal coagulation pH with decreasing temperature has also been observed with coagulation of humic substances.

Jiang *et al.* (1996) compared the performance of three coagulants for the removal of humic substances, under different temperatures. The coagulants were polyferric sulfate, ferric sulfate, and alum. The performance of polyferric sulfate in terms of DOC removal was not substantially different at 18°C compared with 4°C. However, the performances of alum and ferric sulfate were worsened significantly at lower temperatures.

Residual metal concentrations were also higher at lower temperatures for alum and ferric sulfate, compared with the polyferric sulfate, although the residual metal concentrations could be minimized by increasing the pH at low temperatures, as described above.

Fitzpatrick *et al.* (2004) investigated the performance of different coagulants at temperatures ranging from 7°C to 27°C. Three different polyaluminum chloride compounds, together with aluminum sulfate, and ferric sulfate were tested. Alum and ferric sulfate produced the smallest sized flocs. However, ferric sulfate showed less variation of floc size with temperature, compared with the aluminum based coagulants. This could make ferric sulfate preferable for year round operation – particularly for sedimentation performance. Alum produced the smallest flocs at lowest temperature, which would impact sedimentation.

All of the aluminum based coagulants demonstrated an increase in floc size with increasing water temperature. The PACl coagulants had the largest flocs at all temperatures, which partially explains why PACl coagulants are often preferred at low temperatures.

One characteristic of hydrolyzing coagulants such as alum and ferric sulfate is that there is essentially no control over the coagulant species formed under the prevailing conditions of concentration and pH during treatment. A difference with pre-polymerized coagulants such as PACl and polyferric sulfate is that they can be tailored for particular raw water conditions of temperature and other parameters, and produced with optimal hydrolysis and polymeric species. Such compounds may retard the rate of hydroxide precipitation and allow coagulation reactions to proceed with the best species for a longer time.

Therefore, one significant advantage of pre-polymerized coagulants in general is that the region of optimum coagulation can be expanded over a wider range of raw water conditions, including temperature.

4.16 RESIDUAL ALUMINUM

Residual aluminum in treated water is undesirable for aesthetic reasons. In 1991 the USEPA set a recommended maximum contaminant level of 0.05 to 0.20 mg/l for Al based on preventing colored water and post-precipitation in water distribution systems.

However, there is an added concern with residual aluminum related to a possible link between aluminum and adverse neurological effects – principally those manifested as Alzheimer's disease.

Alzheimer's disease was first identified in 1906 by a German neurologist named Alois Alzheimer. The disease is an irreversible, progressive disorder. Deterioration in critical areas of the brain may precede symptoms by as much as 20 to 40 years. Once thought to be rare, Alzheimer's disease is now known to be the commonest cause of dementia and, although statistics vary, affects at least 5-percent of the population aged 70 and over (Edwardson, 1995).

The precise causes of Alzheimer's disease are unknown. However, current understanding is that there are multiple causes, some of which may be genetically inherited, whereas others may come from the environment. Aluminum is an element that is abundant in the environment. The link between aluminum and Alzheimer's disease was put forward in 1965 when it was shown that the injection of aluminum compounds into rabbits caused tangle-like formations in nerve cells.

However, it was also shown that these tangles differed in structure and composition from the characteristic tangles observed in the human brain of Alzheimer patients.

The link between aluminum and Alzheimer's disease was also supported by the evidence that aluminum is associated with the plaques and tangles in Alzheimer brains. However, it is also known that the presence of aluminum does not prove a causal relationship – it could be a coincidental secondary relationship.

It had been suggested that the brain content of aluminum was higher in Alzheimer brains, although more recent studies comparing Alzheimer brains to normal brains failed to find any significant difference in aluminum levels. However, it is possible that the content of aluminum in specific types of cells could be different, rather than the total amount – and this could contribute to the disease (Edwardson, 1995).

It is estimated that human ingestion of aluminum from drinking water is a small percentage of the total intake. Studies have shown that 1 to 2-percent of the daily intake of aluminum comes from drinking water. Of this only 0.3 to 0.4-percent of the Al ingested via alum treated water is absorbed by the body (Stauber *et al.* 1999).

Major sources of aluminum absorption by the body are from foods. Anti-caking agents contain aluminum and are used in a wide variety of food preparations. The additives in processed food may contribute as much as 50-percent the average US intake of aluminum (Reiber *et al.* 1995). Coffee and tea also contain substantial quantities of aluminum bound to organic constituents such as phenolic compounds.

Aluminum is also an important constituent of many antacid preparations. Persons on a regular regimen of such antacids could consume as much as 3 g Al/day. Aluminum compounds are also very common in underarm antiperspirants. The transfer of Al from aluminum cooking utensils can also be a major source.

Although the ingestion of aluminum from drinking water is comparatively small, it is still unclear whether residual aluminum should be regulated for health reasons (see for example Walton & Wilcox, 1995; and Reiber *et al.* 1995). Currently, the World Health Organization, USEPA, the European Union, Health Canada and Australia have concluded that the level of scientific knowledge is not at the point where health based limits can be established (Pontius, 2000). However, operational guideline values have been set to limit the aluminum in treated water – for possible health effects but also to address the issues of colored water and post-precipitation in water distribution systems.

On a weight basis, alum consists of about 48% sulphate and almost 10% aluminum. Much of the added aluminum will be removed through the precipitation of aluminum hydroxide floc and other insoluble aluminum compounds, such as aluminum hydroxysulfates and aluminosilicates. The solubility of the various aluminum compounds follow somewhat similar trends with respect to pH. Optimization of the pH of coagulation will consequently reduce the level of soluble aluminum to low concentrations, even with the higher alum concentrations typical of enhanced coagulation.

Despite the higher alum dosages of enhanced coagulation, utilities may actually see a drop in residual aluminum levels, particularly if the utility had not been practicing careful pH control for conventional coagulation, before changing to enhanced coagulation.

Residual aluminum levels will also be affected by the concentrations of substances such as sulfates, silica, orthophosphates, and humic material in the source water, as well as temperature. For example, Jiang *et al.* (1996) showed that the optimum pH to minimize aluminum residuals should be increased, as the water temperature decreases.

Figure 4.55 presents a compilation of residual aluminum from jar tests, using membrane filters ranging from 0.05 μm to 0.45 μm. The results shown are considered more reliable than determining residual aluminum from equilibrium equations. Within the relatively short residence times during water treatment, aluminum hydroxide is not expected to reach equilibrium. The common method of determining "soluble" residuals is also an issue. The use of 0.45 membrane filters allows a significant portion of colloidal material to pass, thereby confounding the equilibrium calculations (Jekel & Heinzmann, 1989).

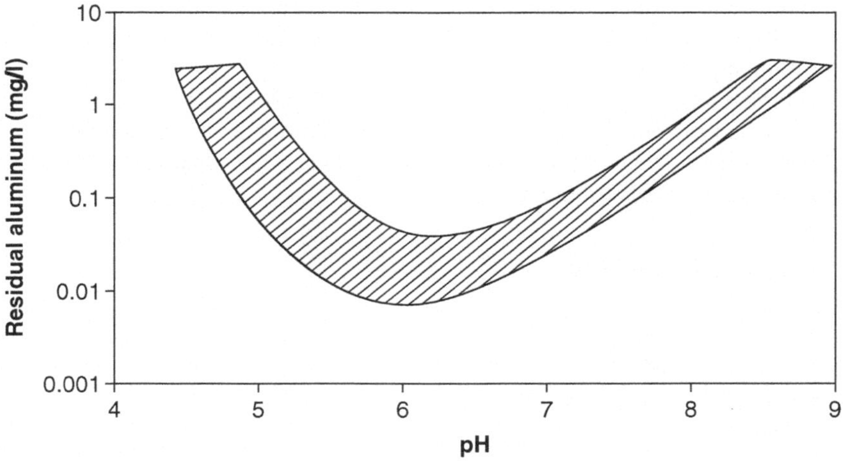

Figure 4.55 Compilation of residual aluminum determined from jar tests using membrane filters ranging from 0.05 to 0.45 μm (adapted from Jekel & Heinzmann, 1989).

The results show that, if the limit for residual Al is set at 200 μg/l, the final pH should be within the approximate range 5.0 to 7.5. If the limit is set at 50 μg/l the approximate range would be 5.5 to 7.0.

The contribution of colloidal material to the aluminum residual emphasizes the importance of achieving low final treated water turbidities, at least less than 0.1 NTU, to minimize final aluminum residuals (Letterman & Driscoll, 1988). When addressing high aluminum residuals, it is also important to determine whether the aluminum is in the particulate form, which would indicate improvements to filter retention, or whether it is soluble, which would require improving the chemistry of coagulation – particularly the pH before filtration.

A complicating factor when determining the conditions for minimum aluminum residuals is the presence of natural organic matter. Because of complexation of aluminum species with humic substances, the residual aluminum is linked to the removal of NOM.

Experiments carried out by Jekel and Heinzmann (1989) demonstrated that at low alum dosages applied to humic waters, residual aluminum concentrations after treatment were relatively high. At higher applied alum dosages, where a larger proportion of the humic substances were removed, residual Al concentrations after treatment were significantly lower. If preozonation is practiced, residual aluminum tends to be even higher in the presence of organic matter.

Vik *et al.* (1985) similarly showed that in the presence of humic compounds, when the alum dosage was increased to achieve maximum removal of humics at optimum pH, the residual aluminum was also minimized. The optimum pH range to minimize aluminum residual increased with increasing alum dosages. The optimum coagulation pH for minimum aluminum residual was slightly different to the optimum pH for the removal of NOM.

Frommell *et al.* (2004) showed that adding orthophosphates during the rapid mix stage, approximately one-minute after alum addition, resulted in significantly reduced residual aluminum concentrations and a residual orthophosphate concentration in the distribution system.

4.17 REFERENCES

Alvarez M., Voorhees J. and Hartman G. (1997). Using color as an indicator to comply with the proposed D-DBP rule. AWWA Water Quality Technology Conference Proceedings.

Amirtharajah A. and Mills K. (1982). Rapid-mix design for mechanisms of alum coagulation. *Jour. AWWA*, **74**(4), 210–216.

Amirtharajah A. (1990). Coagulation: rejuvenation for a classical process. *Water Engineering & Management*, 25–32.

Ammary B. and Cleasby J. (2004). Effect of addition sequence on dual-coagulant performance. *Jour. AWWA*, **96**(2), 90–101.

Angbo L., Mazet M. and Mossa W. (1990). Adsorption of humic substances onto prebuilt flocs. *Water Supply*, **8**, 87–91.

Ayoub G., Koopman B. and Pandya N. (2001). Iron and aluminum hydroxy (oxide) coated filter media for low-concentration phosphorus removal. *Wat. Env. Res.*, **73**(4), 478–485.

Baillod, C., Cressey G. and Beaupre R. (1977). Influence of phosphorus removal on solids budget. *Jour. WPCF*, 131–145.

Barnard J. (1974). Cut P and N without chemicals. *Wat. and Waste Eng.*, 33–36 and 41–44.

Baur R., Benisch M., Muirhead W., Neethling J., Stephens H. and Willey B. (1999). How low can you go? Can EBPR plus filtration meet part-per-billion phosphorus limits? Water Environ. Fed. Annual Conf., WEFTEC, New Orleans.

Becker W. and O'Melia C. (2001). Ozone: its effect on coagulation and filtration. *Wat. Sci. Tech.: Water Supply*, **1**(4), 81–88.

Benedek A. and Bancsi J. (1977). Comparative evaluation of commercial polyelectrolytes for flocculating alum precipitated domestic wastewater. *Prog. Wat. Tech.*, **9**, 33–42.

Benisch M., Baur R., Mengelkoch M., Muirhead W., Neethling J., Stephens H., Willey B. and Cullen N. (2001). An Oregon facility tweaks its enhanced biological phosphorus removal system to cut costs and improve compliance. *WE&T*, 39–44.

Benisch M., Clark D., Neethling J., Fredrickson H. and Gu A. (2007). Can tertiary phosphorus removal reliably produce 10 μg/L? Pilot results from Coeur d'Alene, ID. Proc. IWA/WEF Nutrient Conf., Baltimore, MD.

Bernhardt H. and Clasen J. (1991). Flocculation of micro-organisms. *J. Water SRT – Aqua*, **40**(2), 76–87.

Black A., Rice O. and Bartow E. (1933). Formation of floc by aluminum sulfate. *Ind. Eng. Chem.*, **25**, 811.

Black A. and Willems D. (1961). Electrophoretic studies of coagulation for removal of organic color. *Jour. AWWA*, 589–604.

Black A. and Christman R. (1963). Characteristics of colored surface waters. *Jour. AWWA*, **55**, 753–770.

Black A. and Christman R. (1963). Chemical characteristics of fulvic acids. *Jour. AWWA*, 897–912.

Black A., Singley J., Whittle G. and Maulding J. (1963). Stoichiometry of the coagulation of color causing organic compounds with ferric sulphate. *Jour. AWWA*, **55**, 1347–1366.

Black A. and Walters J. (1964). Electrophoretic studies of turbidity removal with ferric sulphate. *Jour. AWWA*, **56**, 99–110.

Black A. and Chen C. (1965). Electrophoretic studies of coagulation and flocculation of river sediment suspensions with aluminum sulphate. *Jour. AWWA*, **57**, 354–362.

Black A. and Chen C. (1967). Electrokinetic behaviour of aluminum species in dilute dispersed kaolinite systems. *Jour. AWWA*, 1173–1183.

Blackall L., Tandoi V. and Jenkins D. (1991). Continuous culture studies with *Nocardia amarae* from activated sludge and their implications for nocardia foaming control. *J. Water Pollut. Control Fed.*, **63**(1), 44–50.

Blake D., Phelps T., Eckenwiler S., Lee J., Smith K. and Burroughs B. (2015). Chemically enhanced primary treatment the next step in Columbus' wet weather plan. Water Environ. Fed. Annual Conf., WEFTEC, Chicago, IL.

Boller M. (1984). Full scale experience with tertiary contact filtration. *Wat. Sci. Tech.*, **16**, 225–239.

Boulay N. and Edwards M. (2000). Copper in the urban water cycle. *Critical Reviews in Environmental Sci. and Technol.*, **30**(3), 297–326.

Brandes M. (1977). Effective phosphorus removal by adding alum to septic tank. *Jour. WPCF*, 2285–2296.

Bratby J. and Marais G. (1974). Dissolved air flotation. *Filtration and Separation*, **11**(6), 614–624.

Bratby J. and Marais G. (1977). Thickening of brown water sludges by dissolved air pressure flotation. *Water SA*, **3**(4), 202–212.

Bratby J. (1978). Feasibility study concerning removal of nutrients from stabilization ponds (Guara) using metal coagulants and flotation. Contract Report III, CAESB, Brasilia, D.F., Brasil.

Bratby J. (1982). Treatment of raw wastewater overflows by dissolved-air flotation. *Jour. WPCF*, **54**(12), 1558–1565.

Bratby J. (1988). Phosphorus removal: a comparison between activated sludge systems and physical-chemical processes. Annual conference, Associação Brasileira de Engenharia Sanitaria, ABES, São Paulo, Brazil.

Bratby J. (1988). Optimizing manganese removal and washwater recovery at a direct filtration plant in Brazil. *Jour. AWWA*, 71–81.

Bratby J. (1990). The importance of the sequence of chemical addition on the coagulation of manganese bearing waters. 21st Fine Particle Society Meeting. San Diego, CA, 21–25.

Bratby J. (2006). Technical Memorandum for City of Boulder: Copper Jar Testing Study.

Briley D. and Knappe D. (2002). Optimizing ferric sulfate coagulation of algae with streaming current measurements. *Jour. AWWA*, **94**(2), 80–90.

Bronk D., Roberts Q., Sanderson M., Canuel E., Hatcher P., Mesfioui R., Filippino K., Mulholland M. and Love N. (2010). Effluent organic nitrogen (EON): bioavailability and photochemical and salinity-mediated release. *Environmental Science and Technology*, **44**, 5830–5835.

Cannon F. and Yeh T. (2001). Copper removal to ppb residuals via iron coagulants and biosolids storage conditioning. *Jour. Env. Eng. Div. ASCE*, **127**(8), 712–723.

Carlson K., Via S., Bellamy B. and Carlson M. (2000). Secondary effects of enhanced coagulation and softening. *Jour. AWWA*, 63–75.

Carlson K. and Gregory D. (2000). Optimizing water treatment with two-stage coagulation. *Jour. Env. Eng. ASCE*, 556–561.

Carter P., Worrell J., Daigger G., Allen E. and Land G. (2003). Enhanced primary treatment: full scale pilot answers many questions. Water Environ. Fed. Annual Conf., WEFTEC, Los Angeles.

Cavagnaro P., Work S., Bennett E. and Linstedt K. (1978). Enhancement of phosphorus removal through iron coagulation following lime precipitation. *Jour. WPCF*, 95–100.

Chack J., Rubino V., McFarland S., Krasnoff P. and Liubicich J. (1994). Advanced primary treatment bridges the gap. *Water Env. & Tech.*, 49–53.

Chaudhary R., Shao Y., Crosse J. and Soroushian F. (1991). Primary treatment: evaluation of chemical addition. *Water Env. & Technology*, 66–71.

Chaudhuri M. and Engelbrecht R. (1970). Removal of viruses from water by chemical coagulation and flocculation. *Jour. AWWA*, 563–566.

Chen J. and Yeh H. (2006). Comparison of the effects of ozone and permanganate preoxidation on algae flocculation. *Wat. Sci. Tech.: Water Supply*, **6**(3), 79–88.

Chen J., Yeh H. and Tseng I. (2009). Effect of ozone and permanganate on algae coagulation removal – Pilot and bench scale tests. *Chemosphere*, **74**, 840–846.

Cheng W., Chen W. and Yu R. (2010). PACl coagulation for the solid-liquid separation of highly concentrated algae suspensions. *Desalination and Water Treatment*, **16**, 290–297.

Cheng Y., Juang Y., Liao G., Ho S., Yeh K., Chen C., Chang J., Liu J. and Lee D. (2010). Dispersed ozone flotation of *Chlorella vulgaris*. *Bioresource Technology*, **101**, 9092–9096.

Cheng Y., Juang Y., Liao G., Tsai P., Ho S., Yeh K., Chen C., Chang J., Liu J., Chen W. and Lee D. (2011). Harvesting of *Scenedesmus obliquus FSP-3* using dispersed ozone flotation. *Bioresource Technology*, **102**, 82–87.

Christman R. and Chassemi M. (1966). Chemical nature of organic color in water. *Jour. AWWA*, **58**, 723–741.

Chwirka J., Thomson B. and Stomp J. (2000). Removing arsenic from groundwater. *Jour. AWWA*, **92**(3), 79–88.

Chwirka J., Colvin C., Gomez J. and Mueller P. (2004). Arsenic removal from drinking water using the coagulation/microfiltration process. *Jour. AWWA*, **96**(3), 106–114.

Clark D., Copp R. and Nissen J. (1999). Achieving 10 ppb, the state-of-the-art for phosphorus removal. Water Environ. Fed. Annual Conf., WEFTEC, New Orleans.

Clark T. and Stephenson T. (1999). Development of a jar testing protocol for chemical phosphorus removal in activated sludge using statistical experimental design. *Water Research*, **33**(7), 1730–1734.

Committee Report (1967). Research committee on color problems. *Jour. AWWA*, **59**, 1023–1035.

Committee Report (1970). Coagulation and color problems. *Jour. AWWA*, *62*, 311–314.

Cookson J. (1974). Viruses and water supply. *Jour. AWWA*, 707–711.

Cooper P. and Thomas E. (1974). Recent developments in sewage treatment based on physico-chemical methods. *Wat. Pollut. Control*, 505–520.

Cornwell D. (1994). Enhanced coagulation of three surface waters. A critical assessment. Enhanced coagulation research workshop, Dec. 4–6, Charleston, North Carolina.

Culp R. and Stoltenberg H. (1958). Fluoride reduction at La Crosse, Kan. *Jour. AWWA*, 423–431.

Cushing R., Houck C., McGlothlin P., Stevens R., Johnson R. and Schafran G. (2000). Evaluation of alternative coagulation strategies for fluoride control. Proc. AWWA National Conf., Denver, CO, June 11–15.

Czerwionka K., Makinia J., Pagilla K. and Stensel H. (2012). Characteristics and fate of organic nitrogen in municipal biological nutrient removal wastewater treatment plants. *Water Research*, **46**, 2057–2066.

Davis D., Binder G., Bucurel N., Melcer H. and Ifft J. (2014). Advancing wet weather treatment in a big way-NEORSD is getting it done! Water Environ. Fed. Annual Conf., WEFTEC, New Orleans, LA.

Davis D., Binder G., Bucurel N., Melcer H. and Cramer J. (2015). Two for the price of one – combining CEPT with disinfection. Water Environ. Fed. Annual Conf., WEFTEC, Chicago, IL.

deBarbadillo C., Shellswell G., Cyr W., Edwards B., Waite R., Sabherwal B., Mullan J. and Mitchell R. (2010). Development of full-scale sizing criteria from tertiary pilot testing results to achieve ultra-low phosphorus limits at Innisfil, Ontario. Water Environ. Fed. Annual Conf., WEFTEC, New Orleans, LA.

De Haas D. and Dubery I. (1989). Unreliability of cold-stored samples for assessment of chemical precipitates of phosphate in activated sludge. *Water SA*, **15**(4), 257–260.

De Haas D. and Greben H. (1991). Phosphorus fractionation of activated sludges from modified Bardenpho processes with and without chemical precipitant supplementation. *Wat. Sci. Tech.*, **23**, 623–633.

De Haas D., Borain G. and Kerdachi D. (1993). Review of treatment performance at Hammarsdale wastewater works with special reference to alum dosing. *Water SA*, **19**(2), 93–106.

De Haas D., Wentzel M. and Ekama G. (2000–2001). The use of simultaneous chemical precipitation in modified activated sludge systems exhibiting biological excess phosphate removal, Parts 1–7. *Water SA*, **26**(4), 439–504 and *Water SA*, **27**(2), 117–168.

De Godos I., Guzman H., Soto R., García-Encina P., Becares E., Muñoz R. and Vargas V. (2011). Coagulation/flocculation-based removal of algal-bacterial biomass from piggery wastewater treatment. *Bioresource Technology*, **102**, 923–927.

De Hek H., Stol R. and De Bruyn P. (1978). Hydrolysis precipitation studies of aluminum (III) solutions: (3) The role of the sulfate ion. *Jour. Colloid Interface Sci.*, **64**(1), 72–89.

Dentel S. and Gossett J. (1987). Coagulation of organic suspensions with aluminum salts. *Jour. WPCF*, **59**(2), 101–108.

Dentel S. and Gossett J. (1988). Mechanisms of coagulation with aluminum salts. *Jour. AWWA*, 187–198.

Dielmann P., Shepherd W., Leveque E., Semenza S., Clyburn M. and Narayanan B. (2004). Operational strategies and treatment technologies for meeting very low total phosphorus limits. Water Environ. Fed. Annual Conf., WEFTEC, New Orleans.

Drury D., Shepherd W., Narayanan B. (2005). Phosphorus – how low can you go? Water Environ. Fed. Annual Conf., WEFTEC, Washington, DC.

EC Environment Directorate (2002). Eutrophication of waters, role of phosphates, preventive measures – phosphates and alternative detergent builders. WRC synthesis-06.

Edwards G. and Amirtharajah A. (1985). Removing color caused by humic acids. *Jour. AWWA*, 50–57.

Edwardson J. (1995). Aluminum, mercury and Alzheimer's disease. Alzheimer's Disease Society, London, 1–4.

Edzwald J. and Wingler B. (1990). Chemical and physical aspects of dissolved-air flotation for the removal of algae. *J. Water SRT – Aqua*, **39**, 24–35.

Edzwald J. (1994). Coagulation: Applications, basics and concepts of TOC removal. Enhanced Coagulation Research Workshop, Dec. 4–6, Charleston, North Carolina.

Edzwald J., Tobiason J., Parento L., Kelley M., Kaminski G., Dunn H. and Galant P. (2000). Giardia and Cryptosporidium removals by clarification and filtration under challenge conditions. *Jour. AWWA*, 70–84.

Eggers E., Dirkzwager A. and van der Honing H. (1991). Full-scale experiences with phosphate crystallization in a Crystalactor. *Wat. Sci. Tech.*, **23**, 819–824.

Eklund L., Hellstrom B., Hultman B., Lind J. and Nordstrom B. (1991). Swedish full-scale experiments on modified operational modes in removal of nutrients. *Wat. Sci. Tech.*, **24**(7), 97–102.

Ekster A. and Jenkins D. (1996). Nickel and copper removal at the San Jose/Santa Clara water pollution control plant. *Water Env. Research*, **68**(7), 1172–1178.

EPA, United States Environmental Protection Agency (1999). Enhanced coagulation and enhanced precipitative softening guidance manual. EPA 815-R-99–012, May.

Fan M., Brown R., Sung S., Huang C., Ong S. and Leeuwen H. (2003). Comparisons of polymeric and conventional coagulants in arsenic(V) removal. *Water Env. Research*, **75**(4), 308–313.

Farrell J., Salotto B., Dean R. and Tolliver W. (1968). Removal of phosphate from wastewater by aluminum salts with subsequent alum recovery. *Chem. Eng. Progress, Symp. Series*, **64**, 232–239.

Fearing D., Goslan E., Banks J., Wilson D., Hillis P., Campbell A. and Parsons S. (2004). Staged coagulation for treatment of refractory organics. *Jour. Env. Eng. ASCE*, 975–982.

Feld C., Lytle D. and Schock M. (2000). The effect of fluoride on conventional drinking water treatment using aluminum sulfate. Proc. Water Quality Technology Conf., AWWA.

Ferguson J. and King T. (1977). A model for aluminum phosphate precipitation. *Jour. WPCF*, 646–658.

Finger R. (1973). Solids control in activated sludge plants with alum. *Jour. WPCF*, **45**(8), 1654–1662.

Fitzpatrick C., Fradin E. and Gregory J. (2004). Temperature effects on flocculation, using different coagulants. *Water Sci. Tech.*, **50**(12), 171–175.

Fitzpatrick J., Tetrick N., Sengupta S., Martin D., Gelderloos A., Kadava A. and Shaw A. (2012). Accurate wet-weather flow characterization and dynamic modelling help optimize treatment designs and operations. Water Environ. Fed. Annual Conf., WEFTEC, New Orleans, LA.

Frommell D., Feld C., Snoeyink V., Melcher B. and Feizoulof C. (2004). Aluminum residual control using orthophosphate. *Jour. AWWA*, **96**(9), 99–109.

Gates D., Luedecke C., Hermanowicz S. and Jenkins D. (1990). Mechanism of chemical phosphorus removal in activated sludge with Al(III) and Fe(III). ASCE National Conf. on Env. Eng., Arlington, VA.

Gerhardt M., Nascimento D., Witzgall R. and Fan M. (2002). Enhancing copper removal in secondary treatment by chemical addition. Water Environ. Fed. Annual Conf., WEFTEC, Cicago, IL.

Giroult E. (1977). Ensuring the quality of drinking water. *WHO Chronicle*, **31**(8), 316–320.

Gnirss R. and Dittrich J. (2000). Microfiltration of municipal wastewater for disinfection and advanced phosphorus removal: results from trials with differnt small-scale pilot plants. *Wat. Env. Res.*, **72**(5), 602–609.

Government Gazette (1984). Requirements for the purification of waste water or effluent. *Government Gazette, South Africa*, **227**(991), 12–17.

Gregory D. (1998). Enhanced coagulation for treating spring runoff water. *AWWA Opflow*, 12–13.

Gregory J. (1977). Stability and flocculation of colloidal particles. *Eff. Wat. Treat. Jnl.*, **17**(12), 641–643 and 647, 649, 651.

Gros H. and Morgeli B. (1980). Optimal advanced treatment and phosphorus removal by deep sand filtration. *Prog. Water Technol.*, **12**, 315.

Gu A., Liu L., Neethling J., Stensel H. and Murthy S. (2011). Treatability and fate of various phosphorus fractions in different wastewater treatment processes. *Water Sci. Tech.*, **63**(4), 804–810.

Gupta K., Hogge A., Romero I. and Swider C. (1998). A substitute yields savings. *WE&T*, 37–40.

Haarhoff J. and Cleasby J. (1988). Comparing aluminum and iron coagulants for in-line filtration of cold water. *Jour. AWWA*, 168–175.

Hach C. (1977). Measurement of color in water. *Jour. Indian Wat. Wks Asscn.*, **IX**(3), 249–252.

Hall E. and Packham R. (1965). Coagulation of organic color with hydrolyzing coagulants. *Jour AWWA*, **57**, 1149–1166.

Hanna G. and Rubin A. (1970). Effect of sulfate and other ions in coagulation with aluminum (III). *Jour. AWWA*, **62**, 315–321.

Hanner N., Mattsson A., Gruvberger C., Nyberg U., Aspegren H., Fredriksson O., Nordqvist A. and Andersson B. (2004). Reducing the total discharge from a large WWTP by separate treatment of primary effluent overflow. *Water Sci. Tech.*, **50**(7), 157–162.

Hanson A. and Cleasby J. (1990). The effects of temperature on turbulent flocculation: fluid dynamics and temperature. *Jour. AWWA*, 56–73.

Harleman D. and Murcott S. (2001). *Water 21*, 44–48 and 57–59.

Haron M., Wasay S. and Tokunaga S. (1997). Preparation of basic yttrium carbonate for phosphate removal. *Wat. Env. Res.*, **69**(5), 1047–1051.

Hartland Township, MI (1999). Dualsand Pilot Study. Report prepared by McNamee, Porter & Seeley, Inc., October.

Hartig J., Horvath F. and Waybrant R. (1982). Effects of Michigan's phosphorus detergent ban on municipal chemical costs. *Jour. WPCF*, **54**(3), 316–317.

Haruta S., Takahashi T. and Nishiguchi T. (1991). Basic studies on phosphorus removal by the contact aeration process using iron contactors. *Wat. Sci. Tech.*, **23**, 641–650.

He Q., Leppard G., Paige C. and Snodgrass W. (1996). Transmission electron microscopy of a phosphate effect on the colloid structure of iron hydroxide. *Water Research*, **30**(6), 1345–1352.

Healey K., Kerdachi D. and Borain G. (1989). The use of simultaneous precipitation to supplement biological treatment in a nutrient removal (activated) sludge process. Proc. Water Institute of Southern Africa Biennial Conf., Mar. 28–30, Cape Town.

Henderson R., Parsons S. and Jefferson B. (2008). The impact of algal properties and pre-oxidation on solid-liquid separation of algae. *Water Research*, **42**, 1827–1845.

Henderson R., Parsons S. and Jefferson B. (2010). The impact of differing cell and algogenic organic matter (AOM) characteristics on the coagulation and flotation of algae. *Water Research*, **44**, 3617–3624.

Hook G. and Ott R. (2001). The ultimate challenge for technology: 0.02 effluent total phosphorus. Water Environ. Fed. Annual Conf., WEFTEC, Atlanta.

Horn J., Hendricks D., Scanlan J., Rozelle L. and Trnka W. (1988). Removing Giardia cycts and other particles from low turbidity waters using dual-stage flitration. *Jour. AWWA*, 68–77.

Huang C. (1977). Removal of phosphate by powdered aluminum oxide adsorption. *Jour, WPCF*, 1811–1817.

Huang C. and Yeh H. (2009). Study on algae coagulation control. *Wat. Sci. Tech.:Water Supply*, **9**(2), 167–172.

Hundt T. and O'Melia C. (1988). Aluminum-fulvic acid interactions: mechanisms and applications. *Jour. AWWA*, 176–186.

Hunt K., Maxwell M., Morine M., Neher M., Okamura K., Okazaki J., Porter D., Segler K. and Smith T. (2000). Advanced wastewater treatment testing showdown in Henderson, Nevada. Water Environ. Fed. Annual Conf., WEFTEC, New Orleans.

James B., Rabenhorst M. and Frigon G. (1992). Phosphorus sorption by peat and sand amended with iron oxides or steel wool. *Wat. Env. Res.*, **64**(5), 699–705.

James G. (1971). Removal of Iron and Manganese. Water Treatment. Technical Press, London.

Jarvis P., Jefferson B. and Parsons S. (2004). Characterising natural organic matter flocs. *Wat. Sci. Technol.: Water Supply*, **4**(4), 79–87.

Jefferson B., Jarvis P., Sharp E., Wilson S. and Parsons S. (2004). Flocs through the looking glass. *Wat. Sci. Tech.*, **50**(12), 47–54.

Jekel M. and Heinzmann B. (1989). Residual aluminum in drinking-water treatment. *J. Water SRT – Aqua*, **38**, 281–288.

Jenkins S., Benefield L., Keal M. and Peacock R. (1984). Effective manganese removal using lime as an additive. *Jour. AWWA*, **76**(9), 82–86.

Jiang J., Graham N. and Harward C. (1993). Comparison of polyferric sulphate with other coagulants for the removal of algae and algae-derived organic matter. *Water Sci. Tech.*, **27**(11), 221–230.

Jiang J., Graham N. and Harward C. (1996). Coagulation of upland coloured water with polyferric sulphate compared to conventional coagulants. *J. Water SRT – Aqua*, **45**(3), 143–154.

Jiang J. and Graham N. (1998). Pre-polymerised inorganic coagulants and phosphorus removal by coagulation – a review. *Water SA*, **24**(3), 237–244.

Johnson P. and Amirtharajah A. (1983). Ferric chloride and alum as single and dual coagulants. *Jour. AWWA*, 232–239.

Jones R. and Dold P. (2005). Simulation applications made possible by pH modeling. Water Environ. Fed. Annual Conf., WEFTEC, Washington, DC.

Joye D. (1991). Aluminum chloride poly-aluminum chloride enhanced phosphorus removal. *Wat. Env.&Tech.*, 48–49.

Kang S., Hoversten K. and Lund D. (2001). The highest level of phosphorus removal practicable from municipal wastewater treatment plants. Water Environ. Fed. Annual Conf., WEFTEC, Atlanta.

Kastl G., Sathasivan A., Fisher I. and Leeuwen J. (2004). Modeling DOC removal by enhanced coagulation. *Jour. AWWA*, **96**(2), 79–89.

Kavanaugh M., Krejci V., Weber T., Eugster J. and Roberts P. (1978). Phosphorus removal by post precipitation with Fe(III). *Jour. WPCF*, 216–233.

Keaney J., Bratby J., Allenwood M., Simmons D., Pepin R. and Woodard S. (2012). Achieving ultra-low effluent phosphorus during start-up under stress conditions and maintaining performance long term. Water Environ. Fed. Annual Conf., WEFTEC, New Orleans, LA.

Kelkar U. and Schafran G. (1994). Enhanced coagulation: impact on residual characteristics and dewaterability. AWWA Enhanced Coagulation Research Workshop, Dec. 4–6, Charleston, NC.

Kerrison P., McEwen B., Phillips G. and Crook B. (1989). The use of redox potential to control ferric sulphate dosing during phosphate removal. *J. IWEM*, **3**, 397–403.

Kim W., Ludwig H. and Bishop W. (1965). Cation exchange capacity and pH in the coagulation process. *Jour. AWWA*, **57**, 327.

Klein A., Constantine T., Crawford G., Daigger G. and Brauer R. (2005). Technical comparison of nine treatment alternatives to achieve 0.1 mg/l effluent phosphorus. Water Environ. Fed. Annual Conf., WEFTEC, Washington, DC.

Krasner S., Amy G. and Zhu H. (1994). Use of some simple NOM characterization techniques to evaluate enhanced coagulation treatability. AWWA Enhanced Coagulation Research Workshop, Dec. 4–6, Charleston, NC.

Krasner S. and Amy G. (1995). Jar test evaluations of enhanced coagulation. *Jour. AWWA*, 93–107.

Krugel S., Melcer H., Hummel S. and Butler R. (2005). High rate chemically enhanced primary treatment as a tool for wet weather plant optimization and re-rating. Water Environ. Fed. Annual Conf., WEFTEC, Washington, DC.

Kwak D., Kim S., Jung H., Won C., Kwon S., Ahn H. and Lee J. (2006). Removal of clay and blue-green algae particles through zeta potential and particle size distribution in the dissolved air flotation process. *Wat. Sci. Tech.: Water Supply*, **6**(1), 95–103.

Lea W., Rohlich G. and Katz W. (1954). Removal of phosphates from treated sewage. *Sew. & Ind. Wastes*, **26**(3), 261–275.

Lee G. and Jones R. (1986). Detergent phosphate bans and eutrophication. *Environ. Sci. Technol.*, **20**(4), 330–331.

Lee R., Carlson J., Bril J., Cramer J. and Harenda J. (2015). Pilot testing reveals alternative methods to meet Wisconsin's low level phosphorus limits. Water Environ. Fed. Annual Conf., WEFTEC, Chicago.

Leentvaar J., Buning W. and Koppers H. (1978). Physico-chemical treatment of municipal wastewater. Coagulation-flocculation. *Water Research*, **12**, 35–40.

Leentvaar J., Ywema T. and Roersma R. (1979). Optimalization of coagulant dose in coagulation-flocculation of sewage. *Water Research*, **13**(3), 229–236.

Letterman R., Tabatabaie M. and Ames R. (1979). The effect of the bicarbonate ion concentration on flocculation with aluminum sulfate. *Jour. AWWA*, 467–472.

Letterman R. and Vanderbrook S. (1983). Effect of solution chemistry on coagulation with hydrolyzed Al(III). Significance of sulfate ion and pH. *Water Research.*, **17**(2), 195–204.

Letterman R. and Driscoll C. (1988). Survey of residual aluminum in filtered water. *Jour. AWWA*, 154–158.

Levesque S. (2001). Comparison of chemicals to achieve low phosphorus limits. Water Environ. Fed. Annual Conf., WEFTEC, Atlanta.

Licsko I. (2004). Coagulation mechanisms – nano- and microprocesses. *Wat. Sci. & Tech.*, **50**(12), 193–200.

Lin S. and Carlson D. (1975). Phosphorus removal by the addition of aluminum (III) to the activated sludge process. *Jour. WPCF*, **47**(7), 1978–1986.

Lind C. (1996). Reducing total and dissolved organic carbon: comparing coagulants. *Env. Tech.*, 54–58.

Linderoth J., Cervenka F., Pospech L. and Novak L. (2004). Influence of PIX-XL2 dosing on operation of biological treatment line with activated sludge system. *The Water Mirror*, **1**, 12–17.

Linenfelser B. and Fevig S. (2006). Copper effluent limit compliance – a multi-pronged approach. Presentation to RMWEA Annual Conference, Colorado.

Liu D., Li F. and Zhang B. (2009). Removal of algal blooms in freshwater using magnetic polymer. *Wat. Sci. Tech.*, **59**(6), 1085–1091.

Liu H., Jeong J., Gray H., Smith S. and Sedlak D. (2012). Algal uptake of hydrophobic and hydrophilic dissolved organic nitrogen in effluent from biological nutrient removal municipal wastewater treatment systems. *Environmental Science and Technology*, **46**, 713–721.

Lötter L. (1991). Combined chemical and biological removal of phosphate in activated sludge plants. *Wat. Sci. Tech.*, **23**, 611–621.

Luedecke C., Hermanowicz S. and Jenkins D. (1989). Precipitation of ferric phosphate in activated sludge: A chemical model and its verification. *Wat. Sci. Tech.*, **21**, 325–327.

Lynch W. and Potter L. (1981). Start-up of a physical-chemical treatment plant. *Jour. WPCF*, **53**(3), 318–324.

Lytle D., Magnuson M. and Snoeyink V. (2004). Effect of oxidants on the properties of Fe(III) particles and suspensions formed from the oxidation of Fe(II). *Jour. AWWA*, **96**(8), 112–124.

Ma J., Li G., Chen Z., Xu G. and Cai G. (2001). Enhanced coagulation of surface waters with high organic content by permanganate preoxidation. *Water Sci. and Technol.: Water Supply*, **1**(1), 51–61.

Ma J. and Liu W. (2002). Effectiveness and mechanism of potassium ferrate (VI) preoxidation for algae removal by coagulation. *Water Research*, **36**, 871–878.

Ma J., Fang J., Wang L., Guo J. and Chen Z. (2006). Effect of preozonation on characteristics of algae cells and algae-derived organic matter (AOM) with respect to their removal by coagulation. *Water Sci. and Technol.: Water Supply*, **6**(4), 145–152.

Mackrle S. (1962). Mechanism of coagulation in water treatment. *J. San. Eng. Div. ASCE*, SA 3, 1.

Maher C., Neethling J., Murthy S. and Pagilla K. (2015). Kinetics and capacities of phosphorus sorption to tertiary stage wastewater alum solids, and process implications for achieving low-level phosphorus effluents. *Water Research*, **85**, 226–234.

Mamais D., Jenkins D. and Pitt P. (1993). A rapid physical-chemical method for the determination of readily biodegradable soluble COD in municipal wastewater. *Water Research*, **27**(1), 195–197.

Mamais D., Kalaitzi E. and Andreadakis A. (2007). Foaming control in activated sludge treatment plants by polyaluminum chloride addition. Proc. 10th Internat. Conf. on Environ. Science and Technology, Greece, 5–7 Sep.

Mangravite F., Buzzell T., Cassell E., Matijević E. and Saxton G. (1975). Removal of humic acid by coagulation and microflotation. *Jour. AWWA*, 88–94.

Manwaring J., Chaudhuri M. and Engelbrecht R. (1971). Removal of viruses by coagulation and flocculation. *Jour. AWWA*, 298–300.

Marchioretto M. and Reali M. (2001). Ozonation followed by coagulation/flocculation and flotation as post treatment of the effluent from an anaerobic baffled reactor treating domestic sewage. *Wat. Sci. Tech.*, **43**(8), 99–106.

Marion S. and Thomas A. (1946). Effect of diverse ions on the pH of maximum precipitation of aluminum hydroxide. *J. Colloid Sci.*, **1**, 221.

Matijević E. (1973). Cooloid stability and complex chemistry. *Jnl. Colloid Interface Sci.*, **43**(2), 217–245.

Matsushita T., Matsui Y., Inoue T. (2004). Irreversible and reversible adhesion between virus particles and hydrolyzing-precipitating aluminum: a function of coagulation. *Water Sci. Technol.*, **50**(12), 201–206.

Maurer M. and Boller M. (1999). Modelling of phosphorus precipitation in wastewater treatment plants with enhanced biological phosphorus removal. *Wat. Sci. Tech.*, **39**(1), 147–163.

McDermott J. (1974). Virus problems and their relation to water supplies. *Jour. AWWA*, 693–698.

McGarry M. (1970). Algal flocculation with aluminum sulfate and polyelectrolytes. *Jour. WPCF*, **42**(5), 191–201.

McGarry M. and Tongkasame C. (1971). Water reclamation and algae harvesting. *Jour. WPCF*, **43**(5), 824–835.

McNeill L. and Edwards M. (2003). Degradation of drinking water treatment plant infrastructure from enhanced coagulation. *Jour. Infrastructure systems, ASCE*, 145–156.

Melcer H., Krugel S., Butler R., Carter P. and Land G. (2005). Alternative operational strategies to control pollutants in peak wet weather flows. Water Environ. Fed. Annual Conf., WEFTEC, Washington, DC.

Melcer H., Robinson W., Jue P., Doty J., Yanasak J. and Land G. (2009). The application of PAX compounds for Microthrix Parvicella foam control in BNR systems. Proc. WEF Specialty Conf. Nutrient Removal, Washington, DC.

Melcer H., Klein A., Land G., Butler R., Carter P., Ciolli M. and Lilienthal R. (2011). Revisiting a wet weather option optimizing chemically enhanced primary treatment. *WE&T*, 57–61.

Melcer H., Davis D., Xiao S., Shaposka H., Ifft J., Bucurel N. and Land G. (2012). Wet weather flow treatment with a difference: novel ideas for applying chemically enhanced primary treatment with high rate disinfection. Water Environ. Fed. Annual Conf., WEFTEC, New Orleans, LA.

Melcer H., Carter P., Butler R. and Moore Z. (2014). Bench- to pilot- to full-scale verification of CEPT assisted dual use primary clarifier technology at King County's Brightwater treatment plant. Water Environ. Fed. Annual Conf., WEFTEC, New Orleans, LA.

Metsch V., Bantz I. and Hahn H. (1985). Phosphorus removal by pre- and simultaneous coagulation/precipitation effects on biological wastewater and sludge treatment. Proc. Int. Conf. Manage Strategies for Phosphorus in the Environ., Lisbon.

Mohapatra M., Anand S., Mishra B., Giles D. and Singh P. (2009). Review of fluoride removal from drinking water. *Jour. Env. Management*, **91**, 67–77.

Morris J. and Knocke W. (1984). Temperature effects on the use of metal-ion coagulants for water treatment. *Jour. AWWA*, 74–79.

Mueller P., Danzer J., Wable M., Carlson M., Daigger G. and Poling M. (1999). Advanced chemical phosphorus removal by direct filtration without tertiary clarifiers. Water Environ. Fed. Annual Conf., WEFTEC, New Orleans.

Murthy S. and Novak J. (1999). Factors affecting floc properties during aerobic digestion: implications for dewatering. *Water Environ. Res.*, **71**(2), 197–202.

Murthy S. and Novak J. (2001). Influence of cations on activated sludge effluent quality. *Water Environ. Res.*, **73**(1), 30–36.

Murthy S., Takacs I., Dold P. and Al-Omari A. (2005). Examining the bioavailability of chemically removed phosphorus at the Blue Plains Advanced Wastewater Treatment Plant. Water Environ. Fed. Annual Conf., WEFTEC, Washington, DC.

Najm I., Tate C. and Selby D. (1998). Optimizing enhanced coagulation with PAC: a case study. *Jour. AWWA*, **90**(10), 88–95 (and discussions: 91, 4, April, 4 and 189).

Narasiah K., Morasse C. and Lemay J. (1991). Nutrient removal from aerated lagoons using alum and ferric chloride – a case study. *Wat. Sci. Tech.*, **23**, 1563–1572.

Narayanan B., Karam W., Leveque E., Besett T. and Baadsgarrd M. (2000). A new approach for defining the limits of chemically enhanced primary treatment. Water Environ. Fed. Annual Conf., WEFTEC, Anaheim, CA.

Narayanan B., de Leon C., Radke C. and Jenkins D. (2010). The role of dispersed Nocardioform filaments in activated sludge foaming. *Water Environ. Res.*, **82**(6), 483–491.

Narkis N. and Rebhun M. (1977). Stoichiometric relationship between humic and fulvic acids and flocculants. *Jour. AWWA*, 325–328.

Nasser A., Weinberg D., Dinoor N., Fattal B. and Adin A. (1995). Removal of hepatitis A virus (HAV) poliovirus and MS2 coliphage by coagulation and high rate filtration. *Wat. Sci. Tech.*, **31**(5–6), 63–68.

Neethling J., Benisch M., Clark D. and Gu A. (2007). Phosphorus speciation provides direction to produce 10 µg/L. Proc. IWA/WEF Nutrient Conf., Baltimore, MD.

Newman D., Melcer H., Davis D., Pepe L., Winn R. and Nascimento D. (2013). At the nexus of process and design: optimizing a wet weather treatment system. Water Environ. Fed. Annual Conf., WEFTEC, Chicago, IL.

Novak J., Becker H. and Zurow A. (1977). Factors influencing activated sludge properties. *Jour. Env. Eng. Div. ASCE*, **103**(EE5), 815–828.

NYCDEP (2000). Village of Stamford/NYCDEP tertiary wastewater treatment demonstration project comparing continuously backwashed upflow dual sand filtration and microfiltration technologies. New York City Department of Environmental Protection, May.

Ødegaard H. (1992). Norwegian experiences with chemical treatment of raw wastewater. *Wat. Sci. Tech.*, **25**(12), 255–264.

Ødegaard H. (1995). Optimization of flocculation/flotation in chemical wastewater treatment. *Wat. Sci. Tech.*, **31**, 3–4, 73–82.

Ødegaard H. (2001). The use of dissolved air flotation in municipal wastewater treatment. *Wat. Sci. Tech.*, **43**(8), 75–81.

Oden S. (1919). Humic acids. Kolloid chem. Beihefte, 11, 75.

Oh H., Lee S., Park M., Kim H., Kim H., Yoon J., Kwon G. and Yoon B. (2001). Harvesting of *Chlorella vulgaris* using a bioflocculant from *Paenibacillus* sp. AM49. *Biotechnology Letters*, **23**, 1229–1234.

Packham R. (1963). The coagulation process – a review of some recent investigations. *Proc. Soc. Wat. Treat. Exam.*, **12**, 15.

Packham R. (1964). Studies of organic color in natural waters. *Soc. Wat. Treat. Exam.*, **13**, 316–334.

Packham R. (1965). Some studies of the coagulation of dispersed clays with hydrolyzing salts. *J. Colloid Sci.*, **20**, 81.

Parker D., Barnard J., Daigger G., TeKippe R. and Wahlberg E. (2001). *Water 21*, June, 49–56.

Philp D. (1985). Phosphorus removal at the Lower Molonglo water quality control centre. *Jour. WPCF*, **57**(8), 841–846.

Pivokonsky M., Kloucek O. and Pivokonska L. (2006). Evaluation of the production, composition and aluminum and iron complexation of algogenic organic matter. *Water Research*, **40**, 3045–3052.

Pizzi N. and Rodgers M. (1997). Preliminary testing for control of DBPs in Cleveland. *Jour. AWWA*, **89**(8), 78–87.

Pommerenk P. and Schafran G. (2002). Effects of prefluoridation on removal of particles and organic matter. *Jour. AWWA*, **94**(2), 99–108.

Pommerenk P. and Schafran G. (2005). Adsorption of inorganic and organic ligands onto hydrous aluminum oxide: evaluation of surface charge and the impacts on particle and NOM removal during water treatment. *Environ. Sci. Technol.*, **39**, 6429–6434.

Pontius F. (2000). Regulations for aluminum in drinking water. *Jour. AWWA*, 18–22 &183.

Posselt H., Reidies A. and Weber W. (1968). Coagulation of colloidal hydrous manganese dioxide. *Jour. AWWA*, **60**(1), 48–68.

Posselt H., Anderson F. and Weber W. (1968). Cation sorption on colloidal hydrous manganese dioxide. *Envir. Sci. & Tech.*, **2**(12), 1087–1093.

Rao V., Symons J., Ling A., Wang P., Metcalf T., Hoff J. and Melnick J. (1988). Removal of hepatitis A virus and rotavirus by drinking water treatment. *Jour. AWWA*, 59–67.

Rebhun M., Galil N. and Narkis N. (1985). Kinetic studies of chemical and biological treatment for renovation. *Jour. WPCF*, **57**(4), 324–331.

Reese S., Nozaic D., Pryor M., Rajogopaul R., Trollip D. and Smith R. (2001). Enhanced coagulation: a viable option to advance treatment technologies in the South African context. *Wat. Sci. Technol.: Water Supply*, **1**(1), 33–41.

Reiber S., Kukull W. and Standish-Lee P. (1995). Drinking water aluminum and bioavailability. *Jour. AWWA*, 86–100.

Roels T., Dauwe F., Van Damme S., Wilde K. and Roelandt (2002). The influence of PAX-14 on activated sludge systems and in particular on *Microthrix parvicella*. *Wat. Sci. Tech.*, **46**, 1–2, 487–490.

Rook J. (1977). Chlorination reactions of fulvic acids in natural waters. *Env. Sci. and Tech.*, **11**(5), 478–482.

Rossetti S., Tomei M., Nielsen P. and Tandoi V. (2005). *Microthrix parvicella*, a filamentous bacterium causing bulking and foaming in activated sludge systems: a review of current knowledge. *FEMS Microbiology Reviews*, **29**(1), 49–64.

Rubin A. and Blocksidge H. (1979). Coagualtion of montmorillonite suspensions with aluminum sulfate. *Jour. AWWA*, 102–109.

Ruehl K. (1999). Easing the pain of meeting the D/DBP rule with enhanced coagulation. *Water Eng. & Mgmt*, 22–25.

Sandbank E. and Shelef G. (1988). Harvesting of algae from high-rate ponds by flocculation-flotation. *Wat. Sci. Tech.*, **19**(12), 257–263.

Sattayatewa C., Dubanowitz N., Pagilla K., Sharp R., Pitt P., White C. and Bruton T. (2009). DON and CON in seven BNR wastewater treatment plants' processes and effluents. Proceedings of the Water Environment Federation Conference, WEFTEC, Orlando, FL.

Schaefer W. (1990). Environment ministry gets tough on phosphorus. *World Water*, 53–56.

Schneider O. and Tobiason J. (2000). Preozonation effects on coagulation. *Jour. AWWA*, **92**(10), 74–87.

Sedlak D., Bedsworth W., Jenkins D., Kang S. and Murin J. (2000). Assessing methods of removing metals from wastewater: a review os data and methodologies. Water Env. Research Foundation, WERF 97-CTS-4, Alexandria, VA.

Shannon E. and Verghese K. (1976). Utilization of alumized red mud solids for phosphorus removal. *Jour. WPCF*, **48**(8), 1948–1954.

Shapiro J. (1957). Chemical and biological studies on the yellow organic acids of lake water. *Limnol. Oceanog.*, **2**, 161.

Shelton S. and Drewry W. (1973). Tests of coagulants for the reduction of viruses, turbidity and chemical oxygen demand. *Jour. AWWA*, 627–635.

Shiao S. and Akashi K. (1977). Phosphate removal from aqueous solution from activated red mud. *Jour. WPCF*, 280–285.

Shin J. and O'Melia C. (2004). Stoichiometry of coagulation revisited. AWWA 2004 Annual Conference Proceedings.

Shindala A. and Stewart J. (1971). Chemical coagulation of effluents from municipal waste stabilization ponds. *Wat. and Sew. Works*, **118**(4), 100.

Singhal A. (1980). Phosphorus and nitrogen removal at Cadillac, Michigan. *Jour. WPCF*, **52**(11), 2761–2770.

Singley J., Harris R. and Maulding J. (1966). Correction of color measurements to standard conditions. *Jour. AWWA*, **58**, 455–457.

Smith S. and Edwards M. (2002). Bench scale evaluation of innovative arsenic removal processes. *Jour. AWWA*, **94**(9), 78–90.

Smith S., Takács I., Murthy S., Daigger G. and Szabó A. (2008). Phosphate complexation model and its implications for chemical phosphorus removal. *Water Environ. Res.*, **80**(5), 428–438.

Sobeck D. and Higgins M. (2002). Examination of three theories for mechanisms of cation-induced bioflocculation. *Water Research*, **36**, 527–538.

Sollo F., Larson T. and Mueller H. (1978). Fluoride removal from potable water supplies. University of Illinois Water Resources Center, WRC Research Report No. 136, Sep.

South Florida Water Management District – SFWMD (2000). Chemical treatment followed by solids separation advanced technology demonstration project. Final report, December.

Southern African Branch of the Institute of Water Pollution Control (1973). A guide to the Design of Sewage Purification Works. Institute of Water Pollution Control.

Spatzierer G., Ludwig C. and Matsche N. (1984). Biological phosphorus removal in combination with simultaneous precipitation. IAWPRC conference, Paris, Sep., 203–216.

Standard Methods for the Examination of Water and Wastewater (2005). 21st Edition, APHA, AWWA, WEF.

Stauber J., Florence T., Davies C., Adams M. and Buchanan S. (1999). Bioavailability of Al in alum-treated drinking water. *Jour. AWWA*, **91**(11), 84–93.

Stumm W. and Morgan J. (1962). Chemical aspects of coagulation. *Jour. AWWA*, **54**(8), 971–991.

Stumm W. and O'Melia C. (1968). Stoichiometry of coagulation, *Jour. AWWA*, 514–539.

Stumm W. and Morgan J. (1970, 1996). Aquatic Chemistry – An Introduction Emphasizing Chemical Equilibria in Natural Waters. John Wiley and Sons, Inc., New York.

Sugden A. (1985). Phosphate removal – a new packaged plant. *IMIESA*, 40–41.

Sukias J. and Craggs R. (2011). Digestion of wastewater pond microalgae and potential inhibition by alum and ammoniacal-N. *Wat. Sci. Tech.*, **63**(5), 835–840.

Swift J., Lange D., Porter R., Alexander K. and Wilson J. (2006). Evaluation of compliance strategies to meet stringent effluent copper limits. Water Environ. Fed. Annual Conf., WEFTEC, Dallas, TX.

Switzenbaum M., DePinto J., Young T. and Edzwald J. (1980). A survey of phosphorus removal in lower Great lakes municipal treatment plants. *Jour. WPCF*, **52**(11), 2628–2633.

Symons J., Bellar T., Carswell J., DeMarco J., Kropp K., Robeck G., Seegar D., Slocum C., Smith B. and Stevens A. (1975). National Organics Reconnaissance Survey for Halogenated Organics in Drinking Water. *Jour. AWWA*, **67**(11), 634–647. (Update: 67, 12, 708–709, 1975).

Taljard M. (1981). The integration of physico-chemical and biological processes for the production of reclaimed water. *IMIESA*, 45–49.

Tanaka K., Aoki M., Takahashi S., Chida S., Yasuda T., Takagi K., Yumoto H. and Kasakura K. (1991). Study on development of phosphorus removal process by contact filtration. *Wat. Sci. Tech.*, **23**, 739–745.

Takács I., Murthy S. and Fairlamb P. (2004). Modeling chemical phosphorus precipitation using equilibrium chemistry. Water Environ. Fed. Annual Conf., WEFTEC, New Orleans.

Takács I., Murthy S. and Fairlamb P. (2004). Chemical phosphorus removal model based on equilibrium chemistry. Proceedings of the IWA World Water Congress, Marrakech, Sept. 19–24.

Takács I., Murthy S., Smith S. and McGrath M. (2005). Chemical phosphorus removal to extremely low levels – experience of two plants in the Washington, D.C. area. Proceedings of the IWA Specialized Conference: Nutrient Management in Wastewater Treatment Process and Recycle Streams, Krakow, Poland, Sept. 19–24.

Takács I., Johnson B., Smith S., Szabó A. and Murthy S. (2011). Chemical P removal – from lab tests through model understanding to full-scale demonstration. 11th IWA Specialised Conference on Design, Operation and Economics of Large Wastewater Treatment Plants, Budapest, Hungary.

Tebutt T. and Christoulas D. (1975). Performance relationships for primary sedimentation. *Water Research*, **9**, 347.

Tenney M. and Stumm W. (1965). Chemical flocculation of microorganisms in biological waste treatment. *Jour. WPCF*, **37**(10), 1370.

Thakur U., Dhabadgaonkar S. and Deshpande W. (1977). Chemical treatment of sewage. *Indian J. Environ. Hlth*, **19**(1), 16–29.

Thomas C. and Slaughter R. (1992). Phosphate reduction in sewage effluents: some practical experiences. *J. IWEM*, **6**, 158–164.

Tobiason J., Reckhow D. and Edzwald J. (1995). Effects of ozonation on optimal coagulant dosing in drinking water treatment. *J. Water SRT – Aqua*, **44**(3), 142.

Tryland O. (1976). Humic water dissolves metallic iron. *Vatten*, **3**, 271–273.

Tsuno H., Hidaka T. and Nakamoto M. (2004). Development of pre-coagulation and bio-filtration process for advanced treatment of sewage. *Water Sci. Tech.*, **4**(1), 13–22.

Vik E., Carlson D., Eikum A. and Gjessing E. (1985). Removing aquatic humus from Norwegian lakes. *Jour. AWWA*, 58–66.

Wahlberg E., Wang J., Merill M., Morris J., Kido W., Swanson R., Finger R. and Phillips D. (1998). Primary sedimentation: it's performing better than you think. Water Environ. Fed. Annual Conf., WEFTEC, Orlando, FL.

Walkowiak M., Tse R., Mavis J., Rippon D., Towey A., Douzinas R., Gerhardt M. and Parker D. (2002). Demonstration of copper and nickel removal technologies to protect San Francisco Bay. Water Environ. Fed. Annual Conf., WEFTEC, Chicago.

Walton J. and Wilcox D. (1995). Aluminum bioavailability debated. *Jour. AWWA*, 4&108. (Reiber *et al.* response: Jour. AWWA, Oct., 108–109).

White M., Thompson J., Harrington G. and Singer P. (1997). Evaluating criteria for enhanced coagulation compliance. *Jour AWWA*, **89**(5), 64–77.

Wiechers H. (1986). Chemical phosphate removal from municipal wastewaters: current practice and recent innovations. *Water SA*, **12**(4), 197–202.

Wiley P., Campbell J. and McKuin B. (2011). Production of biodiesel and biogas from algae: a review of process train options. *Water Environ. Res.*, **83**(4), 326–338.

Wilson T., Bizzarri R., Burke T., Langdon T. and Courson C. (1975). Upgrading primary treatment with chemicals and water treatment sludge. *Jour. WPCF*, **47**(12), 2820–2833.

Xagoraraki I. and Harrington G. (2004). Zeta potential, dissolved organic carbon, and removal of Cryptosporidium oocysts by coagulation and sedimentation. *Jour. Env. Eng., ASCE*, 1424–1432.

Yao C. and Yan X. (1989). Removal of turbidity and colour by alum. *Shanghai Water and Wastewater*, **2**, 1–5.

York D. and Drewry W. (1974). Virus removal by chemical coagulation. *Jour. AWWA*, 711–716.

Zafer M., Daniel P., Steiguer A. and Wong D. (1994). It's not easy being Green Lake. *WE&T*, 64–68.

Zhao X., Zhang Y., Li X., Liu C. and Zhu L. (2010). Algae removal efficiencies of AS/PDMDAAC coagulants. *Jour. AWWA*, **102**(7), 119–128.

Zheng X., Zhang Y., Jiang L., Liu X. and Chen L. (2002). The development and application of chemically and biologically enhanced primary treatment process for municipal wastewater treatment. Water Environ. Fed. Annual Conf., WEFTEC, Chicago.

Zhou S., Wilson T., Huang J., Xu Y., Zhang S., Zhang C. and Liu X. (2004). Optimization of chemical enhanced primary treatment process. Water Environ. Fed. Annual Conf., WEFTEC, New Orleans.

Chapter 5

Treatment with polymers

5.1 INTRODUCTION

It was shown in Chapter 3 that polyelectrolytes contain functional groups along their backbone which may or may not carry a charge. If the polymer is charged, the groups may be such as to give an anionic character to the chain, a cationic character or an ampholytic character, where both anionic and cationic charged sites are present. The intensity of charge carried by the polyelectrolyte is dependent on the degree of ionization of the functional groups, or on the degree of copolymerization or substitutions: both of these are usually collectively described as the extent of hydrolysis. The extent of polymerization of the polyelectrolyte is characterized by the molecular weight. High molecular weights signify long chains, whereas low molecular weights indicate short polyelectrolyte chains.

Functional groups along the polyelectrolyte chain, besides the possibility of carrying a charge, are also sites which possess the property of being adsorbed. Therefore, it is understandable that destabilization by polyelectrolytes could involve a mechanism combining both charge effects and effects due to adsorption.

Polyelectrolytes are effective in enhancing the rate of orthokinetic flocculation (see Chapter 7) when added to a system already destabilized with, say, metal coagulants, as shown below (adapted from Kitchener, 1972):

$$\begin{matrix}\textit{Stable}\\ \textit{Colloid}\end{matrix} + \xrightarrow[\text{Coagulant}]{\text{Inorganic}} \begin{matrix}\textit{Unstable}\\ \textit{Colloid}\end{matrix} \xrightarrow[\text{Time}]{\text{Conditioning}} \textit{Flocs} + \xrightarrow[\text{Conditioning}]{\text{Polyelectrolyte}} \begin{matrix}\textit{Large}\\ \textit{Flocs}\end{matrix}$$

Polyelectrolytes may also be effectively applied as primary coagulants, as shown below (Michaels & Morelos, 1955):

$$\begin{matrix}\textit{Stable}\\ \textit{Colloid}\end{matrix} + \xrightarrow[\text{(Conditioning time)}]{\text{Polyelectrolyte}} \textit{Flocs}$$

There are also instances where polyelectrolytes are effective in precipitating substances dissolved in solution.

It is clear from the above that the destabilization mechanism operative with polyelectrolytes is complex and cannot be collectively ascribed to one particular phenomenon. For a given system, there may be a predominance of charge effects, or adsorption, or chemical reactions at the functional groups. The predominance or relative combination of each phenomenon depends on the characteristics of the system and of the added polyelectrolyte.

This Chapter will discuss, in turn, mechanisms of destabilization by polyelectrolytes; the use of polyelectrolytes as primary coagulants; their use as flocculant aids where they are added to a system already destabilized; and as sludge conditioners where they are added to a system where dehydration reactions are required.

5.2 MECHANISMS OF DESTABILIZATION

5.2.1 General

As stated above, there is no single mechanism of destabilization by polyelectrolytes which may be considered applicable in all instances. However, it is possible to set down two principal mechanisms which in some instances may operate conjointly, whereas in others the predominance of one over the other is fairly simple to identify. In some cases, the two phenomena may operate in opposition. The two mechanisms are based on:

(1) A bridging model, where polyelectrolyte segments are adsorbed on the surfaces of adjacent colloids thereby binding them together;
(2) A model whereby ionic polyelectrolytes, bearing a charge of opposite sign to the suspended material, are adsorbed and thereby reduce the potential energy of repulsion between adjacent colloids. Associated with this second model is the phenomenon described in Chapter 2 where oppositely charged polyelectrolytes, by their presence in the double layers of particulate material, reduce the interaction energy. With (1) it is seen that (2) may support the action of the other, or oppose it, depending on whether there is strong electrostatic attraction between the polyelectrolyte and colloid surface. With (2) it is appreciated that if (1) contributes to destabilization, it will always support the action of the other.

The two mechanisms introduced above are referred to as the bridging mechanism and the electrostatic patch mechanism respectively. Although, for convenience, they will be described separately below, it should be reiterated that the overall mechanism of destabilization may in some instances be governed by both phenomena.

5.2.2 The bridging mechanism

The bridging mechanism of destabilization by polyelectrolytes has been an accepted phenomenon for some time (Healy & La Mer, 1962; Michaels, 1955; Ruehrwein & Ward, 1952). The principal basis for acceptance lies in the ability of charged polyelectrolytes to destabilize particles bearing the same charge (Fleer & Lyklema, 1974; Rebhun & Wacks, 1965; Michaels, 1955). To support this, electron micrographs have clearly identified polyelectrolyte bridges between particles (Audsley, 1963).

The stages in the bridging mechanism are summarized below, and schematically represented in Figure 5.1 (Akers, 1972):

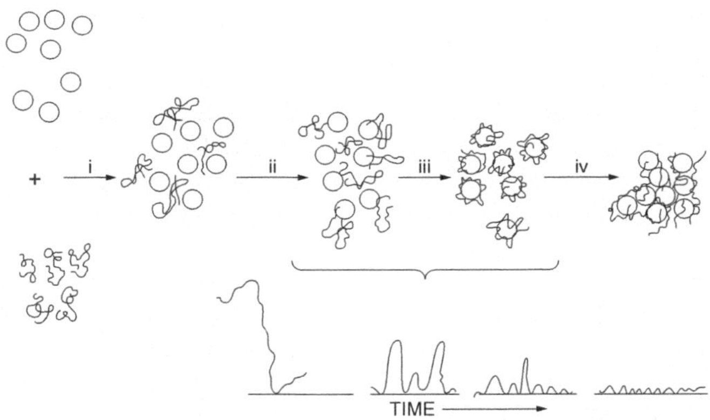

Figure 5.1 Stages in the bridging mechanism: (i) Dispersion; (ii) Adsorption; (iii) Compression or settling down (see inset); (iv) Collision (adapted from Akers 1972).

- Dispersion of polyelectrolyte in the suspension;
- Adsorption at the solid-liquid interface;
- Compression or settling down of the adsorbed polyelectrolyte and
- Collision of adjacent polyelectrolyte coated particles to form bridges and thereby increasingly larger flocs.

Each of these stages is discussed in detail below. Note that an additional step is the breakup of flocs formed by the above sequence of steps. Floc breakup is discussed further in Chapter 7.

5.2.2.1 Dispersion in the suspension

Because of their high molecular weights, polyelectrolytes in solution exhibit high viscosities and low diffusion rates. Since adsorption of polyelectrolytes to particle surfaces is essentially irreversible, for adsorption to occur evenly on all particles it is essential that the polyelectrolyte be dispersed evenly throughout the suspension. Since the adsorption reaction is usually much faster than the rate of diffusion, it is necessary to disperse the polyelectrolyte throughout the suspension as quickly and efficiently as possible. This is accomplished during the rapid mixing stage (Chapter 6) where a short vigorous mixing environment is required to effect dispersion in the shortest practicable time. Ensuring that the polymer solution is adequately diluted is also important. The rate of adsorption of polymer to particle surface is likely transport limited, and depends on this initial dispersion phase (Gregory & Lee, 1990).

5.2.2.2 Adsorption at the solid-liquid interface

After polyelectrolytes have diffused to the solid-liquid interface, adsorption of initially one functional group will occur while the rest of the chain is for a while momentarily free and extends into the solution (Packham & Sheiham, 1977) as depicted schematically in Figure 5.1. As time proceeds, the chain becomes successively attached at more points along its length due to continuous Brownian movement, until eventually there are no dangling ends extending into the solution phase. The configuration is now one of a number of pendant loops extending some distance out from the particle surface.

The mechanism of adsorption of polyelectrolyte segments to a solid surface depends on both the chemical characteristics of the polyelectrolyte and the adsorbent surface. For example, adsorption could be due to cation exchange, as reported for adsorption of cationic polyelectrolytes on clay particles, or to electrostatic linkages, hydrogen bonding or ionic bonding, for cases of anionic polyelectrolyte adsorption to clays (Holmes & Toth, 1957; Martin & Aldrich, 1955; McLaren, 1954; McLaren *et al.* 1958; Michaels & Morelos, 1955; Mortenson, 1957, 1959, 1961; Packter, 1957; Ruehrwein & Ward, 1952).

As mentioned in Chapter 2, polymer adsorption is described by the Simha-Frisch-Eirich (S.F.E.) adsorption isotherm (Birkner & Morgan, 1968). The Langmuir isotherm theoretically represents the S.F.E. isotherm for the special case of $\beta = 1$ (where β = the number of segments per polymer molecule attached to the adsorbent surface). However, polymer adsorption often follows the Langmuir isotherm, even when estimated values of β are, for example, of the order 470 to 1660 (Committee Report, 1971). Instances where the Langmuir isotherm may be inadequate are at low polyelectrolyte concentrations.

In the case of polyelectrolytes and particles carrying charges of like sign, adsorption may be strongly dependent on ionic strength. For example, for the case of hydrolyzed polyacrylamide and negatively charged kaolinite clay surfaces, adsorption has been found to be strongly dependent on the calcium ion concentration (Black *et al.* 1965). In general, with calcium ions a critical concentration of approximately 10^{-3} M is often found, which corresponds to hardness in excess of 100 mg/1 as $CaCO_3$. In some cases, the presence of cations such as Ca^{++} may have the effect of complexing with polyelectrolyte functional groups, thereby aiding adsorption of negative particles by anionic polymers by a cation bridging effect (Loganathan & Maier, 1975; Sommerauer, 1968).

Ionic strength may influence adsorption by two effects:

(1) Reduce repulsion between similarly charged particle surfaces and polyelectrolyte segments, thus permitting adsorption to occur
(2) Reduce the size of the polyelectrolyte coil, thus permitting more polyelectrolyte chains to be accommodated on the particle surface.

Adsorption is effectively irreversible (Silberberg, 1962–1968). Although the energy of adsorption at any particular site may be low and under normal circumstances may indicate a probability of desorption, the chance of desorption occurring simultaneously at a number of sites is low. Although under normal conditions of destabilization with polyelectrolytes, desorption of the whole chain from the particle surface does not occur, displacement may be induced by, for example, suitably adjusting pH conditions or adding a sufficiently high concentration of surface competitor, such as sodium tripolyphosphate (Kitchener, 1972).

5.2.2.3 Compression of adsorbed chains

The probability of a successful bridge forming between two adjacent particles depends on the configuration of the adsorbed polyelectrolyte at the surface. As stated previously (and schematically shown in Figure 5.1) from the onset of adsorption, progressively greater numbers of polyelectrolyte segments will become adsorbed on the particle surface and, consequently, the polyelectrolyte chain will become compressed to the surface. The polyelectrolyte layer requires a finite time to become compressed and initially, will consist of long loops.

It is during this initial period, where loops are longest and extend the furthest into solution, that bridging is most effective. With extensive adsorption of segments to the surface, the polyelectrolyte chain progressively assumes a flatter configuration until the extent of double layer repulsion exceeds the size of the loops. This effect is particularly pronounced for dilute colloidal suspensions where a statistically longer time is available for compression of polyelectrolyte loops. A possible solution is to add polyelectrolyte in stages thus ensuring that large loops are retained for periods sufficient for bridging to occur.

The configuration of the adsorbed polyelectrolyte chain depends upon the size of the polyelectrolyte (molecular weight); its structure; flexibility; charge density (% hydrolysis); interaction energy between polyelectrolyte and particle surface; chemical nature and physical spacing of adsorption sites on the particle surface; and the competition between polyelectrolyte and other adsorbing molecules in solution (Committee Report, 1971).

In general, where the bridging mechanism is predominant, an increase in molecular weight and chain length gives rise to an increase in the optimum polymer/

solid ratio. This is because longer chains do not lie as flat on the surface as shorter chains and, therefore, individual chains do not occupy as many adsorption sites. Therefore, bridging is more efficient with polyelectrolytes of higher molecular weight, of several million (Gregory, 1993) since loops will tend to extend further into solution. This is evident from the fact that, with appropriate systems, for a given applied polyelectrolyte concentration, the higher the molecular weight the higher is the efficiency of destabilization (assessed in terms of say, size of flocs, settling rates, filterability and so on). As stated previously, the upper limit of molecular weight is governed by practical considerations of dissolution.

Flexibility of polyelectrolyte segments influences the size of the pendant loops: a relatively inflexible polyelectrolyte chain will tend to be attached by only a few segments and the loops will tend to be longer than for a relatively flexible chain.

The higher the percentage hydrolysis and charge density, the greater is the repulsion between adjacent segments and the more extended is the polyelectrolyte chain for a given molecular weight. The more extended the polyelectrolyte the further should the pendant loops extend into solution and the more effective should be the bridging mechanism. However, unlike molecular weight there is an upper limit to the charge density, beyond which electrostatic repulsion (in the case of polyelectrolyte and solid surface of similar charge characteristics) will effectively retard adsorption. For example, Michaels (1954) found that with polyacrylamides a 30% degree of hydrolysis was optimal; this value providing a compromise between extended loops and polyelectrolyte surface repulsion.

Ionic strength considerations to a certain extent follow a similar reasoning: as stated earlier, increasing ionic strength tends to increase the efficiency of polyelectrolyte adsorption. Furthermore, increasing ionic strength reduces the range of inter-particle repulsion (after polyelectrolyte adsorption) so that particles have a statistically high opportunity of approaching close enough together for bridging to take place before extensive compression of polyelectrolyte segments occurs. However, on the other hand, the greater the ionic strength, the less extended is the polyelectrolyte chain and the shorter is the range of extension of polyelectrolyte loops. Again, optimum conditions will be governed by a balancing of the above factors.

5.2.2.4 Bridge formation

After adsorption has taken place as described in the preceding sections, polyelectrolyte loops extending into solution from the particle surfaces will further become adsorbed onto adjacent particles thus forming a number of bridges. The strength of the flocs thus formed depends on the number of bridges formed, which in turn is dependent on the number of loops available. From preceding discussions, the number of loops available for mutual adsorption between adjacent particles, before extensive compression takes place and the loops are within the bounds of the respective double layers, is dependent on a number of factors related to both the suspension and the added polyelectrolyte.

A factor of crucial importance during bridging is the availability of adsorption sites on particles to accommodate polyelectrolyte loops from neighbouring particles. Such availability depends on the concentration of polyelectrolyte added. If an excess of polyelectrolyte is added, too many adsorption sites per particle will be occupied and bridge formation is prevented: the particles effectively become restabilized.

Some workers postulated that optimum flocculation occurs when the fraction of surface covered by polyelectrolyte θ, is one half, i.e. $\theta = 0.5$ (La Mer *et al.* 1956–1957; Smellie & La Mer, 1956, 1958).

However, many workers have found that maximum flocculation occurs at θ values much less than 0.5, typically ranging from 0.09 to 0.33 (Black *et al.* 1965; Kragh & Langston, 1962). It has also been suggested that the parameter θ is relatively unimportant in determining the degree to which a suspension will be destabilized when other factors such as particle concentration, inter-particle interactive forces, ionic strength, polymer configuration in the adsorbed state, and intensity and duration of mixing are considered (Committee Report, 1971). However, for many systems a stoichiometric polyelectrolyte dosage-particle concentration relationship often does exist (Black & Vilaret, 1969; Packham, 1967; Shea *et al.* 1971) indicating the predominance of adsorption during destabilization reactions.

Intense mixing may disrupt polyelectrolyte bridges and give rise to desorption and/or rearrangement of looped chains on the particle surface. Therefore, although efficient mixing is required when polymers are added to suspensions, the mixing should not be too violent nor for too long a period. The addition of dilute polyelectrolyte solutions is clearly of advantage. The requirements for rapid mixing are dealt with in Chapter 6.

5.2.3 The electrostatic patch mechanism

For the case of non-ionic and anionic polyelectrolytes applied to a negatively charged colloidal dispersion, a destabilization mechanism described by the bridging model adequately accounts for the phenomena taking place. Indeed, in many cases it would be difficult to account for the phenomena by a mechanism other than some form of bridging (Gregory & Sheiham, 1974). However, for the case of charged polyelectrolytes applied to dispersions with particles carrying surface charges of opposite sign, the bridging model is often inadequate. Such systems include cationic polyelectrolytes applied to a negative colloidal dispersion and could include anionic polyelectrolytes applied to dispersions destabilized with metal coagulants which may be positively charged (Sarker & Teot, 1973).

Kasper (1971) and Gregory (1973) questioned the exclusivity of the bridging model. For systems where strong electrostatic attraction between polyelectrolyte and particle surface exists, they proposed an electrostatic patch mechanism. Rather than adsorption of polyelectrolyte at only a few sites, with the remainder of the chain extending into solution in the form of closed loops, virtually complete adsorption of added polyelectrolyte onto the particle surface takes place with such systems. The adsorbed polyelectrolyte chains thus form a charge mosaic with alternating regions of positive and negative charge as shown schematically in Figure 5.2. Destabilization occurs when the charge mosaics of adjacent particles align to provide strong electrostatic attraction. Conceptually, the electrostatic patch model as described above bears a resemblance to the electrical double layer model of colloid destabilization, where the adsorption of counter-ions in the Stern layer brings about a reduction in the potential energy of repulsion between particles.

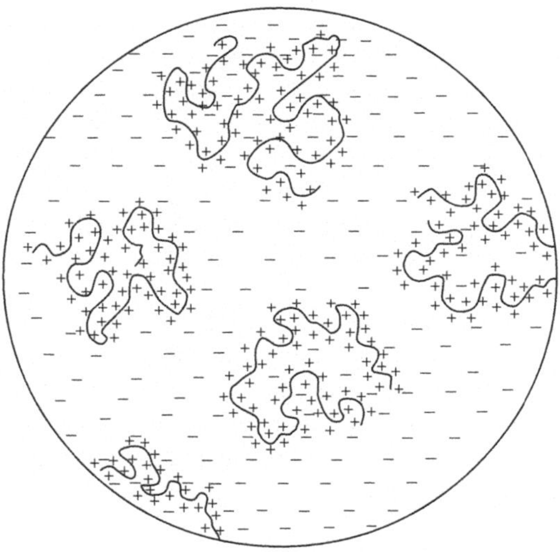

Figure 5.2 Possible arrangement of adsorbed polycations on a particle with low negative surface charge density (from Gregory, 1973).

For high energies of adsorption where $kT < E_{SS} < 10\,kT$ (E_{SS} = energy of interaction between polyelectrolyte segments and particle surface, k = Boltzmann's constant and T = absolute temperature), as exist with polyelectrolyte segments attracted to oppositely charged surfaces, the possibility of bridging via long loops or tails has been suggested as unlikely (Kasper, 1971), even for polyelectrolytes of high molecular weight. However, there are instances where high molecular weights do give rise to increased destabilization efficiency in such systems. This can be explained by a consideration of kinetic effects (Gregory & Sheiham, 1974), as follows:

Adsorption of a polyelectrolyte chain takes place by a series of individual attachments over a finite period of time. As more segments are adsorbed, the configuration of the adsorbed chain becomes progressively flatter and loops become less extensive until some equilibrium state is reached. In the interim between the initial stages of adsorption to the attainment of the equilibrium configuration, there is an interval of time within which polyelectrolyte loops may extend sufficiently far into solution for bridging to take place between adjacent particles.

For a bridging mechanism to be effective, destabilization would need to be virtually complete within the time required for loops to be compressed and be ineffective for bridging. This would be greatly influenced by the collision frequency of particles and, therefore, is directly related to particle concentration. At high particle concentrations a given particle will collide with many others while polyelectrolyte loops are still in a relatively extended state.

Evidence for the above mechanism has been demonstrated with the destabilization of silica and latex particles using cationic polyelectrolytes (Gregory, 1973; Gregory & Sheiham, 1974; Gregory & Lee, 1990). At low particle concentrations, the destabilization mechanism is as described by the electrostatic patch model. Here, since bridging is of negligible consequence, the molecular weight of the polyelectrolyte would have little importance: in view of the difficulty of dissolving high molecular weight products, a polyelectrolyte of low molecular weight is probably preferable. Furthermore, since charge effects are predominant, the use of polyelectrolytes of high charge density should prove most effective.

At high particle concentrations, such as is the case with sludge conditioning, considerable bridging effects are evident and cationic polyelectrolytes with high molecular weights should prove beneficial. In contrast, the results of Gregory's work showed that bridging is negligible for particle concentrations less than of the order 10^{14} particles/l. He points out that typical river waters contain less than of the order 10^{12} particles/l. Therefore, in such applications, if polyelectrolytes as primary coagulants are considered, the use of high molecular weight cationic polyelectrolytes may not be justified.

This was confirmed by Leu and Ghosh (1988) who studied the destabilization of montmorillonite (bentonite) clay suspensions, at turbidities of around 25 NTU, with cationic polyelectrolytes. They found that charge density was important, rather than molecular weight, in selecting the best polyelectrolyte.

Eriksson and Alm (1990) investigated the flocculation of negatively charged latex particles with cationic polyelectrolytes, with low, medium and high charge densities, and two molecular weights of 10^5 and 10^6. They found that at low polymer charge density, a bridging mechanism resulted in compact, shear resistant flocs, but with poor reflocculation ability after floc rupture.

In contrast, at higher charge densities, an electrostatic patch mechanism was operative that produced fractal flocs with less shear resistance, but with good reflocculation ability after floc rupture.

Table 5.1 summarizes the characteristics of the two mechanisms described respectively by the bridging model and electrostatic patch model.

Table 5.1 Summary of characteristics of mechanisms.

Parameter	Influence of Indicated Parameter According to Model	
	Bridging	**Electrostatic Patch Model**
Molecular weight of polyelectrolyte	The higher the M.W. the more effective the bridging. Upper limit dictated by dissolution	In strict accordance with model, M.W. should have no effect. Improved performance evident in some cases with higher M.W. due to onset of bridging
Charge density, or percentage hydrolysis	Bridging more effective at higher charge density due to larger loops. Upper limit dictated by electrostatic repulsion	The higher the charge density, the more pronounced will be the charge mosaic and the more effective will be destabilization
Polyelectrolyte of similar charge to particle surface	Destabilization by bridging mechanism	Electrostatic patch mechanism inoperative
Polyelectrolyte of opposite charge to particle surface	Destabilization by bridging mechanism possible	Destabilization by electrostatic patch mechanism possible
Non-ionic polymer	Destabilization by bridging mechanism	Electrostatic patch mechanism inoperative
Ampholytic polyelectrolyte	Destabilization by bridging mechanism possible	Destabilization by electrostatic patch mechanism possible
Effective surface charge at optimum conditions for destabilization	Usually not zero	Not necessarily zero
Effective surface charge with excess polymer adsorption	Possible reversal of charge by excessive adsorption. Destabilization still possible	Possible reversal of charge by excessive adsorption. Destabilization still possible

Polymer concentration	Optimum destabilization when surface site coverage, $\theta \leq 0.5$	Optimum destabilization not limited by residual adsorption sites. However, θ probably < 0.5
Excess polymer addition	At excessive dosages, particles restabilized by complete site coverage or charge reversal	At excessive dosages, possible restabilization from charge reversal
Mixing	Important for polymer to be adsorbed evenly on particles. Mixing should be short and vigorous at the time of polymer addition. If mixing too violent or for too long a period desorption and/or rearrangement of adsorbed chains could give rise to restabilization	
Particle concentration	Bridging most effective at higher particle concentration. Longer time available for compression at low concentrations	Bridging mechanism likely at high concentrations because of kinetic effects. Electrostatic patch mechanism possible at low concentrations ($<10^4$/l)
Ionic strength	High ionic strength aids adsorption by reducing electrostatic repulsion for polyelectrolyte and particles of opposite charge; Reduces size of polymer coils and permits adsorption of more segments; Aids destabilization by reducing electrostatic repulsion between particles of like charge; Impedes bridging by reducing polymer loops.	High ionic strength probably aids destabilization by reducing repulsive interaction energy between polyelectrolyte adsorbed particles.

5.3 POLYELECTROLYTES AS PRIMARY COAGULANTS

5.3.1 General

There are many instances where polyelectrolytes have been employed as primary coagulants, effectively replacing the use of metal coagulants for this purpose. Such applications include treatment of waters predominantly turbid or colored with humic substances, removal of microorganisms and various industrial applications. In cases where replacement of metal coagulants by polyelectrolytes is possible, the potential advantages are as follows (Beardsley, 1973):

- Soluble metal ion species carry-over from sedimentation basins is prevented
- There is little need for extensive pH adjustment (and subsequent readjustment)
- Carry over of light flocs such as those formed with metal coagulants is minimized
- Sludge volumes for ultimate disposal are reduced
- Improved sludge thickening and dewatering characteristics
- Reduction in the amount of soluble anions.

However, there are disadvantages with the use of polymers as primary coagulants:

- Polymers tend to be very specific to particular types of material. Therefore, overall water quality in some cases may not be acceptable for all parameters
- Many polymers have a low resistance to oxidizing agents
- Many polymers have a relatively short storage life

The following sections present various examples where polyelectrolytes have been successfully applied as primary coagulants.

5.3.2 Turbidity removal using polyelectrolytes

Polyelectrolytes have been applied successfully as primary coagulants in a number of cases to reduce the turbidity of surface waters: Robinson (1974) investigated the use of polyelectrolytes to clarify highly turbid waters (maximum turbidity approximately 5,000 JTU). Using solid and liquid cationic, nonionic and anionic types, he found that the cationic and nonionic types gave very good results, comparable to aluminum sulfate, whereas the anionic types did not produce the required degree of clarification. With the cationic types, dosages ranged from 1 to 4 mg/l depending on the raw water turbidity (range 190 to 2,000 JTU).

A number of workers have reported on the success of using cationic polyelectrolytes as primary coagulants for direct filtration applied to relatively low turbidity waters (Shea *et al.* 1971; Adin & Rebhun, 1974; Habibian & O'Melia, 1975; Terrell, 1977; Letterman *et al.* 1979; Yeh & Ghosh, 1981; Carns & Parker, 1985; Tanaka & Pirbazari, 1986). In general, cationic polymers allow higher filtration rates through coarser media than metal coagulants and, in many cases provide better turbidity removals. Adin and Rebhun observed that the working layer during filtration (where, above the working layer the filter media is saturated with deposit, whereas below it is relatively clean) was broad, poorly defined, and advanced more rapidly when using alum. In contrast, cationic polyelectrolytes produced a better defined layer that advanced at a slower rate. This means that breakthrough was delayed when using polymers.

Letterman *et al.* found that flocculation before direct filtration serves to maximize filter production in terms of water produced per filter run, and delays turbidity breakthrough.

Yeh and Ghosh found that flocculation should be for a short time and at relatively high G values to optimize direct filtration performance. They also suggested that restabilization following over-dosing of the polymer is due to charge reversal rather than to saturation of bridging sites. This follows the reasoning for the electrostatic patch mechanism discussed earlier.

Habibian and O'Melia found that for the polyethylenimene cationic polyelectrolytes studied by them, molecular weight was unimportant in determining the optimum polymer dosage. They also found that with trickling filter effluents, which contain a high proportion of colloidal material, higher stoichiometric amounts of polymer are required. Flocculation would be beneficial in this case, to extend filter runs.

Black *et al.* (1965, 1969) reported a stoichiometric reaction between polyelectrolyte dosage (diallyldimethylammonium chloride) and particle concentration: From Figure 5.3, the optimum dosage/kaolinite clay concentration ratio was of the order 0.002 g/g. Such stoichiometry is to be expected and follows from the overriding influence of adsorption during destabilization with polyelectrolytes; whether the mechanism is governed by the bridging or electrostatic patch model. A possible exception, where stoichiometry may not be evident, would be if the mechanism was principally that of double layer repression as could exist with polyelectrolytes and particles of opposite sign. However, even though this effect is almost certainly present in such cases, there is no evidence to suggest that adsorption would not still be the predominant phenomenon promoting destabilization.

Although stoichiometric relationships as above may be gleaned from the results of several workers, it should be noted that initial turbidity is not sufficient information for an *a priori* estimation of the polyelectrolyte dosage required. The presence of other substances in a particular water may exert an appreciable polyelectrolyte demand over and above that required for turbidity removal. Such substances, which include organic color, anionic detergents, lignins, orthophosphates and so on, react with polyelectrolytes to form soluble complexes or colloidal precipitates (Birkner, 1967).

For turbidity removal, therefore, the polyelectrolyte dosage is over and above that required for these substances. Furthermore, the adsorption of chemical species to particles producing turbidity, such as clays, will reduce the number of adsorption sites for polyelectrolyte and/or alter the adsorption characteristics of polyelectrolyte to the particles.

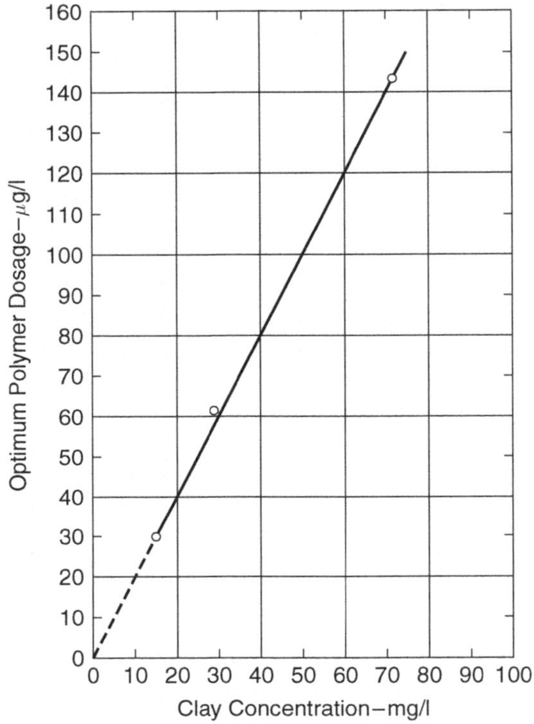

Figure 5.3 Stoichiometry of destabilization of kaolinite clay suspensions with cationic polyelectrolyte PDADMA (from Black *et al.* 1965).

For a given particle mass concentration the optimum dosage depends on the type of material present. For example, smaller particle suspensions require higher dosages of polyelectrolyte at optimum destabilization, because of the higher total surface area. Furthermore, the larger the range of particle sizes of a given suspension, the wider is the range of polyelectrolyte dosage giving good performance. For example, with latex particles the surface area characteristics are generally more uniform and optimum destabilization occurs at a relatively precise polyelectrolyte dosage, when compared with, say, a kaolinite clay suspension. From the results of Birkner and Morgan (1968) stoichiometry is also evident with latex suspensions: using a cationic polyelectrolyte (polyethylenimine), at optimum conditions for destabilization, the dosage: particle number ratio is of the order 2×10^{-12} mg.

Morrow and Rausch (1974) found that cationic polyelectrolytes (polydiallyldimethylammonium chloride) were able to replace metal coagulants successfully for both low and high turbid waters. They pointed out that the often reported unsatisfactory performance of polyelectrolytes when used alone, may be due to too low velocity gradients during the rapid mixing stage. Optimum results were obtained using velocity gradients greater than 400 s^{-1}. The lower the polyelectrolyte dosage, the higher should be the velocity gradient. For example, at a dosage of 2 mg/l, the optimum rapid mix velocity gradient was 500 s^{-1}. Using 1 mg/l the optimum velocity gradient was 750 s^{-1}.

As stated earlier, there is an upper limit to velocity gradient. For example, Morrow and Rausch found that velocity gradients of 1000 s^{-1} resulted in poor overall solids capture, probably due to shearing of polyelectrolyte segments from particles.

Terrell (1977) described the evolution of changing from alum to cationic polymers as filtered water quality goals changed gradually from 1.1 NTU in the 1950s through to 0.02 NTU in 1976, at the Ithaca conventional water treatment plant. Alum based coagulation gave rise to wide seasonal swings in filtered water quality, which was largely resolved by changing to a combination of three polymers: a low charge density cationic, followed by a high charge density cationic, with a nonionic starch polymer used as filter aid. Raw water turbidities ranged from 2 to 176 NTU, with monthly averages varying from 11 to 84 NTU.

Deeble (1976) describes the use of cationic polyelectrolytes (polyamines) in clarifying a number of turbid (and colored) waters. With turbidities ranging from 5 to 30 mg/l SiO_2 (and color 30 to 140 mg/l Pt-Co) dosages of polyelectrolyte from 1.5 to 9.0 mg/l reduced the turbidity and color to approximately 1 mg/l SiO_2 and 5 mg/l Pt-Co units respectively.

Robinson (1979) reported on the successful application of cationic polyelectrolytes to pre-treat highly turbid river water (average 350 NTU with peaks as high as 5,000 NTU). Raw water turbidities of 4,000 to 5,000 NTU were reduced to less than 100 NTU in the pre-sedimentation basins, greatly facilitating subsequent conventional treatment. Together with turbidity, organic color was removed from 10 to 20 units down to approximately 1 unit.

Results obtained by Nozaic *et al.* (2001) illustrate the different dosage-response characteristics when using alum and cationic polyelectrolytes as primary coagulants. Data reproduced in Figure 5.4 show a narrow range of dosage for optimum turbidity removal with the PDADMAC polyelectrolyte used by these workers. Higher dosages resulted in restabilization due to charge reversal. In contrast, alum did not require such precise control. Basically higher dosages improved performance. Nozaic *et al.* found that adequate control of the polymer dosage was afforded by streaming current for a wide range of raw water turbidities.

The minimum turbidity attainable with the polyelectrolyte was approximately 0.2 NTU. With alum, filtered turbidities below 0.05 NTU were obtained. Alum also had the advantage that organics removal through enhanced coagulation was possible. This was not the case when using the polyelectrolyte as primary coagulant.

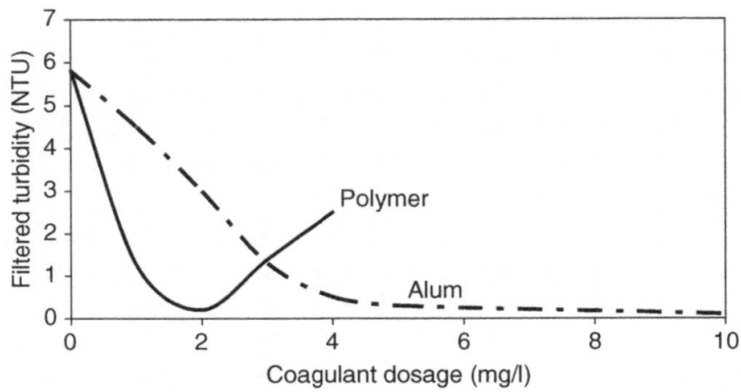

Figure 5.4 Dosage response curves for alum and cationic polyelectrolyte (data from Nozaic *et al.* 2001).

5.3.3 Organics removal using polyelectrolytes

For polyelectrolytes to be generally acceptable as primary coagulants for potable water treatment, it is likely that they should be capable of removing organic material as well as turbidity. By conducting tests with a wide variety of commercial cationic polyelectrolytes and assessing their effectiveness in precipitating fulvic acid (the predominant humic fraction in colored waters) Packham (1972) found that the polyelectrolytes varied widely in their ability to remove organic color.

As seen in Figure 5.5, removals ranged from 3% to 87% with optimum dosages (at maximum color removal for each product) from 2 to 16 mg/l. Although some polyelectrolytes were effective in precipitating fulvic acids (notably polyethylenimine and the product *Primafloc C7*), in almost all cases the flocs formed were very poor, settled slowly and gave rise to high final supernatant turbidities.

It was suggested that the poor flocculation characteristics were because most polyelectrolytes do not form insoluble hydrolysis products. Therefore, they lack the advantage of metal coagulants applied to colored or low turbid waters where hydrolysis products increase particle number concentrations thereby improving flocculation kinetics. An exception was the product *Floccotan*, a tannin based natural polyelectrolyte, which did exhibit acceptable floc formation characteristics. This product reacts with water to form an insoluble hydrolysis product. However, *Floccotan* effected only an average color removal.

By adding montmorillonite to the systems tested, Packham found that with added clay concentrations of at least 5 mg/l, the floc characteristics were markedly improved. Hence, by adding clay to colored and/or low turbid waters, satisfactory color and/or turbidity removals may be achieved by adding the appropriate polyelectrolyte. From Packham's results it appears that the polyelectrolyte dosage may be from 6 to 10% of the corresponding metal coagulant dosage.

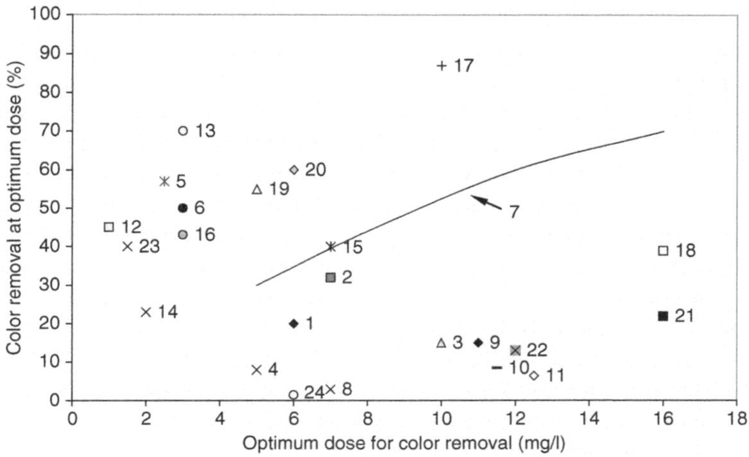

Figure 5.5 Precipitation of fulvic acid with cationic polyelectrolytes. Results of screening tests with 24 commercial polymers (fulvic acid 10 mg/l and pH 5.5): (1) BTICL80; (2) BTICL40; (3) BTIC100; (4) BTIC110; (5) Catafloc; (6) Cirrasol Z; (7) Floccotan; (8) Katafloc; (9) Magnafloc R225; (10) Magnafloc LT22; (11) Magnafloc LT24; (12) Nalfloc N607; (13) Polyethyleneimine; (14) Priamfloc C3; (15) Primafloc C5; (16) Primafloc C6; (17) Primafloc C7; (18) R177; (19) RD2180; (20) RD2181; (21) Reten 205; (22) Reten 210; (23) Superfloc 521; (24) Wisprofloc P. (After Packham 1972).

Amy and Chadik (1983) tested organics removal (measured as the removal of trihalomethane formation potential, THMFP, through 0.45 μm filters) using four different cationic polyelectrolytes on seven different natural waters. They achieved THMFP removals ranging from 37 to 65% depending on the water and the polymer. For four of the waters, a low molecular weight polyalkyl polyamine polymer was best. For two of the waters an ultra-low molecular weight quaternary polyamine was best. For one of the waters, a low molecular weight quaternary ammonium polymer was best.

Amy and Chadik also tested four synthetic waters: two with humic acid, with and without the addition of kaolin clay; and two with fulvic acid, with and without kaolin. For the humic acid water, reductions achieved were over 90%, with or without kaolin. For the fulvic acid water, removals achieved were 79% with kaolin, and 61% without kaolin. The polyelectrolyte dosage required for the fulvic acid water was 50 mg/l, compared with 30 mg/l for the humic acid water. They interpreted the results as follows:

- The higher removals of humic acid compared with fulvic acid were due to the lower charge density of the humic acid molecules
- The higher removals with kaolinite clay were achieved because of the nucleation sites for floc formation provided by the clay particles.

Narkis and Rebhun (1975 and 1977) described the destabilization of humic substances with cationic polyelectrolytes (polyethylenimine) as essentially the result of chemical reactions between polyelectrolyte segments and negatively charged carboxylic and phenolic groups on the humic acid molecules. Both humic and fulvic acid solutions react with the polyelectrolyte to form colloidal precipitates which may be observed as turbidity. With incrementally increasing dosage, optimum conditions of polyelectrolyte dosage required for precipitation were identified by maximum turbidity. Such optima occur at electrophoretic mobilities slightly more or less than the isoelectric point. However, the optimum dosage corresponding to the above usually does not correspond to that required for floc formation.

By adding further polyelectrolyte, bridging between adjacent precipitated particles occurred and settleable and/or filterable flocs were formed. Apparently, the polyelectrolyte required for floc formation need not necessarily be the same as that required for preliminary precipitation. Therefore, the use of sequential polyelectrolyte addition may be advantageous.

The hypothesis of chemical interactions as described above for destabilization of humic substances by polyelectrolytes is supported by the often reported stoichiometry of the reaction. For example, Figure 5.6 (Edzwald *et al.* 1977) demonstrates the stoichiometry between initial humic acid concentration and polyelectrolyte dosage required under optimum conditions of destabilization.

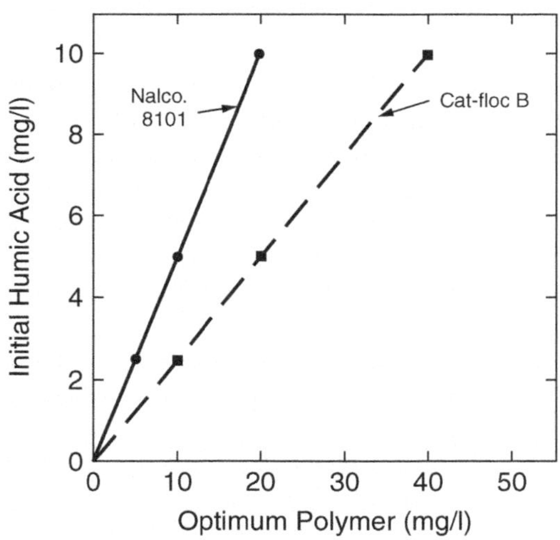

Figure 5.6 Stoichiometry between initial humic acid concentration and optimum cationic polyelectrolyte dosage (from Edzwald *et al.* 1977).

As stated previously, subsequent floc formation after precipitation of humic substances with cationic polyelectrolytes may be improved by increasing the particle concentration with clay addition. With such mixed systems the polyelectrolyte reacts first with the humic substances and only after precipitation is complete will destabilization of clay particles occur (Narkis & Rebhun, 1975, 1977).

Precipitation of humic acids by cationic polyelectrolytes is markedly dependent on pH. Some workers have found a non-linear relationship such that with a decrease in pH from 6.0 to 2.8, the optimum polyelectrolyte dosage required for precipitation decreased from 3.0 to 0.8 mg/l. This effect was attributed to a decrease in the degree of ionization of the humic molecules thus resulting in lower negative charges. Therefore, the lower the pH, the lower is the required polyelectrolyte dosage (Narkis & Rebhun, 1975, 1977).

Several workers have shown that charge neutralization plays an important part in the destabilization reactions between cationic polyelectrolytes and humic substances (Narkis & Rebhun, 1977; Glaser & Edzwald, 1979). Therefore, it is generally accepted that cationic polyelectrolytes with higher charge densities would be more effective for the removal of humic substances. For the same reason one expects optimum maximum dosages, beyond which restabilization and reduced removals result.

Kvinnesland and Ødegaard (2004) showed that with a poly-epichlorohydrin-dimethylamine (poly-Epi-amine) cationic polymer, a polydiallyldimethyl-ammonium chloride (PDADMAC) cationic polymer, and a low molecular weight (300,000 Dalton) cationic polyacrylamide copolymer (LMW CPAAM), there were steady increases in organic color removal through 0.1 μm filters, up to specific dosages of 2.0 to 2.5 mg/mgC, beyond which there were sharp reductions in removals. An exception was the use of chitosan, which followed the same increase in removals with dosage, but continued with improving performance up to a specific dosage of approximately 4 mg/mgC. In general, the higher charged polymers performed better in terms of dosage, than the lower charged polymers.

The charge densities of each of the polymers were poly-Epi-amine: 7.0 meq/g; PDADMAC: 6.2 meq/g; LMW CPAAM: 4.2 meq/g; and chitosan: 4.5 meq/g. Maximum color removals (through 0.1 μm filters) were achieved at specific cationic charge dosages of approximately 11 to 13 μeq/mgC, depending on the polymer. Again, chitosan exhibited increasing removals up to approximately 18 μeq/mgC. The removal efficiencies evaluated in terms of increasing specific cationic charges, performed similarly for the different polymers, except for chitosan as noted above.

The results cited above were at pH 6.0. Kvinnesland and Ødegaard showed that at pH 4, PDADMAC performed significantly better than chitosan. The explanation is that the decreased dissociation of carboxyl groups on the humic molecules reduces the anionic charge density, and polymers able to interact with the increased hydrophobic character of the molecules would perform better. Since PDADMAC has more non-polar characteristics than chitosan, it appears to follow that it would perform better at lower pH values.

These workers also showed no beneficial effect of molecular weight on the removal of humic substances.

The results above related to the destabilization reactions of humic substances with cationic polymers. In terms of flocculation, for ultimate removal, the results were somewhat different. The extent of floc formation, evaluated by the separability of the flocs through 11 μm filters, was not explained in the results of Kvinnesland and Ødegaard by optimal specific cationic charge dosages. The optimal dosages for flocculation were within a narrower range than for destabilization.

Glaser and Edzwald (1979) had observed similar behavior and had suggested that aggregation of the destabilized molecules occurs through cross linkages arising from electrostatic and specific chemical interactions. Efficient cross linkages require a more precise ratio between the humic substances and the polymer, when compared with the destabilization reactions that precede aggregation.

The results of Kvinnesland and Ødegaard demonstrated that polymers with the highest cationic charge densities produced the best flocculation. An exception was chitosan that formed flocs almost as well as poly-Epi-amin, the polymer with the highest charge density. It was suggested that the higher stiffness of the chitosan molecule may have promoted flocculation.

5.3.4 Algae removal and harvesting using polyelectrolytes

Chapter 4 discussed algae removal using metal coagulants. Concerns with the quality of the recovered algae because of high metal concentrations has promoted the use of alternative coagulants including synthetic and natural polymers.

Polyelectrolytes have proved effective in destabilizing algae: suspensions of *Chlorella* and *Scenedesmus* have been effectively destabilized using cationic polyelectrolytes (Cohen *et al.* 1958; Golueke & Oswald, 1965; Tilton *et al.* 1972; Haarhoff & Cleasby, 1989). Using polyethylenimine, there appeared to be a significant increase in effectiveness with increasing molecular weights up to 2,000 (determined by viscosity measurements). However, beyond 2,000 no further improvement was evident. The polyelectrolyte dosage required for algal suspensions (*Chlorella ellipsoidia*) can be much higher than for bacterial suspensions – approximately 30 times the dosage requirements of *E. Coli.* (Tilton & Murphy, 1972). These workers attributed the higher dosage requirements for algae to structural differences of the cell envelopes of the organisms.

Tests with suspensions of *Chlorophyta* and a cationic polyamine polyelectrolyte showed that optimal coagulation occurred when the cationic polyelectrolyte covered approximately 50% of the algae cell surface. This surface coverage corresponded with the zero point of charge measured by electrophoretic mobility. A stoichiometric relationship between algae concentration and cationic polyelectrolyte dosage was demonstrated. The tests also indicated that the optimum pH with this algae species and polyelectrolyte was approximately 3.0. At lower pH the algae cells exhibit their lowest surface charge and the cationic polyelectrolyte will be highly charged and extended, promoting effective destabilization (Tenney *et al.* 1969).

As discussed in Chapter 4, the algal growth phase also influences destabilization. Tests with the same alga (*Chlorophyta*) and cationic polyelectrolyte showed that the least amount of polyelectrolyte was required during the late log and early declining growth phases. During these phases of growth there is an accumulation of extracellular material that comprises long chain polysaccharides, proteins and nucleic acids that act as flocculants, creating bridges between algae cells (Tenney *et al.* 1969).

The negatively charged algal extracellular polymeric material reacts with cationic polyelectrolytes by a charge neutralization/precipitation mechanism. After these neutralization reactions, the cationic polymer then functions as a flocculant of the precipitated particles. Cationic polymers with higher charge densities have performed better than those with lower charge densities (Haarhoff & Cleasby, 1989). Flocculation applied to the water after polymer addition improved initial removals and reduced the ripening period required for subsequent filtration. The rate of head loss development through the filter was also reduced.

Several workers have shown that beyond an optimum dosage, increasing dosages of cationic polyelectrolyte cause charge reversal and restabilization of the algae suspensions (Tenney *et al.* 1969; Tilton *et al.* 1972; Bernhardt & Clasen, 1991; de Godos *et al.* 2011).

In some cases a combination of polymers may be effective. Figure 5.7 shows jar test results on an algae suspension with a high charge polyamine cationic polymer at the indicated dosages followed by 1 mg/L of a high molecular weight cationic polyacrylamide polymer. Without the second polymer algae removals were inadequate.

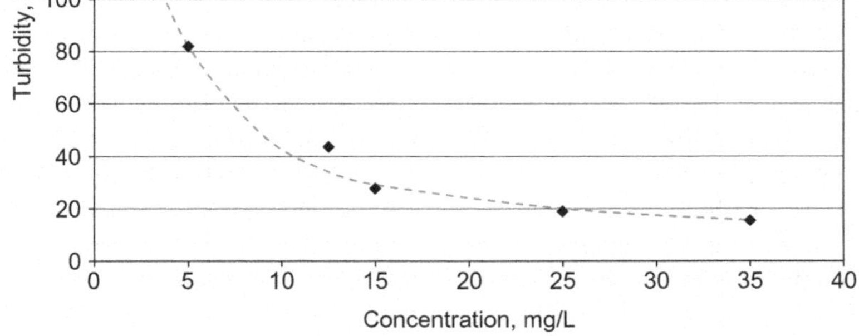

Figure 5.7 Jar test results using a combination of a high charge polyamine polymer with 1 mg/L of a high molecular weight polyacrylamide cationic polymer.

Tests on an algal suspension comprising *Chlorella sorokiniana*, *Scenedesmus obliquus*, *Chlorococcum sp.* and others, together with a bacterial consortium were coagulated with various cationic polyacrylamide polymers and chitosan. All polymers exhibited an optimal dosage, beyond which deterioration in algae removal was evident. The chitosan product was not as effective as the synthetic cationic polyacrylamide products. Chitosan is a natural cationic polymer, usually prepared in acetic acid. It is possible that the progressively reduced pH at higher chitosan dosages may have influenced the results obtained (de Godos *et al.* 2011).

Coagulation tests on several freshwater algae species: *Spirulina*, *Oscillatoria*, *Chlorella* and a brackish species: *Synechocystis* with chitosan as coagulant showed that coagulation effectiveness is very sensitive to pH, reaching a maximum at pH 7 for the freshwater species, but lower for the brackish species. At the optimum pH of 7 for the freshwater species, the optimum dosage of chitosan varied from approximately 2 mg/L to 10 mg/L, depending on the algal concentration. In general higher algae concentrations required higher chitosan dosages, although the dosage-concentration relationship was not consistent (Divakaran & Pillai, 2002).

Chitosan was also applied to algae removal from shrimp culture ponds because of safety concerns with chemicals added to the cultures. The nitrogenous residue developed during intensive shrimp cultures can be assimilated by algae growth and removal from the culture ponds. In this case the dominant algae was the diatom *Chaetoceros* sp. A novel pH adjustment regime was adopted which maximized the efficacy of the chitosan, with approximately 80% removal of the suspended algae: at 40 mg/L chitosan dosage the pH was first adjusted to 6.5 which is the isoelectric point where strong attraction between the bridging chitosan molecules and algae cells occurs. However this low pH is not appropriate for the shrimp aquaculture and the pH was raised to 8.5, after destabilization takes place. Such a sequential pH effect has been found effective elsewhere (see Chapter 4 and manganese removal). By first destabilizing the algae cells at the optimum pH for chitosan, then adjusting the pH to a value appropriate for the aquaculture, algae removals were maximized with removals of over 80%, compared with approximately 30% when chitosan was added only at pH 8.5 (Lertsutthiwong *et al.* 2009).

A mixture of chitosan and magnetite (Fe_3O_4) powder was used to coagulate *Microcystis aeruginosa* (Liu *et al.* 2009). The optimal dosage for the algal suspension was 2 mg/L chitosan with 4 mg/L magnetite. Initial concentrations of algae were approximately 4.8×10^6 cells/mL and turbidity 125 NTU. After treatment, concentrations were 0.0028×10^6 cells/mL and turbidity 0.5 NTU. Control tests with chitosan alone or magnetite alone showed that the combined dosage of chitosan plus magnetite was far more effective with removals of over 99%, compared with less than 30 to 40% with chitosan or magnetite alone.

Another approach to avoid the contamination of recovered algae biomass and the high cost of coagulant chemicals is to use the flocculating ability of bacteria, in a similar fashion to the phenomena occurring in activated sludge systems. This was demonstrated by the flocculation of the marine algae *Pleurochrysis carterae*

(Lee *et al.* 2009) and *Chlorella vulgaris* (Oh *et al.* 2001) by colonies of bacteria. Recovery efficiencies of over 90% were achieved. The type of carbon (acetate, glucose, or glycerine) or excess carbon substrate did not influence results. Substrate was important to develop the bacterial biomass, but it was subsequent substrate depletion that promoted the environmental stresses and the development of extracellular polymeric substance (EPS) responsible for the incorporation of algae cells into the bacterial flocs (Lee *et al.* 2009).

5.3.5 Pathogen removal using polyelectrolytes

Polyelectrolytes have been shown by several workers to be effective in the destabilization and removal of microorganisms. In general, cationic polyelectrolytes are required for effective destabilization; anionic or nonionic types usually produce markedly inferior results.

Tenney and Stumm (1965) showed that cationic polyelectrolytes were effective in destabilizing bacterial and bacterial-algal mixed suspensions. Treweek and Morgan (1977) similarly found that cationic polyelectrolytes (polyethylenimine) were effective in destabilizing suspensions of *E. Coli*. Polyelectrolytes of the latter type, with a wide range of molecular weights (600 to 60,000 Daltons), were strongly adsorbed to the bacterial cells. However, efficient destabilization was only evident with the high molecular weight samples (35,000 and 60,000 Daltons).

The mechanism of destabilization evident in the latter experiments was identified as that described by the electrostatic patch model. The poorer results obtained with lower molecular weight polyethylenimine species was hypothesized to be due to the small size of individual molecules and consequently their inability to neutralize the negative *E. Coli.* surface and thus create positive patches. Therefore, with the low molecular weight species, the mechanism of destabilization was attributed to adsorption and thus reduction of double-layer repulsion.

As seen in Chapter 4, significant virus removal may be effected using metal coagulants. Some workers have also found adequate virus removal using cationic polyelectrolytes. For example, Thorup *et al.* (1970) studied the removal of bacteriophage T2 and type 1 poliovirus. Anionic and nonionic polyelectrolytes were found to be relatively ineffective whereas cationic polyelectrolytes effected removals of up to 96%. The removal was strongly dependent on ionic strength: for at least 95% removal of T2 virus and a polyelectrolyte concentration of 1 mg/l, 0.005 M Na^+, or 0.003 M K^+, or 0.003 M Ca^{++}, or 0,005 M Mg^{++} were required. With solutions containing the above ionic concentrations and with turbidities represented by 5 mg each of bentonite and infusorial earth, floc formation was good.

With poliovirus, removals recorded by these workers were inferior to those with T2 virus: up to 36% only. Again removal was significantly ionic strength dependent.

Amirhor and Engelbrecht (1975) studied the removal of bacterial virus MS2 against *E. Coli.* with diatomaceous earth filtration. They found that uncoated diatomaceous earth filter media are ineffective in virus removal. However, by adding a cationic polyelectrolyte (polyamine type, molecular weight 60,000) either as a precoat to the diatomaceous earth media or applied to the suspension prior to filtration, significant virus removal was obtained.

There was a sharp optimum polyelectrolyte dosage evident in their experiments of 0.25 mg/l for 1.4×10^8 pfu/ml. Beyond this dosage restabilization of the viruses occurred and removals correspondingly decreased. Similar results to the above were obtained by Chaudhuri *et al.* (1974) with removal of bacterial viruses T4 and MS2 through uncoated and cationic polyelectrolyte coated diatomaceous earth media.

5.3.6 Wastewater treatment by coagulation with polyelectrolytes and CEPT

Chapter 4 discussed wastewater treatment by coagulation with metal coagulants, with or without the addition of polymer flocculants, and by chemically enhanced primary treatment, CEPT. In some cases the relatively large amount of metal hydroxide sludge formed by this technique can be a disadvantage. An alternative is to use polyelectrolytes as the coagulant for chemically enhanced primary treatment, CEPT. In general cationic polymers have been successfully used for suspended solids removals in CEPT operations. The advantages associated with using polymers for CEPT are simplicity in operation and a significantly reduced sludge production. However, disadvantages could include issues with residual polymer and, depending on the polymer used, the possibility of NDMA precursors and other toxic compounds in the final effluent which need to be considered and monitored for (see Chapter 3).

Fettig *et al.* (1990) investigated the feasibility of cationic polyelectrolytes for the physical-chemical treatment of domestic wastewaters. The most effective polymer for this purpose was a highly charged cationic polyelectrolyte. Nonionic and anionic polyelectrolytes were not effective. The cationic polyelectrolyte was as effective as alum in removing TSS and COD from the wastewater. The optimum polymer dosage was approximately 20 mg/l. At higher polymer dosages, removals were reduced suggesting restabilization of the particulate material. This was attributed to charge reversal and/or steric stabilization.

Despite the favorable results with TSS and COD removals, the polymer was ineffective with phosphorus removal. Therefore, the use of a metal coagulant such as alum has a significant advantage in this regard, as seen in Chapter 4.

Ebeling *et al.* (2005) screened 19 different nonionic, anionic and cationic polymers for the treatment of the backwash water from microscreens treating fish aquaculture effluents. The backwash water had the following average characteristics: TSS 1015 mg/l; $cBOD_5$ 548 mg/l; soluble reactive phosphorus, SRP (orthophosphate) 12.3 mg/l; total nitrogen 78 mg/l; pH 7.4. None of the nonionic polymers were effective. Five out of eight anionic polymers were ineffective. The remaining three had limited effectiveness, with TSS and turbidity removals of 35 to 45%. From their results it appears that the anionics need to have a combination of high charge density and high molecular weight to have any destabilizing effect.

Only one of the nine cationics had no effect. This product had a very high charge density but a very low molecular weight. One other product had poor performance, with a TSS removal of 48%. This product, described as a polyacrylamide, had a low charge density and very high molecular weight. The other cationics produced turbidity removals of 84 to 98%. These products ranged through very high charge density and very low molecular weight; low charge density and high molecular weight; and medium to high charge density and very high molecular weight. These polymers were described as polyamine, copolymer of quaternary acrylate salt and acrylamide, and cationic polyacrylamide. Optimum dosages ranged from 1.0 mg/l to 50 mg/l, depending on the polymer. The main conclusion was that, without a jar testing campaign (see Chapter 8), it was not possible to identify one specific chemical family, ionic charge, or molecular weight that would predict the successful performance of a polymer as primary coagulant for this wastewater, although it was evident that some sort of cationic polymer was required.

The best polymers in terms of TSS removal were described as cationic polyacrylamides with high cationic charge and very high molecular weight. The optimum dosage in each case was 20 mg/L, reducing an average TSS of approximately 900 mg/L to approximately 12 mg/L. Without polymer, the settled TSS averaged 210 mg/L. Three polyamine types were also effective, although not as effective as the polyacrylamides. These were described as very high cationic charge but very low molecular weight. On average the optimum dosages were 18 mg/L, reducing an average TSS of approximately 960 mg/L to approximately 16 mg/L. Without polymer, the settled TSS averaged 190 mg/L.

Mels *et al.* (2001) applied dissolved air flotation using a cationic copolymer of acrylamide and quaternary ammonium salt, and a high molecular weight cationic polyacrylamide polymer for CEPT of raw wastewater. An in-line static mixer was used to add the polymer to the influent stream and a pipe flocculator with a retention time of approximately 25 seconds before entering the flotation unit. Using the polymer with the highest molecular weight, approximately 80-percent of the influent particulate COD and turbidity were removed using 7.5 mg polymer per g influent COD, or 5 mg polymer per 100 NTU influent turbidity. An effective control strategy was to pace the polymer dosage on influent turbidity. Thickened float concentrations of approximately 10-percent dry solids were obtained.

Averill *et al.* (2001) conducted pilot trials to assess the viability of cationic polymer aided settling of combined sewer overflows. They applied an approximately constant active polymer dosage of 21 mg/L and achieved removals of TSS: 97%; BOD: 80%; COD: 66%. However, they also found toxicity effects in the effluent for rainbow trout and daphnia. These results highlighted the effectiveness of cationic polymer-only treatment for CEPT but also the detrimental effects of coagulant over-dosing and the importance of coagulant dosage control. Over-dosing in this case caused acute effluent toxicity.

Li *et al.* (2003) conducted batch flocculation and settling tests on combined sewer overflows using cationic polyacrylamide polymers. In all cases the polymer dosages were optimized at 5 mg active polymer per g influent TSS. A cationic emulsion polymer with a molecular weight of 13 million produced the best results from the polymers tested with over 80% TSS removal. Polymer addition reduced the non-settleable fraction of the influent TSS from approximately 70% to between 20 and 50%. Jar test trials showed that slow mixing times from 0 to 60 seconds showed no effect on TSS removal performance. Slow mixing was carried out at a G value of 20 s^{-1}. However, in the jar tests, rapid mixing was carried out at a G of 180 s^{-1} for 50 seconds. This suggests that the relatively lengthy rapid mix stage also provided the required flocculation and that further "slow-mix" flocculation did not improve performance.

Wood *et al.* (2005) and Marsalek *et al.* (2005) applied polymers to clarify urban stormwater runoff. A liquid cationic polymer dosage of 4 mg/L with conventional clarification achieved a TSS removal of 77% at surface overflow rates of up to 43 m/h (25,000 gpd/ft^2). Corresponding removals of other constituents were: cBOD$_5$: 24%; COD: 47%; TP: 41%; TKN: 31%; Cr: 53%; Cu: 55%; Mn: 60%; Pb: 28%; Zn: 52%. The heavy metal removals resulted in contaminated sludge which required special disposal. Polymer addition did not increase the acute toxicity of the treated effluent, based on Microtox and 96-hour rainbow trout bioassays. On average, the dosage of 4 mg/L polymer represented a mass dosage of approximately 30 mg polymer per g TSS. The authors do not clarify whether the polymer dosage is as active polymer or as the liquid product. Polymer was added in-line using a static mixer directly upstream of the clarifier inlet. Therefore, no separate flocculation was provided.

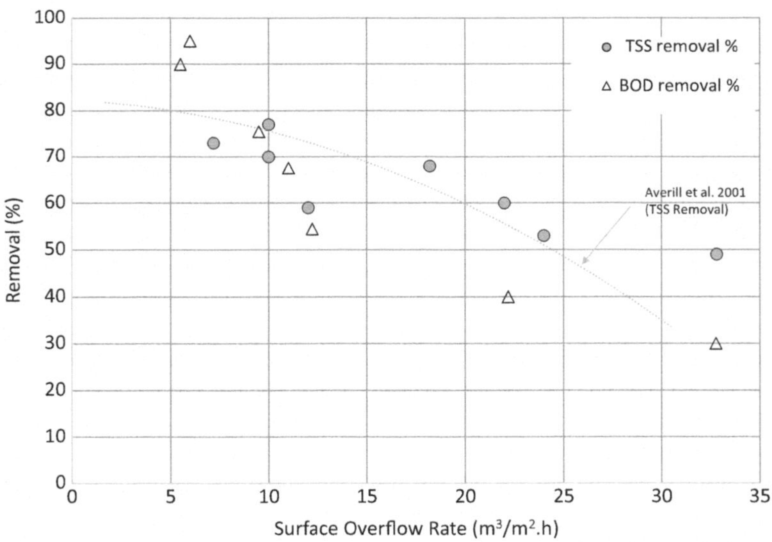

Figure 5.8 Effect of surface overflow rate (SOR) on TSS and BOD removals. (Adapted from Li *et al.* 2009. 1 m^3/m^2 · h = 589 gpd/ft^2).

Li *et al.* (2009) investigated the settling rates of combined sewer overflow suspended solids coagulated with cationic polymers. All the polymers tested were high molecular weight, low cationic charge cationic polyacrylamide polymers. The best polymer was an emulsion type optimized at a dosage of 5 mg active polymer per g TSS. A pilot scale combined sewer overflow settling basin was tested at various surface overflow rates (SOR) at the optimized polymer dosage. Figure 5.8 shows the results obtained by Li *et al* in Windsor, Ontario. Also shown in Figure 5.8 are the results obtained by Averill *et al.* (2001) for TSS removal in a full-scale demonstration combined sewer overflow treatment facility in Toronto using polymer. Li *et al.* also conducted CFD modelling for design of the full scale combined sewer overflow settling basin facilities. The final design was based on a peak SOR of 20 m/h (12,000 gpd/ft^2). The pilot and CFD modelling indicated that TSS and BOD removals at 20 m/h would be approximately 60% and 45%, respectively.

5.3.7 Activated sludge bulking and foaming control and enhanced bioflocculation

The activated sludge process produces biological flocs that, through the action of extracellular polymeric material effectively flocculates dispersed material and forms settleable flocs, producing relatively clear supernatants after secondary sedimentation. However, in some cases, due to an excess or lack of filamentous organisms that assist in floc formation, secondary effluent quality degrades and dispersed floc particles, or excessive sludge blanket depths cause unacceptable increases in effluent suspended solids.

The addition of polymer to mixed liquor prior to secondary clarifiers, with or without the addition of weighting agents, has often proved to be beneficial in terms of increasing zone settling velocities and compacting sludge blankets. For example, Singer *et al.* (1968) demonstrated approximately 70-percent reductions in SVI, from approximately 450 mL/g to less than 140 mL/g by using approximately 3 g/kg cationic polymer added to bulking activated sludge.

Al-Jasser (2009) showed that with polymer, the stirred sludge volume index (SSVI) of activated sludge reduced from the range 156 to 191 mL/g to approximately 40 mL/g. Therefore, allowable surface overflow rates (SOR) and solids loading rates (SLR) can be increased. This has significance particularly during high influent flow periods and during plant upsets, such as bulking sludge incidents. Polymer addition also complements the natural flocculation propensity of activated sludge and enhances the capture of dispersed solids from the supernatant thereby lowering effluent TSS (Patoczka *et al.* 1998; Demel *et al.* 1988).

Figure 5.9 shows one example where initial zone settling rates were increased from 1.8 m/h (1060 gpd/ft^2) without polymer addition, to 6.0 m/h (3530 gpd/ft^2) with the addition of 3.3 g/kg dry solids of a cationic polyacrylamide polymer to activated sludge mixed liquor treating an industrial waste. The mixed liquor concentration was 6100 mg/L (Vanderhasselt & Verstraete 1999). In this case the clarifier capacity could be potentially tripled by polymer addition. Vanderhasselt *et al.* (1999) showed that the best polymer dosing control strategy was feed-forward control based on the sludge load (flow × concentration) at the polymer dosing point. A feed-forward control based on load coupled with a feed-back trim based on sludge blanket depth was also effective.

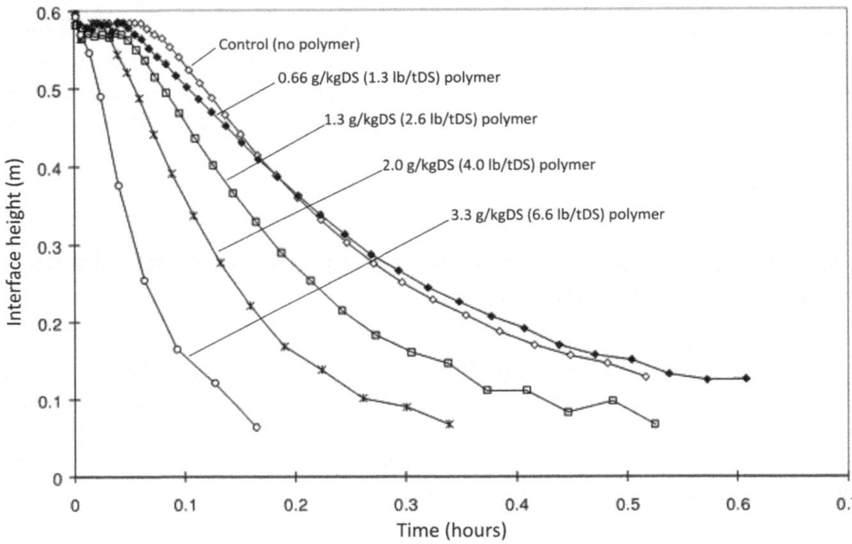

Figure 5.9 Results of polymer addition on settling characteristics of an industrial activated sludge (adapted from Vanderhasselt & Verstraete, 1999).

Figure 5.10 and Table 5.2 show the results of adding polymer to activated sludge mixed liquor treating predominantly domestic wastewater. Small scale interface settling tests showed that with cationic polyacrylamide polymer addition, the initial zone settling velocity was essentially tripled. The depth occupied by the settled sludge was reduced by approximately 40-percent (Bratby 1979).

It was pointed out in Chapter 4 that a major wastewater treatment plant operating challenge arises from biological foaming, usually characterized by viscous, chocolate-colored stable foam at the surface of biological reactors and secondary clarifiers. The problem is particularly severe when the design of the reactors includes sub-surface passages and prevents the free passage of the floating foam from one compartment to the next, or to the clarifiers. Shao *et al.* (1997) demonstrated how a bulking and foaming activated sludge, caused principally by nocardioforms and *Microthrix parvicella*, was controlled by the addition of

0.5 mg/L of polyacrylamide cationic polymer to the mixed liquor channel feeding the secondary clarifiers.

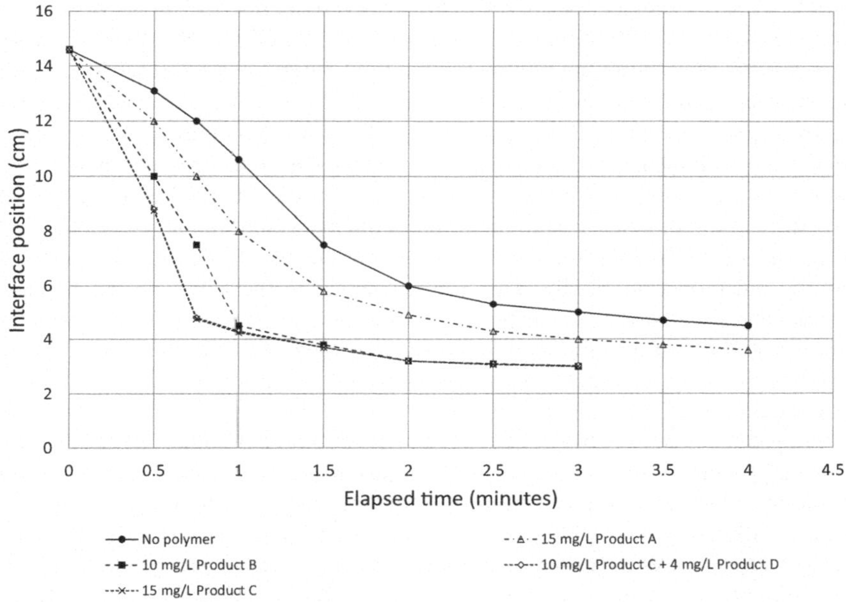

Figure 5.10 Results of polymer addition to activated sludge in jar test (Bratby 1979).

Table 5.2 Results of settling tests with activated sludge (see Figure 5.10).

Product	Description	Dosage mg/L	Dosage g/kgDS (lb/tDS)	Settling Velocity m/h (gpd/ft²)
None	Activated sludge alone	0	0	3.6 (2140)
Product A	Cationic PAM (MW 5×10^4)	15	10 (20)	4.8 (2820)
Product B	Cationic PAM (MW 5×10^6)	10	7 (14)	7.1 (4180)
Product C + Product D	Product D: Anionic PAC (MW 15×10^6)	10 + 4	9.5 (19)	10.3 (6050)
Product C	Cationic PAM (MW 5×10^4)	15	10 (20)	10.8 (6350)

Initially, the total nocardioform filament count was 49×10^6 intersections/g VSS, increasing to 78×10^6 intersections/g VSS within 5 days after the start of polymer addition. Within a further 8 days of polymer addition, the count reduced again to 43×10^6 intersections/g VSS. During this time, the percentage of aeration basin surface covered by foam started at 85%, increased to 96% then decreased again to 83%. By continuing polymer addition at 5 mg/L for a further 6 days, the total filament count did not reduce markedly, at 41×10^6 intersections/g VSS, but the foam surface coverage did reduce to 16%. These workers postulated that it was the dispersed filament organisms that were causing the foaming issues, and not the floc-bound filaments, conforming with the observations of Blackall *et al.* (1991). The effect of the polymer was to incorporate the dispersed filaments into the floc. Therefore, while the total filament count did not change drastically, the extent of foaming did reduce markedly. This was attributed to the action of the polymer in reducing the amount of dispersed filaments. An earlier foaming incident demonstrated that if accumulated foam was first removed from the reactor basins (in their case using vacuum trucks) recovery from the foaming incident when using the same dosage of 0.5 mg/L polymer was within 3 days.

Narayanan *et al.* (2010) confirmed the postulation of Shao *et al.* that the foaming propensity of nocardioforms such as *Gordonia amarae* came less from floc-bound filaments but more from free-floating dispersed filaments. With sub-surface withdrawal from biological reactors, dispersed, free-floating filaments tend to proliferate. With surface withdrawal, floc-bound filaments predominate, as pointed out by Blackall *et al.* (1991). By dosing either 0.5 mg/L cationic polymer (based on influent flow) or 2 g/kg MLSS · d of PACl, the total nocardioform filament counts did not reduce, but the dispersed filament counts did reduce by incorporating the dispersed filaments into the activated sludge flocs, converting the system to floc-bound filaments and reducing the ability to generate foam.

Apart from polymer addition there are other additives that have been used alone, or in conjunction with polymer. Talc (alumino-silicate powder) has been found to be effective as a weighting agent. Vanderhasselt and Verstraete (1999) found that with an activated sludge treating domestic wastewater, the settling velocity increased in different experiments by a factor of 1.7 to 2.1 times the control with no talc. The SVI reduced from 155 mL/g to 120 mL/g with talc addition. The improvements found with polymer addition on the same sludge were more pronounced and had the added benefit of not increasing overall sludge production.

Seka *et al.* (2001) applied a multi-component additive comprising talc, cationic polymer and a cationic quaternary ammonium compound (cetyltrimethyl ammonium bromide) to improve settling issues due to filamentous bulking. Addition of the compound reduced the 30-minute settled volume of the sludge from 800 mL/L to 680 mL/L. The biocidal action of the quaternary ammonium component reduced the numbers of filamentous bacteria.

Magnetite has been applied successfully as a ballast for activated sludge settling both with and without polymer addition. Magnetite (Fe_3O_4) is applied as a powder with a size range of 10 to 30 μm, and has the advantage that it has a very high specific gravity, at approximately 5.2 and is recoverable from the waste activated sludge using a shear mill and magnetic drum. It is a hydrophobic powder and therefore is readily incorporated in the flocs. The main benefit found with magnetite addition is the improvement in settling characteristics. Parallel pilot reactors with similar filamentous bacterial growths showed an average SVI without magnetite addition of 195 mL/g, compared with 75 mL/g with magnetite addition. Sludge settling parameters found in one study are reproduced in Table 5.3.

Table 5.3 Summary of Vesilind settling parameters with and without magnetite and polymer (Jimenez *et al.* 2013).

Magnetite/MLSS Ratio	Polymer Added?	V_0 (m/h)	K (L/g)
No magnetite added (control)	No	2.2	0.37
0.5 to 0.75	No	7.7	0.26
0.5 to 0.75	Yes	66.4	0.39
0.8 to 1.3	No	22.9	0.25
0.8 to 1.3	Yes	77.8	0.31

The results in Table 5.3 show that polymer assists by incorporating the magnetite more efficiently into the flocs. Without polymer, magnetite addition can increase settling by as much as 10-times. With polymer addition, magnetite can increase settling by as much as 36-times.

5.4 POLYELECTROLYTES AS FLOCCULANT AIDS

Polyelectrolytes are widely accepted as flocculant aids, both in water and wastewater treatment applications. In water treatment they are usually added with metal coagulants to achieve some desired effect: Here the objective is not primarily that of destabilization *per se* (which is effected by the metal coagulants) but rather of supplementing the orthokinetic flocculation process and altering floc characteristics in terms of increasing ultimate floc size, density, permeability, compressibility, shear strength, settleability and filterability. In general, the optimum polyelectrolyte dosage required for each of these criteria would be similar. In biological wastewater treatment applications they may be added to supplement the natural biological flocculation characteristics thereby assisting overloaded secondary sedimentation basins.

For some processes, the addition of flocculant aids is almost a prerequisite. A notable example is ballasted sedimentation in which microsand particles (specific gravity 2.65 and sizes ranging from 0.1 to 0.3 mm) act as a ballasting agent. The sand particles are incorporated in the flocculated material, thereby increasing separation rates. In water treatment applications, primary coagulant is typically added first, followed by a polymer flocculant aid and the microsand. Primary coagulant dosages depend on the water or wastewater to be treated. Polymer dosages typically vary from less than 0.5 mg/l to 1.5 mg/l, although higher dosages, up to 10 mg/l have been reported.

Thompson *et al.* (2005) report on the application of ballasted sedimentation to treat a highly organic water source. The enhanced coagulation treatment pH was approximately 4.7. Following pH adjustment, primary ferric coagulant is first added in a mixing tank with a nominal retention time of 2 minutes. Flocculant aid and microsand are then added in a 2-minute retention flash mixing tank. The flow then passes to slower mixed flocculation-maturation basins with a 6-minutes nominal retention time. Separation of the ballasted flocs then occurs in tube settling tanks with an overflow rate of 1.6 m/h (0.65 gpm/ft^2).

Young and Edwards (2003) found that the rate of floc formation during the "maturation" process was essentially the same with or without the microsand. However, the ballasted flocs were more resistant to shear and breakup, probably due to the tighter, rounded shape of the flocs, compared with the more fractal nature of the unballasted flocs. Timing of the addition of the microsand to water dosed with metal coagulant and polymer was not critical, except that when polymer was added after the microsand, increases in turbidity resulted. Therefore, best results were obtained when the ballasting agent was added after the coagulant and polymer had been added. There was also an optimum amount of microsand addition for a given raw water turbidity, coagulant and polymer dosage. Excesses of microsand cannot be incorporated in the flocs, and remain unflocculated in the effluent or treated water

When polyelectrolytes are added to suspensions destabilized with primary coagulants, best results are generally obtained when the addition is some time after primary coagulant addition. This procedure is probably preferable because primary floc particles formed on destabilization are given the opportunity to grow to an appreciable size before polyelectrolyte is added and adsorption occurs, thereby resulting in lower polyelectrolyte requirements. However, there are some instances where flocculant aid has been most effective when added ahead of metal coagulants; examples include the use of activated silica and sodium alginate, where such a procedure may serve to accentuate chemical interaction between polyelectrolyte and primary coagulant (Packham, 1967). In several cases, addition of flocculant aid before metal coagulants gives rise to relatively higher residual aluminum or iron concentrations in the treated water, probably due to incomplete utilization and precipitation of the metal coagulant.

However, the apparent success occasionally reported when polyelectrolytes are added before metal coagulants could be explained partly by inefficient initial dispersion of the flocculant aid and thereby a delayed action effect (Packham, 1967). On the other hand, for cases where polyelectrolyte and particles are of opposite sign, addition of polyelectrolyte before metal coagulant could serve to reduce the surface charge of particles on adsorption thereby subsequently reducing the quantity of metal coagulant required to effect charge reduction.

An interesting example of polyelectrolyte addition before metal coagulant is that reported by Cooper *et al.* (1974): For the reduction of organic solids and COD from tannery lime-sulphide unhairing wastes, and operating at pH values of 8.0 or greater (to prevent H_2S liberation) it was found that addition of an anionic polyelectrolyte as primary coagulant, followed by a cationic polyelectrolyte and subsequently by a metal coagulant gave superior results to all other combinations tested. With the latter procedure, COD was reduced by 56% and organic solids by 98%.

The latter results notwithstanding, in most cases flocculant aid is normally added to greatest benefit after metal coagulant addition. In such instances anionic polyelectrolytes are often found to be effective, probably because primary floc particles formed by metal-hydroxide species are positively charged under most conditions (Gregory, 1977). Polyelectrolytes as flocculant aids do not necessarily lead to a reduction in metal coagulant concentration since their function is not primarily that of destabilization. However, the benefits derived in terms of improved settleability, filterability and reduced sludge volume generally offsets the increased chemical cost. In cases where reduced metal coagulant dosages on flocculant aid addition have been reported, the reason may be that original metal coagulant dosages were dictated by factors such as settleability requirements, rather than only by destabilization.

The following presents various selected examples of polyelectrolytes applied as flocculant aids. The examples cited are by no means exhaustive but are intended to serve merely as an indication of some of the conditions where such applications have been successful.

Cohen *et al.* (1958) investigated the use of several polyelectrolytes as flocculant aids using aluminum sulfate as primary coagulant applied to turbid waters. With an anionic polyelectrolyte (hydrolyzed polyacrylamide), although it functioned well as a flocculant aid, no significant turbidity removal was evident without prior aluminum sulfate addition. Dosage of the polyelectrolyte was very critical: excess dosages gave rise to restabilization. Using an organic cationic polyelectrolyte, apart from its effectiveness as a flocculant aid, it could be effectively used as a primary coagulant, removing not only live organisms such as algae but also dissolved substances. A nonionic polyelectrolyte (of a form similar to carboxymethylcellulose) was found to be considerably less effective than either the anionic or cationic forms used.

Black and Hannah (1961), investigating (kaolinite) turbidity removal with aluminum sulfate, assessed the efficacy of the following polyelectrolytes as flocculant aids: activated silica, potato starch, anionic polyacrylamide, a cationic type, anionic carboxymethylcellulose, guar gum, a cationic natural polymer, and caustic hydrolyzed potato starch. In general the anionic aids worked best when particle charges (determined by electrophoretic mobility measurements) had been reduced to near zero with aluminum sulfate. The cationic aids were most efficient with low alum dosages. The eight aids all could be made to improve or inhibit destabilization/floc formation by appropriate selection of dosages of alum and flocculant aid.

Mueller and Burbank (1964), studying river turbidity removal with ferric sulfate, evaluated the use of three polyelectrolytes as flocculant aids: a nonionic, cationic and anionic polyelectrolyte (characteristics unspecified). Although all flocculant aids could give rise to improvements under appropriate conditions of metal coagulant and aid dosages, the nonionic aid (probably a polyacrylamide) was found to be the most effective in terms of settling rate and residual turbidity.

Black and Christman (1961), applying flocculant aids to accelerate floc formation during the lime-soda softening process, used four polyelectrolytes: activated silica, anionic polyacrylainide, nonionic potato starch and a cationic organic polyelectrolyte. As expected, for each of the aids a distinct optimum dosage was found. Each of the aids improved floc formation characteristics, although the anionic polyacrylamide appeared to offer acceptable floc removal but at a much lower optimum dosage (0.025 mg/l as compared to approximately 1.0 mg/l for the other aids). Interestingly, with the nonionic (potato starch) aid, the addition of 17 mg/l aluminum sulfate as primary coagulant had a marked detrimental effect on floc formation.

Huck *et al.* (1977) applied polyelectrolytes to aid flocculation of lime precipitated heavy metal (Fe, Zn, Cu, Mn, Pb) hydroxide flocs. The polyelectrolytes used had molecular weights of around 9.0×10^6 (determined by gel permeation chromatography). Although the type of polyelectrolyte used was unspecified, they found that the degree of hydrolysis, which varied from 2 through 37%, had no effect on flocculation. From their results, it appears that the optimum polyelectrolyte dosage is of the order 1/500 times the metal concentration. At the optimum dosage, the degree of floc formation and settleability were markedly improved.

Benedek and Bancsi (1977) applied a wide variety of polyelectrolytes to evaluate the properties most appropriate when applied to phosphorus precipitated domestic wastewater using aluminum sulfate as primary coagulant. They found that, in contrast to normal manufacturers' recommendations of the highest hydrolyzed polyelectrolyte species for alum precipitated phosphorus flocs, the polyelectrolyte giving the most superior results in terms of floc formation and settling velocity was an anionic polyacrylamide of molecular weight greater than 8.0×10^6, and 10 to 20% hydrolysis. Under optimum conditions, overflow rates during primary sedimentation could be increased in some cases from approximately 0.75 m/h (440 gpd/ft^2) with alum alone up to 7.0 m/h (4,100 gpd/ft^2). The criterion used to monitor limiting overflow rate was 90% phosphorus removal. The aluminum dosage varied from 110 mg/l to 190 mg/l.

Shine and Rasmussen (1977) demonstrated that using a cationic polymer as flocculant aid also allowed partial substitution of the metal coagulant dosage for treatment of Lake Michigan water, with raw water turbidities ranging from approximately 2 to 5 NTU. The cationic polymer dosage found optimal throughout the year was 0.3 mg/l. The combination of alum plus cationic polymer resulted in lower sedimentation tank effluent turbidities, although filtered turbidities were similar, with and without polymer. Filter runs were longer, however, with polymer addition. In general, the alum dosage could be reduced by 4 mg/l with polymer addition. The benefits of alum plus polymer were more pronounced at higher raw water turbidities.

Lawson (1977) demonstrated the benefits of adding polymer to maintain stable operation in sludge blanket clarifiers treating water from Lake Ontario. With alum alone, approximately 25 mg/l was used, together with 15 mg/l kaolin clay. Treatment was satisfactory except during severe temperature changes or increases in treatment flow. During severe temperature drops, denser cold water destroyed the blankets and required emergency operation to maintain satisfactory treatment. By using approximately 5 mg/l alum and 4 mg/l clay, together with 1 mg/l cationic polymer, operation of the sludge blanket clarifiers was much more stable, even during temperature changes, and settled water turbidities lower.

Morrow (1976) describes another water plant with sludge blanket clarifiers treating Lake Ontario water, using the following chemicals: cationic polymer (described as a polyquaternary ammonium polymer) at approximately 2 mg/l; a *hectorite* clay at approximately 4 mg/l; and alum at 2 to 5 mg/l. This replaced previous chemical dosages of alum 18 to 24 mg/l; starch 1 to 2 mg/l; and clay at 7 to 9 mg/l.

5.4.1 Polymers as filter aids

Polyelectrolytes are commonly added to improve filter performance in both water and wastewater treatment applications, and in conventional configurations with sedimentation, as well as direct filtration applications. There are essentially three alternative configurations of coagulant addition: the first is metal coagulant alone, added during rapid mixing and, after flocculation, passed either directly or after sedimentation to the filter; the second is metal coagulant added as above but with a flocculant/filter aid added (e.g. non-ionic polyelectrolyte or activated silica) just before filtration; and the third is metal coagulant eliminated and replaced with a cationic polyelectrolyte serving as primary coagulant.

The removal processes within granular media filters comprise three stages: a working-in stage, a working stage and a breakthrough stage. During the working-in stage (also referred to as the filter ripening period) the turbidity of the filtered water decreases rapidly until it reaches a stable low value. The presence of the working-in stage indicates that for efficient attachment, floc particles must provide an initial coating of the media. The working-in stage is usually shorter when using metal coagulants, compared with using cationic polyelectrolytes as primary coagulants in direct filtration applications. In some cases the working-in stage is shortened by adding polymer to the backwash water. This also has the benefit of conditioning the backwash solids. Yapijakis (1982) demonstrated these benefits by adding 0.05 to 0.15 mg/l nonionic polymer to the backwash water to pre-condition the filters for the following runs.

The working stage, considered to commence when the working-in stage produces a stable low value, is the main phase during filtration. During this stage, polyelectrolytes as primary coagulants often produce a better effluent quality for all depths than metal coagulants in direct filtration applications, when turbidity removal is the performance criterion.

If the filter run is subsequently not terminated for reasons of head loss, the filtrate turbidity begins to rise which indicates the breakthrough stage. The onset of breakthrough is usually faster using metal coagulants, compared with cationic polyelectrolytes as primary coagulants in direct filtration applications.

When polymers are used as primary coagulants, a cationic polymer is usually required at typical dosages ranging from 0.1 to 5 mg/l.

However, in some instances it is necessary to use metal coagulants as primary coagulants to obtain efficient removal of water constituents. In this case, polymers are often used as filter aids. When polymers are used as flocculant (or filter) aids in conjunction with metal coagulants, a nonionic or slightly anionic polyelectrolyte is usually best. Typical dosages range from 0.05 to 0.5 mg/l.

For tertiary filtration to polish effluents from wastewater treatment plants, higher molecular weight cationic polyelectrolytes are often used at typical dosages ranging from 0.05 to 0.15 mg/l. If lower molecular weight polymers are used, dosages can be significantly higher. Pilot plant trials, or careful full scale experimentation to determine the correct polymer and dosage is the best approach. Typically one would start at a low dosage (0.025 mg/l with the higher molecular weight polymers) and gradually increase the dosage by 0.025 mg/l increments to determine the optimal dosage to minimize effluent turbidity.

For tertiary filtration in wastewater treatment plants practicing chemical phosphorus removal, anionic or nonionic polymers may be the best as filter aids. Typical dosages range from 0.05 to 0.1 mg/l. However, as before, careful pilot or full-scale experimentation is the best approach to determine the best type and dosage of polymer, and to avoid serious filter blinding.

5.5 POLYMERS AS SLUDGE CONDITIONERS

Sludge processing and disposal strategies from water or wastewater treatment plants often include some form of mechanical thickening and dewatering operations. Devices used for thickening include gravity thickeners, dissolved-air flotation thickeners, gravity belt thickeners, rotary drum thickeners, and centrifuges. Devices for dewatering include atmospheric drying beds, filter presses, filter belt presses, and centrifuges.

Because of the relative fragility of biological flocs and the high degree of bound water, some form of chemical conditioning is often required prior to thickening and dewatering processes. Inorganic chemical conditioners could include aluminum chlorohydrate, ferric chloride and lime. However, polymers are widely used as sludge conditioners because of their effectiveness in dewatering, the toughness of flocs produced and the fact that, unlike metal coagulants, they do not significantly add to the quantity of sludge to be ultimately disposed of. With biological wastewater treatment sludges, polyelectrolytes used are almost exclusively cationic, although this is not an unbroken rule.

It is possible to overdose with polymer conditioners. This is not always evident since the range of effective conditioning is quite wide with concentrated suspensions. However, at higher than optimal polymer dosages, a worsening of dewatering behavior is often noticed. Christensen *et al.* (1993) attributed this to the increased liquid phase viscosity due to residual polymer. They reported an increase in filtrate viscosity that could form the basis of a control system, since there was a marked increase in viscosity at dosages just above the optimum for dewatering, determined by capillary suction time (CST) and specific resistance measurements (see Chapter 8).

Investigations conducted by Dentel *et al.* (1995) confirmed that filtrate viscosity may be a viable control parameter that required further research to evaluate the practicability of this approach. Of the methods tested by them, the streaming current detector offered the most reliable method of automated control of polymer dosage.

With sludges from water treatment plants, a similar general scheme of processing is evident: Again, the thickening stage may include gravity sedimentation or flotation, often aided by chemical conditioners; and the dewatering stage includes processes such as belt filter presses, pressure filtration, centrifuges, and drying beds.

Chemical conditioners are also required for dewatering water treatment plant sludges. Again, with their inherent advantages, polyelectrolytes have been used with notable success. Here, there is no generality of the type of polyelectrolyte most effective since it depends very much on the type of raw water, its constituents, and the coagulants used during treatment. However, high molecular weight anionic polyelectrolytes with a moderate (15–30%) degree of hydrolysis have often proved to be effective.

The following presents a few selected examples to illustrate chemical conditioning of water and wastewater sludges with polyelectrolytes. As before, the examples are not by any means exhaustive.

Gale and Baskerville (1970) compared the performance of 17 polymer products, together with aluminum chlorohydrate. They applied each product to four different sludges: one raw sludge, one anaerobically digested sludge, and two activated sludges. Evaluations were carried out on the basis of equal cost per ton dry solids.

In one case (one of the activated sludges) aluminum chlorohydrate provided the best dewatering (measured by capillary suction time, CST). For the raw sludge, performance of aluminum chlorohydrate was similar to some of the best polymers. For the digested sludge, aluminum chlorohydrate was much worse than the best polymers (at dosages determined by equivalent costs). The overriding conclusion derived from this work was that the best product for one type of sludge may not be the best for another type of sludge, or even an ostensibly similar sludge at another location. This emphasizes the value of laboratory and pilot testing to determine the best approach for conditioning a given sludge (see Chapter 8).

During investigations on the thickening characteristics of activated sludge by dissolved-air flotation, a cationic polyacrylamide with a medium to high charge density (degree of hydrolysis 10 to 30%) and molecular weight 10 to 12×10^7 was applied at a concentration of 2 g/kg TSS. For a given solids loading rate and height of float above the water level, float solids concentrations achieved were consistently higher (by a factor of 1.35) when compared to no polyelectrolyte addition. Furthermore, polyelectrolyte addition reduced the total depth of float solids thereby allowing greater float accumulation for thickening (Bratby, 1978).

Effluent solids concentrations from the flotation unit were lower when using polyelectrolytes (mean value 6.7 mg/l compared to 37 mg/l with no polyelectrolyte addition). It appeared that polymer strengthened biological flocs were better able to resist break up within the relatively turbulent environment of the flotation unit

Experiments carried out with a belt filter press, using the same sludge, identified the most effective polyelectrolyte in terms of aiding dewatering and prevention of clogging the belt fabric, to be the same as that used for flotation; however, the dosage required was higher (4 g/kg dry solids).

For cases where flotation thickening precedes a dewatering facility (such as belt pressing), the advantage of polyelectrolyte usage is that since float solids concentration is increased, for a given set of operating conditions the volume of float solids to be handled by the dewatering facility (which governs the size of the facility) is reduced. Presuming that the total polyelectrolyte dosage does not increase significantly, a saving in total cost is achieved due to the reduced capital cost of the facility.

Novak *et al.* (2001) determined that it is the amount of proteins in solution that determines the ease of dewatering and the polymer demand of digested wastewater sludges. Anaerobic digestion releases large amounts of protein into solution, derived from anionic biopolymers. The presence of larger amounts of this protein causes deterioration of the dewatering behavior of anaerobic sludges and increases the polymer demand. Aerobic digestion does not release as many proteins and the sludge quality deteriorates to a lesser extent.

Novak *et al.* suggested that it is the reduction and solubilization of iron that leads to the release of biopolymer proteins from the flocs. The major role of conditioning chemicals, both inorganic conditioners or organic polymers, is to coagulate the biopolymers in solution, especially the protein.

Novak *et al.* (1998) investigated the role of inorganic cations on the effectiveness of polymer conditioning of wastewater sludges. They found that when the monovalent-to-divalent (M:D) cation ratio was greater than 2:1 on a milliequivalent basis, dewatering deteriorated and the polymer demand increased. The soluble protein level also increased as the M:D ratio increased.

Addition of sodium (Na^+) was more detrimental to performance compared with the addition of potassium (K^+). The addition of magnesium (Mg^{2+}) improved dewatering characteristics substantially when the background Na and K levels were greater than 10 and 0.1 meq/l respectively. The beneficial dosage of Mg was

7 to 12 meq/l. The Mg exchanged and displaced the monovalent cations in the floc. At lower Na and K levels, Mg addition was less effective since dewatering properties were already good.

O'Brien and Novak (1977) investigated the effects of pH and mixing on polyelectrolyte conditioning of chemical sludges (iron, iron-lime, alum and alum-lime sludges) from several water treatment plants. From their studies, they suggested the following rules for initial polyelectrolyte selection:

- Cationic polyelectrolytes operate most efficiently at pH values of 7.0 or lower.
- Nonionic polymers and the lower percent hydrolysis anionic types function effectively over a pH range 6.5 to 8.5. In general, as the pH increases, the effectiveness of polyelectrolytes decreases.
- A 50% hydrolysis anionic polyelectrolyte is most effective at pH values over 8.5. Below this pH value, the effectiveness correspondingly reduces.
- Over- or under-mixing reduces polyelectrolyte effectiveness greatly. The degree of mixing depends on the solids concentration of the sludge.
- The lower the solids concentration the longer is the optimum rapid mixing time.

Novak and Langford (1977) investigated the use of polyelectrolytes for improving chemical sludge dewatering on sand beds. In their study a high molecular weight anionic polyelectrolyte was applied to various water treatment plant sludges. The sludges used were essentially the same as reported by O'Brien and Novak above. They found that with polyelectrolyte conditioning, dewatering rates were greatly improved and bed penetration minimised.

Bratby and Marais (1977), applying dissolved-air flotation to thicken sludge from a water plant treating humic colored (brown) waters, used a cationic polyacrylamide (potable water grade) with a 10% degree of hydrolysis and molecular weight in excess of 7×10^6. The dosage applied was 0.5 g/kg dry solids. During treatment, humic substances were removed by 10 mg/l sodium aluminate followed by lime to adjust the pH to 5.5 after subsequent addition of 50 mg/l aluminum sulfate.

Because of the gelatinous nature of the flocs formed during the treatment process, the sludge concentration from the bottom of the sedimentation tanks never exceeded 0.25%. With polyelectrolyte conditioning and flotation thickening, waste sludge was thickened to concentrations in excess of 12%. Including polyelectrolyte conditioning at this plant increased the overall chemical costs by 1.43% of the existing chemical costs.

Without polyelectrolyte addition, although bubble-particle attachment did occur to a certain extent during flotation, the union was relatively short lived and, on the whole, the effluent was a very poor quality. A characteristic of alum precipitated brown humic water is the extreme fragility of the floc particles and a marked reluctance for reformation (within the time allowed during flotation) after being sheared apart by the relatively turbulent conditions created by precipitated bubbles in the flotation unit. Polyelectrolyte serves to considerably strengthen flocs and, after bubble-particle attachment, to resist dislodgement by shearing.

5.6 REFERENCES

Adin A. and Rebhun M. (1974). High rate contact flocculation-filtration with cationic polyelectrolytes. *Jour. AWWA*, 109–117.

Akers R. (1972). Factors affecting the rate of the polyelectrolyte flocculation reaction. *Filtration and Separation*, 423–425 and 466.

Akers R. (1972). Flocculation. Inst. Chem. Engnrs., London.

Al-Jasser A. (2009). Enhancement of sludge settling with chemical additives. *Water Env. Res.*, **81**(9), 849–857.

Amirhor P. and Engelbrecht R. (1975). Virus removal by polyelectrolyte aided filtration. *Jour. AWWA*, 187–192.

Amy G. and Chadik P. (1983). Cationic polyelectrolytes as primary coagulants for removing trihalomethane precursors. *Jour. AWWA*, 527–531.

Audsley A. (1963). Scientific Report, N.C.L./DEP, 5, D.S.I.R., London, 1963.

Averill D., Chessie P., Henry D., Kok S., Marsalek J. and Seto P. (2001). Field experience with chemically-aided settling of combined sewer overflows. *Proceedings of NOVATECH: 4th International Conference on Innovative Technologies in Urban Drainage*, Lyon, France.

Beardsley J. (1973). Use of polymers in municipal water treatment. *Jour. AWWA*, 85–90.

Benedek A. and Bancsi J. (1977). Comparative evaluation of commercial polyelectrolytes for flocculating alum precipitated domestic wastewater. *Prog. Wat. Tech.*, **9**, 33–42.

Bernhardt H. and Clasen J. (1991). Flocculation of micro-organisms. *J. Water SRT – Aqua*, **40**(2), 76–87.

Birkner F. (1967). Discussion to cationic polyelectrolytes as prime coagulants in natural water treatment. *Jour. AWWA*, 169–182.

Birkner F. and Morgan J. (1968). Polymer flocculation kinetics of dilute colloidal suspensions. *Jour. AWWA*, 175–191.

Black A. and Christman R. (1961). Electrophoretic studies of sludge particles produced in lime-soda softening. *Jour. AWWA*, 737–747.

Black A. and Hannah S. (1961). Electrophoretic studies of turbidity removal by coagulation with aluminum sulphate. *Jour. AWWA*, **53**, 438–452.

Black A., Birkner F. and Morgan J.J. (1965). Destabilization of dilute clay suspensions with labelled polymers. *Jour. AWWA*, **57**, 1547–1560.

Black A. and Vilaret M. (1969). Effect of particle size on turbidity removal. *Jour. AWWA*, 209–214.

Blackall L., Tandoi V. and Jenkins D. (1991). Continuous culture studies with *Nocardia amarae* from activated sludge and their implications for nocardia foaming control. *J. Water Pollut. Control Fed.*, **63**(1), 44–50.

Bratby J. and Marais G. (1977). Thickening of brown water sludges by dissolved-air (pressure) flotation. *Water SA*, **3**(4), 202–212.

Bratby J. (1978). Aspects of sludge thickening by dissolved-air flotation. *Wat. Pollut. Control*, **77**(3), 421–432.

Bratby J. (1979). Emergency chemical upgrading of South WWTP. Contract Report, CAESB, Brasilia, D.F., Brazil.

Carns K. and Parker J. (1985). Using polymers with direct filtration. *Jour. AWWA*, 44–49.

Chaudhuri M., Amirhor P. and Engelbrecht R. (1974). Virus removal by diatomaceous earth filtration. *Jour. Env. Eng. Div., ASCE*, **100**(EE4), 937.

Christensen J., Sorenson P., Christensen G. and Hansen J. (1993). Mechanisms for overdosing in sludge conditioning. *Jour. Env. Eng., ASCE*, **119**(1), 159–171.

Cohen J., Rourke G. and Woodward R. (1958). Natural and synthetic polyelectrolytes as coagulant aids. *Jour. AWWA*, 463–478.

Committee Report (1970). Coagulation and color problems. *Jour. AWWA*, **62**, 311–314.

Committee Report (1971). State of the art of coagulation. *Jour. AWWA*, 99–108.

Cooper J., Happich W., Bitcover E., Mellon E. and Filachione E. (1974). Effect of flocculants on the sedimentation of organic solids in tannery unhairing effluents. *Proc. 29th Ind. Waste Conf., Purdue Univ.* 510–518.

Culp R. (1977). Direct filtration. *Jour. AWWA*, 375–378.

De Godos I., Guzman H., Soto R., García-Encina P., Becares E., Muñoz R. and Vargas V. (2011). Coagulation/flocculation-based removal of algal-bacterial biomass from piggery wastewater treatment. *Bioresource Technol.*, **102**, 923–927.

Deeble B. (1976). Cationic polyamines as primary coagulants. *Jnl. Inst. Water Eng. and Sci.*, **30**(3), 138–146.

Demel I. and Möbius C. (1988). Improving the settling of activated sludge by chemical additives. *Wat. Sci. Tech.*, **20**(1), 283–286.

Dentel S., Abu-Orf M. and Griskowitz N. (1995). Polymer characterization and control in biosolids management. Water Environment Research Foundation, 91-ISP-5, Alexandria, Virginia, U.S.A.

Divakaran R. and Pillai V. (2002). Flocculation of algae using chitosan. *Jour. Applied Phycology*, **14**, 419–422.

Ebeling J., Rishel K. and Sibrell P. (2005). Screening and evaluation of polymers as flocculation aids for the treatment of aquacultural effluents. *Aquaculture Engineering*, **33**(4), 235–249.

Edzwald J., Haff J. and Boak J. (1977). Polymer coagulation of humic acid waters. *Jour. Env. Eng. Div., ASCE*, **103**, EE6, 989–1000.

Eriksson L. and Alm B. (1990). Effect of cationic polyelectrolyte characteristics, ionic strength and concentration of anionic particles on flocculation kinetics and floc properties. 21st Annual Meeting of the Fine Particle Society, San Diego, CA, Aug.

Fettig J., Ratnaweera H. and Ødegaard H. (1990). Synthetic organic polymers as primary coagulants in wastewater treatment. Water Supply, 8, Jönköping, 19–26.

Fleer G. and Lyklema J. (1974). Polymer adsorption and its effect on the stability of hydrophobic colloids - III. The flocculation process as studied with the silver iodide - polyvinyl alcohol system, *Jour. Colloid-Interface Sci.*, **46**, 1.

Gale R. and Baskerville R. (1970). Polyelectrolytes in the filtration of sewage sludge. *Filtration and Separation*, 37–52.

Glaser H. and Edzwald J. (1979). Coagulation and direct filtration of humic substances with polyethylenimine. *Env. Sci. Tech.*, **13**(3), 299–305.

Golueke C. and Oswald W. (1965). Harvesting and processing sewage grown planktonic algae. *Jour. WPCF*, **37**, 471–498,

Gregory J. (1973). Rates of flocculation of latex particles by cationic polymers. *Jour. Colloid Interface Sci.*, **42**(2), 448–456.

Gregory J. and Sheiham I. (1974). Kinetic aspects of flocculation by cationic polymers. *Br. Polym. J.* **6**, 47–59.

Gregory J. (1977). Stability and flocculation of colloidal particles. *Eff. Wat. Treat. Jour.*, **17**(12), 641–643 and 647, 649, 651.

Gregory J. and Lee S. (1990). The effect of charge density and molecular mass of cationic polymers on flocculation kinetics in aqueous solution. *J. Water SRT–Aqua*, **39**(4), 265–274.

Gregory J. (1993). The role of colloid interactions in solid-liquid separation. *Wat. Sci. Tech.*, **27**(10), 1–17.

Haarhoff J. and Cleasby J. (1989). Direct filtration of *Chlorella* with cationic polymer. *Jour. Env. Eng., ASCE*, **115**(2), 348–366.

Habibian M. and O'Melia C. (1975). Particles, polymers and performance in filtration. *Jour. Env. Eng. Div., ASCE*, EE4, 567–583.

Healy L. and La Mer V. (1962). The adsorption flocculation reactions of a polymer with an aqueous colloidal dispersion. *J. Phys. Chem.*, **66**, 1835.

Holmes R. and Toth S. (1957). Physico-chemical behaviour of clay conditioner complexes. *Soil Sci.*, **84**, 479.

Huck P., Murphy K., Reed C. and LeClair B. (1977). Optimization of polymer flocculation of heavy metal hydroxides. *Jour. WPCF.*, 2411–2418.

Jimenez J., Woodard S., Vareika M. and Parker D. (2013). Improving simultaneous nitrogen removal performance through magnetite addition. *Water Environ. Fed. Annual Conf.*, WEFTEC, Chicago, IL.

Kasper D. (1971). Theoretical and experimental investigations of the flocculation of charged particles in aqueous solutions by polyelectrolytes of opposite charge. Ph.D. Thesis, California Institute of Technology, Pasadena, CA.

Kitchener J. (1972). Principles of action of polymeric flocculants. *Brit. Polym. J.*, **4**, 217–229.

Kragh A. and Langston W. (1962). The flocculation of quartz and other suspensions with gelatin. *Jour. Colloid Sci.*, **17**, 101.

Kvinnesland T. and Ødegaard H. (2004). The effects of polymer characteristics on nano particle separation in humic substances removal by cationic polymer coagulation. *Wat. Sci. Tech.*, **50**(12), 185–191.

La Mer V. and Smellie R. (1956). Flocculation, subsidence and filtration of phosphate slimes (I. General). *Jour. Colloid Sci.*, **11**, 704.

La Mer V. and Smellie R. (1956). Flocculation, subsidence and filtration of phosphate slimes. (II. Starches and agents for improving flocculation, subsidence and filtration of phosphate slimes). *Jour. Colloid. Sci.*, **11**, 710.

La Mer V., Smellie R. and Lui P. (1957). Flocculation, subsidence and filtration of phosphate slimes (IV. Flocculation by gums and polyelectrolytes and their influence on filtration rates). *Jour. Colloid Sci.*, **12**, 230.

La Mer V., Smellie R. and Lui P. (1957). Flocculation, subsidence and filtration of phosphate slimes (V. The optimum filtration rate as a function of solid content and specific area). *Jour. Colloid Sci.*, **12**, 566.

Lawson D. (1977). Polymer cuts cost of Rochester water. *The American City and County*, 97–98.

Lee A., Lewis D. and Ashman P. (2009). Microbial flocculation, a potentially low-cost harvesting technique for marine microalgae for the production of biodiesel. *J. Appl. Phycol.*, **21**, 559–567.

Lertsutthiwong P., Sutti S. and Powtongsook S. (2009). Optimization of chitosan flocculation for phytoplankton removal in shrimp culture ponds. *Aquacultural Engineering*, **41**, 188–193.

Letterman R., Sama R. and DiDomenico E. (1979). Direct filtration using polyelectrolyte coagulants. *Jour. AWWA*, 332–338.

Leu R. and Ghosh M. (1988). Polyelectrolyte characteristics and flocculation. *Jour. AWWA*, 159–167.

Li J., Dhanvantari S., Averill D. and Biswas N. (2003). Windsor combined sewer overflow treatability study with chemical coagulation. *Water Qual. Res. J. Canada*, **38**(2), 317–334.

Li J., Drca P., Manzon, Harold H., El-Gendy A., Biswas N. and McCorquodale J. (2009). Development of retention treatment basin with polymer flocculation for CSO treatment. *Proceedings of the Water Environment Federation Conference*, WEFTEC, Orlando, FL.

Liu D., Li F. and Zhang B. (2009). Removal of algal blooms in freshwater using magnetic polymer. *Wat. Sci. Tech.*, **59**(6), 1085–1091.

Loganathan P. and Maier W. (1975). Some surface chemical aspects in turbidity removal by sand filtration. *Jour. AWWA*, 336–342.

Marsalek J., He C., Rochfort Q., Exall K., Wood J., Krishnappan B., Seto P. and Chessie P. (2005). Upgrading the north Toronto combined sewer overflow (CSO) storage and treatment facility. *NATO Science Series: IV: Earth and Environmental Sciences*, **43**, 111–121.

Martin J. and Aldrich D. (1955). Influence of soil exchangeable cation ratios on the aggregating effects of natural and synthetic soil conditioners. *Soil Sci. Soc. of Amer.*, **19**, 50.

McLaren A. (1954). The adsorption and reactions of enzymes and proteins on kaolinite (I). *Jour. Phys. Chem.*, **58**, 129.

McLaren A., Peterson G. and Barshad I. (1958). The adsorption reactions of enzymes and proteins on clay minerals (IV. Kaolinite and Montmorillonite). *Soil Sci. Soc. of Amer.*, **22**, 239.

Mels A., Rulkens W., van der Meer A., van Nieuwenhuijzen A. and Klapwijk A. (2001). Flotation with polyelectrolytes as first step of a more sustainable wastewater treatment system. *Wat. Sci. Tech.*, **43**(11), 83–90.

Michaels A. (1954). Aggregation of suspensions by polyelectrolytes. *Ind. Eng. Chem.*, **46**, 1485.

Michaels A. and Morelos O. (1955). Polyelectrolyte adsorption by Kaolinite. *Ind. and Eng. Chem.*, **47**, 1801.

Morrow J. and Rausch E. (1974). Colloid destabilization with cationic polyelectrolytes as affected by velocity gradients. *Jour. AWWA*, 646–653.

Mortenson J. (1957). Adsorption of hydrolyzed polyacrylonite on Kaolinite (I. Effect of exchange cation and anion). *Soil Sci. Soc. Amer.*, **21**, 385.

Mortenson J. (1959). Adsorption of hydrolyzed polyacrylonite on Kaolinite (II. Effect of solution electrolytes). *Soil Sci. Soc. Amer.*, **23**, 199.

Mortenson J. (1961). Adsorption of hydrolyzed polyacrylonite on Kaolinite, Ninth National Conf. Clays and Clay Minerals, 530.

Mueller K. and Burbank N. (1964). Effects of polyelectrolyte coagulant aids on removal of missouri river turbidity with ferric sulphate. *Jour. AWWA*, **56**, 333–346.

Narayanan B., de Leon C., Radke C. and Jenkins D. (2010). The role of dispersed nocardioform filaments in activated sludge foaming. *Water Environ. Res.*, **82**(6), 483–491.

Narkis N. and Rebhun M. (1975). The mechanism of flocculation processes in the presence of humic substances. *Jour. AWWA*, 101–108.

Narkis N. and Rebhun M. (1977). Stoichiometric relationship between humic and fulvic acids and flocculants. *Jour. AWWA*, 325–328.

Novak J. and Langford M. (1977). The use of polymers for improving chemical sludge dewatering on sand beds. *Jour. AWWA*, 106–110.

Novak J., Love N., Smith M. and Wheeler E. (1998). The effect of cationic salt addition on the settling and dewatering properties of an industrial activated sludge. *Water Env. Res.*, **70**(5), 984–996.

Novak J., Muller C. and Murthy S. (2001). Floc structure and the role of cations. *Wat. Sci. Tech.*, **44**(10), 209–213.

Nozaic D., Freese S. and Thompson P. (2001). Longterm experience in the use of polymeric coagulants at Umgeni Water. *Wat. Sci. Tech.: Water Supply*, **1**(1), 43–50.

O'Brien J. and Novak J. (1977). Effects of pH and mixing on polymer conditioning of chemical sludge. *Jour. AWWA*, 600–605.

Oh H., Lee S., Park M., Kim H., Kim H., Yoon J., Kwon G. and Yoon B. (2001). Harvesting of *Chlorella vulgaris* using a bioflocculant from *Paenibacillus* sp. AM49. Biotechnology Letters, **23**, 1229–1234.

Packham R. (1967). Polyelectrolytes in water clarification (including discussions). *Proc. Soc. Wat. Treat. Exam.*, **16**, 88–111.

Packham R. (1972). The laboratory evaluation of polyelectrolyte flocculants. *Br. Polym. J.*, **4**, 305–315.

Packham R. and Sheiham I. (1977). Developments in the theory of coagulation and flocculation. *Jnl. Inst. Water Eng. Sci.*, **31**(2), 96–108.

Packter A. (1957). Interaction of Montmorillonite clays with polyelectrolyte. *Soil Sci.*, **83**, 355.

Patoczka J., Lauria J., Scheri J. and Shehhan J. (1998). Stress testing of final clarifier and polymer use for flow maximization. *Water Environ. Fed. Annual Conf.*, WEFTEC, Orlando, FL.

Rebhun M. and Wacks A. (1965). *Int. Congr. Pure Appl. Chem.*, (Moscow) A88.

Robinson C. (1974). Polyelectrolytes as primary coagulants for potable water systems. *Jour. AWWA*, 252–257.

Robinson C. (1979). Cationic polyelectrolytes reduce organic matter in turbid surface waters. *Jour. AWWA*, 226–227.

Ruehrwein R. and Ward D. (1952). Mechanism of clay aggregation by polyelectrolytes. *Soil Sci.*, **73**, 485.

Sarker N. and Teot A. (1973). Coagulation of negatively charged colloids by anionic polyelectrolytes and metal ions. *Jour. Colloid Interface Sci.*, **43**(2), 370–381.

Seka A., Wiele T. and Verstraete W. (2001). Feasibility of a multi-component additive for efficient control of activated sludge filamentous bulking. *Wat. Res.*, **35**(12), 2995–3003.

Shao Y., Starr M., Kaporis K., Kim H. and Jenkins D. (1997). Polymer addition as a solution to nocardia foaming problems. *Water Environ. Res.*, **69**(1), 25–27.

Shea T., Gates W. and Argaman Y. (1971). Experimental evaluation of operating variables in contact flocculation. *Jour. AWWA*, 41–48.

Shine D. and Rasmussen R. (1977). Polymers reduce cost and improve efficiency in water treatment plant. *Water & Sew. Works*, 34–35.

Silberberg A.: *J. Phys. Chem.*, 1962, **66**, 1872; 1962, **66**, 1844; *J. Chem. Phys.*, 1967, **46**, 1105; 1968, **48**, 2835.

Singer P., Pipes W. and Hermann E. (1968). Flocculation of bulked activated sludge with polyelectrolytes. *Jour. WPCF*, **40**(2), R1–R9.

Smellie R. and La Mer V. (1956). Flocculation, subsidence and filtration of phosphate slimes (III. Subsidence behaviour). *Jour. Colloid Sci.*, **11**, 730.

Smellie R. and La Mer V. (1958). Flocculation, subsidence and filtration of phosphate slimes (VI. A quantitative theory of filtration of flocculated suspensions). *Jour. Colloid Sci.*, **23**, 589.

Sommerauer D. (1968). The role of complex formation in the flocculation of negatively charged sols with anionic polyelectrolytes. *Kolloid Zeitschrift*, **225**, 147.

Tanaka T. and Pirbazari M. (1986). Effects of cationic polyelectrolytes on the removal of suspended particulates during direct filtration. *Jour. AWWA*, 57–65.

Tenney M. and Stumm W. (1965). Chemical flocculation of microorganisms in biological waste treatment. *Jour. WPCF*, **37**, 1370–1388.

Tenney M., Echelberger W., Schuessler R. and Pavoni J. (1969). Algal flocculation with synthetic organic polyelectrolytes. *Appl. Microbiol.*, **18**(6), 965–971.

Terrell D. (1977). Organic polymers replace alum and improve water quality in Ithaca. *Jour. AWWA*, 263–266.

Thompson D., Moore R. and Hackworth D. (2005). Ballasted sedimentation boosts floc settling. *AWWA Opflow*, 14–17.

Thorup R., Nixon F., Wentworth D. and Sproul O. (1970). Virus removal by coagulation with polyelectrolytes. *Jour. AWWA*, 97–101.

Tilton R., Murphy J. and Dixon J. (1972). The flocculation of algae with synthetic polymeric flocculants. *Water Research*, **6**, 155–164.

Treweek G. and Morgan J. (1977). Polymer flocculation of bacteria: The mechanism of E. Coli aggregation by polyethylenimine. *Jour. Colloid Interface Sci.*, **60**(2), 258–273.

Vanderhasselt A., Clercq B., Venderhaegen B., Vanrolleghem P. and Verstraete W. (1999). On-line control of polymer addition to prevent massive sludge washout. *Jour. Env. Eng. Div., ASCE*, 1014–1021.

Vanderhasselt A. and Verstraete W. (1999). Short-term effects of additives on sludge sedimentation characteristics. *Wat. Res.*, **33**(2), 381–390.

Wood J., He C., Rochfort Q., Marsalek J., Seto P., Yang M., Chessie P. and Kok S. (2005). High-rate stormwater clarification with polymeric flocculant addition. *Wat. Sci. Tech.*, **51**(2), 79–88.

Yapijakis C. (1982). Direct filtration: polymer in backwash serves dual purpose. *Jour. AWWA*, 426–428.

Yeh H. and Ghosh M. (1981). Selecting polymers for direct filtration. *Jour. AWWA*, 211–218.

Young J. and Edwards F. (2003). Factors affecting ballasted flocculation reactions. *Water Env. Res.*, **75**, 263–272; and Discussions 76, 284–285, 2004.

Chapter 6
Rapid mixing

6.1 INTRODUCTION

Rapid mixing refers to that stage in the overall coagulation-flocculation process where coagulant chemicals are added into the stream to be treated. The function of rapid mixing should not be considered to be one merely of efficient coagulant dispersion. The rapid mixing stage is possibly the most important operation in the process since it is here that destabilization reactions occur and where primary floc particles are formed, the characteristics of which markedly influence subsequent flocculation kinetics.

From an engineering survey of the quality of water treatment at over twenty water treatment plants, Cleasby *et al.* (1989) concluded that efficient rapid mixing was essential and one of the most important unit operations in water treatment. In fact, flocculation appeared to be less important than rapid mixing for successful high rate filtration.

The importance of the efficiency of a rapid mixing facility is evident if the order of time of destabilization reactions is considered: The time for mononuclear complexes, such as $AlOH^{2+}$ and $FeOH^{2+}$, to form is of the order 10^{-10} s. Polynuclear complexes require from 10^{-2} to 1.0 seconds to form. The time required for adsorption of these species to particles is of the order 10^{-10} seconds (O'Melia, 1969). The time needed for adjusting the structure of the double layer is of the order 10^{-8} seconds and for a Brownian collision (diffusion) of the order 10^{-7} to 10^{-3} seconds (Overbeek, 1977).

The time required for the formation of metal hydroxide precipitates and the entrapment of colloidal particles by a sweep mechanism is of the order 1 to 7 seconds (Letterman *et al.* 1973). The longest order of time scale is the time required for charge adjustment at particle surfaces. This time may vary from as little as 10^{-6} to as long as 10^{4} seconds (Overbeek, 1977).

Amirtharajah and Mills (1982) pointed out that the metal coagulant hydrolysis products that are formed within the time range 0.01 to 1.0 seconds are the most important for effective destabilization.

From the above, it is appreciated that in many instances, traditional 30 to 60 second retention times during rapid mixing are unnecessary and that flocculation efficiency may not improve beyond rapid mix times of approximately 5 seconds or less (Griffith & Williams, 1972). Indeed, beyond a certain optimum rapid mix time, a detrimental effect on flocculation efficiency may result (Letterman *et al.* 1973). For a particular water and type of coagulant applied, the optimum retention time during rapid mixing is dependent on the velocity gradient and coagulant dosage applied.

When carrying out coagulation tests in the laboratory, it will be found that optimum solids removal will be achieved by a particular combination of coagulant concentration and pH. The conditions of the laboratory experiments are such that complete and instantaneous dispersion of the coagulant chemicals into the sample of 'water' will be achieved readily (see Chapter 8). However, in practice, by applying the optimum coagulant dosage and pH as found in the laboratory, it is often found that the same degree of process efficiency cannot be achieved. This is often likely due to an inadequate appraisal of the mode of destabilization necessary for the particular type of 'water' in question; and consequently the installation of a rapid mixing tank which cannot provide the requisite mixing environment.

6.2 REQUIREMENTS FOR RAPID MIXING DEVICES

6.2.1 General

The types of rapid mixers most commonly installed in practice are represented schematically in Figure 6.1 (Vrale & Jorden, 1971). Such mixers are given the general name back-mix reactors and usually are designed to provide a 10 to 60 second retention time with a root mean square velocity gradient, G, of the order 300 s^{-1} (Morrow & Rausch, 1974). (Note that the concept of G is discussed in more detail in Chapter 7). In many instances where back-mix reactors have been installed for coagulant addition, they have been abandoned or not used extensively due to the poor results often attained (Ryder, 1977).

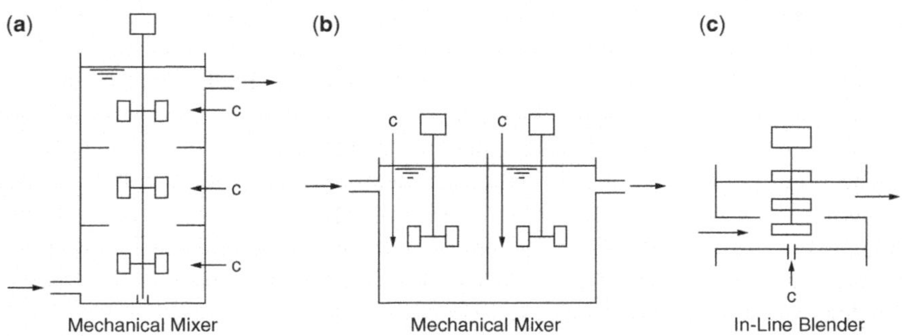

Figure 6.1 Schematics of types of mechanical agitation devices used for rapid mixing (after Vrale & Jorden, 1971).

The basic characteristic (and in many instances, the shortcoming) of back-mix devices is that the particles within them have a distribution of residence times. That is, some of the particles will be short-circuited out almost immediately whereas some will stay in the reactor for longer periods than the nominal time. In the case of an adsorptive mechanism of destabilization using metal coagulants, this can be undesirable for the following reasons:

- Due to long residence times extensive hydrolysis of some of the applied metal coagulant to the relatively inefficient $Fe(OH)_3$ or $Al(OH)_3$ will take place;
- Poor overall destabilization of the particles will occur due to either incomplete adsorption or extensive adsorption of metal hydroxide species.

What is needed is for the same (best) metal hydroxide species to be available to every particle simultaneously. A plug flow type of rapid mixer will be seen to approach this "ideal" situation the most satisfactorily since by virtue of its nature, all elements of liquid within it will theoretically have the same retention time.

For cases where an adsorptive mechanism of destabilization is not operative, the following considerations are important:

- If it is found that one needs to operate at say a $pH < 3$ with Al(III) or Fe(III) then a "Gouy-Chapman" type of destabilization will be effective and one will only need minimal rapid mixing since this type of destabilization is reversible (Overbeek, 1977).
- If an entrapment or 'sweep-floc' type of destabilization is found to be necessary then a back-mix reactor might be desirable since it has been found favorable for the formation of metal-hydroxide precipitates. Back-mix reactors may also be desirable if one desires nucleation effects to take place, such as in the water-softening process. However, note that the practice of phosphorus precipitation in wastewater treatment is usually more effective with some form of plug flow type rapid mixing environment.
- Amirtharajah and Mills (1982) found no significant differences in performance between different rapid mixing velocity gradients, when the sweep mechanism of coagulation was operative.
- This was also concluded from the work of Clark *et al.* (1994) who found that in the sweep floc regime of coagulation, flocculation and sedimentation performance were not sensitive to the type or the intensity of rapid mixing. However, the plug-flow type of mixers (such as the in-line jet mixer discussed later in this Chapter) did require less coagulant when compared with back-mix reactors.
- For sludge conditioning using metal coagulants or polymers, excessive turbulence would lead to floc breakup and hence an increase in particle surface area – this indicates that a laminar flow type reactor would be best.
- Using polyelectrolytes, where the method of destabilization is adsorptive in nature, then a plug flow environment, imparting a velocity gradient appropriate to the application, would be best.

6.2.2 Comparison of back-mix and plug-flow reactors

Vrale and Jorden (1971) were among the first investigators to identify that the *G* value is not a sufficient parameter to characterize the quality of mixing. Their work concentrated particularly on the most appropriate type of rapid mix unit for an adsorptive mechanism of destabilization by metal-hydroxide species. For this purpose they devised the five experimental rapid mix units shown in Figure 6.2:

- Unit 1 was designed to investigate the mixing effect within a pipe.
- Unit 2 was designed to simulate a back-mix reactor.
- Unit 3 was selected because it had been proved elsewhere to afford a very rapid and high degree of mixing. It was intended that this unit should be the criterion for 'ultimate' rapid mixing.
- Unit 4 was designed as a combination of Units 1 and 3. It was employed in order to determine the effect of the time lag before intense mixing occurred.
- Unit 5 was designed to simulate Unit 3 without the excessive head loss of Unit 3, and consequential shearing intensity.

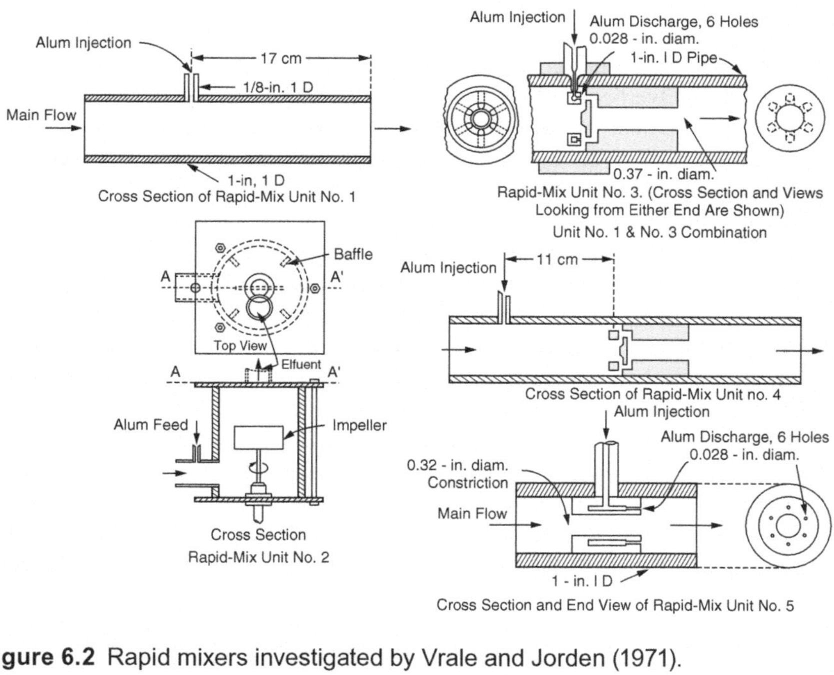

Figure 6.2 Rapid mixers investigated by Vrale and Jorden (1971).

The results obtained from their tests are illustrated in Figure 6.3 and the conclusions they came to with regard to the most efficient rapid mixing system are as follows: (Note: the criterion chosen for the efficiency of rapid mixing was the apparent aggregation rate K_{app} of the particles in a slow mix flocculating reactor, where

$$\frac{1}{T_t} = \frac{1}{T_0} + K_{app} \cdot t \tag{6.1}$$

and
T_t = supernatant turbidity after time t of flocculation;
T_0 = initial turbidity).

- Unit 1 showed a constant rate of aggregation – *irrespective of the flow rate in the pipe.*
- The results obtained for Unit 2, the back-mix reactor, indicate that this is the worst possible method of rapid mixing with a hydrolyzing metal; increasing the impeller speed tended to decrease its efficiency and, conversely, at very low impeller speeds incomplete mixing was obtained.
- Unit 4 showed a marked increase in aggregation rate with increased flow rate, until at a certain flow rate the rate of aggregation began to drop.
- This drop in efficiency at higher flow rates was also evident in Unit 3. It was reasoned that because of the very high head loss (and consequent shearing intensity) in these two units, then at the higher flow rates breakup of the destabilized particles occurred.
- Unit 5 was found to give the best results since its efficiency reached a peak (similar to the peak in the next best Unit 3) but did not decrease as did the others – it maintained a constant efficiency even at very high flow rates. The efficiency of Unit 5 reached a maximum when the hydraulic flow rate through the unit was increased to a value giving an average velocity gradient G of approximately 1000 s^{-1}, thereafter remaining constant.

Vrale and Jorden suggested that Units 3 and 4 did not perform as well as Unit 5 because breakup of primary floc particles at higher flow rates had a detrimental effect on subsequent flocculation reactions. However, from Figure 6.3 it is evident that for Units 3 and 4, at velocity gradients between 1000 to 2000 s^{-1} found optimal for Unit 5, the efficiencies of these two units were minimal. It would appear, therefore, that there are further factors to be considered besides excessive velocity gradients since efficiencies for Units 3 and 4 were maximal at velocity gradients as high as 7,500 s^{-1}.

Even though the points of maximum shearing intensity (or velocity gradient) occurred at the plates, it is evident that the lower velocity gradient at the point of expansion was the most effective with respect to rapid mixing efficiency. There are two possible reasons for this:

- First, the liquid velocity in the system was a maximum at the plates and through the constriction such that the residence time is minimal;
- Second, even though there was a high shearing intensity at the plate, the flow was fairly uniform being at a point of contraction.

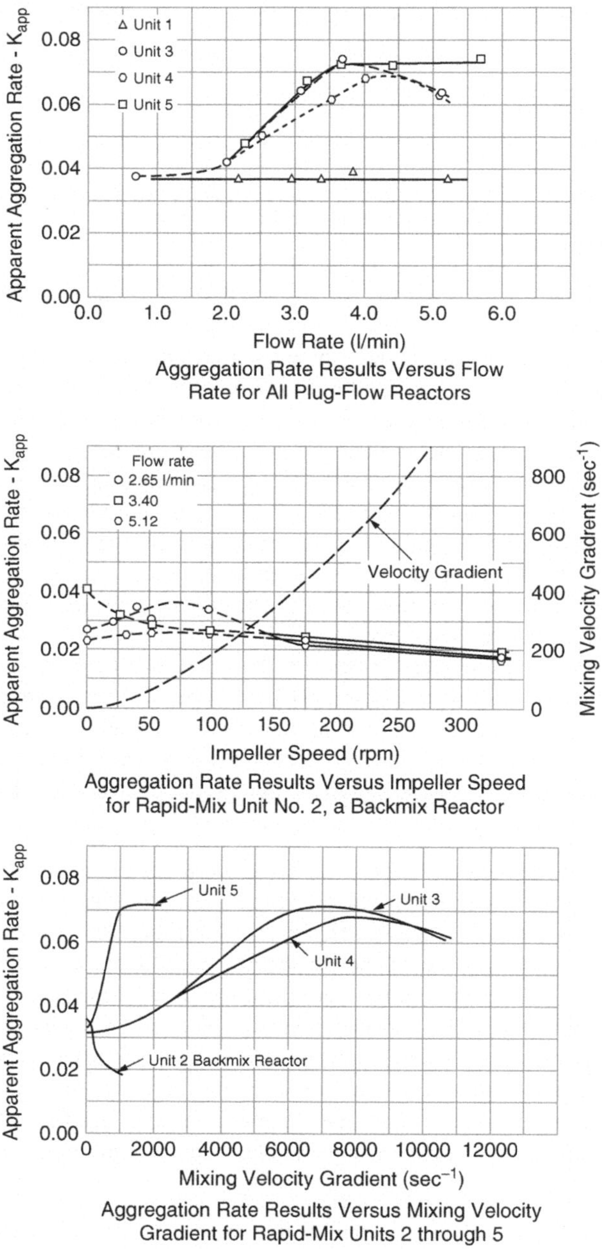

Figure 6.3 Results obtained using rapid mixers shown in Figure 6.2 (from Vrale & Jorden, 1971).

Effective mixing only occurs at the point of expansion where turbulent eddies are set up. This argument is supported by Vrale and Jorden's data where maximum efficiencies for all three Units 3 through 5 occur at approximately the same flow rate. The head loss, h_L, at the point of expansion must be similar in each case, whereas the overall head losses for Units 3 and 4, including the losses through the constrictions upstream of the point of expansion in each case, are far greater. (Note that velocity gradient, G, is proportional to $h_L^{1/2}$ – see Chapter 7).

6.2.3 Velocity gradient requirements

The work of Vrale and Jorden, outlined above, is especially useful in identifying the fact that *a high velocity gradient alone is not sufficient guarantee that efficient rapid mixing is achieved.* The quality of the mixing at the point of coagulant addition and the manner in which the small coagulant flow contacts the whole cross-section of process flow as instantaneously as possible probably has more significance in many applications.

Although some workers have pointed out the desirability of intense rapid mixing (i.e., high G values) to maximize efficient use of coagulant, there is an upper limit for a particular application. Too high velocity gradients during rapid mixing can delay floc formation during subsequent flocculation (Camp, 1968; Ferguson & King, 1977). The magnitude of the delay depends on the velocity gradient applied: using a G of 12 500 s^{-1}, no floc was visible up to a 45 minute flocculation period. With $G = 4400\ s^{-1}$, pinpoint floc was observed after 10 minutes flocculation. However, with rapid mixing at a G of 1000 s^{-1}, then applying a G of 12 500 s^{-1} and then flocculating, flocs formed successfully and were visible after 18 minutes flocculation. This emphasises further the importance of the correct initial rapid mixing procedure, the success of which strongly influences the efficiency of downstream processes.

Some examples of rapid mixing experiences are as follows:

- With an in-line rapid mix (plug flow) unit, using a hydrolyzing metal coagulant, a general design guideline appears to be 1000 s^{-1} under low flow conditions. Higher G values can be allowed to develop at higher flows. Vrale and Jorden (1971) found that coagulation efficiency did not reduce at higher G values for the type of rapid mixing device represented by Unit 5 in their work.
- If a back-mix reactor were used for the same application, the efficiency would decrease as G values are increased from near zero values, based on the findings of Vrale and Jorden.
- Some workers have recommended the use of G values between 400 and 1000 s^{-1} when using cationic polyelectrolytes as primary coagulants in the treatment of turbid waters (Morrow & Rausch, 1974). These workers suggested that, in some instances, the reported failure of using polyelectrolytes as primary coagulants may be due to inadequate velocity gradients being applied at the rapid mix stage. However, too high velocity gradients, greater than

approximately 100 s^{-1}, may give rise to primary floc breakup and "armouring" of the flocs with polyelectrolyte segments, thus effectively inhibiting further flocculation. Of interest in their work is that if rapid mixing occurs for too long a period (over 1 to 2 minutes) or, if high G values are reduced too rapidly, floc formation is inhibited during subsequent flocculation.

In the light of previous comments pointing out the inadequacy of solely relying on a value of velocity gradient, G, in characterizing a rapid mix facility, the guideline previously presented is worth reiterating: *the quality of the mixing at the point of coagulant addition and the manner in which the small coagulant flow contacts the whole cross-section of process flow as instantaneously as possible probably has the most significance in many applications.*

The importance of high efficiency rapid mixing has been demonstrated in many applications, including recent publications pointing out the importance of efficient rapid mixing to achieve very low effluent phosphorus concentrations. In this case efficient and effective rapid mixing is important to present fresh, newly formed precipitates for surface complexation/adsorption of slowly sorbed P species. Even if an application is not as critical as one to achieve very low residual phosphorus concentrations, an approach that assigns the same level of importance to rapid mixing would generally be prudent. The findings of Vrale and Jorden (above) and the quality and intensity of mixing achieved in their *Unit 5* should serve as a conceptual guide for design.

6.2.4 Rapid mixer retention time

It was implied above that rapid mixing for extended periods may give rise to a deleterious effect during subsequent flocculation. For a particular application there does in fact appear to be an optimum rapid mix period for which the rate of flocculation is maximised. This time appears to be principally a function of rapid mix velocity gradient and metal coagulant concentration. The higher the coagulant concentration and the higher the velocity gradient, the lower in general is the required rapid mix period. In one study, the optimum rapid mix period at a G of 1000 s^{-1}, ranged from 9 seconds to 2.5 minutes for alum concentrations ranging from 100 to 10 mg/l, respectively (Letterman *et al.* 1973). In contrast, other workers have found that turbidity removal, after flocculation and settling, was not influenced by rapid mix detention periods in the range 5 to 60 seconds, even though the velocity gradient applied was relatively low, at 350 s^{-1} (Griffith & Williams, 1972).

The best way of determining the appropriate rapid mixing time for a particular water is to conduct laboratory scale and/or pilot scale tests, as described in Chapter 8.

6.2.5 Tapered rapid mix velocity gradient

With coagulation-flocculation facilities there is often an interruption of up to several minutes from the high velocity gradients during rapid mixing to the low velocity gradients during flocculation. The only intermediate mixing in such instances arises from the velocity gradients existing within the connecting conduit or channel. Intermediate G values in these cases are determined from the head loss along the conduit or channel.

A sharp decline in velocity gradient (from, say, 1000 down to 100 s^{-1}) can have a detrimental effect on subsequent (final) water quality. By striving to maintain a given final water quality, coagulant dosages with interrupted mixing have been reported as high as 30 to 40 percent greater than if tapered rapid mixing were provided (Kawamura, 1976). Some workers found that when using cationic polyelectrolytes as primary coagulants, if the rapid mix velocity gradients were suddenly lowered to low velocity gradients during flocculation of from 5 to 100 s^{-1}, small flocs formed which neither grew nor settled during and after flocculation. This effect was not evident if tapered rapid mixing was employed to lower velocity gradients gradually (Morrow & Rausch, 1974).

6.2.6 Coagulant feed concentration

Several workers have reported the effects of the concentration of coagulant solution fed to the rapid mix system (Griffith & Williams, 1972; Jeffcoat & Singley, 1975; Jenkins & Lee, 1977; Kawamura, 1973, 1976). Too high a concentration for a given rapid mix device may result in poor water quality after flocculation and solid-liquid separation. When mixing a small flow of coagulant solution with a much larger stream of raw water, dilute coagulant solutions disperse more rapidly and uniformly than concentrated solutions.

Coagulant solutions that are too low (less than 0.1% to 0.3%) may result in poor performance due to pre-hydrolysis of metal species before addition to the raw water stream. Serious clogging of the solution feed lines could also result.

The solution strength for aluminum sulphate should be greater than 1%. The pH of the solution should also be less than 3.0 to avoid undue scaling in the chemical feed lines. For ferric chloride the solution concentration should be greater than 5% with a solution pH less than 2.0 (Kawamura, 2000).

The feed concentration of polyelectrolytes is usually dictated by manufacturers' recommendations. However, as a rough guide solid grade products are normally recommended to be within the range 0.02 to 0.1%. With liquid grade products, dilution to 100:1 is often recommended. Stock solution concentrations of solid grade products are normally made up with viscosities of the order 500 cps. Higher solution concentrations present handling problems due to the correspondingly higher viscosities. Stock solution concentrations are normally recommended at 0.2 to 0.5%.

6.2.7 Sequence of chemical addition

Multiple point addition is sometimes beneficial when dosing polyelectrolytes. It is possible that such a procedure assists the bridging mechanism of destabilization by retaining loops for a sufficient period. It may also help overcome inadequate rapid mixing. However, from preceding discussions, in general it may be preferable to design a rapid mixing facility with polyelectrolyte addition at one point only provided that an appropriate velocity gradient is applied.

It is generally understood that where alkaline reagent (e.g., lime) or acid needs to be added with the coagulant to adjust raw water pH to the optimum final pH, such chemicals must be added before or in some cases at the same time as the coagulant. It should not be added after coagulant addition since destabilization reactions which commence at the time of coagulant addition are not necessarily reversible.

However, there are exceptions to this. In some cases, adding polymer prior to the metal coagulant has improved performance. In other cases, adding lime to alum coagulated water improved subsequent sedimentation, although the lime increased the pH beyond the optimum for coagulation (Hudson, 1981).

An example illustrating the importance of not rigidly adhering to preconceived notions of treatment is with coagulation applied to manganese bearing waters, where ferric chloride was added first at background pH, then lime was added after destabilization reactions had occurred to bring the pH to 9.5, for efficient precipitation of manganese using chlorine (Bratby, 1988).

6.3 DESIGN OF RAPID MIXING DEVICES

6.3.1 General

There are several types of rapid mixing devices which may be installed in a given coagulation-flocculation process system. Essentially they may be subdivided into three principal groups: back-mix reactors; partial back-mix reactors; and plug-flow in-line reactors.

When designing a rapid-mixing facility, a difficulty arises when different flow rates are to be designed for. As pointed out in an earlier section, for a particular water, coagulant type, dosage and rapid mix device, there is an associated value of velocity gradient, G, which produces best results. With rapid mix devices relying on hydraulic considerations alone, this optimal G value may be difficult to maintain at different flow rates. With rapid mixers relying on impellers to produce velocity gradients, the situation is simplified since an adjustable rotational speed to the impeller can serve to maintain the required intensity of mixing. However, as discussed earlier, such mixers fall into the back-mix or partial back-mix category and may not produce the appropriate mixing environment.

This section discusses the design of several specific rapid mix devices. In several instances it is difficult to categorize whether a particular device may be described as a partial back-mix type or a true plug-flow type. For the purposes of this section, therefore, rapid mixers will be categorised as back-mix reactors, in-line reactors where velocity gradients are variable, and fixed G in-line reactors. Bearing in mind the discussions of preceding sections, the following should assist as a preliminary guide in choosing an appropriate device for a particular system.

6.3.2 Backmix reactors

The configuration of back-mix reactors was schematically presented earlier. Figure 6.4 shows a typical installation using a single impeller. As stated previously there are a few instances (e.g., in the lime-soda softening process) where a back-mix reactor may be suitable. However, for most applications, indications are that a plug-flow type system is best. With installations such as Figure 6.4, the shear gradients applied may be adjusted to cater for varying flow rates by varying the speed of rotation of the impeller. Since velocity gradients produced by such proprietary equipment depend very much on the type of impeller, design is best carried out using data supplied by the manufacturer.

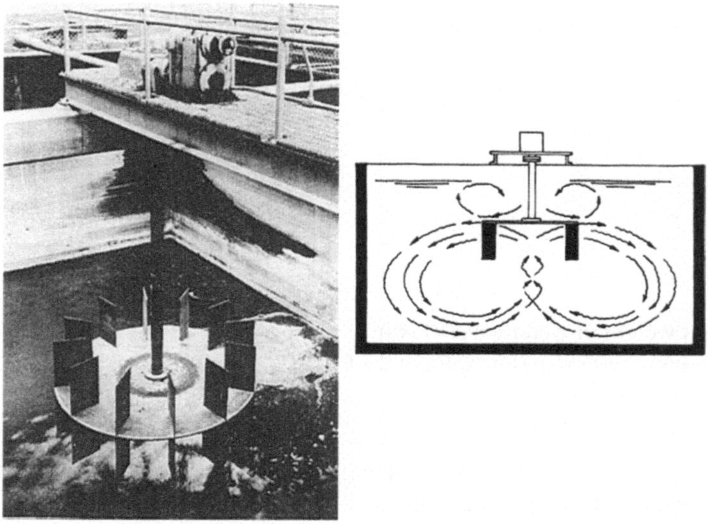

Figure 6.4 Typical back-mix type rapid mixing installation.

6.3.3 In-line mixers without controlled velocity gradient

There are a number of hydraulic appurtenances included in this category. These include static mixers in pipes or channels; pipe bends; sudden expansion within a pipe; orifice plate within a pipe, diffusion grids, and hydraulic jump in channel flow. Figure 6.5 shows pictures of static mixers. Because of the proprietary nature of these, design for a given mixing intensity should be carried out jointly with the manufacturer.

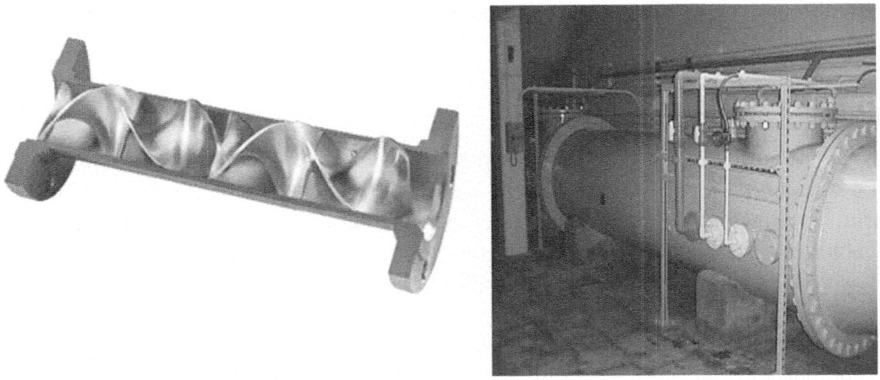

Figure 6.5 Static mixers (courtesy: (left) Chemineer Inc.; (right) Komax Systems Inc.).

The remaining appurtenances are described in more detail in the following sections:

For each of these devices the average velocity gradient is determined by the following general expression (see also Chapter 7):

$$G = \sqrt{\frac{\gamma \cdot h_L}{T \cdot \mu}} \quad s^{-1} \tag{6.2}$$

where

γ = unit weight of water (=9,810 N/m^3)
μ = absolute viscosity (for water at 20°C = 10^{-3} N · s/m^2)
h_L = head loss (m)
T = retention time (seconds).

The required head loss for a given velocity gradient, G, is given by a rearrangement of the above equation, i.e.,:

$$h_L = \frac{G^2 T \mu}{\gamma} \quad m \tag{6.3}$$

6.3.3.1 Pipe bend

The turbulence occurring downstream of a pipe bend, or series of bends, has been effectively used for mixing chemicals (see Figure 6.6). For a number of 90° bends in series, an approximation to the total head loss is given by the sum of the head losses for each bend (Ziemke, 1961). Many hydraulic texts are available to calculate the overall head loss in pipe bends.

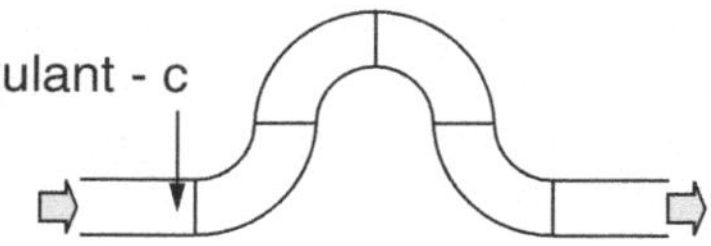

Figure 6.6 Pipe bends.

To compute average velocity gradient, G, from the head loss, an effective mixing length has to be assumed to compute retention time. Strictly speaking, although the flow pattern is only completely re-established 50 to 100 diameters downstream of the bend, 80% of the total head loss occurs within approximately 5 diameters (Ito, 1960). It is a reasonable approximation to assume a mixing length of 7.5 diameters for determining retention time.

6.3.3.2 Sudden expansion within a pipe

At a sudden expansion in a pipe (see Figure 6.7) a backwash of eddies will be formed at the corner of the enlargement giving rise to a loss of head. The magnitude of the head loss is derived from Bernoulli's equation and is given by:

$$h_L = \frac{(v_1 - v_2)^2}{2g} \tag{6.4}$$

where
v_1 = velocity of flow before enlargement (m/s);
v_2 = velocity of flow after enlargement (m/s).

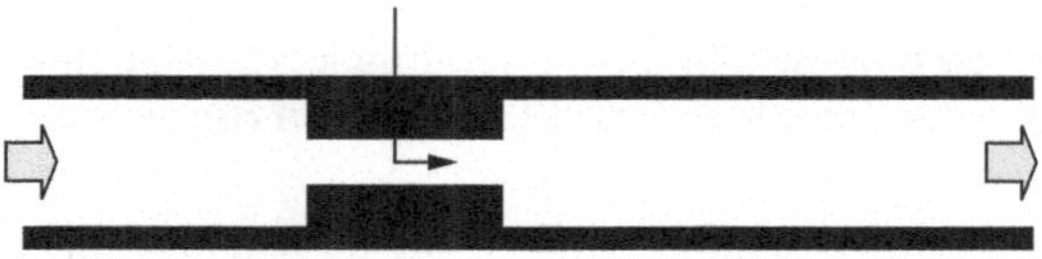

Figure 6.7 Sudden expansion within a pipe.

To compute the retention time for Equation 6.4, it can be assumed that the mixing length in the pipe downstream of the enlargement is approximately 7.5 times the pipe diameter downstream of the enlargement, and that the velocity is the average of the velocity before the enlargement and in the pipe after the enlargement.

6.3.3.3 Orifice plate within a pipe

Head loss within a pipe may be induced by situating an orifice plate at the point desired for mixing (see Figures 6.8a and 6.8b). This mixing device has been used effectively for mixing polymers in sludge lines for conditioning before thickening or dewatering (Bratby *et al.* 1993). The magnitude of the head loss at an orifice in pipe flow is given by (Simpson, 1968; Crane, 1988):

$$h_L = K\frac{v^2}{2g} \tag{6.5}$$

where

$$K = 2.8(1-\beta^2)\cdot\left(\frac{1}{\beta^4}-1\right) \tag{6.6}$$

v = approach pipe velocity, upstream of the orifice (m/s);
β = (diameter orifice)/(diameter pipe upstream of orifice)

(a) (b)

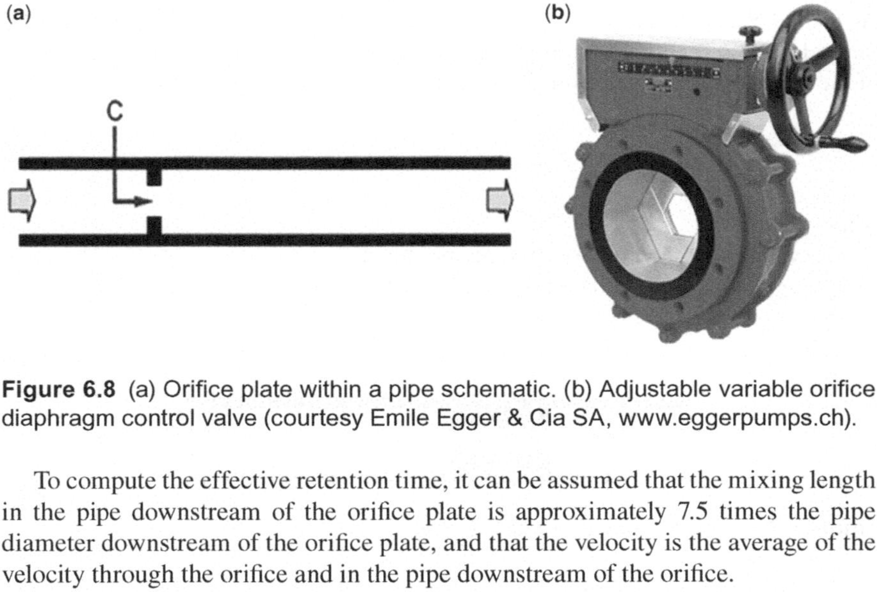

Figure 6.8 (a) Orifice plate within a pipe schematic. (b) Adjustable variable orifice diaphragm control valve (courtesy Emile Egger & Cia SA, www.eggerpumps.ch).

To compute the effective retention time, it can be assumed that the mixing length in the pipe downstream of the orifice plate is approximately 7.5 times the pipe diameter downstream of the orifice plate, and that the velocity is the average of the velocity through the orifice and in the pipe downstream of the orifice.

6.3.3.4 Diffuser grids in channel

Diffuser grids installed in open channels have been used for dosing chemicals, particularly where a specific means for efficient rapid mixing has not been previously provided (see Figure 6.9a). One issue that has been encountered with this method is clogging of the small orifices, particularly when metal hydroxides are used and metal hydroxide precipitate build-ups occur. If possible, a means for removal of the grid for cleaning the orifices is normally provided. The problem of precipitate fouling is exacerbated by the need to carry sufficient dilution water to ensure that the grid functions efficiently as a manifold device, to distribute the coagulant flow evenly across the channel cross-section.

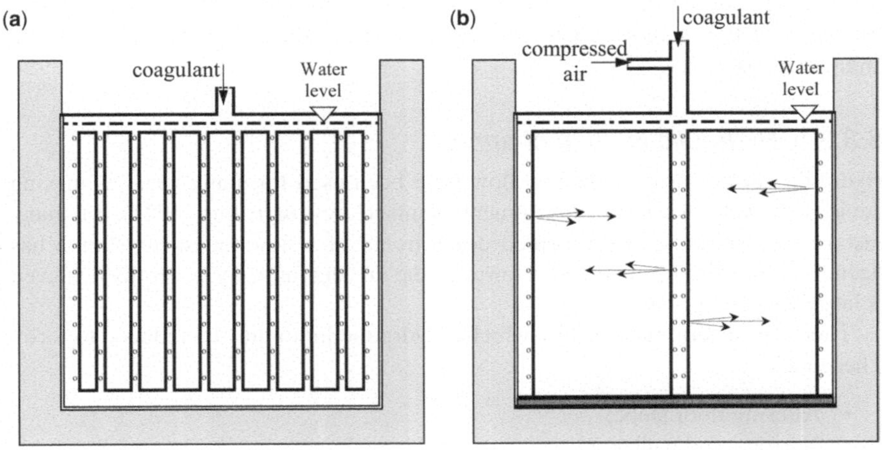

Figure 6.9 (a) Diffuser grid in channel. (b) Diffuser grid with compressed air dispersion in channel.

The velocity of the individual jets through the orifices is generally low, particularly since high dilutions of the metal coagulant are avoided to prevent excessive precipitate fouling. Orifice sizes are determined based on manifold hydraulics and the total flow of coagulant solution. The number and diameter of the grid tubes can be selected on the basis of maintaining proper manifold flow distribution, and creating reasonable head loss and mixing energy in the channel. The channel head loss through the grid can be calculated as follows:

$$h_L = K\frac{(V^2 - v^2)}{2g} \tag{6.7}$$

where

V = approach velocity (m/s)
v = velocity through the grid openings (m/s)
K = headloss coefficient = 1.43

One solution that has been adopted to avoid problems with coagulant flow distribution and orifice clogging, is to use low-pressure compressed air to induce jet flow through a smaller number of orifices. By this means the difficulty of creating sufficient headloss within the channel by the diffuser grid to create an adequate *G* value is avoided. The jet action of the compressed air as it exits the orifices together with the coagulant flow helps to cover the whole channel cross section with coagulant in an instantaneous fashion. Figure 6b shows a schematic of this concept. In some cases, depending on the size of the channel, the jet manifold pipes can be installed one on either side, against the walls of the channel, instead of against the walls and in the center of the channel, as shown.

6.3.3.5 Hydraulic jump in channel

Hydraulic jumps in open channel flow have been used for many years as mixing devices in water/wastewater treatment plants (Levy & Ellms, 1927). In many instances a hydraulic jump formed downstream of a flow measuring flume has been used for mixing chemicals. However, the mixing intensity generally achieved is largely serendipitous.

There are several methods by which a hydraulic jump may be induced to form. These include:

- Altering floor slope;
- Narrowing the channel;
- Including a sluice gate.

The first two methods are incorporated in flow measuring flumes. The third method may provide the most control as a mixing device (see Figure 6.10).

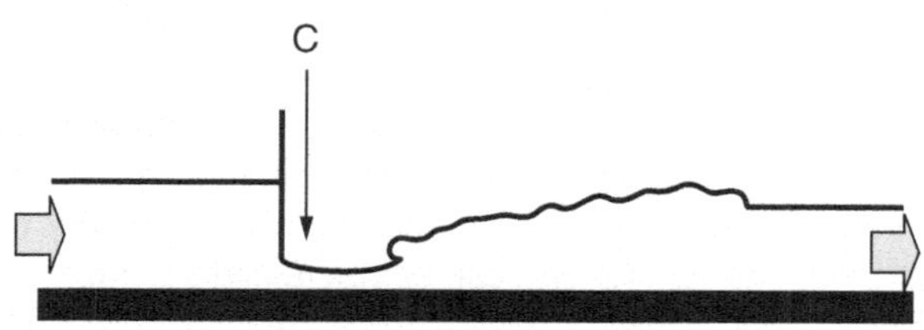

Figure 6.10 Hydraulic jump in channel.

When measuring flumes are also to be used as chemical mixing devices, the chemical solution is often dripped down from a distribution pipe with orifices, suspended transversely above the water level upstream of the jump. One often observes streaks of chemical on the surface of the flow, until dispersion occurs in the turbulent jump. There is no guarantee that the surface layers are completely mixed in the whole depth of flow in the jump. To ensure some penetration of the chemical into the depth of flow, the chemical diffuser pipe should be located as high as possible – at least 0.3 m (1 ft) above the water level – to increase the velocity of the chemical streams.

Another measure that might improve mixing throughout the depth is to install the maximum flume throat width possible for the range of flows expected at the plant. This ensures that the depth of flow through the critical section upstream of the jump is as shallow as possible.

Hydraulic jumps on horizontal floors are of several distinct types and can be classified according to Froude number, F_1, of the incoming flow (Chow, 1959):

- For $F_1 = 1$, the flow is critical and hence no jump can form.
- For $F_1 = 1$ to 1.7, the water surface shows undulations and the jump is called an "undular jump".
- For $F_1 = 1.7$ to 2.5, a series of small rollers develop on the surface of the jump, but the downstream water surface remains smooth. The velocity throughout is fairly uniform and energy loss low. This jump is called a "weak jump".
- For $F_1 = 2.5$ to 4.5, there is an oscillating jet entering the jump bottom to surface and back again with no periodicity. This jump is called an "oscillating jump".
- For $F_1 = 4.5$ to 9.0, the downstream extremity of the surface roller and the point at which the high velocity jet tends to leave the flow occur at practically the same vertical section. The action and position of this jump are least sensitive to variations in tailwater depth. This jump is called a "steady jump".
- For $F_1 > 9.0$, the high velocity jet grabs intermittent slugs of water rolling down the front face of the jump generating waves downstream and a rough surface can prevail. This jump is called a "strong jump".

A hydraulic jump will form in a channel if the Froude number F_1 of the approach flow, the approach flow depth, y_1, and the downstream depth, y_2 satisfy the equation:

$$\frac{y_2}{y_1} = \frac{\sqrt{1 + 8 \cdot F_1^2} - 1}{2} \tag{6.8}$$

where

$$F_1 = \frac{v_1}{\sqrt{g y_1}} \tag{6.9}$$

and

v_1 = approach velocity (m/s).

The loss of energy in the jump is given by

$$h_L = \frac{(y_2 - y_1)^3}{4 y_1 y_2} \tag{6.10}$$

The length of the jump, L, is a function of Froude number and expressed in terms of the ratio L/y_2. Figure 6.11 presents empirical data for L/y_2 (Chow, 1959).

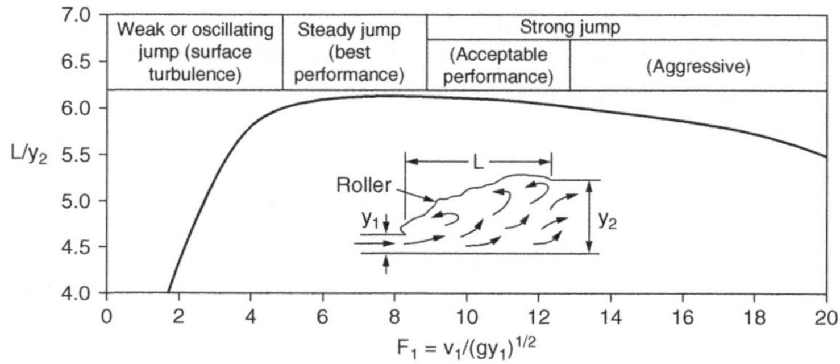

Figure 6.11 Length of hydraulic jump in terms of sequent depth y_2 of jumps in horizontal channels (after Chow, 1959).

To determine the velocity gradient, *G*, provided by the hydraulic jump, the jump length, L, and the average velocity through the jump can be used to compute the average retention time. The average velocity through the jump can be assumed to be the average of v_1 and v_2.

6.3.4 In-line mixers with controlled velocity gradient

The previous rapid mixing devices are generally classed as "wake" devices. The mixing energy is developed downstream of some disturbance in the flow and the dissipating eddies are relied upon to effect mixing. The disadvantage of such devices is that the mixing energy changes as the water flow changes, and no control over *G* is possible.

Figure 6.12 is an example of a rapid mix device where the mixing energy velocity gradient can be adjusted, independent of the flow. This is an example of proprietary devices available. Such devices may be equipped with variable speed drives thus catering for different flow rates.

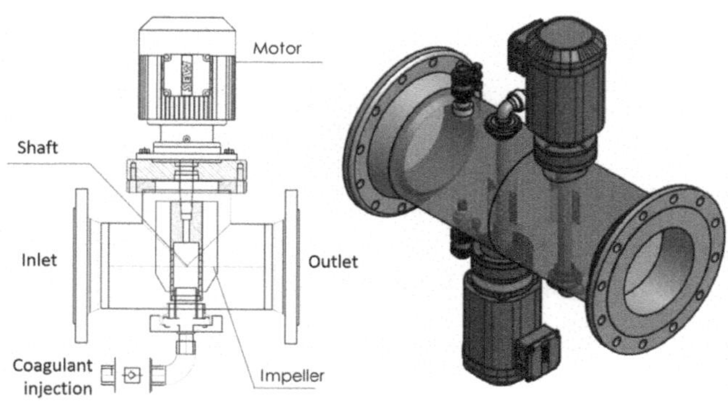

Figure 6.12 Example of in-line mixer with varying velocity gradient (courtesy Aquen Aqua-Engineering GmbH, www.aquen.de).

Figures 6.13a through 6.13c show a concept of in-line jet mixing, where the effective velocity gradient is determined by the discharge rate through a nozzle (Kawamura, 1976, 2000; Chao & Stone, 1979; Monk & Trussell, 1991).

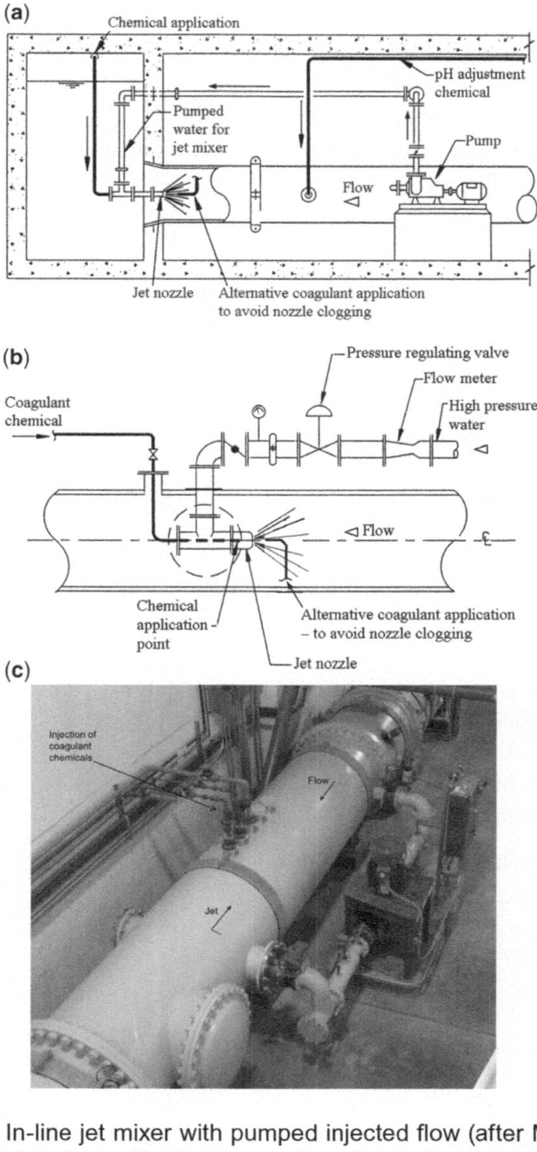

Figure 6.13 (a) In-line jet mixer with pumped injected flow (after Monk & Trussell 1991). (b) In-line jet mixer with plant pressurized flow (after Monk & Trussell 1991). (c) Example installation of an in-line jet mixer.

Design of an in-line jet mixer is as follows:
Velocity gradient, G is given by:

$$G = \sqrt{\frac{W}{\mu}} \qquad (6.11)$$

where

μ = absolute viscosity of liquid (for water = 10^{-3} N · s/m^2 at 20°C);
W = the dissipation function, is given by power dissipated, P, divided by volume, V (see also Chapter 7).

Hence, $W = P/V$ where V is the volume of influent pipe over which rapid mixing is effective.

Monk and Trussell (1991) estimated V from the spray cone angle and diameter d (m) of influent pipe. Therefore, for a 90° spray, $V = 0.13 \cdot d^3$, where d is the internal diameter of the pipe.

From the above, P is given by:

$$P = G^2 \cdot \mu \cdot V \qquad \mathrm{N \cdot m/s\ (= J/s)} \qquad (6.12)$$

The power dissipated by the nozzle flow is also given by:

$$P = \frac{1}{2g} C_d \cdot a \cdot v^3 \cdot \gamma \qquad (6.13)$$

where

C_d = discharge coefficient of orifice (approximately = 0.75);
a = area of orifice (m^2);
v = velocity of flow through nozzle (m/s);
γ = unit weight of liquid (=9,810 N/m^3 for water)
g = weight per unit mass (=9.81 N/kg).

Since $v = q/a$ (where q = flow rate through nozzle, m^3/s), then:

$$G = \sqrt{\frac{C_d \cdot a \cdot v^3 \cdot \gamma}{2g \cdot \mu \cdot V}} \quad \mathrm{s^{-1}} \qquad (6.14)$$

and:

$$a = \frac{1}{G}\sqrt{\frac{C_d \cdot q^3 \cdot \gamma}{\mu \cdot V \cdot 2g}} \quad \mathrm{m^2} \qquad (6.15)$$

As a rule of thumb, Kawamura (2000) recommends that the flow through the nozzle should be 2 to 5% the plant flow rate, to ensure satisfactory diffusion of the chemical into the full flow stream. The nozzle is best selected with an approximate 90° full cone spray pattern. During operation, the velocity gradient can be varied by increasing or decreasing the flow through the nozzle. The G value achieved is essentially independent of the plant flow rate.

It will be noted that a conflict arises between the recommended nozzle flow for good mixing, and the maximum dilution of metal coagulant required to avoid premature hydrolysis and precipitate fouling of the nozzle. For this reason the schematics shown in Figures 6.13a and 6.13b include alternative coagulant introduction points, with discharge within the spray cone, but downstream of the nozzle discharge. By this means, the coagulant flow can be determined based on the desired dilution, and the nozzle flow by the requirements for good mixing.

6.4 REFERENCES

Amirtharajah A. and Mills K. M. (1982). Rapid-mix design for mechanisms of alum coagulation. *Jour. AWWA*, 210–216.

Bratby J., Baugh C. and Ambrose W. A. (1993). Water Plant Sludge Dewatering by Recessed Chamber Filter Press. AWWA/WEF Joint Residuals Conf., 5–8, Phoenix, Arizona.

Bratby J. (1988). Optimizing manganese removal and washwater recovery at a direct filtration plant in Brazil. *Jour. AWWA*, 71–81.

Camp T. R. (1968). Floc volume concentration. *Jour. AWWA*, 656–673.

Chao J.-L. and Stone B. G. (1979). Initial mixing by jet injection blending. *Jour. AWWA*, 570–573.

Chow V. T. (1959). Open Channel Hydraulics, McGraw-Hill Student Edition.

Clark M. M., Srivastava R. V., Lang J. S., Trussell R. R., McCollum L. J., Bailey D., Christie J. D. and Stolarik G. (1994). Selection and Design of Mixing Processes for Coagulation. AWWA Research Foundation, Denver, Colorado.

Cleasby J. L., Dharmarajah A. H., Sindt G. L. and Baumann E. R. (1989). Design and Operation Guidelines for Optimization of the High-Rate Filtration Process: Plant Survey Results. AWWA Research Foundation, September, Denver, CO, U.S.A.

Crane Co. (1988). Flow of Fluids Through Valves, Fittings and Pipe, Technical Paper No. 410, King of Prussia, PA.

Ferguson J. F. and King T. (1977). A model for aluminium phosphate precipitation. *Jour. WPCF*, 646–658.

Griffith J. D. and Williams R. C. (1972). Application of jar test analysis at Phoenix, Arizona. *Jour. AWWA*, **64**(12), 825–830.

Hicks T. C. (1972). Standard Handbook of Engineering Calculations. McGraw-Hill Co., Inc.

Hudson H. E. Jr. (1981). Water Clarification Processes-Practical Design and Evaluation, Van Nostrand Reinhold Company, New York, NY.

Ito H. (1960). Pressure losses in smooth pipe bends. *Jour. Basic Eng., Trans. ASME*, 131–143.

Jeffcoat W. B. and Singley J. E. (1975). The effect of alum concentration and chemical addition times on coagulation. *Jour. AWWA*, 177–181.

Jenkins D. and Lee F. M. (1977). Investigations into the lime precipitation of raw municipal wastewater. *Prog. Wat. Tech.*, **9**, 495–507.

Kawamura S. (1973). Coagulation considerations. *Jour. AWWA*, **65**(6), 417–423.

Kawamura S. (1976). Considerations on improving flocculation. *Jour. AWWA*, 328–336.

Kawamura S. (2000). Integrated Design and Operation of Water Treatment Facilities. John Wiley & Sons, Inc., New York, NY.

Letterman R. D., Quon J. E. and Gemmell R. S. (1973). Influence of rapid mix parameters on flocculation. *Jour. AWWA*, 716–722.

Levy A. G. and Ellms J. W. (1927). The hydraulic jump as a mixing device. *Jour. AWWA*, **17**(1), 1–26.

Licsko I. (2004). Coagulation mechanisms – nano- and microprocesses. *Wat. Sci. Tech.*, **50**(12), 193–200.

Monk R. D. G. and Trussell R. R. (1991). Design of mixers for water treatment plants: rapid mixing and flocculators. In *Mixing in Coagulation and Flocculation*, AWWA Research Foundation, Denver, Colorado.

Morrow J. J. and Rausch E. G. (1974). Colloid destabilization with cationic polyelectrolytes as affected by velocity gradients. *Jour. AWWA*, 646–653.

O'Melia C. R. (1969). A review of the coagulation process. *Public Wks.*, **100**, 87.

Overbeek J. Th. G. (1977). Recent developments in the understanding of colloid stability. *Jour. Colloid Interface Sci.*, **58**(2), 408–422.

Ryder R. A. (1977). State of the art in water treatment design, instrumentation and analysis. *Jour. AWWA*, 612–620.

Simpson L. L. (1968). Process piping systems: sizing piping for process plants. *Chem. Eng.*, 192–214.

Vrale L. and Jorden R. M. (1971). Rapid mixing in water treatment. *Jour. AWWA*, **63**(1), 52–58.

Ziemke P. C. (1961). Pressure drop in pipe bends and elbows. *Jour. AWWA*, **53**, 329–330.

Chapter 7

Flocculation

7.1 INTRODUCTION

In Chapter 6 it was shown that rapid mixing is the stage in the overall coagulation-flocculation process where destabilization reactions occur and where primary, pin-point flocs commence growth. A necessary adjunct to destabilization is the inducement of primary particles to approach close enough together, make contact and progressively form larger agglomerates. This stage in the overall process is termed flocculation.

There are two stages in the flocculation process: The first, given the name perikinetic flocculation, arises from thermal agitation (Brownian movement) and is a naturally random process. Flocculation during this stage commences immediately after destabilization and is complete within seconds since there is a limiting floc size beyond which Brownian motion has no or little effect.

Furthermore, although the potential energy barrier existing between colloidal particles may be overcome by the thermal kinetic energy of Brownian movement, as the particles progressively coalesce the magnitude of the energy barrier increases approximately proportional to the area of the floc, so that eventually perikinetic flocculation of such potentially repellent particles must cease (Wilson & French, 1978).

The rate of flocculation or the rate of decrease in the number of particles of a suspension due to perikinetic flocculation may be described by a second order rate law. For example, in a turbid water containing 10^6 particles per ml, the particle number concentration would be reduced by half within a period of about 6 days, provided all the particles were completely destabilized and the particles did not become too large and be outside the range of Brownian movement. The theoretical development of perikinetic flocculation is dealt with in Section 7.1.

The second stage in the flocculation process is given the name orthokinetic flocculation and arises from induced velocity gradients in the liquid. Such velocity gradients may be induced by setting the liquid in motion by (a) passage around baffles or mechanical agitation within a flocculation reactor; (b) the tortuous path through interstices of a granular filter bed; (c) by differential settlement velocities within a settling basin, and so on. The effect of velocity gradients within a body of liquid is to set up relative velocities between particles thereby providing opportunity for contact.

For a given flocculating system, the principal parameter governing the rate of orthokinetic flocculation is the velocity gradient applied. The degree or extent of flocculation is governed by both applied velocity gradients and time of flocculation. These two parameters influence the rate and extent of particle aggregation and the rate and extent of breakup of these aggregates.

The theoretical development and practical applications of the orthokinetic flocculation phenomenon are dealt with in Section 7.3.

Figure 7.1 shows that there are several variations in the layout of the overall coagulation-flocculation process (Culp, 1977):

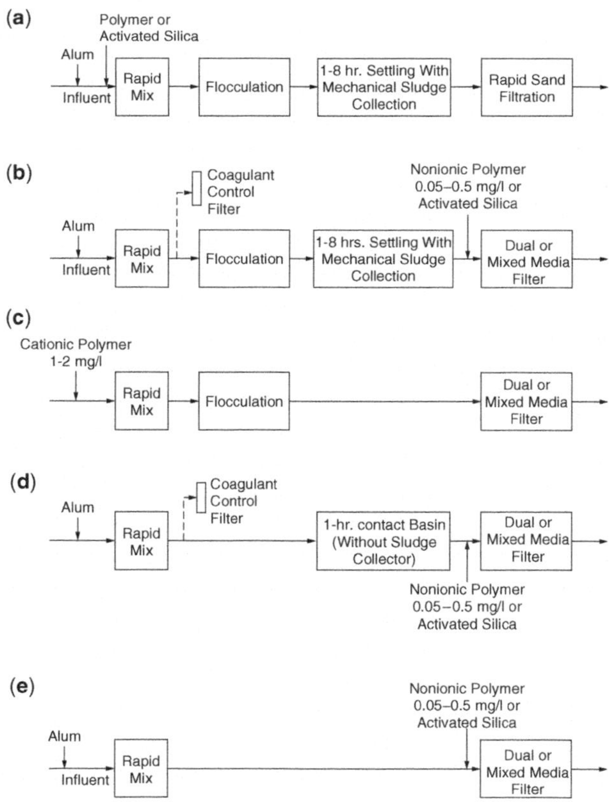

Figure 7.1 Alternative water treatment flow sheets (after Culp, 1977).

(a) Destabilization (rapid mix) flocculation-sedimentation-filtration;
(b) As for (a) but with the inclusion of a coagulant-control filter, addition of flocculant aid to the filter influent and dual or mixed-media filters;
(c) Direct filtration utilizing a flocculation basin with either a cationic polyelectrolyte added before rapid mixing (as shown) or alternatively, alum added before rapid mixing and flocculant aid before filtration;
(d) Direct filtration with a contact basin between the rapid mix stage and filtration. (The objective of the basin is not to provide flocculation or settling but rather a lag time to provide opportunity for response to the control filter);
(e) Direct filtration with only rapid mixing before filtration.

Alternatives (a) to (c) above require the inclusion of a flocculation reactor or basin whereas alternatives (d) and (e) rely on velocity gradients produced within a filter bed.

7.2 PERIKINETIC FLOCCULATION

Von Smoluchowski (1916, 1917) developed a model for perikinetic flocculation where the frequency of collisions was obtained from the diffusional flux of particles towards a single stationary particle. He assumed that particles diffused towards the stationary particle in a radial direction. The number of particles diffusing radially inwards through the surface of a sphere centred on the stationary particle is proportional to the Brownian diffusion coefficient of the particles, the surface area of the sphere, and the particle concentration gradient in the radial direction. Since the diffusion is assumed to be radial the number of particles diffusing radially inwards is equal to the number of collisions with the central particle. Assuming that the central particle also experiences diffusion, von Smoluchowski's equation for perikinetic flocculation is:

$$I_{ij} = 4\pi D_{ij} R_{ij} n_i n_j \tag{7.1}$$

where

I_{ij} = number of contacts per unit time between particles of radius R_i and R_j.
D_{ij} = mutual diffusion coefficient of particles i and j (approximately $D_i + D_j$)
R_{ij} = radius of interaction of the two particles, i.e. the distance between centres of two particles forming a lasting contact, i.e. $R_{ij} = R_i + R_j$.
n_i, n_j = respective number concentration of i and j particles.

If velocity gradients exist in the body of liquid of a magnitude dv/dz greater than approximately 5 s^{-1}, and if particle sizes are larger than 1 micrometer, the effect of perikinesis is negligible and orthokinetic flocculation need only be considered.

7.3 ORTHOKINETIC FLOCCULATION

7.3.1 Theoretical development

Once perikinetic flocculation is complete, the only way in which appreciable contact between particles can be promoted is to induce shear motion in the liquid. This is achieved by inducing velocity gradients whereby particles achieve mutual contact by movement with the surrounding liquid. The process of floc aggregation by this procedure is termed orthokinetic flocculation.

The greater the velocity gradients induced in the liquid the more particle contacts there will be within a given time. However, the greater the velocity gradients the smaller will be the ultimate floc size due to a continuous breakdown of the larger flocs. Hence, for a given velocity gradient there will be a limiting flocculation time beyond which floc particles will not grow further. The lower the velocity gradient the longer will be the time needed to reach the optimum floc size, but the larger will be the final floc size.

Von Smoluchowski (1916, 1917) formulated a mathematical model for the orthokinetic flocculation of colloidal particles under laminar flow conditions.

Two particles, *i* and *j*, are suspended in a liquid moving in the *x* direction in laminar flow. The velocity of the *j* particle is *v* and the velocity gradient in the *z* direction is *dv/dz*. The centre of the *j* particle is the origin of a set of Cartesian coordinates (see Figure 7.2).

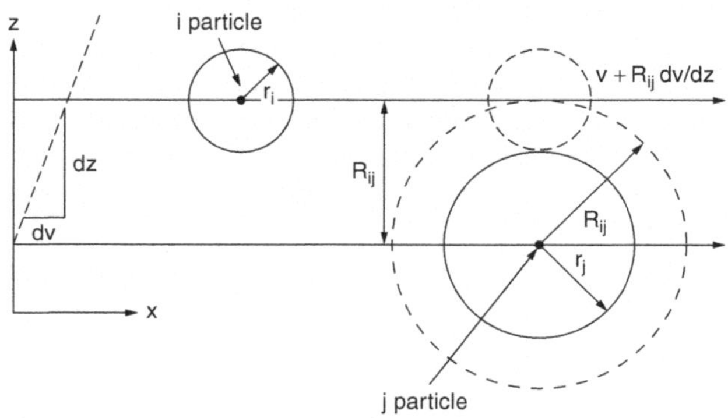

Figure 7.2 Contact between two particles, *i* and *j*, moving under the influence of a velocity gradient *dv/dz*.

For the i particle to come into contact with the j particle it must be brought close enough for London-van der Waal forces of attraction to "pull" and hold the two particles together. The locus of the *i* particle or any number of *i* particles satisfying this requirement is a sphere centred at the origin of the coordinate system, with a radius $R_{ij} = (r_i + r_j)$.

The sum of the radii of the spheres R_{ij} defines the range of influence of London-van der Waal forces for the individual i and j particles. The larger the particles, the narrower, in a relative sense, is the zone of influence of London-van der Waal forces beyond the surface of the particle; the closer therefore is the diameter of the particle to the "effective" diameter (see Figure 7.3).

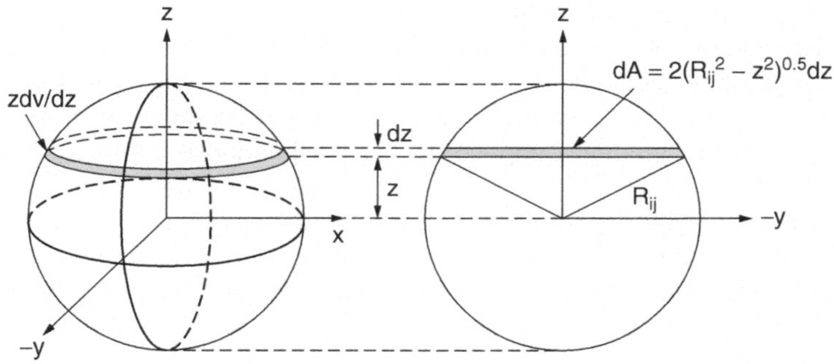

Figure 7.3 Zone of influence around central *j* particle.

If the particles follow the stream lines that would develop in their absence, the number of i particles making contact with the j particles would equal the product of the rate of flow q into the sphere, and the number of i particles per unit volume of fluid.

The velocity of flow at distance z from the j particle = $z \cdot dv/dz$ (relative to the j particle).

The area through which this flow takes place = $dA = 2\sqrt{R_{ij}^2 - z^2 dz}$.

Hence, the rate of flow into the area dA is given by:

$$dq = 2z\frac{dv}{dz}\sqrt{R_{ij}^2 - z^2 dz} \tag{7.2}$$

If the number of i particles per unit volume is n_i, the number of "contacts" H_i of i particles with the central j particle in unit time is:

$$H_i = 2\int_{z=0}^{z=R_{ij}} n_i dq = 2n_i \int_{z=0}^{z=R_{ij}} 2z\frac{dv}{dz}\sqrt{R_{ij}^2 - z^2 dz} \tag{7.3}$$

For a constant velocity gradient over the range of integration:

$$H_i = 4n_i\left(\frac{dv}{dz}\right)\int_0^{R_{ij}} z\sqrt{R_{ij}^2 - z^2 dz} = 4n_i\frac{dv}{dz}\left(\frac{R_{ij}^3}{3}\right) = \frac{4}{3}n_i\left(\frac{dv}{dz}\right)R_{ij}^3 \tag{7.4}$$

For n_j central j particles, the number of contacts between all i particles and j particles per unit time is given by:

$$H_{ij} = \frac{4}{3} n_i n_j R_{ij}^3 \frac{dv}{dz} \tag{7.5}$$

where dv/dz = velocity gradient in laminar flow.

The application of von Smoluchowski's orthokinetic flocculation equation to the design of flocculation tanks is limited because his theoretical treatment is based on laminar flow conditions. The more usual condition existing for the practicable velocity gradients applied during flocculation is a turbulent regime (Camp, 1969).

Since the local velocity gradients are not known in turbulent mixing, Camp and Stein (1943) replaced the velocity gradient term dv/dz by a measurable average value termed the *root mean square velocity gradient*, G, which is defined as follows (Camp, 1946):

$$G = \sqrt{\frac{W}{\mu}} \tag{7.6}$$

where

μ = absolute viscosity and
W = dissipation function = total power dissipated divided by the volume of chamber or conduit etc. = mean value of the work of shear per unit volume per unit time.

Replacing dv/dz in Equation 7.5 by G, Camp and Stein's equation for orthokinetic flocculation is given by:

$$H_{ij} = \frac{4}{3} n_i n_j R_{ij}^3 \cdot G \tag{7.7}$$

The principal objection to the use of the average velocity gradient, G, is that it does not describe the length scale over which local velocity gradients extend. In other words, velocity gradients which extend for a given mixing length do not influence the flocculation of particles which are larger than this distance.

Fair and Gemmell (1964) extended Von Smoluchowski's Equation (7.5) to describe the rate of change in number concentration of flocs of a given size R_k:

$$\frac{dn_k}{dt} = \frac{2}{3}\left[\sum_{\substack{i=1\\ j=k-1}}^{k-1} n_i n_j R_{ij}^3 - 2n_k \sum_{i=1}^{\infty} n_i R_{ik}^3\right]\frac{dv}{dz} \tag{7.8}$$

The first term in the brackets describes the formation of particles of radius R_k from a range of particles, i.e. from primary particles (i and $j = 1$) to ($k - 1$) fold particles. For example, the joining of a (k – 1) fold particle and a primary particle produces a k-fold particle with radius R_k. The second term in the brackets depicts the elimination of k-fold particles due to their collisions with all other particles, i.e. with the full range of particles from $i = 1$ (primary particles) to infinity.

In all the equations above, floc breakup was not considered. Furthermore, all collisions were assumed to result in lasting contact.

Harris and Kaufman (1966) and Harris, Kaufman and Krone (1966) modified Equation 7.8 to include floc breakup and the possibility of non-lasting collisions. They considered the case of $k = 1$, which describes the rate at which primary (1-fold) particles are removed from the system. Their expression for the rate of change in primary particle concentration is as follows:

$$\frac{dn_1}{dt} = -\frac{\alpha \cdot a^3}{\pi}\delta \cdot \phi \cdot n_i \frac{dv}{dz} \tag{7.9}$$

where

α = fraction of collisions which result in aggregation;
a = ratio of collision radius of a floc to its physical radius;
ϕ = floc volume fraction;
δ = size distribution function, given by:

$$\delta = \frac{\sum_{i=0}^{p-1} n_i (i^{1/3} + 1)^3}{\sum_{i=0}^{p} i n_i}$$

The maximum floc size permitted is p-fold. Harris *et al.* presented other distribution functions for various floc formation and breakup models. Hudson (1965) adopted a different approach by assuming that in a continuous flow system a simplified bimodal floc size distribution is valid. In a system composed of only primary particles and large flocs, assuming size variations within each group are small compared to the differences between the groups, the removal of primary particles is hypothesized to occur primarily by their collisions with flocs. Accordingly, Equation 7.9 is simplified to:

$$\frac{dn_1}{dt} = -\frac{4}{3} n_1 n_F {R_F}^3 \frac{dv}{dz} \tag{7.10}$$

where

R_F = radius of flocs;
n_F = floc number concentration.

Flocculation as described by the above equations probably occurs whenever applied velocity gradients are linear over a distance of a particle diameter. In a turbulent regime, however, flocculation kinetics are better described by turbulent diffusion.

Argaman and Kaufman (1968, 1970) proposed a diffusion model for orthokinetic flocculation, analogous to that of von Smoluchowski for perikinetic flocculation. It is based on the hypothesis that particles suspended in a turbulent regime experience a random motion resembling gas molecules. In their analysis a simplified bimodal floc size distribution comprising primary particles and large flocs was assumed. This assumption was also made by Hudson (1965). Argaman and Kaufman experimentally verified the validity of the bimodal distribution by analysing floc size measurements. Their expression for the rate of collision of primary particles and flocs is given by:

$$H_{1F} = 4\pi \cdot K_S R_F^{\,3} n_1 n_F u^2 \tag{7.11}$$

where

K_S = proportionality coefficient expressing the effect of the turbulence energy spectrum on the effective diffusion coefficient;
R_F = radius of floc;
n_1, n_F = number concentration of primary particles and flocs, respectively;
u^2 = mean square velocity fluctuation, which is related to the root mean square velocity gradient, G, and is a measure of the intensity of turbulence.

Although Equation 7.11 was derived for orthokinetic flocculation using the concept of turbulent diffusion, it is equivalent in form to the orthokinetic flocculation equation of von Smoluchowski using the concept of laminar flow velocity gradients. In fact all the equations derived for particle aggregation kinetics during orthokinetic flocculation are similar in form to Equation 7.5. The major differences lie in the constant terms and the definitions of velocity gradient.

It was recognised by Argaman and Kaufman that two opposing processes are responsible for the changes in particle concentration during flocculation: aggregation of primary particles and small flocs to form larger flocs (as described by Equation 7.11); and the breakup of flocs into smaller fragments. The breakup mechanism assumed by Argaman and Kaufman is described by the rate of formation of primary particles by shearing from the floc surface, i.e.

$$\frac{dn_1}{dt} = B \cdot R_F^{\,2} \frac{n_F}{R_1^{\,2}} u^2 \tag{7.12}$$

where B = breakup constant

Parker *et al.* (1970, 1972) on reviewing alternative mechanisms for floc breakup (i.e. erosion of primary particles from the floc surface, or floc deformation and subsequent splitting) concluded that the surface erosion concept used by Argaman and Kaufman was the more plausible for the biological floc system studied by them.

Parker *et al.* discussed the maximum floc sizes under given mixing intensities. They showed that the translation of a given mixing intensity to floc breakup depends on an *energy cascade* effect. Large-scale eddies contain most of the kinetic energy produced by mixing devices. The energy in the larger eddies is expended in producing progressively smaller eddies until eventually the energy is dissipated by viscous friction. The influence of a particular eddy scale depends on the size of the particles. Floc particles smaller than a given eddy size will be unaffected by that size.

The *Kolmogorov* microscale, η identifies a dividing point, above which inertial energy has a disruptive influence on floc survival, and below which energy is dissipated and there is little influence on flocs. The *Kolmogorov* microscale is calculated as: $\eta = \sqrt{\nu/G}$ where ν = kinematic viscosity (cm^2/s) and G = root mean square velocity gradient.

For example, with the viscosity of water 0.01 cm^2/s and an applied G value of 30 s^{-1}, η = 0.018 cm or 0.18 mm. Therefore, flocs larger than 0.18 mm will be subject to floc breakup during flocculation.

Combining Equations 7.11 and 7.12, the rate of change in concentration of primary particles is given by:

$$\frac{dn_1}{dt} = -4\pi \cdot \alpha \cdot K_S R_F^3 n_1 n_F u^2 + B\frac{R_F^2}{R_1^2} n_F u^2 \tag{7.13}$$

where α = fraction of particle collisions resulting in lasting aggregation.

Argaman and Kaufman applied Equation 7.13 to the case of a single, completely mixed continuous flow tank reactor and, at steady state, obtained the following equation:

$$\left(n_0 - n_1\right) = \left(4\pi \cdot \alpha \cdot K_S R_F^3 n_1 n_F u^2 - B\frac{R_F^2}{R_1^2} n_F u^2\right) \cdot T \tag{7.14}$$

Rearranging Equation 7.14:

$$\frac{n_0}{n_1} = \frac{1 + 4\pi \cdot \alpha \cdot K_S R_F^3 n_F u^2 T}{1 + \dfrac{B \cdot R_F^2 n_F u^2 T}{n_0 R_1^2}} \tag{7.15}$$

where

n_0 = number concentration of primary particles at time $T = 0$

n_1 = number concentration of primary particles at time T

n_0/n_1 is referred to as the performance parameter.

7.3.2 Working equation

In order to develop a working equation incorporating experimentally measurable parameters, Argaman and Kaufman made a number of assumptions as follows:

(1) *The average floc size is closely related to the mean square fluctuating velocity as follows*:

$$R_F = \frac{K_2}{\bar{u}^2} \tag{7.16}$$

where K_2 is a constant

They showed experimentally that, for practical purposes, this assumption is valid.

(2) *The mean square velocity fluctuations can be estimated from the root mean square velocity gradient as follows*:

$$\bar{u}^2 = K_P G \tag{7.17}$$

where

K_P = performance parameter dependent on the type of stirring arrangement and referred to as "stirrer performance coefficient".

This assumption was supported by experimental observations using a hot-film anemometer for a range of G values from 30 to 240 sec^{-1}. For a stake and stator stirrer arrangement they found $K_P = 0.0458\ cm^2/sec$. For a turbine stirrer $K_P = 0.0233\ cm^2/sec$.

The difference in K_P for the different stirrers supports the contention that G is a parameter whose influence is dependent on one particular stirrer arrangement and reactor configuration.

(3) *Assuming spherical particles, the total volume of flocs in a reactor is given by*:

$$\phi = \frac{4}{3}\pi \cdot n_F R_F^3 \tag{7.18}$$

where ϕ = floc volume fraction.

(4) *The aggregation-collision ratio, α, is a constant.* They point out, however, that as the time of flocculation increases the flocs become more compact and of a more regular shape, and therefore the probability of successful adhesion on collision decreases.

Substituting Equations 7.16, 7.17, and 7.18 into Equation 7.15 and describing a flocculation constant, K_F, as:

$$K_F = 3\alpha\phi \tag{7.19}$$

and a breakup constant, K_B, as:

$$K_B = \frac{3}{4\pi} \cdot \frac{B \cdot \phi \cdot K_P^2}{n_0 R_1^2 K_2} \tag{7.20}$$

the flocculation equation becomes:

$$\frac{n_0}{n_1} = \frac{1 + K_A GT}{1 + K_B G^2 T} \tag{7.21}$$

where

$K_A = K_F K_S K_P$
K_B and K_A are constants for a single tank
T is the retention time.

7.3.3 Flocculation reactors in series

By making a further assumption that with a number of reactors, m, in series the constants K_A and K_B retain the same values, a general working equation for the i-th tank in the series is given by:

$$\frac{n_{i-1}}{n_i} = \frac{1 + K_A GT/m}{1 + K_B \dfrac{n_0}{n_{i-1}} G^2 T/m} \tag{7.22}$$

where T is the overall retention time for the series.

The overall performance of an m-tank series system, with equal velocity gradients in each tank, is therefore given by:

$$\frac{n_0}{n_m} = \frac{\left(1 + K_A GT/m\right)^m}{1 + K_B G^2 T/m \sum_{i=0}^{m-1} \left(1 + K_A GT/m\right)^i} \tag{7.23}$$

By carrying out experiments using a synthetic turbid water with the number of reactors progressively increasing from one to four, Argaman and Kaufman demonstrated the validity of Eq. 7.23. To illustrate the interrelationship between the various parameters, Figure 7.4 shows the root-mean-square velocity gradient, G, plotted against total retention time, T, for a series of reactors m (1 to 4) and for various values of performance parameter n_0/n_m (Argaman, 1971).

The marked beneficial effect of compartmentalization is evident, especially where higher performances are desired. For example, if 75% removal of primary particles ($n_0/n_m = 4$) is desired, a single reactor would require a retention time of

2,000 seconds, whereas four reactors in series would require 800 seconds, or less than half the overall retention time.

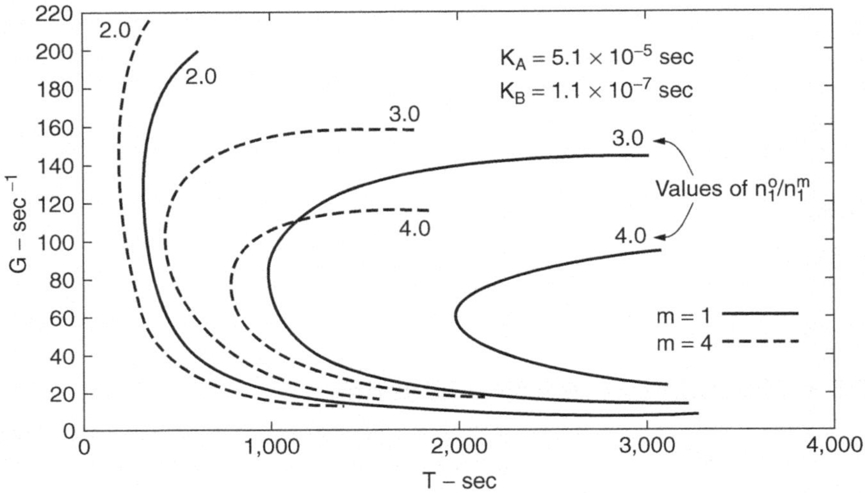

Figure 7.4 Performance of flocculation system as related to *G*, *T* and *m*. (From Argaman, 1971).

7.3.4 Adequacy of *G* and *GT* as design parameters

As stated previously, an objection to the concept of the average velocity gradient, *G* is that it does not describe the length scale over which local velocity gradients extend. Several workers have emphasized the limitations of the Camp and Stein (1943) concept of *G* (for example, Han & Lawler, 1992; Kramer & Clark, 1997; Lawler, 1993; McConnachie, 1991; Pedocchi & Piedra-Cueva, 2005).

Although the concept of *G* is problematic, it does characterize reasonably well the collision rates occurring in isotropic turbulence. Pedocchi and Piedra-Cueva point out the inadequate terminology of *G* as *velocity gradient* – but assign lesser importance to the terminology provided it is understood that the definition is: $G = \sqrt{W/\mu}$. These workers accept that Camp and Stein's pioneering concepts are still plausible in turbulent conditions of flocculation.

However, it must be remembered that there are inaccuracies in applying a certain G value for design, without also considering the reactor geometries and the type of mixing device. *G* is not a universal parameter.

Gregory (2006) has noted that the dimensionless term *GT* (i.e. GxT), often referred to as the Camp number (after Camp, 1955), has been and still is of practical importance. For example, Bernhardt and Schell (1993) found that within certain limits of G, high or low G values can be exchanged for high or low T values to maintain a certain GT, and still maintain flocculation efficiency.

However, there are limitations with the GT concept that must be understood. Of particular note in Figure 7.4 is the fact that the product GT does not adequately identify the conditions for a given flocculation performance. For example, from Figure 7.4, at a desired performance $n_0/n_m = 3$ using four reactors in series, there is a minimum retention time of 2,000 seconds below which that particular performance cannot be achieved regardless of the magnitude of G. The inadequacy of the GT product as a design parameter is because the phenomenon of floc breakup is not accommodated. An exception is where G values are low or, alternatively, K_B is low (very strong flocs) and Equation 7.23 approaches $n_0/n_m = (1 + K_A GT/m)^m$. In this case, performance for a particular flocculation system is directly dependent on the product GT.

7.3.5 Experimental determination of flocculation parameters

The equations describing both the aggregation and breakup phases of the flocculation process have been set out in the preceding sections. These equations contain certain "constants" which must be evaluated for each particular water. One method is to carry out continuous type pilot plant trials. However, this has the disadvantage that relatively complicated and costly apparatus and lengthy testing procedures are required. However, many workers consider it essential that the experimental determination of flocculation parameters be carried out using pilot plant studies (for example, Argaman, 1971).

A further possibility is the use of batch testing procedures. However, it has been stated that the application of batch test data is appropriate for plug flow reactors only (O'Melia, 1967). Despite this, if it were shown possible to determine the appropriate parameters of flocculation by a simple batch testing procedure, a powerful means would be available to design engineers and researchers for the design of continuous completely mixed flocculation systems.

By correlating batch test data with those from continuous type tests, the theoretical link between batch or plug flow behaviour, and continuous test performance has been demonstrated (Bratby *et al.* 1977). By taking cognizance of the appropriate (different) formulations, batch test data can be used for the design of in-series completely mixed flocculation reactors.

In Equation 7.23, if K_A and K_B remain constant for all tanks in a series of completely mixed reactors, then these constants should theoretically remain unchanged when the number of tanks is increased to infinity, that is, for plug flow or batch conditions.

Batch test results are governed by a relationship of the form given by Equation 7.13. Substituting Equations 7.16, 7.17 and 7.18 into Equation 7.13 and expressing

$K_A = K_F K_S K_P$ and
$K_B = 3(B\phi K_P^2)/(4\pi \cdot n_0 R_1^2 K_2)$ as before,
the equivalent flocculation equation for a batch system is given by (Bratby *et al.* 1977):

$$\frac{dn_1}{dt} = -K_A n_1 G + K_B n_0 G^2 \tag{7.24}$$

Integration of Equation 7.24 yields

$$n_1 = \frac{K_B}{K_A} n_0 G + \frac{n_0\left(1 - \frac{K_B}{K_A}G\right)}{\exp(K_A GT)}$$

Rearranging

$$\frac{n_0}{n_1} = \left[\frac{K_B}{K_A}G + \frac{\left(1 - \frac{K_B}{K_A}G\right)}{\exp(K_A GT)}\right]^{-1} \tag{7.25}$$

The experimental apparatus and procedures recommended for determining the "constants" K_A and K_B by a batch test procedure are described in detail in Chapter 8.

Chapter 8 also presents values of K_A and K_B determined for various waters by different researchers.

7.3.5.1 Inconstancy of K_B

Although K_A and K_B have both been described above as constants, this is strictly only valid for K_A. Due to an inadequacy of the assumed breakup mechanism in the above analyses, K_B may assume a constant value for only one particular G value. However, this does not deter from the value of the relationships describing flocculation performance since the same variation in K_B is evident in both batch and continuous performance (see Chapter 8). For convenience, therefore, when analyzing data from batch tests, the value of K_B may be empirically related to G by an expression of the form

$$K_B = k_1 \ln G + k_2 \tag{7.26}$$

where k_1 and k_2 are constants for a particular water.

The following results were obtained for various waters (Bratby *et al.* 1977; Bratby, 1981):

- *For a synthetic (kaolin clay) turbid water using alum as coagulant*: $K_A = 2.5 \times 10^{-4}$; $k_1 = -0.87 \times 10^{-7}$; $k_2 = 7.72 \times 10^{-7}$.
- *For a humic coloured water with zero turbidity and using alum as coagulant*: $K_A = 3.1 \times 10^{-4}$; $k_1 = -3.04 \times 10^{-7}$; $k_2 = 15.2 \times 10^{-7}$.
- *For a natural lake water using 15 mg/l alum as coagulant*: $K_A = 1.28 \times 10^{-4}$; $k_1 = -4.12 \times 10^{-7}$; $k_2 = 20.6 \times 10^{-7}$.
- *For a natural lake water using 3 mg/l alum and 0.15 mg/l anionic polymer*: $K_A = 0.77 \times 10^{-4}$; $k_1 = -2.68 \times 10^{-7}$; $k_2 = 15.1 \times 10^{-7}$.

7.4 DESIGN OF FLOCCULATION BASINS

7.4.1 General

For a given destabilized water and flocculation device, the principal parameters governing the effective degree of flocculation are the retention time and velocity gradients applied. These parameters influence the rate and extent of particle aggregation and the rate and extent of breakup of these aggregates. The kinetics of the flocculation process are dependent not only on the type of raw water, but also on the mode of destabilization, that is, on the type of primary coagulant used, e.g. aluminium sulphate, ferric chloride, some form of polyelectrolyte, etc., and any flocculant aid added. Before design of a flocculation system, therefore, a thorough testing program must be undertaken. The first stage is to identify the most appropriate chemical(s) to achieve the required degree of clarification (see Chapter 8). At the end of this stage one will presumably have a number of alternatives. The second stage is to assess the alternatives in terms of the efficiency of flocculation and subsequent separation using sedimentation, flotation, or filtration, etc.

7.4.2 Types of flocculation chambers and devices

There are several methods by which velocity gradients may be induced in a body of liquid. These include:

- Baffled chambers
- Granular media beds
- Diffused air
- Spiral flow
- Reciprocating blades
- Rotating blades.

Each of these methods is described below.

7.4.2.1 Baffled chambers

Baffled flocculation chambers, probably among the earliest types of flocculation basins, may be installed either with over-and-under baffles (an example of which is shown in Fig. 7.5) or around the end baffles. The latter configuration is often preferable for several reasons: it is easier to construct; it is easier to drain and clean; and scum trapping is easier to avoid (Hudson & Wolfner, 1967).

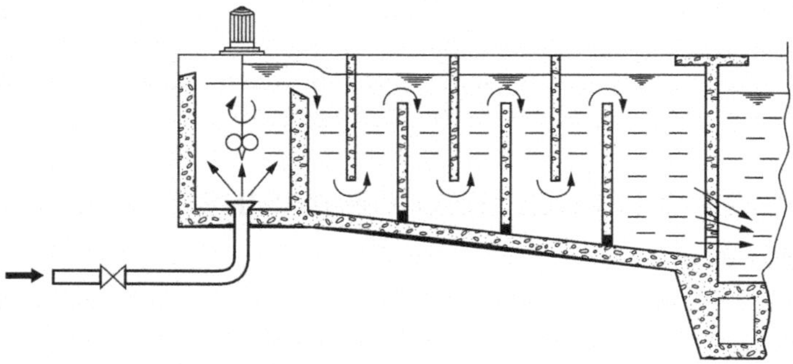

Figure 7.5 Baffled flocculation chamber.

Advantages of baffled chambers are that they are virtually maintenance free and short circuiting is minimised (Kawamura, 1973). However, disadvantages are as follows: (1) Since most of the head loss occurs at the 180-degree bends, velocity gradients are probably excessive at the bends and too low along the straight sections (Camp, 1955); (2) With such devices, velocity gradients are wholly dependent on the flow rate through the basin and, therefore, offer no degree of control; (3) Deposition of silt and flocculated material in the basin, particularly at low flows, is difficult to avoid.

For baffled flocculation basins, the average velocity gradient, G, is computed from Equation 7.6:

$$G = \sqrt{\frac{W}{\mu}} \quad s^{-1} \tag{7.6}$$

where

μ = absolute viscosity (=10^{-3} N · s/m^2 at 20°C)

and W, the dissipation function (=dissipated power per unit volume, P/V) is given by:

$$W = \frac{\gamma \cdot h_L}{T} \quad \text{N/m}^2 \cdot \text{s} \tag{7.27}$$

where

γ = weight/unit volume of liquid (N/m^3)
and $\gamma = \rho \cdot g$

where ρ = density of water (=1,000 kg/m^3)
and g = weight per unit mass = 9.81 N/kg
h_L = head loss (m)
T = retention period (s)

Hence,

$$G = \sqrt{\frac{\gamma \cdot h_L}{T \cdot \mu}} = \sqrt{\frac{\rho \cdot g \cdot h_L}{T \cdot \mu}} = \sqrt{\frac{g \cdot h_L}{T \cdot \nu}} \quad s^{-1} \tag{7.28}$$

where

ν = kinematic viscosity (=10^{-6} m^2/s at 20°C)
The method of calculation is as follows:

- Decide on the G-profile through the compartments (for example: 80; 40; 20 s^{-1})
- Decide on the total retention time at average flow (for example: 30 minutes)
- Decide on the number of tanks, and the number of channels per tank (for example: two tanks operating in parallel, with six passes, or channels per tank)
- Assign the number of channels to each G value (in this example, presumably 2 channels for each G value)
- Knowing the temperature, or range of temperatures, the kinematic viscosity can be determined
- Knowing the flow rate, or range of flow rates, the retention time, T can be determined
- The required total head loss in each of the sets of channels can be determined using Equation 7.28
- Divide these total head loss values into a suitable number of chambers per channel
- Using the head loss per baffle of $h_L = Kv^2/2g$, where v = velocity through the baffle openings, and K = 1.5, determine the openings for each baffle (keep the openings equal, but vary the number of baffles for the different G values, as recommended by Kawamura (2000)
- A horizontal baffled flocculation system is shown schematically in Figure 7.6.

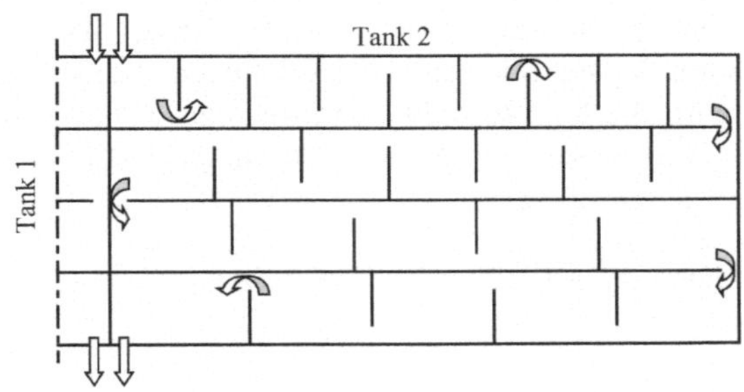

Figure 7.6 Horizontal baffled flocculation tanks.

7.4.2.2 Granular media beds

Granular media flocculators have the advantage of not requiring mechanical equipment although, as for the baffled flocculator, the intensity of mixing is not controllable and depends on the water flow. However, such flocculators do provide good service. Low-cost versions include channels (similar to the baffled tank shown above) but with gravel packing (Ahsan *et al.* 1996). More sophisticated versions include buoyant media that serves the purpose of flocculation as well as pre-filtration (see Figure 7.7).

Figure 7.7 Buoyant granular media flocculator/pre-filter (courtesy US Filter).

Horizontal channel gravel flocculators are susceptible to fouling, particularly with high turbidity waters (Bhole, 1993), although this can be mitigated by maintaining relatively high scouring velocities through the beds, at approximately 5 m/minute (Ayoub & Nazzal, 1989). Ayoub (1996) also determined the flocculation constant K_A for horizontal gravel flocculating beds, and found it to vary linearly with the feed turbidity and the size of the gravel used in the bed. The floc breakup coefficient, K_B could be neglected in their studies. The expression for K_A was as follows:

$$K_A = 0.016 \times 10^{-4} \cdot N_0 + 0.225 \times 10^{-4} \cdot N_0 \cdot D_m$$

where

N_0 = feed turbidity (NTU)
D_m = gravel size (m)

Several applications of granular media flocculators use a vertical arrangement, with some form of air scouring and/or backwash to remove accumulated material. In such cases, the system serves as both flocculator and roughing filter (Polasek, 1988; Tambo & Matsui, 1989; Rajapakse & Ives, 1990; Schulz *et al.* 1994).

The value of G in granular media flocculators can be estimated if pore clogging due to floc retention is ignored, as follows (after Monk & Trussell, 1991):

$$G = \sqrt{\frac{\gamma \cdot h_L}{T \cdot \varepsilon \cdot \mu}} = \sqrt{\frac{\rho \cdot g \cdot h_L \cdot Q}{V \cdot \varepsilon \cdot \mu}} = \sqrt{\frac{g \cdot h_L \cdot Q}{V \cdot \varepsilon \cdot \nu}} \quad \mathrm{s}^{-1} \tag{7.29}$$

where

γ = weight/unit volume of liquid (N/m^3)
h_L = head loss (m)
T = retention or flocculation time = V/Q (s)
Q = flow rate (m^3/s)
V = bulk volume of granular media (m^3)
ε = porosity of the media
μ = absolute viscosity (=10^{-3} N·s/m^2 at 20°C)
γ = weight/unit volume of liquid (N/m^3)
and $\gamma = \rho \cdot g$
ρ = density of water (=1,000 kg/m^3)
g = weight per unit mass = 9.81 N/kg
ν = kinematic viscosity (=10^{-6} m^2/s at 20°C)

The (clean) head loss can be determined from the Kozeny equation, as follows:

$$h_L = \frac{k \cdot \mu}{\rho \cdot g} \cdot \frac{(1-\varepsilon)^2}{\varepsilon^3} \cdot (S)^2 \cdot \frac{L^2}{T} \quad \mathrm{m} \tag{7.30}$$

where

k = dimensionless constant, commonly taken as = 5.0
S = specific surface = grain surface area per unit grain volume = $6/d$ for spheres (d = grain diameter, m)
L = length of granular media bed (m)

7.4.2.3 Other hydraulic flocculators

Velocity gradients may be induced by introducing liquid into a basin at relatively high velocity. These flocculation tanks are also known as hydraulic jet-action flocculators. The mean velocity gradient, G, for each compartment of the basin is computed by assuming that the head loss, h_L is the velocity head through the inlet. Disadvantages with this method are that velocity gradients are not uniformly distributed and passage from one basin to the next in series may result in excessive floc breakup (Camp, 1955), making these types of flocculation tanks less desirable (Kawamura, 2000). Examples of such flocculators are the helicoidal-flow tanks, and the Alabama type flocculator (see Monk & Trussell, 1991).

Pipe flocculators can be appropriate, particularly for relatively small installations although, as for other hydraulic flocculators, the flocculation efficiency is dependent on the flow rate, and the possibility of sludge deposition at low flows would reduce flocculation efficiency (Kurbiel *et al.* 1991). One example is shown in Figure 7.8. The approach for calculating G values is similar to that for *Pipe Bend* in Chapter 6.

Figure 7.8 Example of pipe flocculator.

7.4.2.4 Diffused air

Flocculation in a body of liquid may be achieved by introducing air through diffusers at the bottom. The velocity gradient produced in this case may be controlled to a certain extent by regulating the air flow.

The expression for the flow rate of free air, Q_a, required for a mean velocity gradient G is given by (Camp, 1955):

$$Q_a = G^2 \mu V \frac{\left(\dfrac{H}{2} + 10.33\right)}{H \times 10^5} \quad \mathrm{m}^3/s \tag{7.31}$$

where

μ = absolute viscosity (=10^{-3} N s/m^2 for water at 20°C);
V = volume of liquid in basin (m^3);
H = depth of diffusers (m).

A disadvantage of diffused air flocculators is that with practicable bubble sizes produced by available diffusers, the local velocity gradient produced at the bubble surface is much higher than the mean (and desired) for the whole volume. This maximum local velocity gradient G_{max} is given by:

$$G_{max} = g(\rho - \rho_1)D_b/6\mu \quad s^{-1} \tag{7.32}$$

where

ρ_1 = density of air in bubble (kg/m^3);
ρ = density of liquid (kg/m^3);
D_b = diameter of bubble (m)

Table 7.1 shows values of G_{max} for a range of bubble sizes, for water at 20°C).

Table 7.1 Values of *G* produced by a range of bubble sizes.

Bubble Diameter (mm)	G_{max} (s^{-1})
0.01	16
0.05	82
0.1	163
0.15	245
0.5	816
1.0	1633
2.0	3300

In most cases for lower *G* values, bubble sizes must be extremely small (approximately 10 micrometers in, say, the final compartment of an in-series flocculation system). Although the production of such bubbles is practicably feasible (for example by a dissolved-air precipitation technique) inevitably flotation of suspended flocs would occur which, even if a float removal mechanism were provided at the top of the flocculation unit, would progressively impair the rate of flocculation by reducing the number of flocculating particles.

Kawamura (2000) suggests that of the flocculation methods available, diffused air is the least preferred, because of the reasons cited above, and the low overall efficiency of flocculation achieved.

7.4.2.5 Rotating blades

Examples of rotating blade type flocculation devices are shown in Figure 7.9. Generally, such devices are of two types: horizontal shaft and vertical shaft. Horizontal shaft units have under-water bearings which require maintenance. In some cases the drive shaft passes through to a dry well, which often gives rise to problems of leakage, or the drive may be overhead via V-belts or chains.

Figure 7.9 Examples of rotating blade flocculators. Left: vertical paddle (courtesy Amwell); Center: horizontal (reel) paddle (courtesy WesTech); Right: axial flow (courtesy top-right: Anco; bottom-right: Philadelphia Mixers).

An advantage of horizontal shaft units is that compartmentalization is simple. With adjacent sections along the drive shaft carrying different configurations of agitators, different tapered *G* values are produced for one particular drive shaft rotational speed.

With most designs of horizontal shaft flocculators, the provision of stators within the flocculation basin to minimise rotation of the water with the blades is difficult.

The principal advantages of vertical shaft rotary devices are that underwater bearings are not required; the drive unit is above the water level and the arrangement for stators is simple. A further advantage is that with combined flocculation-sedimentation units (see later) vertical shaft flocculators are simpler to install than horizontal shaft units.

A further rotating flocculation device is the turbine flocculator. A disadvantage of such units is that there is a much wider diversity of *G* values within the flocculation basin for a given mean velocity gradient – being high in the vicinity of the device and low near the walls of the basin. Such devices generally provide lower flocculation performance when compared with devices accommodating as much of the flocculation basin volume as possible (Argaman & Kaufman, 1968, 1970).

Hudson (1981) compared axial flow flocculators and plate turbine devices. He reported on extreme turbulence trailing the blades of the turbine device, whereas there was little turbulence at the blades of the axial flow device. He also pointed out that paddle type flocculation devices rely on the highly localized eddies trailing behind the blades, as well as turbulent vortices in the corners of square basins, and the drag on the basin walls, for most of the power input for flocculation. Figure 7.10 reproduces his findings, and shows that axial flow flocculating devices appear to be the preferred form of mechanical flocculation devices.

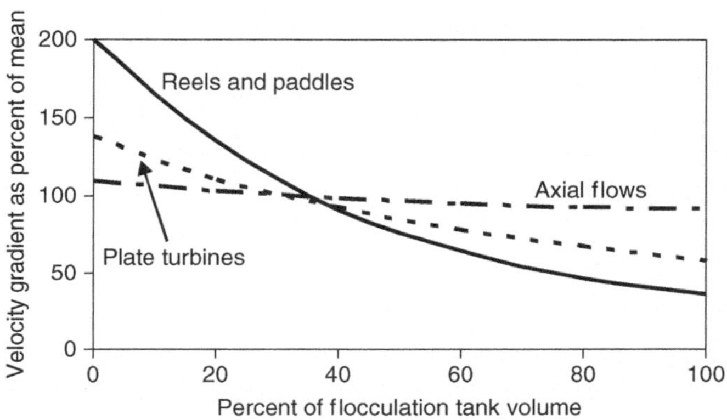

Figure 7.10 Spatial distribution of velocity gradients throughout the tank volume, for different flocculation devices (after Hudson, 1981).

Oldshue and Mady (1979) compared the performances of paddle type flocculators with axial flow (downward flow) impellers. They found that for the same settled water quality, the paddle flocculators required approximately 3-times higher calculated G values, compared with the axial flow impellers. This means that the power draw of the axial flow impellers would be significantly lower for the same degree of flocculation. The efficiency of flocculation by axial flow impellers was more dependent on the rotational speed; more so than the paddle flocculators. With axial flow impellers, higher speeds caused more floc shearing, whereas speeds lower than optimal caused a faster drop-off in flocculation efficiency.

The power drawn by a rotating mixing device is determined by its rotational speed and the geometry of the tank in which it operates. The power consumed by a mixing device is given by (Leentvaar & Ywema, 1980):

$$P = \phi \cdot \rho \cdot n^3 \cdot D^5 \quad \mathrm{N \cdot m/s} \quad (1\ \mathrm{N \cdot m/s} = 0.001\ \mathrm{kW} = 0.0013\ \mathrm{hp}) \tag{7.33}$$

where

ϕ = dimensionless power number
ρ = liquid density (kg/m^3)
n = mixer rotational speed (revolutions per second)
D = diameter of mixer impeller (m).

The root mean square velocity gradient, G is defined as (Camp & Stein, 1943):

$$G = \sqrt{\frac{W}{\mu}} \quad s^{-1} \tag{7.6}$$

where

W = dissipation function ($N/m^2.s$)
= dissipated power per unit volume (=P/V)
V = volume of flocculation tank (m^3)
μ = absolute viscosity (=10^{-3} N s/m^2 for water at 20°C).

Therefore, G is related to the power number by:

$$G = \sqrt{\frac{\phi \cdot \rho \cdot n^3 \cdot D^5}{V \cdot \mu}} \quad s^{-1} \tag{7.34}$$

The concept of power number is useful because for a particular impeller and tank geometry, the scale up of small scale pilot systems should be possible with the same power number, if geometrical similarity is maintained.

Leentvaar and Ywema showed that the power number is essentially independent of Reynolds number, if the latter is higher than approximately 10^4 – which is probably the case in most flocculation applications.

Figures 7.11a through 7.11c show a compilation of flocculation devices and tank geometries investigated by various workers, grouped according to similar power numbers.

Figure 7.12 shows G values versus impeller speed, for the flocculation systems shown in Figures 7.11a through c.

When velocity gradients are induced by mechanical mixing, the value of the dissipation function, W, may be estimated from the drag force of the stirring blades and the distance moved per second. The drag force for any body in motion in a liquid is given by:

$$F_D = C_D \cdot A \cdot \gamma \frac{v^2}{2g} \tag{7.35}$$

where

C_D = drag coefficient;
A = cross-sectional area of the submerged object in a plane perpendicular to the direction of the motion (m^2);
V = relative velocity of the object with respect to the liquid (m/s)
γ = weight/unit volume of liquid (N/m^3).

In order to simplify design of vertical paddle type devices it is desirable that rotor paddles be mounted parallel to the shaft on arms attached to the shaft. The velocity of a paddle with respect to the tank is then proportional to its distance from the shaft.

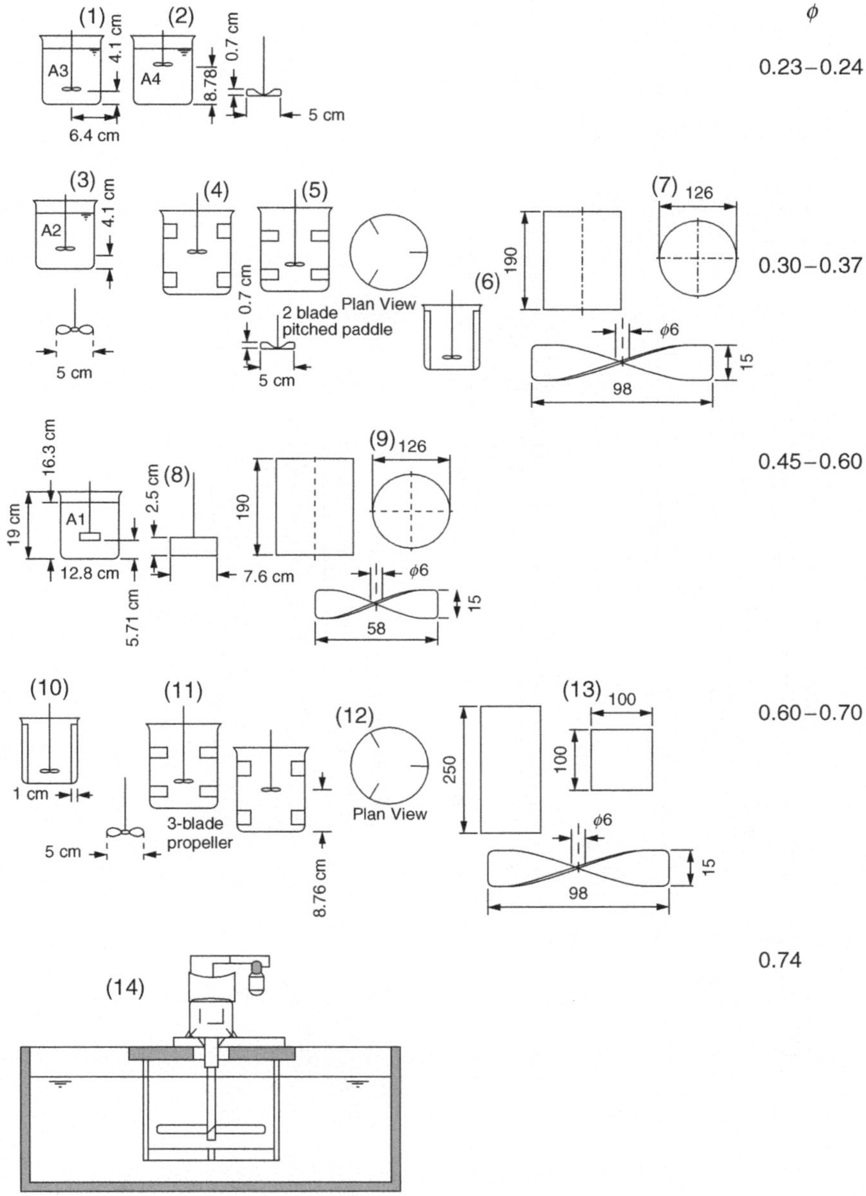

Figure 7.11a Flocculation devices, tank geometries, and power numbers (Units 1–6, 8, 10–12: Lai *et al.* 1975; Units 7, 9, 13: Leentvaar & Ywema, 1980; Unit 14: full scale axial flow unit with cross-baffles (*FMC-Filsan*, Brazil).

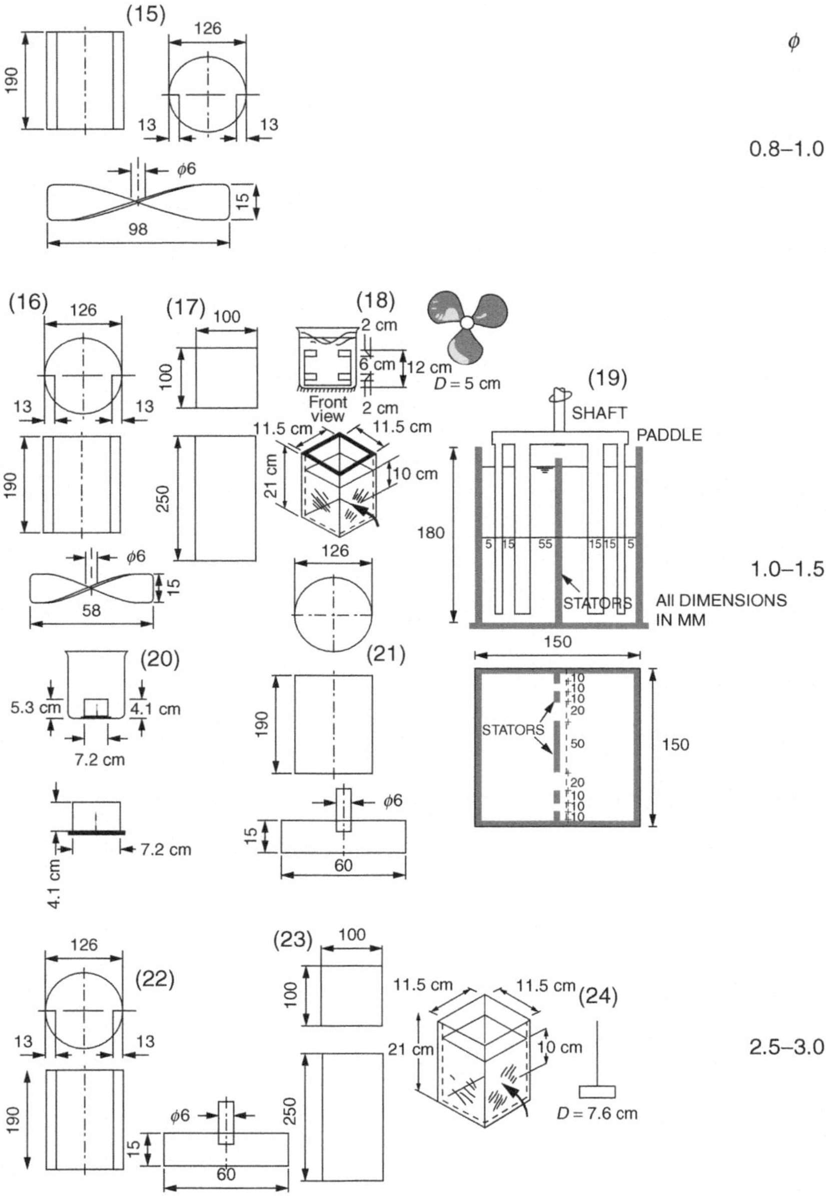

Figure 7.11b Flocculation devices, tank geometries, and power numbers (Units 15–17, 21–23: Leentvaar & Ywema, 1980; Units 18, 24: Cornwell & Bishop, 1983; Unit 19: Bratby, 1977; Unit 20: Lai *et al.* 1975).

	ϕ
(25)	3.0–3.5
(26)–(29)	3.5–4.0
(30)–(34)	5.0–6.0

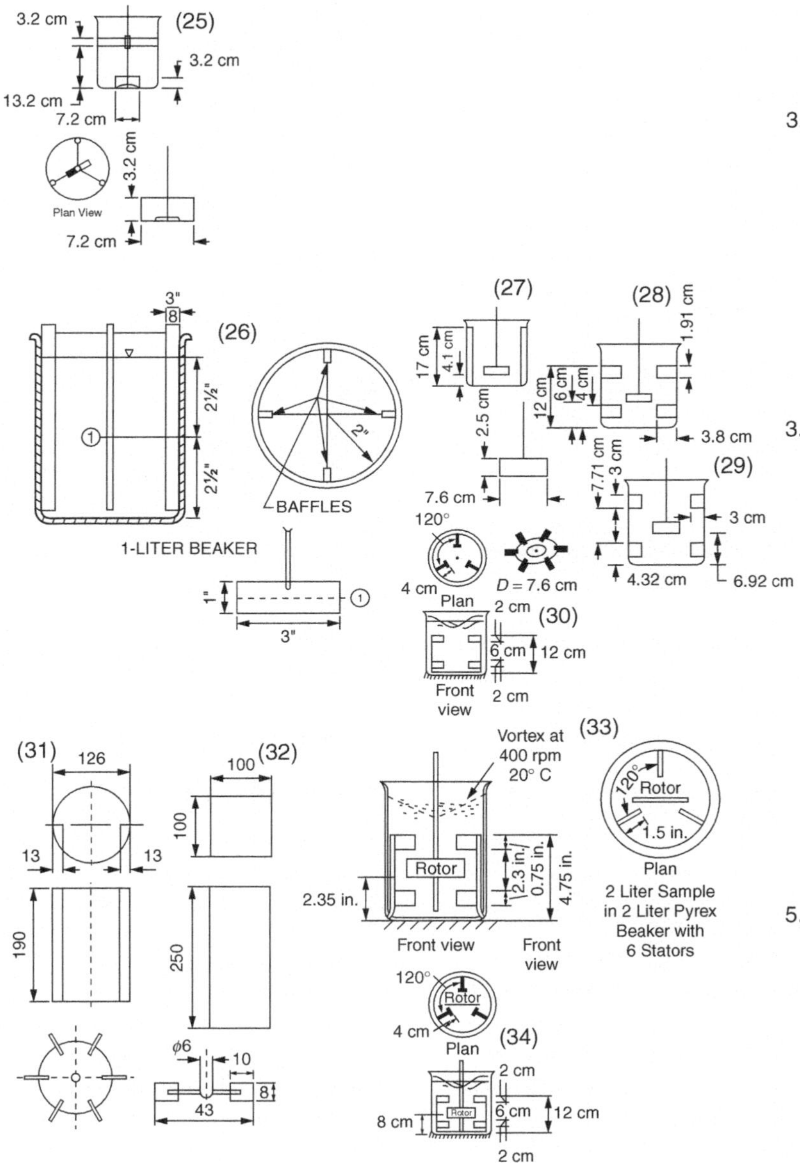

Figure 7.11c Flocculation devices, tank geometries, and power numbers (Units 25, 27–29: Lai *et al.* 1975; Unit 26: Andeu-Villegas and Letterman 1976; Units 30, 34: Cornwell and Bishop 1983; Units 31, 32: Leentvaar and Ywema 1980; Unit 33: Camp, 1968).

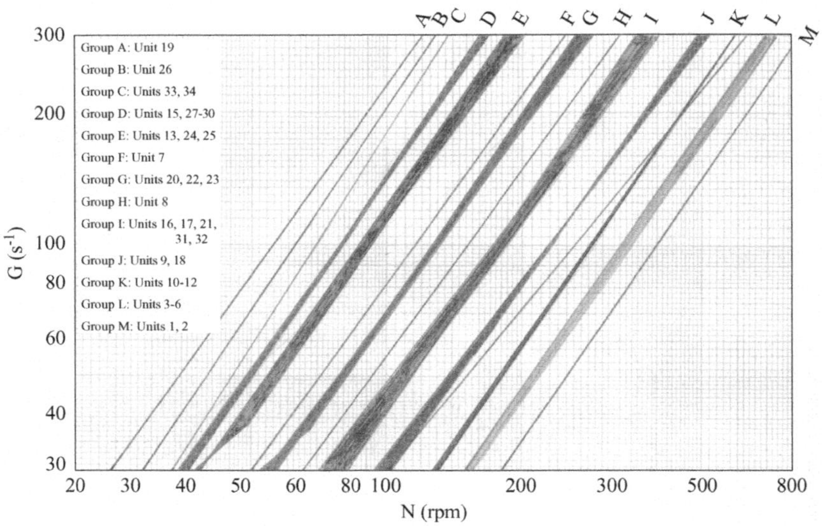

Figure 7.12 *G* values versus impeller speed, for the flocculation devices shown in Figures 7.11a through c.

A flocculation tank must have several rotor paddles at varying distances from the shaft. The area between the paddles should comprise stators fixed to the tank to minimise rolling of the water.

The value of dissipation function W for all the paddles in a tank is given by:

$$W = \sum \frac{F_d \cdot v}{V} = \sum \frac{C_D \cdot A \cdot \gamma \cdot v^3}{2 \cdot g \cdot V} = \frac{1.24 \times 10^5 C_D (1-k)^3 S_S{}^3}{V} \sum A \cdot r_b{}^3 \tag{7.36}$$

where

C_D = drag coefficient = 2.0 for flat blades for rotating devices;
S_S = speed of rotation of shaft (rev/s);
kS_S = speed of rotation of water (rev/s);
V = volume of water in tank (m^3);
A = cross-sectional area of each blade in the plane perpendicular to direction of motion (m^2);
r_b = distance of centre of each blade from shaft (m).

The root mean square velocity gradient is given by:

$$G = \sqrt{\frac{W}{\mu}} = \sqrt{\frac{1.24 \times 10^5 C_D (1-k)^3 S_S{}^3}{V \cdot \mu} \sum A \cdot r_b{}^3} \quad s^{-1} \tag{7.37}$$

where

μ = absolute viscosity (=10^{-3} N s/m^2 for water at 20°C).

The speed of rotation of the water ($k \cdot S_S$) is difficult to pre-determine since rotary stirring devices set the liquid in spiral motion with an average angular velocity less than the velocity of the rotors, and with the velocity beyond the central core diminishing with distance from the centre. In a few plants, the width of the paddles is so great that the water is carried along at approximately the same rate as that of the paddles (Camp, 1955). The only mixing in such cases arises from velocity gradients set up between the outside edges of the paddles and the walls of the tank.

As a guide, k usually falls within the range 0.2 at high impeller speeds (0.03 to 0.09 rev/s) to 0.35 at low impeller speeds (0.015 to 0.05 rev/s) when $\Sigma A \cdot r_b^3$ for stators is equal to $\Sigma A \cdot r_b^3$ for rotors (Camp, 1955). In laboratory scale stake-and-stator systems, k was found to have a constant value of 0.24 for the full range of rotational speeds used in the experiments (Bratby *et al.* 1977).

7.4.2.6 Reciprocating blades

The advantage of reciprocating blade flocculation devices is that the effect of rotating water is virtually eliminated. However, an issue with reciprocating devices is that the stresses on bearings and other mechanical components can be severe. Figures 7.13a and 7.13b illustrate two types of reciprocating blade flocculation devices.

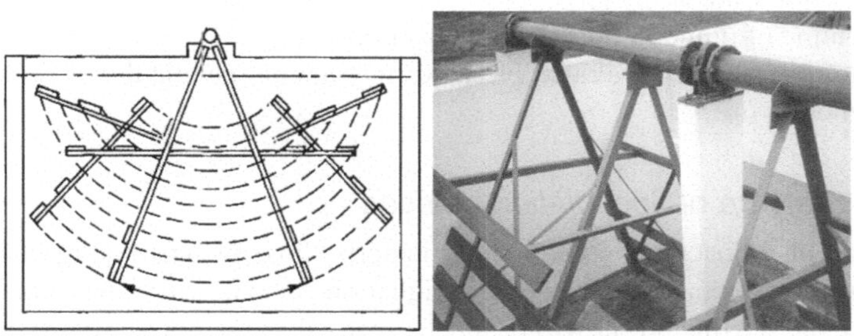

Figure 7.13a Example of reciprocating blade flocculator (courtesy Eimco).

Determination of velocity gradients produced by such devices is carried out using the principle of simple harmonic motion. The relative velocity between blade and liquid changes continuously throughout a cycle. To obtain values for the mean velocity gradient, *G*, it is necessary to compute the mean value of the dissipation function, *W*, throughout a cycle. An expression for *G* is as follows Camp, 1955):

As before,

$$G = \sqrt{\frac{W}{\mu}} \quad s^{-1} \tag{7.6}$$

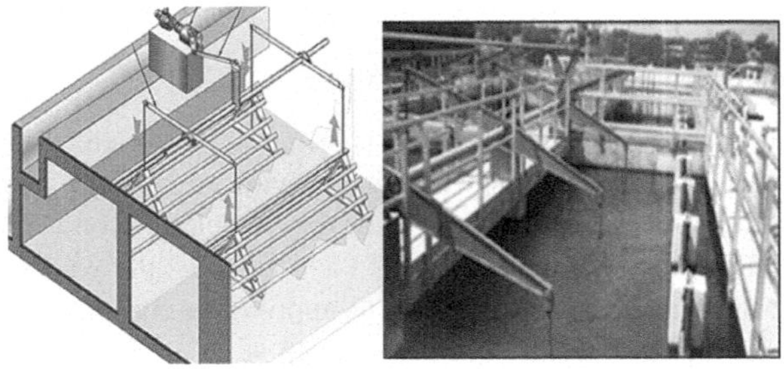

Figure 7.13b Examples of walking beam reciprocating flocculators (courtesy: left – Jim Myers & Sons, Inc.; right – JDV Equipment Corporation).

where

$$W = \frac{6.58 \times 10^3 \cdot C_D \cdot D^3 \cdot S_S{}^3}{V} \sum A \quad \mathrm{N/m^2 \cdot s} \tag{7.38}$$

and

C_D = drag coefficient = 3.0 for flat blades for reciprocating devices;
D = length of stroke of paddles in reciprocating motion (m);
S_S = speed of rotation of drive shaft (rev/s) or number of complete strokes per unit time (s^{-1});
V = volume of liquid in flocculation basin (m^3);
ΣA = total cross sectional area of blades in plane perpendicular to direction of motion (m^2).

7.4.2.7 Start-up of flocculation devices

At start-up of a flocculation device, particularly those using rotating blades, the liquid is stationary and the value of k in Equation 7.36 is zero. If the rotors start at full speed, the starting power will be much greater than the power after the water is rotated at its equilibrium velocity. Drive motors should be oversized, therefore, or provision should be made to bring the paddles up to speed slowly during start-up. With some devices, Camp (1955) showed that the startup loads can be as high as eight times the running load.

7.4.3 Short circuiting in flocculation reactors

As stated in preceding sections, the degree of flocculation achieved within a basin is dependent on the magnitude of the mean velocity gradient, G, and the retention period in the basin. With the types of flocculation devices discussed above, in almost all cases (with the possible exception of baffled chambers) the action of the device gives rise to a certain degree of short circuiting in the basin. That is, some elements of liquid will leave the flocculation basin after a time less than the design retention period, whereas others will remain for longer.

The degree of short circuiting may be determined by introducing a slug of tracer (e.g. dye or salt) at the inlet to the basin and measuring its concentration, with time, t, at the outlet. If mixing within a tank is instantaneous, the concentration of tracer at the outlet, C, is given by (Camp, 1955):

$$C/C_0 = e^{-t/T} \tag{7.39}$$

where

C_0 = initial concentration in the basin assuming instantaneous dispersion;
T = design retention time of basin.

By carrying out experiments on a single basin with a moderate degree of mixing, Camp (1955) found that experimental results correlated reasonably well with Equation 7.34. For different fractions of the nominal retention period, t/T, the respective percentages of tracer slug leaving the tank within a given time as predicted theoretically and measured experimentally were as follows: From integration of Equation 7.39, 22% of the slug theoretically leaves the tank within one-quarter the retention period; 40% in one half; and 63% in a time equal to the retention period. Experimentally, the respective values were approximately 23%, 32% and 47%.

From this it is seen that theoretically, only 37% of flocculating particles remain in the basin for the full design retention period under conditions of complete and instantaneous mixing. This short circuiting effect may be diminished with rotating devices by installing suitably designed stators and minimizing the rolling factor, k, of the water. At the same time, uniform inlet and outlet conditions should be ensured.

When using inlet and outlet baffles, openings of 3 per cent of flow area have been found to minimize short circuiting (Kawamura, 1973). However, the flow through the openings should not give rise to velocity gradients greater than the mean value existing in the flocculation basin, otherwise breakup of previously formed floc could result.

The most effective means of minimizing short circuiting is to install a number of flocculation reactors in series. This is the subject of the following section.

7.4.4 Compartmentalization

Previously, the progressively reduced total retention time for a flocculation system with an increasing number of flocculation basins in series was demonstrated. The minimum possible retention period is achieved when the number of reactors in series is infinity, or with plug flow conditions. This phenomenon is of general occurrence when reaction kinetics are of first order, as is the case for orthokinetic flocculation.

A second beneficial effect of compartmentalization is to minimize the effect of short circuiting. An equation similar in form to Equation 7.39 describes the instantaneous dispersion curve for n compartments of equal size in series (Camp, 1955):

$$\frac{C}{C_0} = \frac{n^n}{(n-1)!}\left(\frac{t}{T}\right)^{n-1} e^{-n \cdot t/T} \tag{7.40}$$

where C_0 and T are based on the total volume of n tanks in series.

Figure 7.14 shows the dispersion curves produced by Equation 7.40. The area under a given curve at a particular time t represents the fraction of tracer slug which has passed through the system. Hence, integrating Equation 7.40, the fraction of tracer slug passing through system in time t is equal to:

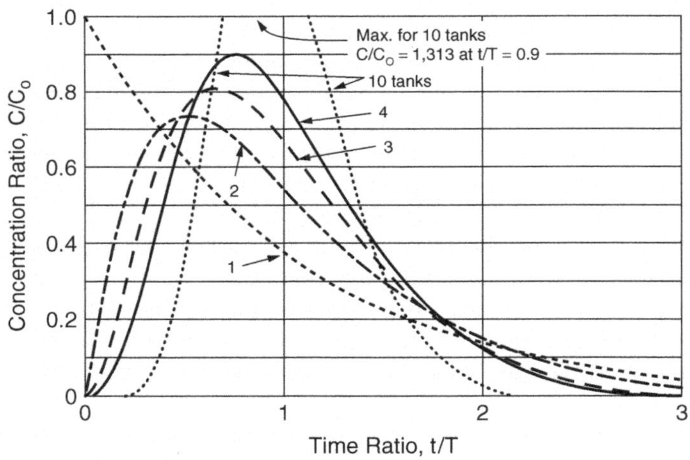

Figure 7.14 Dispersion curves for tanks in series (after Camp, 1955).

$$1 - \frac{n^n \cdot e^{-\frac{t}{T}}}{n!}\left[\left(\frac{t}{T}\right)^{n-1} + \frac{(n-1)}{n}\left(\frac{t}{T}\right)^{n-2} + \frac{(n-1)\cdot(n-2)}{n^2}\left(\frac{t}{T}\right)^{n-3} + \ldots + \frac{(n-1)!}{n^{n-1}}\right] \tag{7.41}$$

It is seen that as the number of compartments increases, the time of maximum response increasingly approaches the nominal retention time, *T*. For example, with three basins or compartments in series, 4% of a tracer slug will pass through the series within one-fourth the retention time and 19% of the slug within one-half the retention time. Note that with $n = 10$, C/C_0 is greater than 1.0 since C_0 is based on the total volume of the series.

A further advantage of compartmentalization during flocculation is that a tapered velocity gradient through the system may be achieved. Higher velocity gradients during flocculation produce lower floc volume concentrations. Therefore, flocs are more compact and less prone to shear disruption. Results for floc volume concentration for *G* values from 80 to 1000 s^{-1} are shown in Figure 7.15 (Camp, 1968).

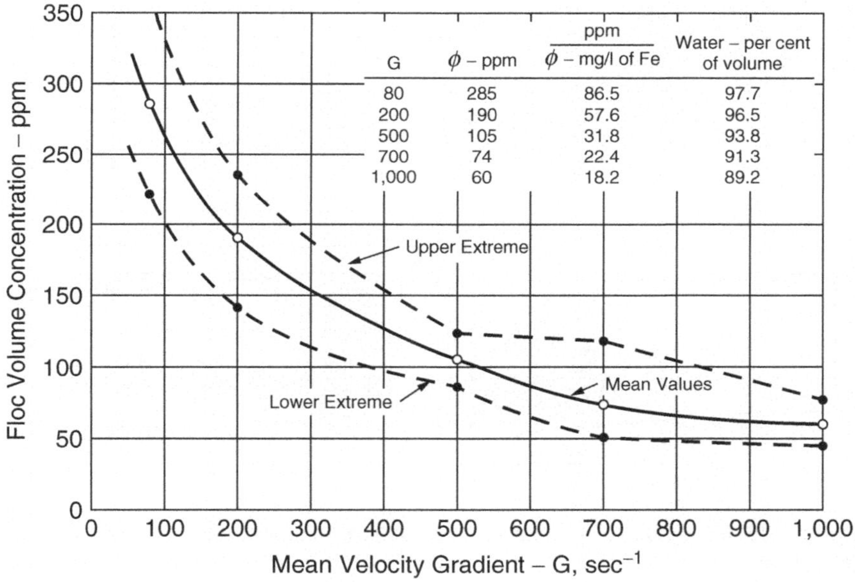

G	ϕ – ppm	$\frac{\text{ppm}}{\phi - \text{mg/l of Fe}}$	Water – per cent of volume
80	285	86.5	97.7
200	190	57.6	96.5
500	105	31.8	93.8
700	74	22.4	91.3
1,000	60	18.2	89.2

Figure 7.15 Floc volume concentrations for 15 mg/l ferric sulfate in tap water at pH 6.0 and after rapid mixing at indicated *G* values for 2 minutes (after Camp, 1968).

By providing a high velocity gradient in the first chamber and successively lower velocity gradients in subsequent compartments, flocs with denser, more robust characteristics will be produced compared to a system where a single low *G* value is assigned to one or a series of tanks. A further beneficial effect with tapered velocity gradients is that the total retention time for a series of reactors is lower than a series of reactors subjected to the same mean velocity gradient in each reactor.

The transfer of flocculating water from one compartment to the next, or of flocculated water from the final compartment to a sedimentation basin, must be carried out so as to prevent excessive breakup of flocs already formed (see later).

7.4.5 Combined flocculation – sedimentation basins

As an alternative to providing separate flocculation and sedimentation units, another option is to combine flocculation and sedimentation in one unit. Figure 7.16 illustrates two types of such combined units:

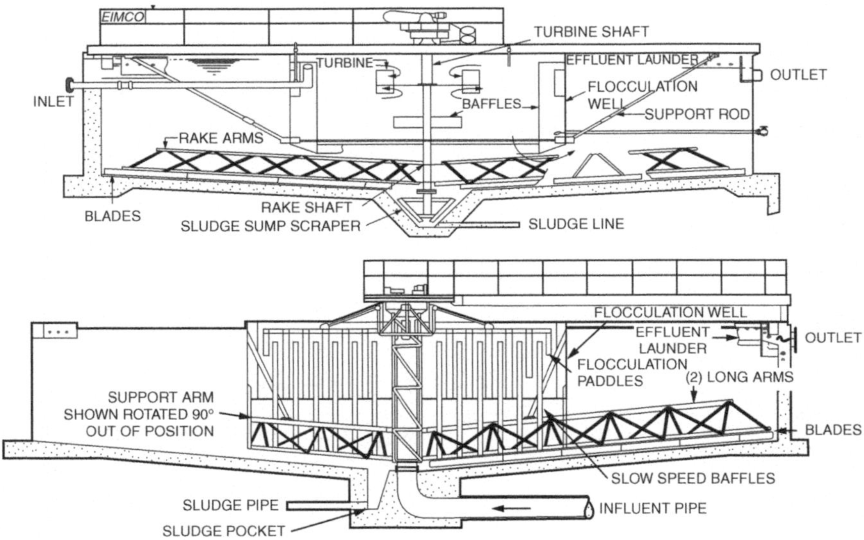

Figure 7.16 Examples of combined flocculation-sedimentation basins (courtesy U.S. Filter).

With the first unit, it is seen that neither compartmentalization nor a tapered velocity gradient are accommodated. As shown in the preceding section, provision of both these facilities is beneficial.

With the second type, typical of larger units, the flow is horizontal and, therefore, a tapered velocity gradient may in principle be accommodated by appropriate design of paddle areas. However, paddles are often designed all of the same size which, theoretically, produces an increasing velocity gradient in the direction of flow (i.e. a reverse tapered velocity gradient profile). As an example, considering such a unit with seven equally spaced paddles of equal cross sectional area in a plane perpendicular to the direction of travel, and taking the volume swept by each paddle as a separate compartment, the ratio of the mean velocity gradient in the last compartment to the first compartment (i.e. considering successive compartments from the centre outwards in the direction of flow) is of the order 1:2.

Short circuiting considerations for such combined units coincide with those for standard sedimentation basins (Camp, 1946).

A disadvantage of the units described in this section, particularly with the larger types, is the possibility of settling occurring in the flocculation section thereby reducing the overall efficiency of flocculation. With centre-feed units, the velocity of flow decreases as floc formation proceeds.

7.4.5.1 Solids contact blanket clarifiers

A special type of combined flocculation-sedimentation device is the solids contact blanket clarifier, examples of which are shown in Figure 7.17.

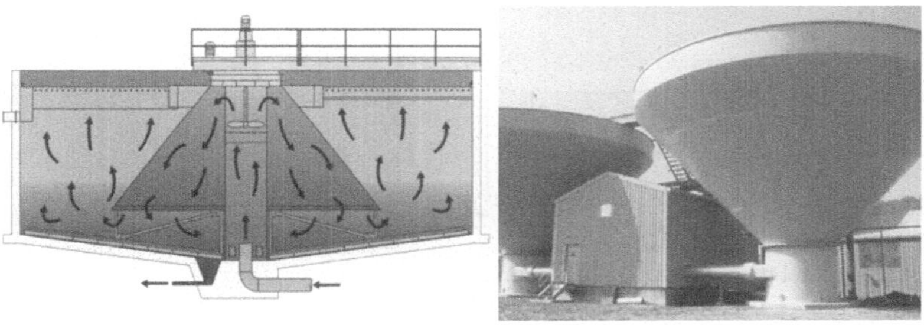

Figure 7.17. Examples of solids contact blanket clarifier (courtesy left: U.S. Filter; right: CBI Walker).

This device is widely used in the United Kingdom in the water treatment field as well as in several industrial applications. Such devices have also been used effectively in tertiary wastewater treatment, particularly with chemical phosphorus removal applications.

The principal features of solids contact blanket clarifier systems are as follows (after Stevenson, 1977):

- A rapid mixing system for coagulants;
- A delay period from the point of coagulant addition to entry to the clarifier;
- An inlet system to mix incoming water with preformed blanket floc, and to maintain the latter in stable suspension;
- A blanket or flocculation zone where primary floc particles formed during destabilization are filtered out and incorporated with preformed flocs;
- A supernatant clear water zone which serves primarily as a buffer to allow surges in blanket volume to be accommodated, and also to provide a degree of post flocculation;
- A clarified water decanting system;
- A blanket level control system (often carried out visually);
- A sludge concentrator, which may also serve as a level controller;
- A sludge bleed control.

Figure 7.18 shows a variation of the blanket clarifier known as the *Pulsator.*

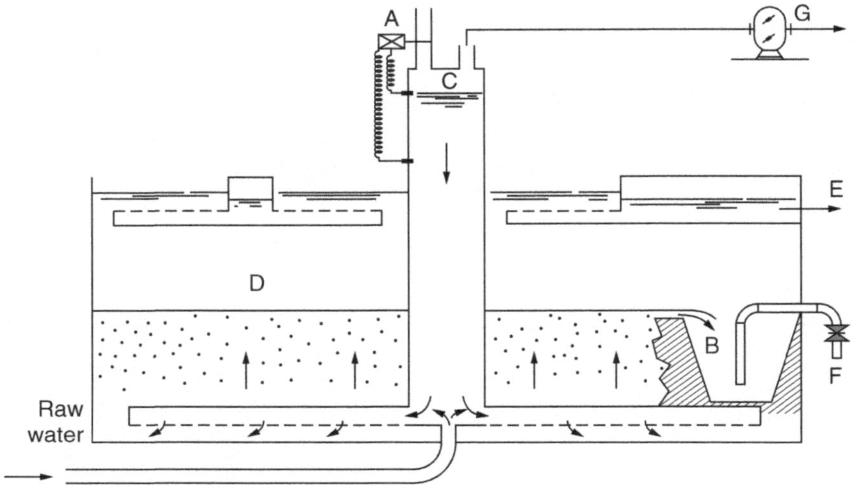

Figure 7.18 Schematic of pulsed sludge blanket (courtesy Degremont).

Here, the mixing energy within the flocculating zone may be varied independently of the base flow. Approximately every four minutes a pulse of water is injected through the inlet at the bottom. Therefore, periodically the G values within the blanket are temporarily increased and then subsequently decreased. Furthermore, flocs settled on the floor of the clarifier are re-dispersed causing them to move upwards for reflocculation within the blanket. Currently these devices also include plate settlers in the clarification zone to improve settled water clarity.

In general, the principle of operation of solids contact blanket clarifiers is that floc particles (formed within the clarifier) with a settling velocity greater than the surface overflow rate will be held in a position of equilibrium at some point where the upflow velocity relative to the particles equals the hindered settling velocity. Thereby, larger particles will form an equilibrium layer within which all smaller particles with lower settling velocities are trapped.

Solids concentration within the blanket increases with depth from the bottom, with the highest concentration being at the top of the blanket. For floc blanket control and to minimize water withdrawal, sludge should be withdrawn from the top of the blanket. Calculating an exact sludge level is difficult and thus a sludge concentrator is normally positioned as near to the top of the blanket as possible. Concentrated sludge is thus withdrawn intermittently from the concentrator. Withdrawing sludge from a single opening in the wall is very unsatisfactory since a very watery sludge results and blanket control is difficult.

Instability of the sludge blanket can occur and is principally due to two reasons:

(1) Poor inlet design, where the momentum of incoming water is not destroyed or, in the case of expanding units, boundary separation occurs. In the latter instance a practicable solution has been to install a series of baffles at the bottom of expanding flow units;

(2) Poor outlet design, where clarified water is collected unevenly from the top of the unit. To alleviate this, sufficient freeboard is usually provided above the top of the blanket. (As a rule-of-thumb, the freeboard depth should be approximately one-half the distance between overflow weirs). Poor inlet and outlet conditions give rise to short circuiting and streaming, with disturbance of the blanket surface and consequent floc carry-over with the effluent.

Referring to Figure 7.19 (after Fair *et al.* 1971), velocity gradients, G, existing in the sludge blanket are given by:

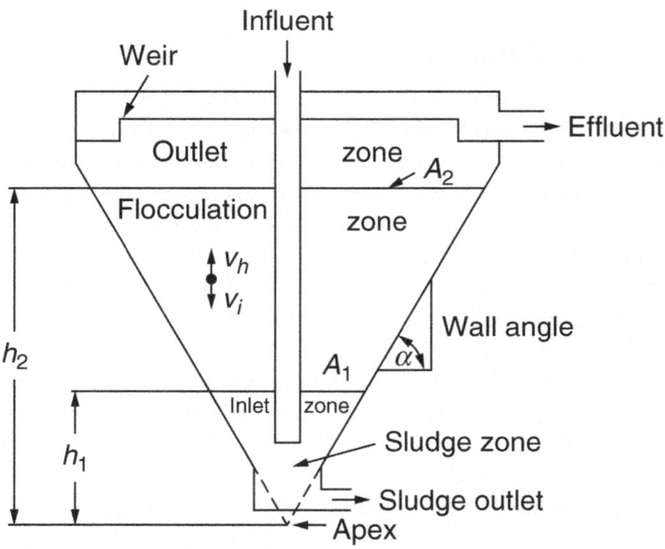

Figure 7.19 Vertical section through conical or pyramidal upflow tank (after Fair *et al.* 1971).

$$G = \sqrt{\frac{\frac{\rho \cdot g}{\mu}(s_s - 1)(1 - f_e)(h_2 - h_1)}{(C/Q)}} \quad \mathrm{s}^{-1} \tag{7.42}$$

The retention time within the blanket, or flocculating zone = $T = f_e C/Q$
where

g = gravity constant;
ρ = mass density of liquid;
μ = viscosity of liquid;
s_s = specific gravity of flocs;
f_e = relative pore space of flocculation zone;
$(h_2 - h_1)$ = blanket or flocculation zone depth;
C = zonal volume or capacity;
Q = rate of flow.

In assessing G at different depths within the blanket, values for h_2 and h_1 are taken as appropriate. In controlling G within the flocculation zone, f_e, h and C are the control variables and are selected to yield a desired floc growth rate and size. A relatively high initial G value (at the bottom of the clarifier) is sought to promote floc growth. High G values at the top of the blanket are avoided to prevent floc breakup with subsequent loss over the effluent weirs.

7.4.6 Transfer of flocculated water

An important consideration with flocculation operations, is the transfer of flocculated water to the solid-liquid separation device, without undue breakup of the flocs formed. Therefore, ports, conduits and baffles must be designed such that at maximum expected flow, the mean velocity gradient during passage from one tank to the next is close to the G value in the preceding tank or compartment.

On the other hand, from a consideration of uniformity in flow pattern, the head loss through port orifices, conduits and baffles must be relatively high with respect to the approaching velocity head, and the number of openings should be as large as possible (Hudson & Wolfner, 1967).

From the above, it is clear that considerations pertaining to minimizing short circuiting, maintaining uniformity of flow, and preventing floc breakup are inter-related – and may in some cases be in conflict. It is often necessary to compromise the above factors during design.

When passing from one compartment to another of in-series flocculation chambers, as well as from the flocculation tank to rectangular sedimentation or flotation tanks, this is often done through dividing walls with ports. The ports provide a small amount of head loss to ensure uniform distribution of flow through the dividing wall. This arrangement is relatively common in water plants.

The head loss through the ports should be about four times higher than the kinetic energy of the approach velocities to ensure uniform distribution of flow. Some workers have recommended the ratio of openings to area of flow in each compartment, in a plane perpendicular to direction of flow, of approximately 3% (Kawamura, 1973).

At the same time, the head loss through the ports should be close to or a little higher than the last flocculation compartment, to avoid undue floc breakup (Hudson, 1981).

For square-edged cylindrical ports, the head loss through each port is given by:

$h_L = 1.7 \times v^2/2g$ where v = velocity through port.

In general, flow contracts into the ports and does not expand to port diameter in the port. Therefore, almost none of the energy in the contracted jet is expended in the port. Instead it is expended downstream of the port in a relatively large volume.

Hudson suggests that the volume of dissipation is a cube given by approximately $6 \times a^3$, where a = distance between ports, as shown schematically in Figure 7.20.

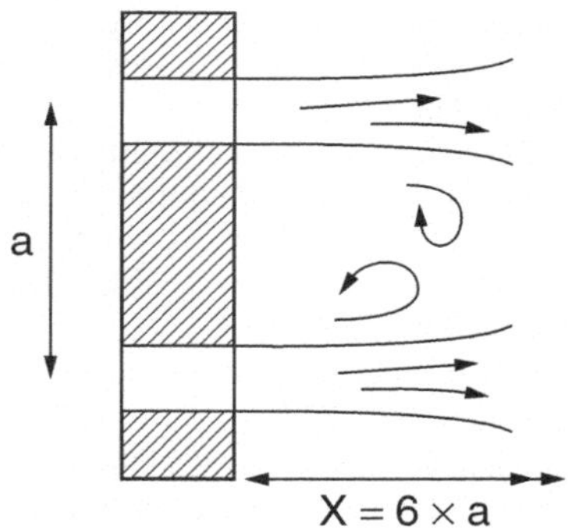

Figure 7.20 Schematic of zone of energy dissipation after ports.

Therefore, as before, $G = \sqrt{\dfrac{\gamma \cdot h_L}{T \cdot \mu}}$

Hence:

$$G = \sqrt{\frac{0.23 \cdot \gamma \cdot Q^3}{N^2 \cdot d^4 \cdot a^3 \cdot g \cdot \mu}} \tag{7.43}$$

where

N = total number of ports
d = diameter of each port
a = spacing between ports
(all consistent units)

7.5 REFERENCES

Adin A. and Rebhun M. (1974). High-rate contact flocculation-filtration with cationic polyelectrolytes. *Jour. AWWA*, 109–117.

Ahsan T., Alaerts G. J. and Buiteman J. P. (1996). Direct horizontal-flow roughing filtration. *J. Water SRT – Aqua*, **45**(5), 262–271.

Argaman Y. and Kaufman W. J. (1968). Turbulence in orthokinetic flocculation. SERL Report no.68–5, Berkeley: SERL, Univ. Calif., Berkeley, Calif., July.

Argaman Y. and Kaufman W. J. (1970). Turbulence and flocculation. *Jour. San. Eng. Div. ASCE*, **96**(SA2), Proc. Paper 7201, 223–241.

Argaman Y. (1971). Pilot-Plant Studies of Flocculation. *Jour. AWWA*, 775–777.

Ayoub G. M. and Nazzal F. F. (1989). Gravel packed baffled channel flocculator. *Jour. Env. Eng. Div. ASCE*, **114**, 6.

Ayoub G.M. (1996). A model for flocculation in a gravel reactor: development and assessment. *J. Water SRT – Aqua*, **45**(2), 76–83.

Bernhardt H. and Schell H. (1993). Effects of energy input during orthokinetic aggregation on the filterability of generated flocs. *Wat. Sci. Tech.*, **27**(10), 35–65.

Bhole A. G. (1993). Performance of static flocculators. *Wat. Sci. Tech.*, **27**(10), 181–194.

Bratby J., Miller M. W. and Marais G. V. R. (1977). Design of flocculation systems from batch test data. *Water SA*, **3**, 173–182.

Bratby J. and Lutzow W. (1977). Flocculation Characteristics of Coloured Water Derived from Table Mountain Catchment. Unpublished Results, Dept. Civil Eng., Univ. Cape Town, South Africa.

Bratby J. (1981). Interpreting laboratory results for the design of rapid mixing and flocculation systems. *Jour. AWWA*, 318–325.

Camp T. R. and Stein P. C. (1943). Velocity gradients and internal work in fluid motion. *J. Boston Soc. Civil Eng.*, **30**(4), 219–237.

Camp T. R. (1946). Sedimentation and the design of settling tanks. *Trans. ASCE*, **III**, 909.

Camp T. R. (1955). Flocculation and flocculation basins. *Trans. ASCE*, **120**(1), 1–16.

Camp T. R. (1961). Applied Hydraulic Design of Treatment Plants and Fluid Mechanics Applied to Waste Treatment Problems. Seminar papers on wastewater treatment and disposal, Boston Soc. Civil Eng., Sanitary Section, Boston, MA.

Camp T. R. (1968). Floc volume concentration. *Jour. AWWA*, 656–673.

Camp T. R. (1969). Discussion to agglomerate size changes in coagulation, *San. Eng. Div. ASCE*, **SA6**, 1210–1214.

Culp R. L. (1977). Direct filtration. *Jour. AWWA*, 375–378.

Fair G. M., Gemmell R.S. (1964). A mathematical model of coagulation. *J. Colloid Sci.*, **19**(4), 360–372.

Fair G. M., Geyer J.C. and Okum D.A. (1971). Elements of Water Supply and Wastewater Disposal. John Wiley and Sons Inc., N.Y.

Gregory J. (1993). The role of colloid interactions in solid-liquid separation. *Wat. Sci. Tech.*, **27**(10), 1–17.

Gregory J. (2006). Particles in Water. Properties and processes. IWA Publishing – CRC Press, Taylor & Francis Group, FL.

Habibian M. T. and O'Melia C. R. (1975). Particles, polymers and performance in filtration. *Jour. Env. Eng. Div., ASCE*, **101**(EE4), 567–583.

Han M. and Lawler D. F. (1992). The (relative) insignificance of G in flocculation. *Jour. AWWA*, 79–91.

Harris H. S. and Kaufman W. J. (1966). Orthokinetic flocculation of polydispersed systems. SERL Report no. 66–2, Berkeley: SERL, Univ. Calif., Berkeley, Calif., July.

Harris H. S., Kaufman W. J. and Krone R. B. (1966). Orthokinetic flocculation in water purification. *Jour. San. Eng. Div., Proc. ASCE*, **92**(SA6), 95–111.

Hudson H. E. (1965). Physical aspects of flocculation. *Jour. AWWA*, **57**(7), 885–892.

Hudson H. E. and Wolfner J.P. (1967). Design of mixing and flocculation basins. *Jour. AWWA*, **59**, 1257–1267.

Hudson H. E. (1981). Water Clarification Processes. Practical Design and Evaluation. Van Nostrand Reinhold, New York.

Hutchison W. and Foley P. D. (1974). Operational and experimental results of direct filtration. *Jour. AWWA*, 79–87.

Ives K. J. (1977). Deep Bed Filters. Chapter 8 in Solid-liquid Separation Equipment Scale-Up. Uplands Press, Croydon, England, U.K.

Kavanaugh M., Eugster J., Weber A. and Boller M. (1977). Contact filtration for phosphorus removal. *Jour. WPCF*, 2157–2171.

Kawamura S. (1973). Coagulation considerations. *Jour. AWWA*, **65**(6), 417–423.

Kawamura S. (1975). Design and operation of high-rate filters. *Jour. AWWA*, 535–544.

Kawamura S. (1976). Considerations on improving flocculation. *Jour. AWWA*, 328–336.

Kawamura S. (2000). Integrated Design and Operation of Water Treatment Facilities. John Wiley & Sons, Inc., New York.

Kurbiel J., Sapulak A. and Schade H. (1991). The use of a turbulent pipe-flow for rapid flocculation of precipitate in the electroplating wastewater treatment. *Wat. Sci. Tech.*, **24**(7), 255–259.

Kramer T. A. and Clark M. M. (1997). Influence of strain-rate on coagulation kinetics. *Jour. Env. Eng. ASCE*, **123**(5), 444–1029.

Lawler D. (1993). Physical aspects of flocculation: from microscale to macroscale. *Wat. Sci. Tech.*, **27**(10), 165–180.

Leentvaar J. and Ywema T. S. J. (1980). Some dimensionless parameters of impeller power in coagulation-flocculation processes. *Water Research*, **14**, 135–140.

Loganathan P. and Maier W. J. (1975). Some surface chemical aspects in turbidity removal by sand filtration. *Jour. AWWA*, 336–342.

McConnachie G. L. (1991). Turbulence intensity of mixing in relation to flocculation. *Jour. Env. Eng., ASCE*, **117**(6), 731–750.

Monk R. D. G. and Trussell R. R. (1991). Design of Mixers for Water Treatment Plants: Rapid Mixing and Flocculators. In: Mixing in Coagulation and Flocculation, Amirtharajah A., Clark M. and Trussell R. (eds), AWWA Research Foundation, Denver, Colorado, p. 380.

Oldshue J. Y. and Mady O. B. (1979). Flocculator impellers: A comparison. *Chem. Eng. Progress*, **75**(5), 72–75.

O'Melia C. R. (1967). Discussion to Orthokinetic flocculation in water purification. *Jour. San. Eng. Proc. ASCE*, **93**(SA5), 120–125.

O'Melia C. R. and Stumm W. (1967). Theory of water filtration. Pub. No.61, *Dept. Env. Sci. and Eng.*, Univ. North Carolina, Chapel Hill.

Parker D. S., Kaufman W. J. and Jenkins D. (1970). Characteristics of biological flocs in turbulent regimes. SERL Report No. 70–5, Berkeley: SERL, Univ. Calif., Berkeley, Calif., July.

Parker D. S., Kaufman W. J. and Jenkins D. (1972). Floc breakup in turbulent flocculation processes. *Jour. San. Eng. Div. ASCE*, **98**(SA1), 79–99.

Pedocchi F. and Piedra-Cueva I. (2005). Camp and Stein's velocity gradient formalization. *Jour. Env. Eng., ASCE*, 1369–1376.

Polasek P. (1988). Granular fluidised layer in waterworks design. *Wat. Sew. & Eff.*, **8**(2), 30–35.

Rajapakse J. P. and Ives K. J. (1990). Pre-filtration of very highly turbid waters using pebble matrix filtration. *J. IWEM*, **4**, 140–147.

Ryder R. A. (1977). State of the art in water treatment design, instrumentation and analysis. *Jour. AWWA*, 612–620.

Schulz C. R., Singer P. C., Gandley R. and Nix J. E. (1994). Evaluating buoyant coarse media flocculation. *Jour. AWWA*, 51–63.

Shea T. G., Gates W. E. and Argaman Y. A. (1971). Experimental evaluation of operating variables in contact flocculation. *Jour. AWWA*, 41–48.

Spink C. M. and Monscvitz J. T. (1974). Design and operation of a 200 mgd direct filtration facility. *Jour. AWWA*, 127–132.

Stevenson D. G. (1977). Coagulation Clarification. In: Solid-Liquid Separation Equipment Scale-Up, Purchas D. (ed.), Uplands Press, Croydon, England, U.K., p. 127.

Stumm W. (1977). Chemical Interaction in Particle Separation. *Env. Sci. and Tech.*, **11**(12), 1066–1070.

Sweeney G. E. and Prendiville P. W. (1974). Direct filtration: an economic answer to a city's water needs. *Jour. AWWA*, 65–71.

Tambo N. and Watanabe Y. (1979). Physical characteristics of flocs. I. The floc density function and aluminium floc. *Water Res.*, **13**, 409–419.

Tambo N. and Matsui Y. (1989). Performance of fluidized pellet bed separator for high concentration suspension removal. *Aqua*, **38**(1), 16–22.

Tate C. H. and Lang J. S., Hutchinson H. L. (1977). Pilot plant tests of direct filtration. *Jour. AWWA*, 379–384.

Tredgett R .G. (1974). Direct-filtration studies for Metropolitan Toronto. *Jour. AWWA*, 103–108.

Von Smoluchowski M. (1916). Drei Vortrage uber diffusion, Brownshe molekular beweging und koagulation von kolloidteilchen. *Physik. Z.*, **17**, 557.

Von Smoluchowski M. (1917). Versuch einer mathematischen theorie der koagulationskinetik kolloid losungen. *Z. Physik. Chem.*, **92**, 155.

Wilson D. J. and French R. H. (1978). Kinetic and equilibrium aspects of floc coagulation. II. Slow mixing criteria. *Sep. Sci. and Tech.*, **13**(2), 95–106.

Chapter 8

Testing and control of coagulation and flocculation

8.1 INTRODUCTION

The efficiency of the coagulation-flocculation process is dependent on many variables. For a particular water these may include:

- Type of coagulant used
- Coagulant dosage
- Final pH
- Coagulant feed concentration
- Type and dosage of chemical additives other than primary coagulant (e.g. polyelectrolytes)
- Sequence of chemical addition and time lag between dosing points
- Intensity and duration of mixing at rapid mix stage
- Type of rapid mix device
- Velocity gradients applied during flocculation stage
- Flocculator retention time
- Type of stirring device used
- Flocculator geometry

To adequately assess the optimum conditions for the overall process it is necessary to use a testing procedure which sequentially maintains each parameter constant while the optimum value of a particular parameter is determined.

The prerequisite for such a series of tests is, if possible, to thoroughly analyze the water and thereby identify the impurities most requiring consideration. In many cases the impurities of major concern are organic material and turbidity, but in others, residual phosphorus, iron, microorganisms, or other parameters may be the governing criteria. The particular application obviously decides the desired final water quality and, therefore, the combination of coagulation-flocculation variables most appropriate.

In some instances, tests must be carried out on a water or wastewater with unknown characteristics. Although an *a priori* estimation of the likely range of process variables in such cases is not possible, the testing procedure remains essentially the same. With some applications final effluent quality is not of major importance; an example is sludge conditioning where the specific resistance or capillary suction time of the sludge after treatment is of concern.

A word of caution is appropriate at this stage. When following the treatment presented in this and earlier chapters, there may be a danger of gaining the impression that waters to which the coagulation and flocculation processes are applied have well defined properties requiring pre-analysis only and fixation of design parameters on this basis. In practice most waters, especially from natural sources, are seasonally variable and, therefore, require testing during all changes in water quality. Plant design should naturally be based on the variation in water quality rather than on one set of data obtained at one particular time.

8.2 OPTIMIZING PRIMARY COAGULANT TYPE, DOSAGE AND PH

8.2.1 General

The starting point in a series of tests serving to characterize a particular water in terms of coagulation/flocculation performance is to determine the type (or types) of primary coagulant, coagulant sequencing, optimum dosage of the coagulant(s), and the pH at which coagulation operates most efficiently. The testing procedure followed is referred to as the jar test procedure.

The jar test is the ideal tool for optimizing coagulation operations. This is particularly evident when one considers the number of papers devoted to the subject over the last 85 years (examples include Langelier, 1921; Black *et al.* 1957; Cohen, 1957; Camp, 1970; TeKippe & Ham, 1970; Packham, 1972; Griffith & Williams, 1972; Hudson, 1973, 1981; Lai *et al.* 1975; Benedek & Bancsi, 1976; Bratby, 1977, 1981; Cornwell & Bishop, 1983; Dentel *et al.* 1988; Clark, 1992; Bernhardt & Schell, 1993; Barthelme, 1994; Stanley & Smith, 1995; Clark & Stephenson, 1999).

However, despite the importance with which these researchers have held jar test procedures, in many cases the full potential of the information that could be gleaned from jar tests, to optimize the operation of treatment plants, is not realized.

Hudson and Wagner (1981) reviewed jar testing techniques and their uses. To quote these researchers: *If procedures that simulate treatment plant conditions are followed, jar testing can produce important information quickly and economically, and the data are directly applicable to plant design, modification, and operation. The procedure offers greater flexibility and economy than the traditional pilot plant test for pretreatment. It has shown excellent scale-up correlations with plant operating records.* It is important to ensure that jar test procedures properly simulate plant facilities, and that chemicals are made up at the time of testing.

8.2.2 Apparatus

The basic apparatus used for assessing coagulation performance is the jar test apparatus, an example of which is shown in Figure 8.1. It consists essentially of a rack of stirrers, driven by one motor, under which the jars are arranged. Jars smaller than 600 ml should not be used since irreproducible results may occur due to difficulties in accurately adding coagulants or coagulant aids. Furthermore, where the test criterion is settling velocity, tall jars of large capacity are to be preferred.

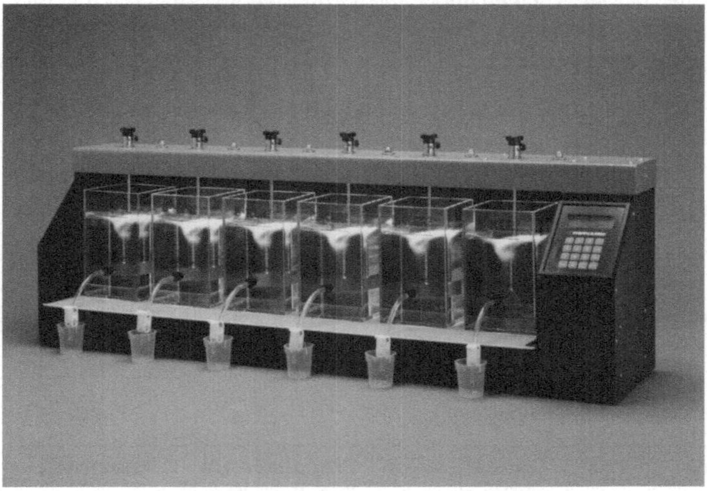

Figure 8.1 Example of jar test apparatus. This particular model includes six paddles and accommodates 2-liter square *gator* jars, or smaller jars. It also includes programmed stirring speeds and durations. (Photo courtesy of Phipps & Bird, Inc., Richmond, VA. © 2015).

A further factor supporting the use of jars of large capacity is that when using polyelectrolytes either as primary coagulants or coagulant aids, the inner surfaces of the jars serve as adsorption sites for the polymer (Habibian & O'Melia, 1975). This effect is accentuated when smaller jars are used and for waters of low turbidities, where the total colloidal surface area may approach the same order of magnitude as the jar inner surface area. For this reason jars of at least 1-liter capacity, and preferably 2-liters, are recommended.

The speed of rotation of the stirrers on the jar test apparatus is variable because a fast speed is used during coagulant addition, and slow speeds for the flocculation phase.

Further desirable features of the apparatus are:

- A light source arranged either above or below the jars e.g. a daylight 'cold' fluorescent tube or LED lighting to avoid appreciable heat transfer)

- A black background to the jars is very helpful to observe the flocs
- Flat bladed stirrers
- Square jars (often referred to as *gator* jars) to improve mixing and reduce the rotation that occurs in circular jars
- A means to allow simultaneous chemical addition to all the jars. Addition of coagulant chemicals should be rapid and the stirring should be such that dispersion occurs in as short a time as possible. (One operator has fabricated a syringe rack that serves this purpose. The rack is held over the jars, one syringe per jar. Then an upper and a lower bar are squeezed together by hand to depress the syringes and inject coagulant into the jars (Mical, 1997). More than one syringe rack could be fabricated for sequences of chemicals).

G values for a number of jar sizes, temperatures and paddle rotation are given later in this Chapter.

8.2.3 Chemical solutions

Stock solutions of each coagulant or coagulant/flocculant aid are prepared such that the volumes corresponding to the dosages to be used may be accurately and conveniently measured. Table 8.1 presents a general guide for suitable stock solution strengths of some chemicals.

Table 8.1 Stock solution strengths.

Chemical	Concentration of Stock Solution	Prepare Fresh Solution after at Least	Comments
Alum ($Al_2(SO_4)_3 \cdot 16H_2O$)	10 g/l	1 month	Renew before solution becomes opalescent
Ferric sulfate ($Fe_2(SO_4)_3 \cdot 9H_2O$)	10 g/l	1 week	Renew before solution becomes opalescent
Ferric chloride ($FeCl_3$)	10 g/l	1 week	Renew before solution becomes opalescent
Polymer (polyacrylamide derivatives)	0.05%	1 week	Solutions should be at least 1 day old Dilute 10 times immediately before use
Sulfuric acid (H_2SO_4)	0.1 N	3 months	
Sodium hydroxide (NaOH)	0.1 N	1 month	

If a number of coagulants are to be tested and compared, it may be convenient to conduct the tests at the same concentration expressed as the metal ion. If analytical grade reagents are used in the laboratory, the corresponding commercial coagulant concentrations for plant applications may be determined from the percentage of metal ion given by the manufacturer. Table 8.2 presents atomic weights of principal elements to assist in the calculations.

Table 8.2 Atomic weights.

Element	Atomic Weight
Al	26.982
Cl	35.453
Fe	55.847
H	1.008
N	14.007
O	15.999
S	32.064
Si	28.086

For example, with a stock solution strength of 10 g/l alum, the metal concentration is determined by the following steps:

- Determine chemical formula (from manufacturer's literature, or from laboratory reagent bottle (for example $Al_2(SO_4)_3 \cdot 16H_2O$)
- Calculate molecular weight: 2×26.982
  $3 \times (32.064 + 4 \times 15.999)$
  $16 \times (2 \times 1.008 + 15.999)$
  $= 630.384$
- Concentration of metal $= 10\ \text{g/l} \times (2 \times 26.982)/630.384$
  $= 0.856\ \text{gAl/l}$

It is convenient to adjust the stock solution concentrations such that, for example, 1 ml of stock solution added to the 1-liter jars produces a dosage of 1 mg/l. For example, for alum (assuming the same chemical formula as above) the stock solution concentration would be 11.68 g/l measured as the chemical $Al_2(SO_4)_3 \cdot 16H_2O$, or 1 g/l as Al.

In the case of expressing dosages in terms of the chemical, then a stock solution of 10 g/l would be suitable – in this case, 1 ml added to the 1-liter jars would produce a dosage of 10 mg/l. If a lower range of coagulant dosages is to be tested, then the stock solution concentration could be suitably diluted. For example, a 10-times dilution would produce a stock solution of 1 g/l as the chemical, and every 1 ml added to the 1-liter jar would produce a dosage of 1 mg/l. However, the diluted stock concentration would have to be prepared fresh for each jar test session to avoid hydrolysis of the coagulant.

An alternative method of evaluating different coagulant products is to apply ranges of dosages that reflect similar costs. In other words, instead of applying similar ranges in terms of mg/l as Al or Fe for example, one could apply the ranges in terms of $/l (depending on the local currency). This provides a direct indication of the cost-effectiveness of different products. One may still express results in terms of mg/l, but the stock solution concentrations, and the applied dosages, would reflect respective costs of the products.

For routine monitoring of plant performance, made-up chemical from the plant may be used, suitably diluted to, say, 10 g/l alum or ferric chloride etc.

The keeping properties of chemical solutions depend principally on the chemical and the concentration. For example, for a 10 g/l alum concentration, the solution may be used up to one month. For the same concentration of ferric sulfate, solutions should not be used for longer than one week. If the solutions become opalescent before this time, they should be discarded. (However, it is usually preferable to prepare fresh stock solutions on a daily basis.)

Sulfuric acid and sodium hydroxide solutions of 0.1 N should be renewed after 3 and 1 month respectively.

With coagulant aids it is best to refer to the manufacturer's literature. However, as an indication, a 0.05% stock solution of a polyacrylamide derivative should not be used for longer than approximately one week and should be at least 1 day old before use. Prior to each day's testing program, sufficient quantity of polyacrylamide stock solution should be diluted to at least 0.005 percent.

8.2.4 Criteria describing process performance

Before coagulation/flocculation tests are carried out it is necessary to decide which criterion (or criteria) of process efficiency is the most appropriate for the particular water under test and for the existing or proposed method of treatment (i.e. sedimentation, flotation, filtration, etc.). The most common criteria are to take supernatant samples after a given period of settling, or the filtrate after passing through filter papers, and analyze for residual turbidity, color, iron, etc. as discussed earlier.

However, as the following list shows, the evaluation of optimum coagulant type, dosage and pH using the jar test procedure may be carried out on the basis of a wide variety of criteria; the applicability of each dependent on the particular circumstances and the process used after coagulation (Baskerville & Gale, 1968; TeKippe & Ham, 1970; Packham, 1972; Hudson & Wagner, 1981):

(1) Settling and assessing supernatant quality
(2) Filtration through a small granular bed filter and assessing filtrate quality
(3) Filtration through filter papers and assessing filtrate quality

(4) Determining the time from coagulant addition to the first appearance of a visible floc
(5) Visually recording or comparing flocs as they are formed
(6) Floc density measurements
(7) Floc volume concentrations determined by removing samples from the jar test apparatus and manually counting and sizing the particles using a stereo-microscope
(8) Volume of settled floc
(9) Residual coagulant concentration
(10) Silting index using a 0.45-micrometer membrane filter at constant pressure, where silting index, SI, is given by

$$\mathrm{SI} = \frac{(\text{Time to filter 2nd 5ml sample}) - (\text{Time to filter 1st 5ml sample})}{\text{Time to filter 1st 1ml sample}}$$

(11) Filterability number, which is the ratio of the time to filter 200 ml of jar test supernatant through a 0.45-micrometer membrane filter, to the time required to filter an equal volume of distilled water
(12) Membrane Refiltration, which involves the formation of a filter cake by passing the jar test suspension through a 0.45-micrometer membrane filter under constant pressure. The filtrate is collected and refiltered through the cake and membrane, the time required to refilter a specified volume being recorded
(13) Inverted gauze filtration
(14) Cation exchange capacity
(15) Surface area concentration
(16) Conductivity
(17) Zeta potential
(18) Streaming current detection
(19) Colloid titration
(20) Cotton plug filters
(21) Electronic particle counting
(22) Specific resistance to filtration
(23) Capillary suction time.

TeKippe and Ham (1970) selected a number of the above criteria and evaluated their respective merits. They formed groups of three categories: those based on colloidal charge reduction (e.g. colloid titration); those measuring floc settleability (e.g. conventional-settled jar test; floc formation time test; visual floc size comparisons; and electronic particle counting); and those measuring floc filterability (e.g. membrane refiltration).

For the suspension used by them, the colloid titration technique was found to be of limited use in determining an optimum coagulant concentration. For the category of settleability, the conventional jar test was found to be the most reliable or practicable. Electronic particle counting, although a useful laboratory research tool and a more sensitive measuring technique than nephelometry or turbidimetry cannot replace the conventional jar test for the control or optimization of coagulant dosages.

The results for the technique of membrane refiltration were found to correlate well with pilot scale filter experiments. Although not a useful technique in cases where the settleability of a suspension is of importance, it may be recommended for control of plants which rely heavily on diatomite filtration or rapid sand filtration. Shortcomings of the test are that, firstly, it does not indicate the clarity of the filtrate (Cleasby, 1969) and, secondly, in the use of multi-media filters, where the formation of a compressible cake is minimized, the technique may not predict actual performance (TeKippe & Ham, 1970).

The procedure for membrane refiltration is as follows: Figure 8.2 shows the apparatus used by TeKippe and Ham. A 250 ml sample is taken immediately after the slow mix period and filtered through a 0.45 μ membrane filter under a vacuum of 100 mm Hg. Approximately 100 ml of the filtrate is collected and refiltered through the cake and membrane and collected in a 100 ml cylinder. The time taken to filter 80 ml of filtrate (i.e. from the 10 ml to 90 ml mark on the graduated cylinder) is recorded as refiltration time. Typical results are shown in Figure 8.3.

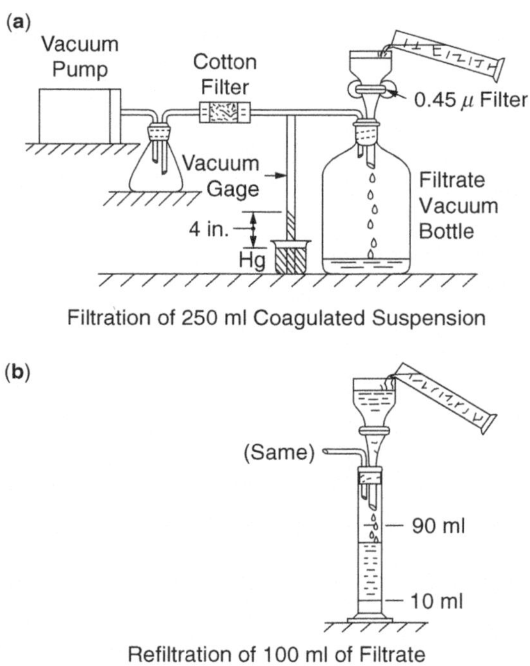

Figure 8.2 Apparatus for refiltration experiments. (From TeKippe & Ham, 1970).

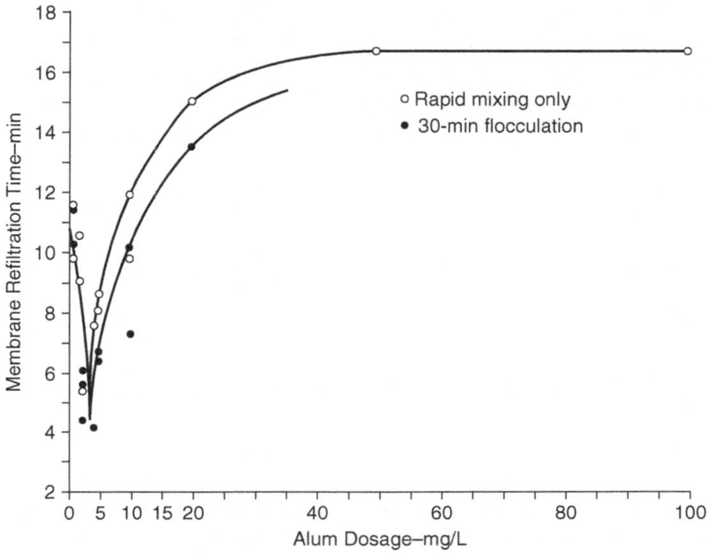

Figure 8.3 Example of membrane refiltration experiments. Each alum dosage was at the optimum pH; filtration was carried out with 47-mm membrane filters with 100-mm Hg vacuum; times recorded were those required to refilter 80 ml. (From Bratby, 1981.)

Figure 8.3 demonstrates the value of the membrane refiltration test: Hudson (1973) mentioned a problem in conventional water plants where excess coagulant hydrolysis products can pass over to filters and rapidly clog them, although adequate settling and turbidity removals are obtained. This was also noticed elsewhere (Bratby, 1986). If the clogging effect is concentrated principally within the surface layers of the filter, the membrane refiltration test should provide a good indication of the coagulant dosage to minimize surface clogging.

Figure 8.3 shows that for the water tested, a well-defined minimum refiltration time was evident at an alum dosage of 3 mg/l. The same trend is evident with and without flocculation, so the hypothesis of excess hydrolysis products at higher dosages is supported (supernatant turbidity progressively reduced with increasing alum dosages). Pilot scale filter experiments subsequently carried out confirmed the optimal alum dosage of 3 mg/l.

Another filtration technique that is very useful (and more straightforward than the membrane refiltration technique) is to use general purpose filter papers, as illustrated in Figure 8.4. The filter paper is folded twice to produce a cone and inserted in a funnel, which is placed in an *Erlenmeyer* flask. Fresh filter papers are used for each test. One filter assembly is placed in front of each jar. After the slow mix period, a sample is taken from the jar and poured through the filter. The first portion of the filtrate should be discarded and the balance used for measuring filtered water turbidity. Hudson and Wagner (1981) found that the results correlate well with filtered water turbidity in the plant. However, the test cannot be used to predict lengths of filter runs. Pilot plant filters would be used for this information.

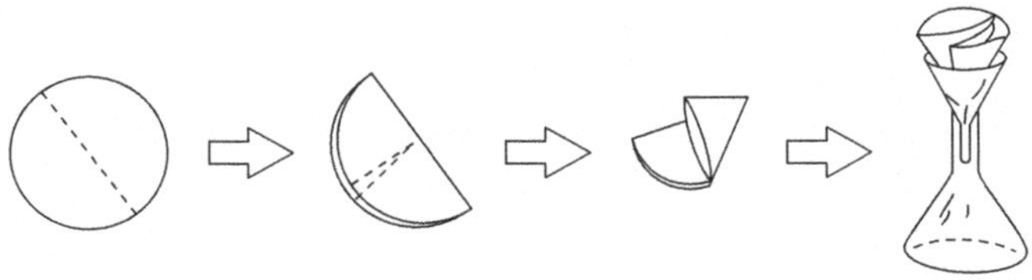

Figure 8.4 Jar test filtration assembly. Filter paper folded twice to fit into funnel. (Suitable papers are 125 mm Whatman No. 40.)

Di Bernardo (1984) also reported that jar test results using filter papers predicted quite accurately the filtered water quality of direct filtration in upward flow filters.

The optimum coagulant dosage for direct filtration may be much lower in some cases (especially for low turbid or colored waters) than for settling requirements. However, several workers have reported that settled jar test results closely predict optimum dosage requirements during filtration, using cationic polymers as primary coagulants (Yao *et al.* 1971; Adin & Rebhun, 1974; Ghosh *et al.* 1975; Habibian & O'Melia, 1975; Glaser & Edzwald, 1979).

The results of Adin and Rebhun are shown in Figure 8.5. The filtration apparatus used by them to optimize dosage comprised sand filters 50 and 150 mm in length. TeKippe and Ham (1970) showed that filters of such small size do correlate well with jar test settling results but not necessarily with larger pilot or full scale filters. However, Habibian and O'Melia (1975), using polyelectrolytes as primary coagulants, found that conventional jar test (settling) results correlated well with large scale pilot filtration results, provided the error attributed to using small capacity jars was eliminated, as explained earlier.

Bratby and Marais (1974) also showed that optimum metal coagulant dosages determined in jar tests coincided with the optima found in pilot scale flotation tests. The optimum pH was also well predicted, although during flotation pH had a more acutely defined optimum, compared with jar test results.

From the preceding discussions, in the initial investigations for the design of a water treatment facility where the jar test is carried out to assess optimum coagulant conditions, it is sometimes useful to assess both supernatant quality after settling, and filtration using filter papers. A fairly reliable indication will thereby be achieved of the likely range of conditions using sedimentation, or direct filtration, or conventional treatment with filtration.

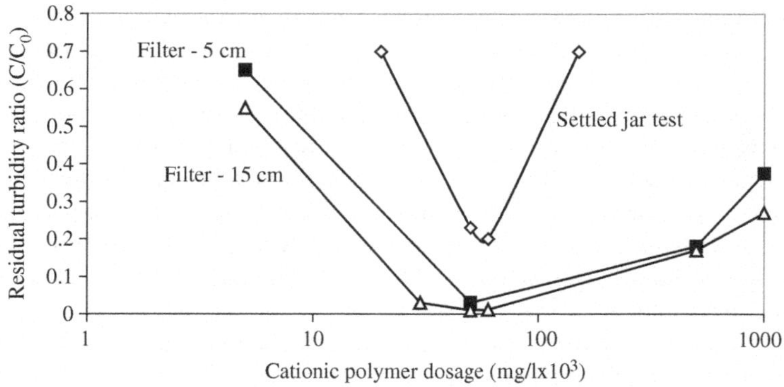

Figure 8.5 Results from jar test (settling) experiments and small scale granular filters gave same optimum cationic polymer dosages. (Data from Adin & Rebhun, 1974.)

8.2.5 Jar test procedure

Whichever criterion for coagulation efficiency is chosen, the experimental procedure remains essentially the same and is set out below. As described earlier, stock solutions of each coagulant to be tested are first prepared. A 1-liter sample (depending on the apparatus available) is placed on the jar test apparatus and the impeller speed set to maximum.

It is convenient to run each batch of tests at a constant coagulant dosage. Therefore, each sample for a particular test run is dosed with acid or alkaline reagent (H_2SO_4 or NaOH etc.) to produce a range of pH values. For example, for a jar test apparatus with six jars, the initial pH in each may be adjusted to 4; 5; 6; 7; 8 and 9 respectively, using a pH electrode immersed in the sample.

Maintaining the impellers at the fast speed, coagulant is then added individually and sequentially to each jar, or simultaneously to all the jars using the syringe rack described earlier, so that coagulant addition is rapid and dispersion virtually instantaneous. This procedure simulates as far as possible typical rapid mixing in actual plants.

The pH recorded as operative for a particular test is the final pH after coagulant addition. It should be noted that post adjustment of pH after coagulant addition to some pre-determined value is not valid since destabilization reactions are usually not reversible – however, there are exceptions to this in specific cases (for example, Bratby, 1990).

On adding the coagulant a stop clock/watch should be started and the fast rate of stirring continued for a set period (the same in the whole series of tests for the particular water). A satisfactory rapid mix time is 15 to 30-seconds. After this period, set the stirrers to slow speed. A satisfactory speed should be pre-determined by trial and error. A guide is that the speed should be slightly greater than the minimum required to maintain flocs in suspension. This speed, once determined for the particular water under test, should be used for the whole series of tests.

Although the jar testing procedure primarily targets coagulation chemistry, there may be value in duplicating rapid mix and flocculation mixing intensities that occur in the plant. For this purpose one could apply the same G values and retention times as the full scale facilities. Several workers have provided information on the G values produced by different impellers used in jar testing apparatus (Lai *et al.* 1975; Hudson & Wagner, 1981; Cornwell & Bishop, 1983; AWWA, 1992). A compilation of the performance of a wide variety of small flocculation devices is presented in Chapter 7.

The following expressions can be used to calculate G values for different jar sizes, rotational speeds and temperatures:

- For 1-liter square *(gator)* jars with 1-liter water, flat blade paddles (60 mm wide × 40 mm high), and anti-rotation baffle, the velocity gradient, G is given by:

  $$G = \left(0.00270 \times \mathrm{T} + 0.127\right) \times \mathrm{R}^{(1.5205-0.0006\times \mathrm{T})}\ \mathrm{s}^{-1}$$

- For 1-liter square *(gator)* jars with 1-liter water, flat blade paddles (75 mm wide × 25 mm high), without an anti-rotation baffle, the velocity gradient, G is given by:

  $$G = (0.00270 \times \mathrm{T} + 0.146) \times \mathrm{R}^{(1.504-0.000261\times \mathrm{T})}\ \mathrm{s}^{-1}$$

- For 1-liter square *(gator)* jars with 1-liter seawater, flat blade paddles (75 mm wide × 25 mm high), without an anti-rotation baffle, the velocity gradient, G is given by:

  $$G = (0.00258 \times \mathrm{T} + 0.144) \times \mathrm{R}^{(1.504-0.000264\times \mathrm{T})}\ \mathrm{s}^{-1}$$

- For 2-liter square *(gator)* jars with 2-liter water, flat blade paddles (75 mm wide × 25 mm high), without an anti-rotation baffle, the velocity gradient, G is given by:

  $$G = (0.00191 \times \mathrm{T} + 0.1028) \times \mathrm{R}^{(1.505-0.000266\times \mathrm{T})}\ \mathrm{s}^{-1}$$

where

T = temperature (°C)
R = rotation (rpm)

After slow stirring for, say, 10 minutes, the stirrers are stopped. Depending on the criterion of destabilization/flocculation efficiency deemed to be the most appropriate, the samples are allowed to settle in the jars for 15 to 30 minutes, or the sample filtered immediately, and so on.

In cases where settling is the criterion of interest the procedure is as follows: After switching off the stirrers, the samples are allowed to stand for a set period of, say,

15 minutes (again, the same for the whole series). After settling, a supernatant sample is taken at the same depth from the water surface, from each jar. In the apparatus shown in Figure 8.1 sampling stopcocks are provided. Care should be taken to discard the first small volume of sample to ensure that a representative sample is obtained.

An alternative is to use a pipette modified for the purpose. Figure 8.6 shows a pipette, the end spout of which has been cut off with a glass cutter, and then bent through 90° using a laboratory bunsen burner. A bent pin is attached, the point of which is 100 mm from the opening. A pipette bulb helps to control the rate of sampling. The pipette is gently immersed in the supernatant, until the point just touches the surface, and the sample slowly extracted. If further samples are to be taken at additional time intervals for settling tests (see the following section) then the pipette is removed slowly so as not to disturb the suspension.

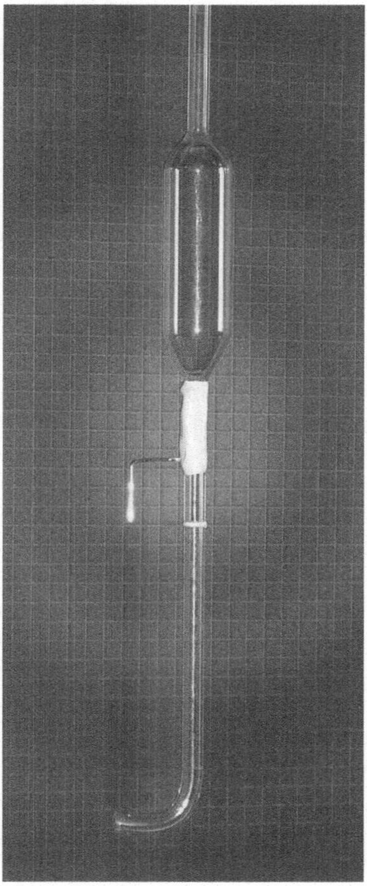

Figure 8.6 Sampling pipette, adapted in the laboratory.

The supernatant samples are then analysed for whatever criterion or criteria of effluent quality is the most appropriate, for example, residual turbidity, color, phosphorus etc.

In the case of turbidity measurements, greater reproducibility may be attained by adding, say, one drop of concentrated reagent grade hydrochloric acid per 100 ml of supernatant sample (to reduce the pH to approximately 2.5) and vigorously shaking the sample. This procedure reduces the effect of particle size on residual turbidity measurements. However, in many cases this may be considered unnecessary since the jar test is comparative in nature.

8.2.6 Analysis of results

Following the above procedures, results are recorded on a log sheet. A suitable log sheet (that can be adapted for the particular conditions) is shown in Figure 8.7.

For convenience a graph is provided on the log sheet so that results can immediately be visualized and evaluated, and the values to be applied to the next batch of jar tests determined.

After a series of jar test experiments, where a range of coagulant dosages and pH values have been applied, a series of curves are plotted of, for example, supernatant turbidity versus pH, each curve at a particular dosage (see Figure 8.8a). Optimization of the results may then be conveniently carried out by presenting the results in the form of two-dimensional contour graphs as shown in Figure 8.8b. The results shown are for supernatant turbidity after settling. However, the same method of presentation may be used for other criteria, such as filtrate turbidity, membrane refiltration times, phosphorus, etc.

Interpretation of results can be complex. With color the jar test procedure described above usually predicts very closely the actual in-plant performance. However, in the case of turbidity, the measurements are complicated by such effects as shape, size, number and refractive index of particles responsible for turbidity. For example, Black and Walters (1964) showed that for equal gravimetric concentrations (approximately 65 mg/l) of three different clay suspensions: montmorillonite, fuller's earth and kaolinite, the respective turbidities measured were 58, 79 and 294. One method of overcoming such difficulties has been to express the dosage necessary to reduce the turbidity or color to one-half its initial value (Packham, 1960) but this parameter has some disadvantages. For treatment plant control the 50% dosage differs markedly from results desirable in water treatment plants. Another method is to use 90% or 95% removal as the criterion (Black & Walters, 1964). However, the disadvantage of this is that with waters of low turbidity or color, small changes in supernatant quality assume inordinate significance. The latter methods are perhaps more applicable to research applications. When optimizing the conditions for a particular water for design purposes, reducing the effluent parameter to some fixed level is necessary to meet treatment and regulatory limits.

Raw water:

Turbidity (NTU) _____ pH _____

Alkalinity (mg/lCaCO$_3$) _____ Temperature (°C) _____

Hardness (mg/l CaCO$_3$) _____ Color (mg/l Pt-Co) _____

UV$_{254}$ _____

Rapid mix (rpm) _____ duration (sec) _____

Flocculation (rpm) _____ duration (min) _____

Settling time (min) _____

Reagent 1 __________ stock soln. _____ g/l

Reagent 2 __________ stock soln. _____ g/l

Reagent 3 __________ stock soln. _____ g/l

Jar No.		1	2	3	4	5	6
Reagent 1	ml						
	mg/l						
Reagent 2	ml						
	mg/l						
Reagent 3	ml						
	mg/l						
pH							
Settled turbidity (NTU)							
Filtered turbidity (NTU)							
(Obs.: 1-best; 6-worst)							

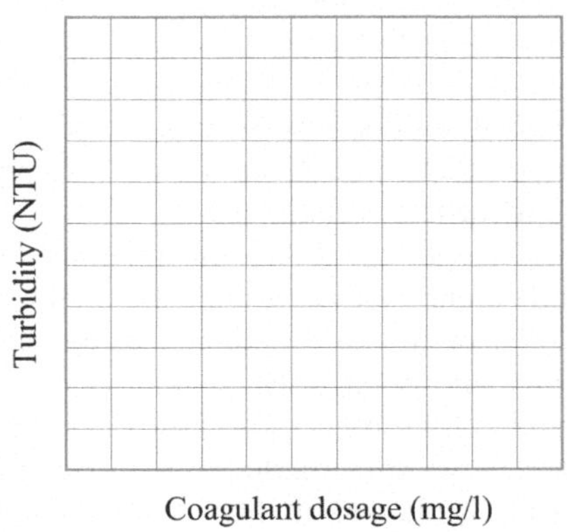

Figure 8.7 Example log sheet for jar test results.

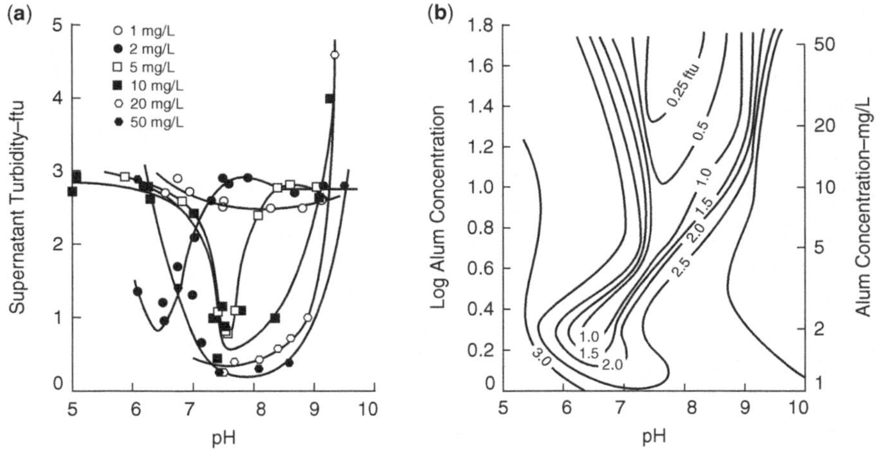

Figure 8.8 Results from jar test (settling) experiments. (Bratby, 1981).

By repeating the procedures described in the preceding sections for a variety of coagulants, the performance of various coagulant types may be evaluated. By comparing supernatant quality (for example) for each type, a final choice of the most appropriate coagulant to use (based on desired effluent quality, economics, availability etc.) may be made.

8.3 USING THE JAR TEST TO EVALUATE SETTLING

The traditional test for the setting characteristics of flocculant suspensions is the column settling test described by Camp (1946) and Conway and Edwards (1961). However one issue with this test, particularly with fragile floc suspensions of the type dealt with here is the transfer of the flocculated suspension from the flocculation reactor to the column without breakup of flocs and without appreciable settling before the test is commenced.

Despite attempts to devise an apparatus to overcome these difficulties, Bratby (1981) found that the results obtained from the column settling test were markedly inferior to those obtained using the jar test reactor and by taking samples from one point only. With flocculant settling samples, column test results should exhibit higher overflow rates for given desired effluent qualities, than results of tests with single point abstraction; the contrary results obtained by Bratby were explained on the basis that the aforementioned difficulties were not overcome. Hudson (1973) had also pointed out the reliability of jar test settling analyses.

Therefore, it is concluded that for settling tests with fragile flocculating suspensions, the jar test reactor with samples taken from one point only yields reliable results for design purposes.

The procedure is as follows:

- After primary coagulant is added (at the correct dosage and pH as determined previously) together with the optimized dosage of flocculant aid, if used, the stirrers are turned down to slow speed and the sample flocculated for 15 to 30 minutes, depending on the water.
- During this time, after a period of 5 minutes slow stirring, two samples are taken at a depth of 100 mm with the bent pipette, or the stopcock, as described earlier (the depth is not critical, as long as it is consistent throughout, and the sampling depth recorded).
- After the flocculation period has elapsed, the stirrers are switched off and the sample allowed to stand. Samples are then taken at the same depth at suitable periods, determined by observation of the suspension under study. For very slow or fast settling suspensions, the time increments can be increased, or decreased, respectively. As an initial guide, the sampling times could be 1.0; 1.5; 2.0; 3.0; 4.0; 5.5; 8.0; 11; 16 and 25 minutes after the stirrers are stopped (based on an approximate $\sqrt{2}$ geometric progression).
- These samples (say 5 to 10 ml volume) are then analyzed for turbidity
- The mean settling velocity of the flocs may then be determined by standard procedures, as follows (see, for example, Camp, 1946):
  - As described above, samples are taken at one depth, Z at different times, t.
  - Any particles with settling velocities great enough to carry them past the sampling point within a time t_i will not be present in the sample taken at time t_i. Hence all particles in the sample have settling velocities less than or equal to Z/t_i $(=v_i)$.
  - If Y_i is the fraction of particles in the original suspension with settling velocity less than v_i, then $Y_i = C_i/C_0$, where C_i is the concentration of particles in the sample taken at time t_i.
  - If Y_0 is the fraction of particles with velocity less than v_0 then the fraction $(1-Y_0)$ of particles will be completely removed.
  - For slower settling particles, they will be removed in the ratio v_i/v_0 and the fractional removal of these particles will be:

    Fractional removal = $\Sigma(v_i/v_0) \times$ (fraction of particles with settling velocity v_i).
  - The total removal is thus $Y_T = (1 - Y_0) + \dfrac{1}{v_0}\int_0^{Y_0} v_i dY$
  - The settling test will yield a distribution curve for the particles in the suspension as shown below. The integral expression corresponds with the shaded area in the Figure, and may be determined graphically.

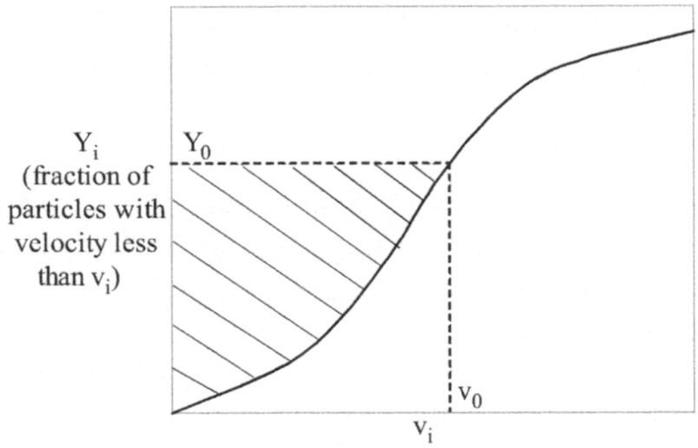

- If the total fractional removal is expressed as a percentage then for various values of v_0 (the theoretical overflow rate) the corresponding percentage removals may be determined.
- A plot of percentage removal versus overflow rate is made.
- For a desired effluent quality, the corresponding overflow rate may be determined, since $C_{eff} = C_{inf} \cdot (1 - \%Removal/100)$
- Because of non-ideal flow in settling tanks, design overflow rates will likely be approximately 25-percent lower than determined by this procedure.

Figure 8.9 shows the results of batch settling tests, relating overflow rates and coagulant dosages required for given effluent water qualities. For a given effluent quality in terms of turbidity, the overflow rate that could be applied increased as the alum dosage increased. At certain alum dosages in each case, further dosage increases did not improve the settling rate.

8.4 EVALUATING FLOCCULANT AIDS

8.4.1 General

Flocculant aids, usually polyelectrolytes, are normally added after destabilization by metal coagulants to speed up the aggregation rate or to strengthen flocs formed by metal coagulants.

Because polyelectrolytes are added in this application to enhance flocculation after destabilization is complete, it is not very meaningful to analyze, for example, supernatant quality, which is usually appropriate when evaluating primary coagulants. Packham (1972) described a procedure based on the settling velocity distribution of the suspension. When evaluating flocculant aids, the advantage in utilizing a method based on settling velocity over other methods such as evaluating floc size, density, permeability, compressibility, shear strength or filterability is that, to quote Packham (1972) *coagulant (flocculant) aids are flocculants which are added to a system that is already flocculating; small differences in timing and in mixing therefore can have very large effects …. floc is very highly compressible and its properties are undoubtedly affected very considerably by any handling operation.* By evaluating settling velocities, it is possible to devise a test procedure which produces flocs under carefully controlled conditions and which eliminates the necessity of any transfer operation. To lend support to Packham's suggested methodology, other researchers (Benedek & Bancsi, 1976, 1977; Mueller & Burbank, 1964) point out that settling velocity determinations give by far the most meaningful results in this application.

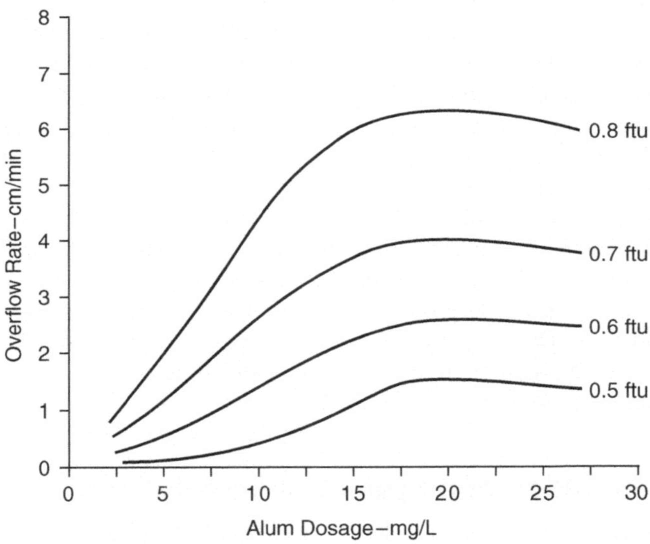

Figure 8.9 Batch settling test results. Optimum pH values used throughout. Average raw water turbidity = 3.24 NTU (Bratby, 1981).

8.4.2 Initial choice of flocculant aid

At the commencement of a testing program using polymeric flocculants, it is likely that a variety of polyelectrolyte samples will be obtained from those manufacturers who can safely guarantee regular delivery in the future. Initial choices of polyelectrolyte brands will probably be influenced by such factors as whether the chemical is manufactured in the particular country or whether it is imported, delivery costs, and so on. Having compiled a reasonable short list of brands, there will be a range of products, each having a particular basic chemical constituent e.g. polyacrylamide, polyamine, etc.

A useful first step in evaluating each range is to take a representative sample of each ionic group. For example, a particular range of polyelectrolyte products is divided into three groups: cationic, nonionic and anionic respectively. With information of charge density from the particular manufacturer, the cationic and anionic groups have charge densities ranging from 5 to 30% and 10 to 40% respectively. By carrying out experiments with a cationic type of charge density 30%; nonionic type of charge density 0% and anionic type of charge density 30%, an indication of the appropriate ionic group for the particular application is obtained. By further carrying out tests for the remainder of the appropriate ionic group i.e. for different molecular weights within that group, compilation of a final short list of products is possible.

A preliminary elimination of unsuitable types of flocculant aids – especially when a large number are to be tested – is most suitably carried out on a subjective but consistent visual basis with the jar test apparatus. The best products within each group can then be compared later. This subjective assessment of the performance of the products can be in terms of floc formation time, floc size, and observed rate of settling. Results can be reported in terms of best (1), second best (2), and so forth (for example, see Bratby, 1981).

After all the products in all the charge groups are tested, and after those products exhibiting no beneficial effect are eliminated, a final comparison of the short-listed products is possible. The most meaningful method of comparison is to calculate dosages for each respective product in terms of equal costs per unit volume of water treated. It is also useful to run a series of tests for a range of cost-equivalent dosages, starting with the highest dosage allowable in terms of human health, for whichever product gives the maximum effect.

8.4.3 Preparation of polyelectrolyte solutions

When preparing different stock solutions of flocculant aid (each of a particular type of polyelectrolyte), a method proposed by Gale (1977) more readily facilitates optimization in terms of performance and cost. All solutions should be prepared such that they are each of the same cost per unit volume. (This same procedure should also be followed when evaluating sludge conditioners.) As a starting point in calculating solution strengths, the maximum recommended strength for the least expensive polyelectrolyte to be tested may be used with the strengths of the remaining polyelectrolytes calculated according to their respective costs.

When adding suitably diluted polyelectrolyte solutions to the jar test samples, a check should be made that the diluted solution concentrations added to samples at no time exceed those recommended by the manufacturer. For example, with many products, the highest concentration recommended before addition to samples is 0.01%.

A suitable method of preparing powder or microbead polyelectrolyte stock solutions is as follows: To a dry 200 ml bottle add the required quantity of polyelectrolyte followed by 3 ml of methylated spirit, methanol or ethanol, in order to wet the material. Pour on rapidly 97 ml distilled water and *immediately* agitate vigorously for 10 to 15 seconds. Continue shaking the contents of the bottle periodically for a further 30 to 60 minutes. With solid grade polyelectrolyte products the highest stock solution concentration should not be greater than 5 g/l. As described above, this concentration is appropriate for the cheapest product in the range of products to be tested. With products of higher cost, the stock solution concentration would decrease.

Volumes of stock polyelectrolyte solutions prepared as above are usually sufficient for most laboratory testing programs. However, if larger volumes are required, 500 ml or 1-liter of distilled water may be placed on a magnetic stirrer, say, and the required quantity of solid grade polymer added slowly to the vortex formed while stirring; addition of the chemical should take place over a one to two minute period. Stirring should continue, after polymer addition, at a moderate rate for one to one-and-a-half hours.

With liquid grade polyelectrolytes, the method of preparation is simpler: The required quantity of polyelectrolyte should be added to a jar and distilled water stirred in by hand to obtain a homogeneous solution. The volume of distilled water added makes up the total volume to 100 ml. The highest concentration stock solution should not exceed 100 g/l.

8.4.4 Experimental procedure

The jar testing technique is as follows:

- After primary coagulant is added (at the correct dosage and pH as determined previously), various dosages of polyelectrolyte are added during the rapid mix stage, approximately one minute after the primary coagulant is added.
- The errors which may result from using jars of small size when evaluating polyelectrolytes have been pointed out earlier.
- After a further 30-seconds rapid mixing the stirrers are turned down to slow speed and the sample flocculated for approximately 15 minutes.
- During this time, after a period of 5 minutes slow stirring, two samples are taken at a depth of 100 mm, as described earlier.
- After the flocculation period has elapsed, the stirrers are switched off and the sample allowed to stand. Samples are then taken at the same depth after periods of 0.5; 1.0; 1.5; 2.0; 2.5; 3.0; 4.0; 6.0; 8.0; 12; 20; 30 and 40 minutes settling.
- Possibly the most appropriate analysis carried out with samples that have been treated with metal coagulants before the addition of a flocculant aid is to measure the metal. However, this is not essential, particularly for less sophisticated laboratories.
- Turbidity in most cases is satisfactory. Absolute turbidity values are likely of interest at the lower end of the scale only. Throughout the settling test, relative turbidity values (as opposed to absolute values) are really all that is required.

- Following the above procedure, the mean settling velocity of the flocs may then be determined following the procedures explained previously.
- By determining the effective settling velocity as above for metal coagulant only and metal coagulant plus various dosages of different polyelectrolytes, a quantitative assessment is possible of the relative merits of each combination. Furthermore, the method may be used to assess the relative merits of various chemical addition times, sequences of addition and so on.

Since the settling velocity is related to other parameters such as floc size, density, permeability, compressibility, etc., the above test has meaning for most intended purposes (Packham, 1972; Gale, 1977).

Figure 8.10 presents an example of settling test results, conducted as described. In this case, beyond a polymer dosage of 0.15 mg/l, there was little improved performance.

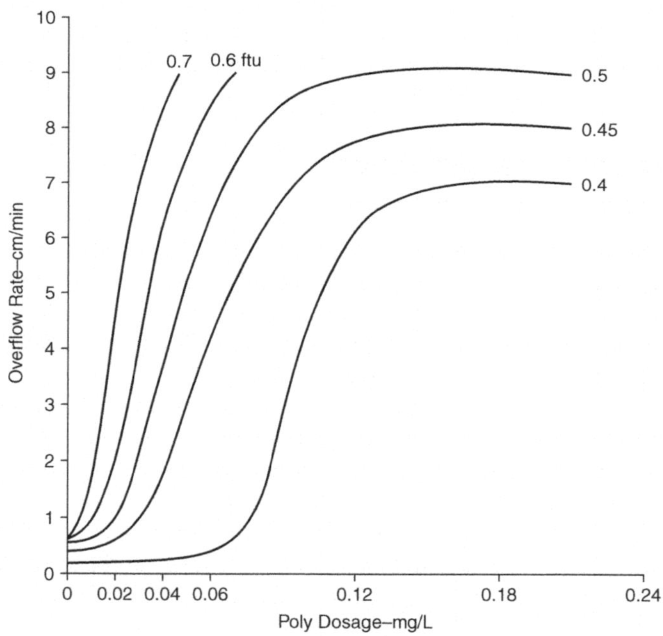

Figure 8.10 Polymer dosage optimized in terms of settling. Raw water turbidity 2.26 NTU; Alum dosage 3 mg/l at pH 6.4; Rapid mix for alum 500 s^{-1} for 10 s; Preflocculation 100 s^{-1} for 10 min; Rapid mix for poly 100 s^{-1} for 30 s; Flocculation 18 s^{-1} for 15 min; Turbidity values are for effluent quality. (From Bratby, 1981.)

Figure 8.11 compares the settling performance of alum only, and alum plus polymer. In both cases, the alum was at the optimal dosage and pH. Rapid mix and flocculation conditions were also optimized. The enhanced overflow rates possible when polymer was used with alum indicated the superiority of this treatment option.

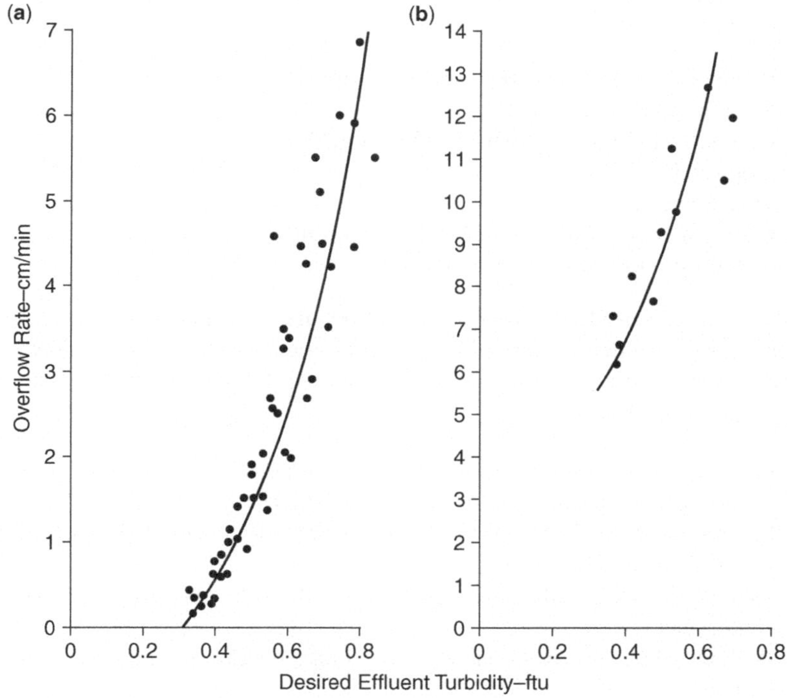

Figure 8.11 Settling characteristics at optimized conditions for alum only, and alum plus polymer. (a) Experiments based on alum dosage of 15 mg/l at pH 7.6; Rapid mix for alum 500 s^{-1} for 10 s; N_0(average) 1.2 NTU; Raw water turbidity 1.45 NTU; G_1, G_2, G_3 80, 25, 12.5 s^{-1}; Total flocculation time 18 min. (b) Experiments based on alum dosage 3 mg/l at pH 6.4; Rapid mix 500 s^{-1} for 10 s; Preflocculation 100 s^{-1} for 10 min; Poly dosage 0.15 mg/l; Rapid mix for poly 100 s^{-1} for 10 s; N_0(average) 0.87 NTU; Raw water turbidity 2.29 NTU; G_1, G_2, G_3 100, 50, 30 s^{-1}; Total flocculation time 12 min. (From Bratby, 1981.)

8.5 EVALUATING SLUDGE CONDITIONERS

8.5.1 General

With sludge conditioning, the testing procedures are oriented towards evaluating two principal properties: (a) filterability of the conditioned sludge and (b) floc strength.

8.5.2 Experimental procedures

The traditional procedure for measuring sludge filterability is to determine specific resistance (Swanwick & Davidson, 1961; Gale, 1977). The procedure can be carried out under vacuum or under pressure. For the vacuum method the apparatus required is as follows:

- Buchner funnel with woven wire mesh insert
- Vacuum pump and gauge
- Graduated cylinders (250 ml is often suitable)
- Stopwatch
- Thermometer
- Filter papers (e.g. Whatman No. 17 or similar)
- Various tubings, stopcocks, clamps etc.

Sludge samples are first conditioned with polymer or other conditioning agent. A blank sludge sample is normally tested, without any conditioner added. The jar test stirrer could be used to add the conditioner, but another method has also provided good results:

- Place the sludge sample volume into a clean jar (a sludge volume of 500 ml is normally sufficient – in this case use a 1-liter jar)
- In a second clean 1-liter jar place the correct volume of conditioner solution for the dosage to be tested
- Take the jar with sludge and rapidly invert the contents into the jar with the conditioner
- Without pause, invert the jar with the sludge-conditioner mix back into the first jar
- Repeat this procedure twice more, giving a total of four mix-pours
- Note that, depending on the sludge, if the number of mix-pours is continued beyond about six to seven, the flocs formed start to break down, indicating excessive agitation of the conditioned sludge.

The test procedure is as follows:

- With the Buchner apparatus assembled, place the woven wire mesh in the funnel, then the filter paper on top. Wet the filter paper with distilled water, and adjust the vacuum to 369 mm mercury (15-inches Hg).
- Pour the conditioned sludge sample into the Buchner funnel (first measuring temperature, and obtaining a sub-sample for solids determination)
- Gradually open vacuum so that vacuum is achieved within about 30 seconds
- Record the filtration rate by measuring the volume of filtrate collected in the graduated cylinder during various time intervals. The intervals do not need

to be constant but can be gradually widened to account for the deceleration of filtration rate

- When 10 or more sets of readings have been taken, turn off the vacuum. Check that there is still liquid in the sludge in the funnel and that the sludge sample hasn't dried (if so, the final readings will be in error)
- Plot Volume (ml) as abscissa and Time/volume (sec/ml) as ordinate, as shown below:

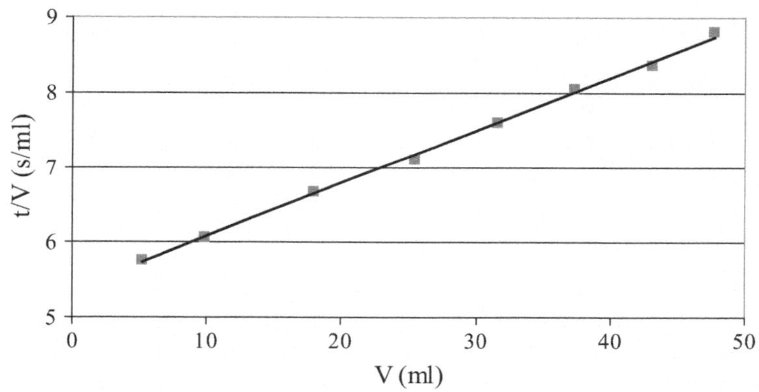

- Determine the slope (b) of the plot
- The apparent specific resistance is given by

$$r = \frac{2A^2Pb}{c\mu} \quad s^2/g$$

where

A = area of filter paper = area of filter cake (cm^2)
P = vacuum or pressure applied (g/cm^2)
b = slope of t/V versus V plot (s/ml^2)
c = mass of solids deposited onto the filter paper per unit volume of filtrate passed through the paper (g/ml) where $c = (C \times 10)/[1000 - (C \times 10)]$ and C = concentration of sludge sample (%)
μ = water viscosity (g/cm.s)

- For the above example, $A = 38.5\ cm^2$; P = vacuum of 369 mm Hg = 500 g/cm^2; $b = 0.071\ s/ml^2$; $c = 0.0332$ g/ml; temperature = 20°C and $\mu = 0.01$ g/cm · s. Therefore, $r = 3.17 \times 10^8\ s^2/g$ (or 3.11×10^{12} m/kg)
- Note that, during the test, if the initial portion of the t/V versus V plot is non-linear, then the values need to be re-calculated based on a starting point at the beginning of the linear portion, as exemplified below:

Experimental Data			Adjusted Data		
t	*V*	*t/V*	*t*	*V*	*t/V*
30	7.2	4.17			
60	13.0	4.62			
90	18.2	4.95	30	5.2	5.77
120	22.9	5.24	60	9.9	6.06
180	31.0	5.81	120	18.0	6.67
240	38.4	6.25	180	25.4	7.09
300	44.6	6.73	240	31.6	7.59
360	50.3	7.16	300	37.3	8.04
420	56.1	7.49	360	43.1	8.35
480	60.7	7.91	420	47.7	8.81

An alternative to using vacuum for the specific filtration test is to use pressure. Since many sludges are compressible, the pressure (or vacuum) applied during the test makes a difference. When evaluating sludges for filter presses for example, it is worthwhile to carry out the test at the same pressure as the press. Bratby *et al.* (1996) used a pressurized filter apparatus, with a circular cut-out of the same filter cloth used in the presses. A high pressure nitrogen gas cylinder was used to pressurize the samples to 1,550 kPa (225 psig). Apart from this, the procedure followed that described above.

An alternative, which has the advantage of rapidity, is the Capillary Suction Time (CST) apparatus developed by the Water Research Centre, U.K. (Baskerville & Gale, 1968) and shown in Figure 8.12. The method relies on the suction pressure created by capillaries within absorbent paper and is, essentially, a constant pressure filtration. Whereas specific resistance is a fundamental parameter, the CST is not. It can, however, be directly correlated with specific resistance for a particular sludge sample.

The CST apparatus automatically measures the time for unit area of paper to become saturated with filtrate when a fixed area of paper is exposed to the suspension. The methodology presented here is that given by Gale (1977):

- Place a filter, paper rough side uppermost, on top of the plexiglass stand, and place the block having the stainless steel probes in it probe-side downwards on top of the filter paper. Light hand pressure will ensure good contact between the probes and the filter paper
- Select either the 10 mm diameter reservoir cylinder for fast sludges, or the 18 mm cylinder for slow sludges, and insert it into the block. Rotate the cylinder, applying slight downward pressure to ensure even contact with the filter paper

- Switch on the apparatus
- Press the reset button and note that the counter is at zero and the reset lamp alight
- Pour the conditioned sludge sample into the reservoir cylinder
- Liquid from the sludge is absorbed by the paper in an elliptical wetting pattern. When the liquid front reaches the first pair of probes, the counter starts and timing commences. When the front reaches the third probe, timing ceases and the counter indicates the CST in seconds.
- Fresh paper must be used for each test
- A double layer of paper can be used to slow down the rate of travel of the liquid front, for very fast draining sludges.

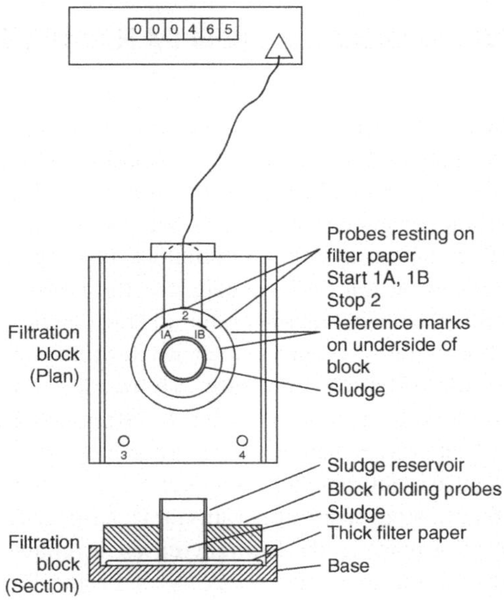

Figure 8.12 Capillary suction time apparatus. (From Gale, 1977).

When initially preparing a short list of products to be tested, and when preparing stock conditioner solutions, identical considerations to those described for the case of flocculant aids are applicable here. When adding suitably diluted conditioner solutions (such that each concentration is half the previous concentration, as before) it is recommended that 100 ml of conditioner solutions of varying strengths are always added to 500 ml of sludge sample. The solutions are added to the sludge at a constant speed of stirring on a laboratory high speed stirrer (1000 rpm) during a 10 second period of mixing – or the mix-pour method described earlier is used. After mixing, a portion of the sample is analyzed for filterability using either specific resistance techniques or the CST apparatus. The remainder of the sample is returned to the stirrer and further periods of mixing of 10, 30 and 60 seconds applied. After each period a sample is taken and the filterability determined as before. Total stirring times applied to each sample to which filterability determinations are made are 0, 10, 40 and 100 seconds (excluding the initial mixing period). By determining filterability for a range of stirring periods, an indication of floc strength is achieved.

As regards initial choice of conditioners, in some cases a reliable indication is available since the most effective conditioners are likely those that are also the most effective as flocculant aids, in the treatment of the same water.

8.6 OPTIMIZING FLOCCULATION PARAMETERS

8.6.1 General

The previous sections of this Chapter have described methods by which primary coagulant type, dosage, pH and flocculant aid are optimized for a particular water. Further tests which could be conducted on the water are those by which rapid mixing and flocculation parameters are optimized. For the design of continuous completely mixed flocculation systems, continuous type pilot plant trials have often been used to glean design data. Because of the considerable expenditure involved in constructing the necessary apparatus and the lengthy testing procedures involved, design of flocculation systems (especially for relatively small plants) has often been based on rule-of-thumb procedures. However, it is possible to develop reliable design information using laboratory batch testing procedures.

The advantages of laboratory batch testing procedures are that normally they involve simple inexpensive equipment and a decreased testing period, compared with the relatively complicated, costly apparatus and lengthy methodology required for continuous type pilot plant trials. In many cases, a reliable laboratory batch testing technique would eliminate the necessity of basing design on subjective asessments, especially for relatively small plants.

However, a problem arises when batch tests are considered for the design of rapid mixing and flocculation basins. Particle aggregation kinetics are of first order; and for this reason the direct application of batch test data in such cases is appropriate for plug-flow reactors only. Whereas batch test data of mixing intensity (i.e. G values) are valid for design, retention time values cannot be taken from batch test results and directly applied to the design of completely mixed continuous systems. With rapid mix units this is usually not a problem if plug-flow type units are adopted. However, it is a problem with completely mixed flocculation devices.

In order to interpret batch test results for use in designing flocculation devices, it is necessary to recognize the appropriate (different) formulations pertaining to batch systems, and to completely mixed, continuous systems.

For a given destabilized water, flocculation reactor geometry and agitation device, the principal parameters governing the effective degree of flocculation are the retention time and velocity gradients applied. These parameters influence the rate and extent of particle aggregation and the rate and extent of breakup of these aggregates. From an analysis of equations describing both the aggregation and breakup phases of the flocculation process, it is possible to obtain design data for continuous, completely mixed flocculation systems using batch test results only.

The development of flocculation kinetic equations and the theoretical link between batch and completely mixed, continuous flocculation performance, was presented earlier in Chapter 7.

The validity of the theoretical link between batch type (or plug flow) behavior and continuous type performance has been verified by comparing batch and continuous pilot scale data (Bratby *et al.* 1977). Results predicted from the batch test data were verified with a three tank in-series continuous pilot plant setup. It was concluded that batch tests are valid for designing continuous flocculation systems – with certain stipulations.

A factor of crucial importance is the applicability of bench scale results to the design of full scale units, and whether such results can be applied directly or whether scaling-up factors are required. Argaman and Kaufman (1968, 1970) and Argaman (1971) found good correlation between results obtained from laboratory scale continuous type experiments and those obtained from a full scale plant treating the same type of water. They concluded that laboratory scale continuous experiments may be successfully extrapolated to full scale design.

The good correlation between batch test results and continuous experiments demonstrated by Bratby *et al.* (1977) indicates that batch data may also be directly applied to full scale design provided the geometry and stirring mechanisms of the laboratory and full scale reactors are similar.

Flocculation batch tests are strictly applicable only to turbulent agitation devices. Examples of relevant appurtenances are baffled chambers, rotating devices (horizontal– and vertical–shaft stake-and-stator types, turbine impellers, etc.), and reciprocating–blade devices. The batch flocculation apparatus is essentially a stake-and-stator device. Oldshue and Mady (1979) compared dangling plate and turbine flocculation devices in a full scale application and found no significant difference in performance for equivalent conditions of mixing intensity. Since the flocculation devices compared by these workers were very different, it is likely that the results obtained in batch tests using a stake-and-stator type of device are generally applicable to other turbulent agitation devices.

Batch tests are not appropriate for contact flocculation, which occurs in solids contact blanket clarifiers, since some continuous form of experimentation would be necessary. It would be difficult, if not impossible, to generate a similar concentrated floc environment in batch reactors.

Bratby *et al.* (1977) demonstrated that the use of small round jars in the jar test, even with a stake-and-stator arrangement, seriously over-predicted flocculation performance obtained in square tanks. It was surmised that the corners in square reactors would produce dead pockets that affect performance. Therefore, although jar test apparatus could likely be used in the tests described below, care should be taken that the jars are square, and that they are properly baffled.

The apparatus and methodology required for the batch tests are described below.

8.6.2 Apparatus

The apparatus previously used for evaluating flocculation performance is shown in Figure 8.13. It consists of a reactor square in plan with a capacity of 3 liters and fitted with stators. Vertical paddles are driven by a motor, the rotational speed of which is varied using a variable speed control. The root mean square velocity gradient, G, in the reactor was determined by installing a torquemeter on the drive shaft with a stroboscopic light arrangement. The relationship between the measured torque and G is given by:

$$G = \sqrt{\frac{2\pi \cdot t \cdot N \cdot g}{60 \cdot V \cdot \mu}} \quad s^{-1} \tag{8.1}$$

where

t = torque measured at drive shaft (gm-cm)
N = speed of rotation of impeller (rpm)
V = volume of water in reactor (cm^3)
G = weight per unit mass (9.81×10^{-3} N/g)
μ = viscosity of water (10^{-7} N · s/cm^2 at 20°C

The measured G values were also predicted well by the procedure of Camp (1955) if k, the ratio of speed of rotation of water to speed of rotation of stirrers is taken as 0.24 (k was found to have a constant value of 0.24 for the full range of rotational speeds applied). Camp's expression is as follows:

$$G = \sqrt{\frac{1.24 \cdot 10^5 \cdot C_D (1-k)^3 S_S{}^3}{V \cdot \mu} \sum A \cdot r_b{}^3} \quad s^{-1} \tag{8.2}$$

where

C_D = drag coefficient = 2.0 for flat blades
S_S = speed of rotation of shaft (rev/s)
kS_S = speed of rotation of water (rev/s)
V = volume of water in reactor (m^3)
A = cross sectional area of each blade in the plane perpendicular to direction of motion (m^2)
r_b = distance of center of each blade from shaft (m)
μ = absolute viscosity of water (=10^{-3} N · s/m^2 at 20°C)

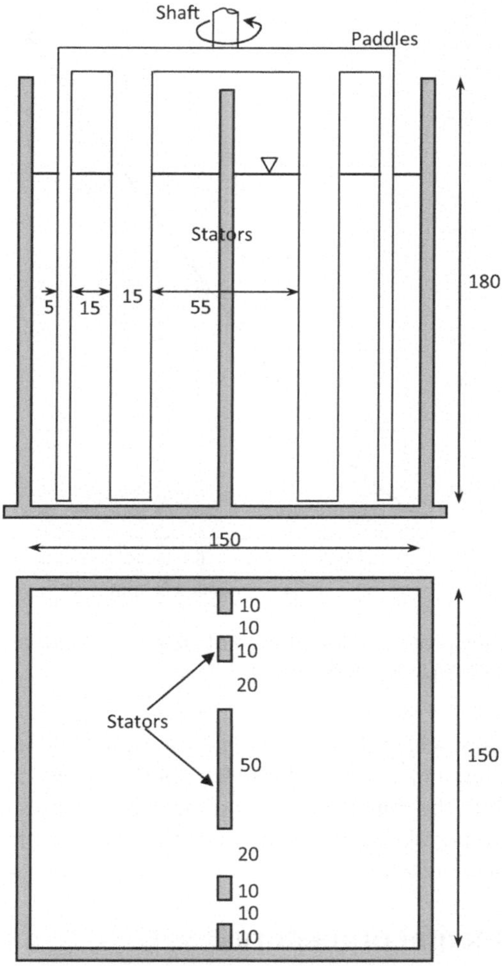

Figure 8.13 Reactor for evaluating flocculation performance. Dimensions in mm (Bratby, 1977, 1981).

Measured values of G with respect to impeller rotational speed (for the reactor shown in Figure 8.13) are shown in Figure 8.14.

Samples can be taken from the apparatus using the sampling pipette shown in Figure 8.6. Alternatively, a sampling tap could be installed 100 mm below the initial water level with a short length of flexible tubing and laboratory pinch clamp. If using a sample tap, corrections need to be made to the sampling depth from the water level to the tap each time a sample is taken to account for liquid drawdown. This correction is not required if the sampling pipette is used.

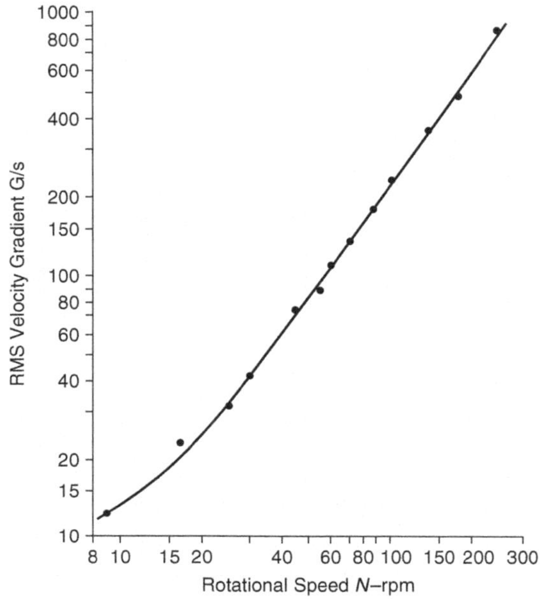

Figure 8.14 G values produced for a range of impeller rotational speeds for reactor shown in Figure 8.13 (Bratby, 1977, 1981).

Chemicals are best added during rapid mixing using a syringe with plunger so that the coagulant aliquot is injected rapidly, approximately at mid-water depth, as near as possible to the impellers. This arrangement ensures, as far as possible, adequate initial mixing of coagulant and flocculant aid chemicals into the whole volume of water in the reactor.

8.6.3 Experimental procedure

The experimental procedure is as follows: The reactor is filled to the 3-liter mark with the raw water to be tested. The impeller is set to the rotational speed giving the *G* value found optimal for rapid mixing for the water under investigation. Without specific information, the impellers could be set to approximately 300 rpm, giving a *G* value of approximately 1,000 s^{-1}.

Immersing a pH probe in the reactor contents, the required volume of alkaline or acid reagent is added to adjust the initial pH to the value found optimal previously. The correct primary coagulant type and dosage (previously determined) are then added. The final pH after coagulant addition should be the optimum pH found previously in the jar test experiments.

After say 30-seconds of rapid mixing, flocculant aid, if required, is added. Note that all chemicals used, their dosages, sequence of addition, time lag between dosing, etc., follow exactly those found optimal in the jar test experiments described previously.

After, say, a further 30-seconds rapid mixing the stirrers are slowed down to the speed corresponding to the flocculation G value set for that particular experiment and stirring is continued for the time also set for that test. After the given time has elapsed, stirring is stopped and the flocs allowed to settle for a given time – the time beyond which no significant improvement in supernatant quality is evident (for example, 30 minutes).

A sample is then withdrawn from the sampling tap and analyzed for some parameter representative of the number of primary particles which remain unflocculated (and, therefore, are present in the supernatant sample). The actual parameter used depends on the type of water under test. For example, for a turbid water the appropriate parameter is turbidity. For a colored water UV spectrophotometric measurements are probably the most appropriate. For a water destabilized with a metal (primary) coagulant and subsequently with a coagulant aid (e.g. a polyelectrolyte) analyzing for the metal (aluminum or iron) may give the most meaningful results.

The apparatus and general procedure is also used to evaluate rapid mix conditions. Figure 8.15 shows an example of results obtained. In this example, a rapid mix G value of 625 s^{-1} was optimal. For rapid mix experiments, performance is best monitored in terms of the ratio (turbidity after rapid mixing, N_0)/(turbidity after flocculation and 30 min settling, N).

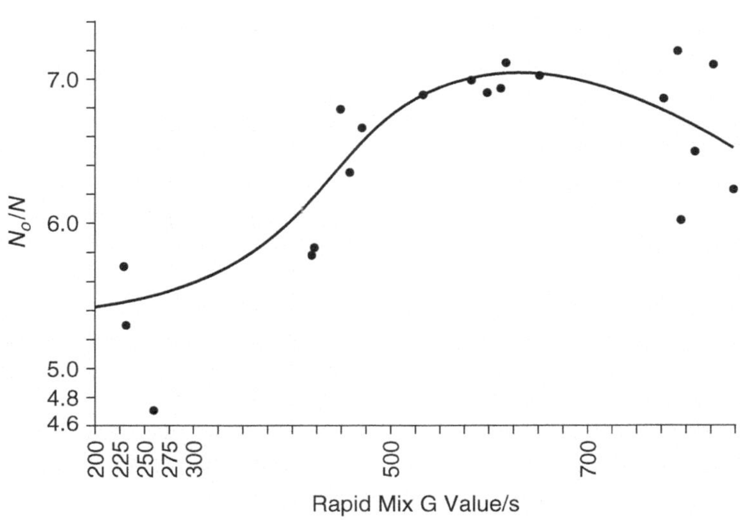

Figure 8.15 Optimizing rapid mix G value in terms of flocculation performance (Bratby, 1981).

Figure 8.16 shows that in the above example, there was no significant difference in flocculation performance for rapid mix times in the range 0 to 120 seconds. Data points are presented as a spread of 8 seconds greater than the recorded time because the time required for adding coagulant chemicals is about 8 seconds for the apparatus shown in Figure 8.13. Beyond approximately 120 s, flocculation performance progressively deteriorated and flocs formed during flocculation were smaller and settled poorly.

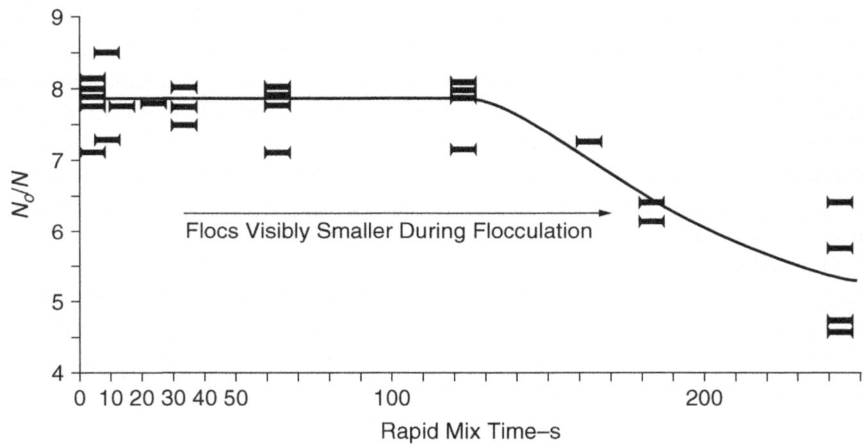

Figure 8.16 Optimizing rapid mix time. (Bratby, 1981).

In the case of adding sequential primary coagulants, or adding flocculant aids, the same methodology should be used to determine the appropriate conditions for pre-flocculation. Figure 8.17 shows that for the example above, pre-flocculation before flocculant aid addition should be approximately 100 s^{-1}. Further tests identified the optimum preflocculation time at 100 s^{-1} to be 10 minutes.

The procedure may be modified slightly to assess the effect of sludge recycling on flocculation rates, where this is of interest. After primary coagulant addition and the rapid mixing stage, known quantities of settled sludge and/or filter backwash are introduced into the sample before flocculation.

For a range of G values and flocculation periods, series of curves of best fit are drawn through plotted values of N_0/N versus flocculation period, T (as shown in Figures 8.18a and b). Note that N_0 = initial supernatant turbidity (or color or metal concentration, etc.) after destabilization and after settling, *but at zero flocculation period*, and N = supernatant turbidity (etc.) after a flocculation period, T, and after settling.

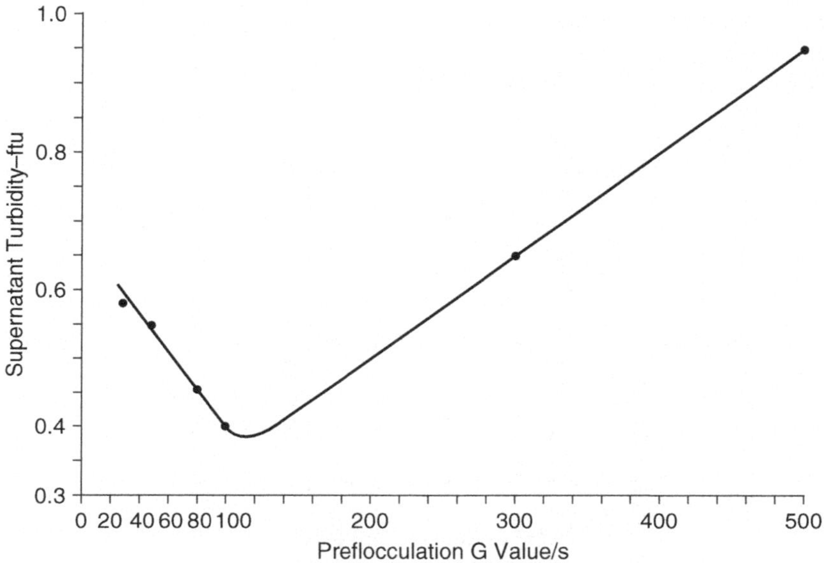

Figure 8.17 Optimizing pre-flocculation *G* value. (Bratby, 1981).

From curves such as those shown in Figures 8.18a and b, the flocculation and breakup constants, K_A and K_B are determined by fitting Equation 7.25 to the curves. Assuming that the number concentration of primary particles, *n*, is proportional to supernatant turbidity (etc.), *N*, equations 7.23 and 7.25 may thus be rewritten as

$$\frac{N_0}{N_m} = \frac{(1 + K_A G \cdot T/m)^m}{1 + K_B G^2 T/m \sum_{i=0}^{m-1} (1 + K_A G \cdot T/m)^i} \tag{8.3}$$

for a continuous, completely mixed in-series system of flocculation reactors, and

$$\frac{N_0}{N} = \left[\frac{K_B}{K_A} G + \left(1 - \frac{K_B}{K_A} G\right) e^{-K_A G \cdot T}\right]^{-1} \tag{8.4}$$

for a batch or plug flow system
where

N_0 and N are as given above;
N_m = supernatant turbidity of *m* th reactor in series of completely mixed flow-through reactors with total retention time in m reactors, *T*;

K_A = flocculation constant;
K_B = breakup constant;
G = root mean square velocity gradient.

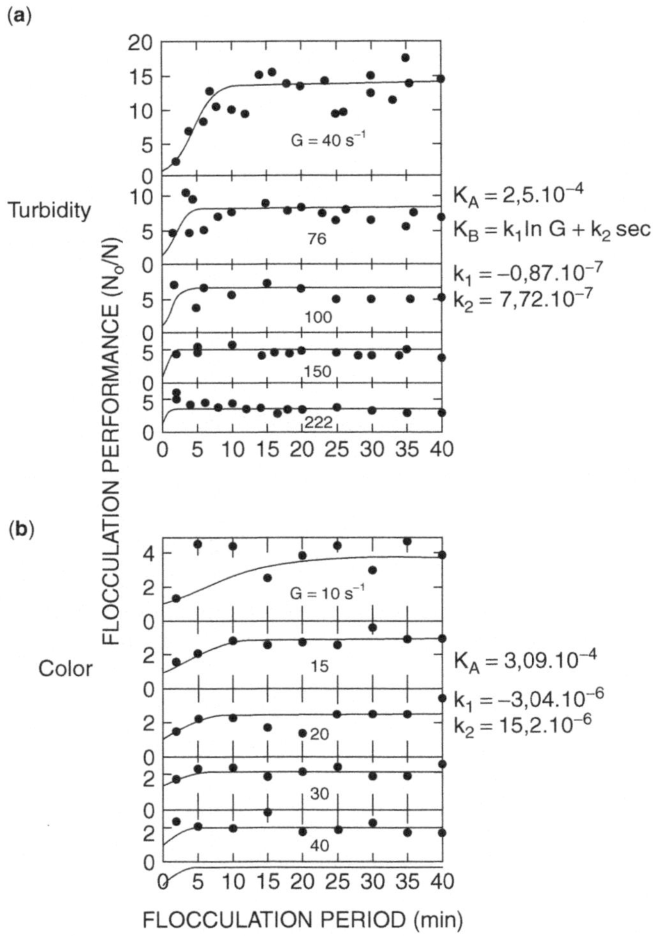

Figure 8.18 Flocculation performance, N_0/N for a range of flocculation times and *G* values for (a) turbid water and (b) colored water. (From Bratby *et al.* 1977).

8.6.4 Analysis of data

There is an apparent inadequacy of the assumed breakup mechanism giving rise to an over-prediction of floc erosion at high *G* values (Bratby *et al.* 1977). This is reflected in the fact that when fitting Equations (8.1) and (8.2) to experimental data,

K_B does not remain constant for each G value applied. In terms of interpreting the laboratory data for design, however, this does not present a problem since the same variation in K_B is evident during continuous operation. In the light of the foregoing, the following methodology is recommended when determining the values of K_A and K_B from experimental data:

(1) From the horizontal portion of the curves (the experiments should be conducted such that a significant horizontal portion is produced) the ratio K_B/K_A is determined from

$$\frac{K_B}{K_A} = \frac{1}{G \cdot N_0/N} \tag{8.5}$$

(2) Taking successive points on the curves of best fit for each G value, values of K_A are determined from a rearrangement of Equation (8.2), i.e.

$$K_A = \frac{1}{G \cdot T} \ln \left[\frac{1 - \frac{K_B}{K_A} G}{\frac{1}{N_0/N} - \frac{K_B}{K_A} G} \right] \tag{8.6}$$

(3) From the K_A values calculated for each point on the curve for each particular G value, the mean value is calculated and, from the values of K_A/K_B for each curve, the values of K_B for each curve are calculated.

(4) Plot the values of K_B as ordinate and ln G as abscissa. The value of K_B at any G value is given by a relationship of the form:

$$K_B = k_1 \ln G + k_2 \tag{8.7}$$

where k_1 and k_2 are constants for a particular water.

For Figure 8.8a, which shows the results for a kaolin clay suspension using aluminum sulfate as coagulant, K_B is given by

$$K_B = (7.72 - 0.87 \ln G) \times 10^{-7} \text{ seconds.}$$

For Figure 8.8b, which shows results for a water with zero initial turbidity but colored with humic substances and using aluminum sulfate as coagulant, K_B is given by

$$K_B = (15.2 - 3.04 \ln G) \times 10^{-6} \text{ seconds.}$$

For the turbid (kaolin clay) suspension, K_A was found to be 2.5×10^{-4} and for the colored water $K_A = 3.09 \times 10^{-4}$.

(5) From the values of K_A and K_B determined as above, the curves are re-plotted and, if necessary, the values of K_A and/or K_B adjusted slightly to give the best fit to all the curves at each G value. Note that the K_A value is constant throughout the G values but K_B may vary with G as demonstrated above.

It should be noted that the empirical relationship between the breakup constant, K_B and velocity gradient, G, given above is for convenience in analysis. The need for such a relationship illustrates the complexity of the breakup mechanism. However, for design purposes, the empirical "fix" described above is satisfactory since the results from batch tests may be applied directly (taking cognisance of the different formulations pertaining) to continuous completely mixed systems (Bratby *et al.* 1977).

Having determined the constants K_A and K_B, the design of a flocculation system for the particular (destabilized) water under test may be made following the procedures described earlier. As stated previously, during the coagulation testing stage a number of chemical combination alternatives will probably be available. These may now be finally optimized, having assessed the optimal flocculation performance of each combination. A final comparative economic assessment is carried out where operating (chemical) costs are weighed against capital (flocculation) costs.

Figure 8.19 shows the settling characteristics of two pretreatment alternatives, tested under optimized flocculation conditions determined as above. Based on optimized flocculation conditions, and the associated overflow rates for given settled water qualities, a rational decision can be made regarding pretreatment to be applied in the plant.

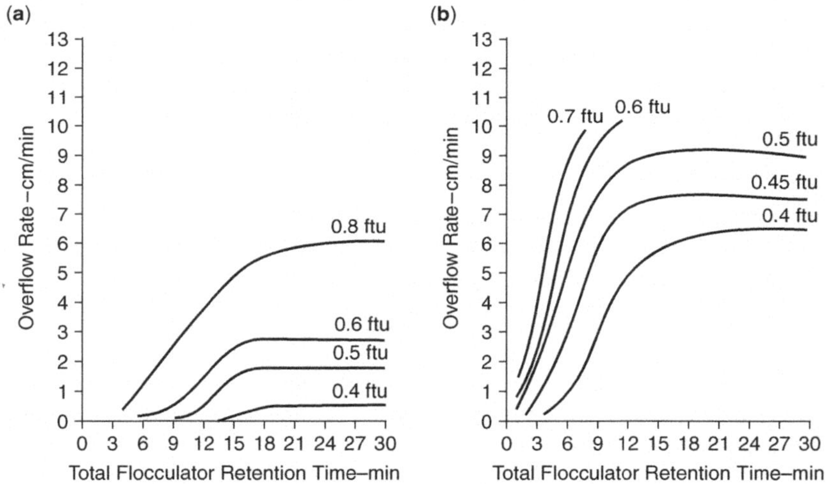

Figure 8.19 Settling characteristics for a range of total flocculator retention times T ($T_1 = T_2 = T_3 = T/3$). (a) Alum used alone, dosage 15 mg/l at pH 7.6; rapid mix 500 s^{-1} for 10 s; N_0(average) 1.9 NTU; raw turbidity (average) 1.5 NTU; G_1, G_2 and G_3 80, 25 and 12.5 s^{-1} resp. (b) Alum with polymer. Alum dosage 3 mg/l at pH 6.4; rapid mix for alum 500 s^{-1} for 10 s; preflocculation 100 s^{-1} for 10 min; poly dosage 0.15 mg/l; rapid mix for poly 100 s^{-1} for 10 s; N_0 (average) 0.88 NTU; raw water turbidity (average) 2.32 NTU; G_1, G_2 and G_3 100, 50 and 30 s^{-1} respectively. (From Bratby, 1981).

Table 8.3 presents the values of K_A and K_B determined by various researchers. The temperature effects investigated by Jin (2005) demonstrate that low temperatures significantly reduce the rate of flocculation, and produce more fragile flocs that are more susceptible to breakup.

Table 8.3 Flocculation kinetic parameters determined by various researchers.

System		K_A	K_B	Reference
Kaolin – Alum		0.45×10^{-4}	1.0×10^{-7}	Argaman (1970)
Kaolin (40 NTU) – Alum		2.5×10^{-4}	$*4.5 \times 10^{-7}$	Bratby (1977)
Natural turbid water – Alum		0.18×10^{-4}	0.8×10^{-7}	Argaman (1971)
River water – Alum	22°C	4.4×10^{-4}	3.0×10^{-7}	Jin (2005)
	4°C	2.1×10^{-4}	8.8×10^{-7}	
	0°C	2.0×10^{-4}	9.8×10^{-7}	
River water – Ferric sulfate	22°C	4.5×10^{-4}	4.2×10^{-7}	
	4°C	2.1×10^{-4}	9.9×10^{-7}	
	0°C	2.0×10^{-4}	10.0×10^{-7}	
Lake water – Alum		1.3×10^{-4}	$*5.3 \times 10^{-7}$	Bratby (1981)
Lake water – Alum + polymer		0.77×10^{-4}	$*5.2 \times 10^{-7}$	
NOM colored water		3.1×10^{-4}	$*40.0 \times 10^{-7}$	Bratby (1977)
Alum-phosphate precipitate		2.8×10^{-4}	3.4×10^{-7}	Ødegaard (1979)
Alum-phosphate precipitate + polymer		2.7×10^{-4}	1.0×10^{-7}	
Lime-phosphate precipitate, pH 11		0.56×10^{-4}	2.4×10^{-7}	

*Average values (K_B values vary with G)

8.7 CONTROL SYSTEMS FOR COAGULATION

8.7.1 Introduction

This chapter has dealt with in some detail the optimization of chemicals to produce required treated water qualities by coagulation operations. An important next step is to apply the optimized dosage(s) and coagulation conditions to the full scale operating plant. This can be a challenge.

Natural waters can be subject to seasonal changes in raw water quality, and required coagulation conditions. These changes could arise from temperature shifts, intense rainfall in the catchment area, very rapid changes in turbidity, seasonal changes in raw water chemistry and organic content, and more.

It is true that the jar test is the best tool for determining what the best coagulation treatment conditions are, under all of these changing raw water conditions – but when raw water quality changes occur very quickly, it can be difficult for operators to glean the information required fast enough to be able to respond in a timely manner.

For this reason, many water and wastewater plants that rely on chemical addition for treatment, have applied on-line automatic control of coagulant dosage.

There are several systems that have been used. These include electrokinetic measurements; floc formation monitors; and redox potential. Of these, redox potential appears to have had limited applications, although some success appears to have been attained with controlling the dosage of ferric sulfate for chemical phosphorus removal in wastewater plants (Kerrison *et al.* 1989). The principle is that the redox controller responds to ferric-phosphate complexes and free ferric to provide a redox potential. A difficulty with this system is that redox control is confounded by fouling of the sensor, as well as background redox readings. For example, whereas control of phosphorus removal appeared to be successful at a trickling filter plant, this was not the case at an oxidation ditch activated sludge plant, where large fluctuations in background redox occurred.

Electrokinetic measurements and floc formation monitoring are discussed in more detail below.

8.7.2 Electrokinetic measurements

In Chapter 2 it was shown that the charge potential at the plane of shear between a moving particle and the surrounding liquid is known as the zeta potential. The usefulness of zeta potential measurements in the control of coagulation–flocculation processes is largely dependent on whether the destabilization mechanism is predominantly due to charge effects, in which case control to a given zeta potential may be feasible. However, in many cases charge effects are not the predominant mechanism, although they may play a role during destabilization.

Therefore, in such instances, not only would adjustment to a given value of zeta potential (for example, zero) not necessarily lead to destabilization but destabilization may not be predicted by any particular value of zeta potential (Ockershausen & Peterman, 1964; Packham & Sheiham, 1977; Stumm & O'Melia, 1968). For example, Black *et al.* (1961, 1965) found that with three separate types of clay turbid water using aluminum sulphate as coagulant, no reliable indication of the optimum conditions for destabilization could be inferred from zeta potential measurements (or more exactly, electrophoretic mobilities – see later).

However, some workers (for example Gupta *et al.* 1975) presenting results of zeta potential measurements have inferred that optimum conditions of coagulant type, dosage and pH occur at the isoelectric point, that is at zero zeta potential. This can be misleading since it considers the destabilization mechanism to be solely that of charge neutralization, which may be a gross simplification in some cases.

Although charge neutralization may not be a prerequisite for destabilization, one may use the occurrence of surface charge response as a means of controlling the performance of certain processes relying on destabilization. For example, in the control of cationic polyelectrolyte dosage prior to a direct filtration facility, the maintenance of a zeta potential of approximately 14 mV successfully controlled the process by using electrokinetic measurements (Letterman & Tanner, 1974).

Bearing in mind that electrokinetic measurements are limited in their ability to predict performance without some prior correlation – using jar test or pilot plant results, success of such a means of control has the prerequisite that the nature of the water quality does not change.

Although operational control of a facility employing cationic polyelectrolytes may be feasible by carrying out electrokinetic measurements, this is not the case with anionic polyelectrolytes applied to predominantly negatively charged suspensions, where such measurements may be ineffective as a means of control. For example, Black *et al.* (1965), investigating the use of polyelectrolytes for the destabilization of a kaolin clay suspension, found that with a cationic polyelectrolyte the response to electrokinetic measurements gave adequate indication of the conditions for destabilization (see Figure 8.20a).

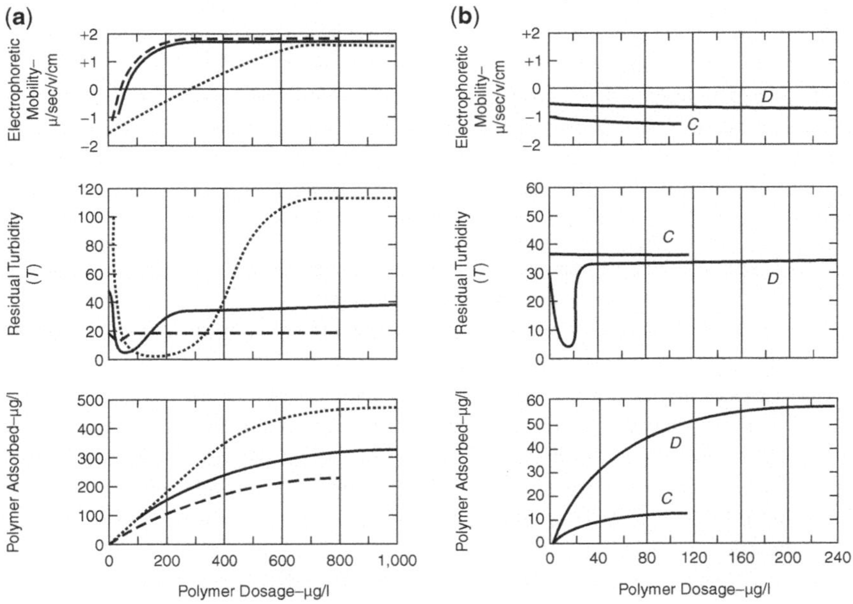

Figure 8.20 (a) Effect of initial clay concentration on destabilization of kaolin clay suspensions with cationic polymer: the dashed, solid and dotted curves represent clay concentrations of 14.9 mg/l, 29.8 mg/l and 73.2 mg/l respectively. (b) Effect of $CaCl_2$ on destabilization of kaolin clay suspensions with anionic polymer: the kaolin suspensions (33.3 mg/l) contain 25 mg/l $CaCl_2$ (curve C) and 250 mg/l $CaCl_2$ (curve D). (From Black *et al.* 1965.)

However, with an anionic polyelectrolyte, although the optimum dosage to give lowest residual turbidity was well defined, there was no corresponding response with electrokinetic measurements (see Figure 8.20b).

Although electrokinetic measurements in research applications can be valuable (for example, Bean *et al.* 1964; Black *et al.* 1961, 1965; Packham & Sheiham, 1977), optimum conditions for destabilization in practical applications are only adequately predicted at present by techniques such as the jar test. The subsequent application of electrokinetic measurements for control purposes is possible after adequate correlation with the results of jar test, pilot filters, or some other control method. However, when waters are subject to significant seasonal changes, as discussed earlier, several jar test correlations would be required.

8.7.2.1 Electrophoresis measurements

Electrophoresis refers to the movement of a charged particle suspended in a fluid induced by an applied electrical force. When a direct current electric field is applied across a suspension containing particles with a net double layer charge, the particles will migrate to the positive or negative pole depending on whether the particles carry a negative or positive charge respectively. Counter-ions surrounding the colloid will migrate in the reverse direction of the particles.

In measurements of electrophoresis, the technique essentially comprises visual observation of the suspension through a microscope and timing a number of particles travelling a given distance under an applied electric field. Figure 8.21 shows examples of electrophoresis instruments. In general the instruments comprise the following (Riddick, 1961; Swope, 1977):

- A stereoscopic microscope with ocular micrometer, 15× eyepieces and 2×, 4×, 6× and 8× (adjusted magnification) objectives
- A special illuminator capable of producing a thin beam of intense blue-white light, with a heat absorbing filter
- A direct-current power supply, continuously variable from 0 to 500 V
- An electrophoresis cell equipped with platinum-iridium electrodes. The electrophoresis cell consists essentially of a clear plastic block with a small polished tube extending through the centre. The tube is connected by two tapered ports to two solution chambers, within which are positioned the platinum-iridium electrodes
- A cell holder consisting of a thick and highly reflective mirror directing the light 45° upward through the tube of the cell
- An interrupted type cumulative reading electrical timer, reading in 0.1 seconds.

Before carrying out measurements of electrophoresis it is necessary to calibrate the particular cell in use to compensate for the effect of electroosmosis (Black & Smith, 1966; Riddick, 1961). A liquid in contact with a surface such as glass may possess a potential with respect to the latter. When an electric field is applied to such a system, there will be movement of the liquid along the walls of the cell in one direction and along the centre of the cell in the other direction. There is a layer at a fixed distance from the walls of the cell where the flow of liquid is zero. It is at this stationary layer where particles should be observed and electrophoretic mobilities measured.

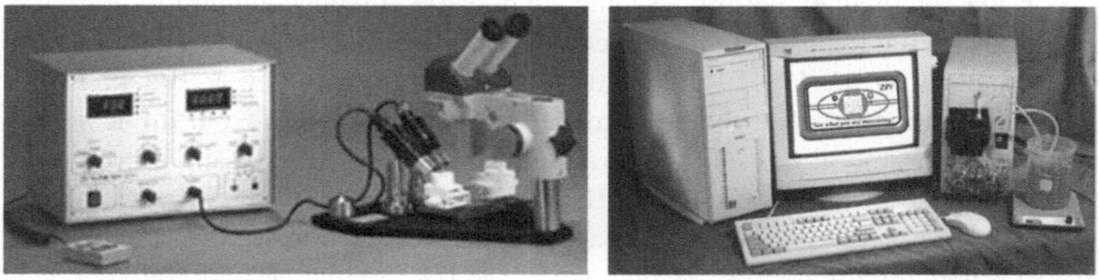

Figure 8.21 Electrophoresis measurement. (Left: courtesy Zeta-Meter, Inc.; Right: courtesy ZPi instruments, inc.).

Black and Smith (1966) describe a procedure for calibrating a rectangular type of electrophoresis cell of the Briggs type, which consists of measuring the mobilities of a standard suspension (polyvinyl chloride microspheres) at different positions in the cell, plotting the mobility profile and thus determining the mean velocity for the cell. The position in the cell at which the mean velocity occurs defines the stationary layer. Riddick (1961) gives a value of 15% tube diameter as defining the position of the stationary layer from the wall of a tube type electrophoresis cell.

Some modern electrophoretic instruments have automatic particle tracking devices, making them suitable for automatic on-line dosing control.

Electophoretic mobility commonly has units of microns/second per volts/centimeter (μ/s/V/cm). μ/s is a measure of the velocity of the tracked particles, and V/cm expresses the electrical field strength. Electrophoretic mobilities may be converted to zeta potential using expressions related to particle size and electrolyte concentration (Edwards & Monke, 1967; Loganathan & Maier, 1975). However, because of difficulties in assigning values to various terms in the appropriate equations, the calculated zeta potential may differ significantly from the true value (Committee Report, 1971). For this reason many workers express results solely in terms of electrophoretic mobility, rather than convert to zeta potentials.

With regard to control of destabilization of clay turbidity, the zeta potential or mobility value giving lowest residual turbidity depends on the type and concentration of clay, primary coagulant and flocculant aid used. In general, when using aluminum sulphate, the values of electrophoretic mobility corresponding to minimum residual turbidity range from −0.5 to 2.0 μ/s/V/cm (Black *et al.* 1961, 1965). However, in some cases, although residual turbidity has a sharply defined minimum at optimal operating conditions, no accompanying response in electrophoretic mobility value is evident. As previously stated, this would severely limit the utility of electrophoresis as a means of control without prior verification that changes in destabilization efficiency are accompanied by appreciable changes in zeta potential.

Using polyelectrolytes to destabilize clay dispersions, electrophoretic mobility values appear to be within a range similar to those using metal coagulants – for example, see Figure 8.20.

For the removal of natural organic color using metal coagulants, optimum conditions giving the lowest residual color generally coincide with mobilities near to zero (Black *et al.* 1961) – but seldom exactly zero (Singley & Black, 1967). Using a cationic polyelectrolyte to remove organic color (humic and fulvic acids), electrophoretic mobilities of approximately zero have also been reported as coinciding with lowest residual color (Narkis & Rebhun, 1977).

Sharp *et al.* (2005) investigated the applicability of zeta potential as a control tool on two waters high in natural organic matter (NOM) content. They found that with both waters the window of zeta potential for minimum residual dissolved organic carbon (DOC) was approximately –5 to +5 mV. This optimum range was the same for both alum and ferric sulfate used as coagulants. These workers concluded that for the waters tested, control by electrokinetic measurements was feasible.

During the lime-soda softening process, using flocculants to aid aggregation of the precipitates formed, the response of electrophoretic mobility to the condition of optimum flocculation of precipitates is dependent on the type of flocculant used (Black *et al.* 1961): With activated silica, data shows that a sharp peak of the already negative mobility occurs at the point of lowest residual turbidity. When using a high molecular weight anionic polyacrylamide, mobility data does not define optimum flocculation conditions. With potato starch, which is a nonionic polymer, mobilities are completely unaffected by flocculant dosage. With a high molecular weight cationic polyelectrolyte, optimum dosage is identified with zero mobility or the point of charge reversal of the suspension.

Destabilization of colloidal hydrous manganese dioxide with calcium occurs when the extent of Ca^{2+} sorption is sufficient to reduce the electrophoretic mobility of MnO_2 from approximately –3 to –1 μ/s/V/cm (Posselt *et al.* 1968). The mobility remains at –1 μ/s/V/cm even at very high calcium dosages, indicating that neither charge reversal nor restabilization occurs.

From the above, one point is clearly evident: In none of the examples cited have the conditions describing optimum performance coincided with zero particle surface charge. This lends support to statements made earlier that the mechanism of destabilization is usually more complex than that of simple charge neutralization.

8.7.2.2 Streaming current measurements

A disadvantage cited of electrophoretic measurements is that they are relatively lengthy and subjective, requiring visual observation and timing of individual particles (Smith & Somerset, 1971; TeKippe & Ham, 1970) – although modern instruments do incorporate automatic particle tracking.

Streaming current measurements, on the other hand, have the advantage of speed and are not as subjective as tests of electrophoresis. Furthermore, with the streaming current technique, results obtained are immediately in terms of the average for the system.

Streaming current devices measure the net residual charge surrounding particles in water. These particles have a net negative surface charge. Coagulants such as alum, ferric salts or cationic polymers surround the particles with cations or positive charges and reduce or reverse the net surface charge. When in control mode, the streaming current monitor alters the coagulant dose until a pre-set end point is reached. This set point is determined by jar tests and confirmatory streaming current measurements. Figure 8.22 shows a laboratory streaming current monitor used to correlate jar test results with streaming current.

Figure 8.22 Laboratory streaming current instrument (Courtesy Chemtrac).

The streaming current technique involves placing a sample in a special cylinder containing electrodes at the top and bottom. A loose fitting piston is then partially submerged in the sample and reciprocated along its axis. Particles in the sample are temporarily immobilized on the piston and cylinder surfaces due to electrostatic and Van der Waal forces. As the water is moved back and forth, mobile counter ions surrounding the particles are sheared from the particles and are moved down to the electrodes.

Movement of liquid and counter-ions through the annular space between piston and cylinder creates a (streaming) current since the piston and bore assume the charge characteristics of the charged particles. The piston is connected to a synchronous motor and is driven at a set number of cycles (approximately 200 strokes per minute) thereby generating an alternating current. By connecting the motor to a synchronous rectifier and the electrodes to an amplifier, the alternating current is read on a direct current meter, with the polarity of output adjusted to indicate surface charge.

The streaming current is related to zeta potential as follows:

$$i \propto ZD/N$$

where

i = streaming current
Z = zeta potential
D = dielectric constant
N = viscosity

As with electrophoresis, although results may be expressed in terms of zeta potential, because of the uncertainty in the numerical values of certain constant terms, results are usually presented in terms of (positive or negative) streaming current. Figure 8.23 shows an on-line instrument used for coagulant control.

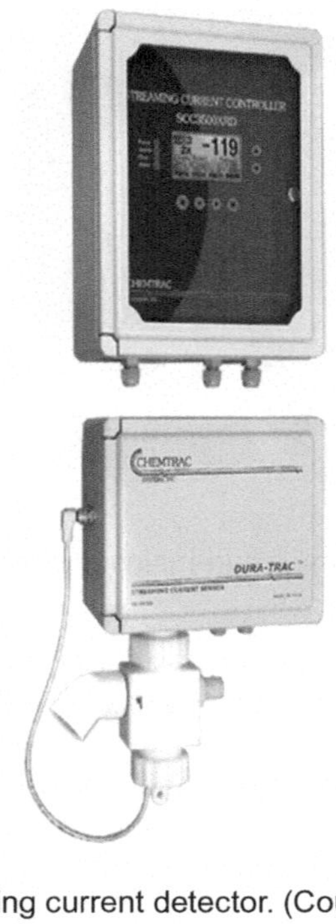

Figure 8.23 On-line streaming current detector. (Courtesy Chemtrac).

Although streaming current measurements are relative and results depend on the particular streaming current detector used, results obtained with turbid and colored waters, and with the lime treatment of raw sewage and of wastewater from the paper industry – all verified with jar tests (Smith & Somerset, 1971) – indicate that optimum conditions for destabilization approximately coincide with the inflection point on the streaming current titration curve produced with the particular coagulant used. However, Briley and Knappe (2002) found that optimum conditions for algae removal with ferric sulfate did not coincide with the inflection point – but with values close to the zero point of charge.

Generally, for clay (turbid) and colored waters, streaming current values corresponding to optimum destabilization conditions range from near zero to –12 μA. However, the optimum setpoint value must be determined by confirmatory jar tests for a particular water.

The sampling point is critical for streaming current detectors (Lind, 1994). If it is too close to the point of coagulant addition, the coagulant may not have enough time for charge neutralization. If it is too far downstream, large flocs will form and results will be distorted by the breakup of the floc particles.

Streaming current has been applied successfully for polymer dosage control for thickening and dewatering operations (Bryant, 1991; Abu-Orf & Dentel, 1997, 1998; Wang, 1999). The sampling point for control is usually the filtrate or the centrate, which still contains residual particulate material. The polymer, or inorganic conditioner is varied on a trial-and-error basis and the streaming current response monitored, until ideal thickening or dewatering conditions are obtained. This value of streaming current then becomes the setpoint. In the work of Abu-Orf and Dentel, optimum dewatering in terms of cake concentration and filtrate or centrate clarity coincided with near-zero point of charge. They found that the streaming current controller responded adequately to changes in sludge flow rate and in changes to the polymer make-up solutions. Proper maintenance to clean the probes was required regularly to maintain satisfactory operation.

Sensor fouling and sample line clogging were issues also reported at water plants by Dentel *et al.* (1989). However, in general these workers reported satisfactory operation if adequate preventive maintenance was practiced.

8.7.2.3 Colloid titration

The colloid titration technique essentially comprises neutralization of a charged colloidal dispersion by an equivalent amount of a colloid of opposite charge (Kawamura *et al.* 1967; Shelton & Drewry, 1973; TeKippe & Ham, 1970; Verma & Chaudhuri, 1978). An advantage of this technique over both electrophoresis and streaming current measurements is the ease and rapidity with which measurements are carried out, but without the need for specialized equipment.

The procedure is as follows:

- add 100 ml sample to a 250 ml Erlenmeyer flask (prepare also a 100 ml distilled water blank)
- add 5 ml of 0.001 N methyl glycol chitosan (a standard positive colloid) to the sample
- add two drops of 0.1% toluidine blue indicator to the contents of the flasks
- titrate with 0.001 N polyvinyl alcohol sulphate (standard negative colloid), using a microburette. The end point is indicated by a subtle color change from light blue to bluish-purple. Since the color change is time dependent, titration should be carried out as quickly as possible. The above procedure should also be carried out for the blank.

Depending on the suspension concentration, the above chemical quantities may need to be adjusted so that sufficient methyl glycol chitosan is added to neutralize the original suspension and leave a measurable excess.

Colloid charge, conveniently expressed as milli-equivalents per litre of positive or negative colloid charge, is given by:

$$\text{Charge (meq/l)} = 1000 \cdot (A - B) \cdot N/V \tag{8.8}$$

where

A = volume of polyvinyl alcohol sulphate (PVSK) added, ml
N = normality of PVSK (e.g. 0,001 N)
B = volume of PVSK added to blank, ml
V = volume of sample, ml.

From comparative data there appears to be good correlation between the results of colloid titration and electrophoretic measurements, up to the point describing zero surface charge (Kawamura *et al.* 1967; Verma & Chaudhuri, 1978). Beyond this point, mobility and colloid charge values diverge markedly. However, whether this divergence is due to inaccuracy in colloid charge or electrophoretic mobility determinations, it does not deter from the value of the colloid titration technique as a method of control of coagulation, in those cases where such a method of control is applicable; As pointed out before, prior correlation with, say, jar test results are in any case necessary.

It appears that with clay turbidity removal, the colloid charges corresponding to optimum conditions of destabilization are within the range -25×10^{-4} to $+300 \times 10^{-4}$ meq/l (Kawamura *et al.* 1967; Verma & Chaudhuri, 1978). In the case of color removal from surface waters, the range appears to be from -200×10^{-4} to $+200 \times 10^{-4}$ meq/l (Kawamura *et al.* 1967).

It should be noted that although colloid titration measurements are carried out readily, the limitations of electrokinetic measurements in general, as discussed earlier, are still applicable. For example, TeKippe and Ham (1970) found that with the coagulation of silica particles with alum, the results from colloid titration experiments had no correlation with either jar test or pilot scale filtration experiments.

This further supports previous statements indicating that the applicability of electrokinetic measurements for controlling the coagulation process depends very largely on the type of suspension. The colloid titration technique is also not readily adaptable for use on-line for automated coagulation control.

8.7.3 Monitoring floc formation

This class of coagulation control monitors floc formation, rather than the surface charge carried by the particles. Research and applications have been ongoing for some time with this approach (for example: Gregory & Nelson, 1984, 1986; Gregory & Kayode, 1988; Gregory, 1988; Guibai & Gregory, 1991; Ching *et al.* 1994; Chou *et al.* 1997; Tambo *et al.* 1997; Gregory & Duan, 1999; Gregory, 2004). In general, the advantage of monitoring floc formation is that optimum coagulation results can be obtained 10 to 50 times faster than by conducting conventional jar tests. The monitoring of floc formation also allows the rate of floc aggregation to be readily determined – information not readily available with jar testing.

A commercial instrument, based on developments at University College, London is shown in Figure 8.24. The principles of operation are as follows (see preceding references):

Figure 8.24 Photometric dispersion analyzer (courtesy Rank Brothers Ltd.).

The monitoring technique is based on measurement of the intensity of light transmitted through a flowing suspension. This is monitored by a detector which provides an output consisting of a large direct current (DC) component and a smaller fluctuating alternating current (AC) component. The DC values are typically of the order volts, and the AC values typically of the order millivolts.

The DC component is a measure of the average transmitted light intensity, which is dependent on the turbidity of the flowing suspension. The fluctuating AC component arises from random variations in the number and size of particles in the light beam. Because the suspension is flowing, the illuminated sample is continuously renewed. Without flow, there is a fixed sample in the light beam and the AC component becomes negligibly small.

The AC signal component is separated from the much larger DC value, and can be amplified by any desired factor. The root-mean-square (RMS) value of the amplified AC signal is derived electronically, and the ratio (RMS/DC) determined. This ratio is referred to as the Flocculation Index (FI), and is strongly influenced by the state of aggregation of the particles. FI always increases as aggregation occurs, and decreases when aggregates or flocs are broken.

A benefit of using the RMS/DC ratio is that it is practically unaffected by fouling of the optical surfaces (Gregory, 2004), which makes the system of considerable value as an on-line monitor.

Plots of FI with time exhibit a sigmoid curve, as shown in Figure 8.25.

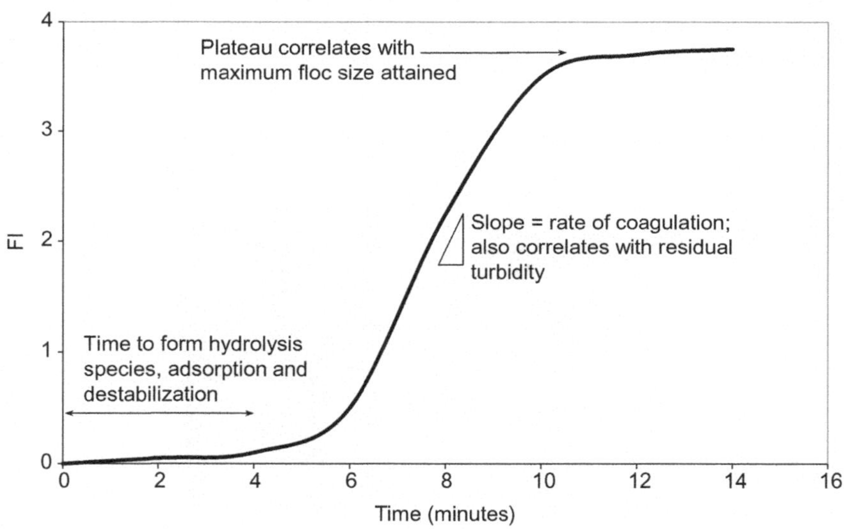

Figure 8.25 Characteristic FI versus time plot (see Gregory, 2004).

In the case of metal coagulants, the lag time exhibited by the FI-time plot represents the time for the formation of hydrolysis species and their adsorption to and destabilization of particles. In the case of cationic polymers, there is generally a longer lag time due to the slower adsorption of polymer molecules to the particles. However, the plateau achieved with polymer is generally higher than with metal coagulants, indicating larger and stronger flocs (Gregory, 2004).

The slope of the rising portion of the FI-time curve indicates the rate of coagulation, and the plateau correlates with the maximum floc size attained. The latter depends on the strength of the flocs, and the intensity of flocculation applied.

Residual turbidity during treatment correlates well with the maximum slope of the FI-time curve, as well as with the plateau value. This allows effective application of the system for monitoring coagulation operations.

Ching *et al.* (1994) showed very good correlation between the rate of increase of FI and residual turbidity at different pH values, using ferric chloride for the coagulation of kaolin suspensions. The results from their series of experiments were also able to detect the occurrence of sweep floc coagulation by a dramatic jump in the rate of increase in FI.

Guibai and Gregory (1991) were able to use DC-time plots to determine the settling rate of flocculated particles in very high turbidity waters. By stopping the flocculator paddles, and continuing to plot DC with time, the time difference between when the stirring was stopped, and the abrupt vertical rise of the DC signal is used to calculate settling velocity, since the depth of the sampling tube below the water surface is known.

The applicability of the technique for monitoring polymer conditioning of sludges for dewatering has been demonstrated by Gregory and Kayode (1988) and Kayode and Gregory (1988). In this case, a specific value such as the FI plateau does not appear to always indicate optimum conditioning. But control was achieved by conditioning to a FI setpoint, determined by trial-and-error. The advantage of the monitor was that responses to polymer dosing were obtained within seconds, allowing rapid feedback control of the polymer dosage.

Another optical floc monitoring approach is shown in Figure 8.26 (Jorden, 2005). As of 2015 this instrument is not available commercially, although various prototypes have been tested successfully. The system relies on the translation of digital images of floc particles into particle size diameters. Floc particle images are captured several times per second and processed in a PC, with proprietary software, to calculate mean particle size.

The system was initially conceived as comprising various optional components. The first is basically a laboratory setup allowing a pre-programmed series of jar tests to be conducted automatically. Up to 50 series of tests may be programmed at a time. A series of automatic chemical dosing syringes allow a wide combination of chemical types and sequences, with pH control. The apparatus comprises essentially a 1-liter reactor vessel, flocculation paddles, a digital camera probe to monitor the development of flocs, and automation hardware and software.

The camera probe monitors floc development continuously so that the increase of floc size with time can be determined. From this the rate of floc growth is determined. From the test series, the maximum floc growth rate determines the optimum coagulation conditions. The optimum determined by this means has been correlated with optimum conditions determined in conventional jar tests, as well as optimum filtered water quality (Ingels *et al.* 2005).

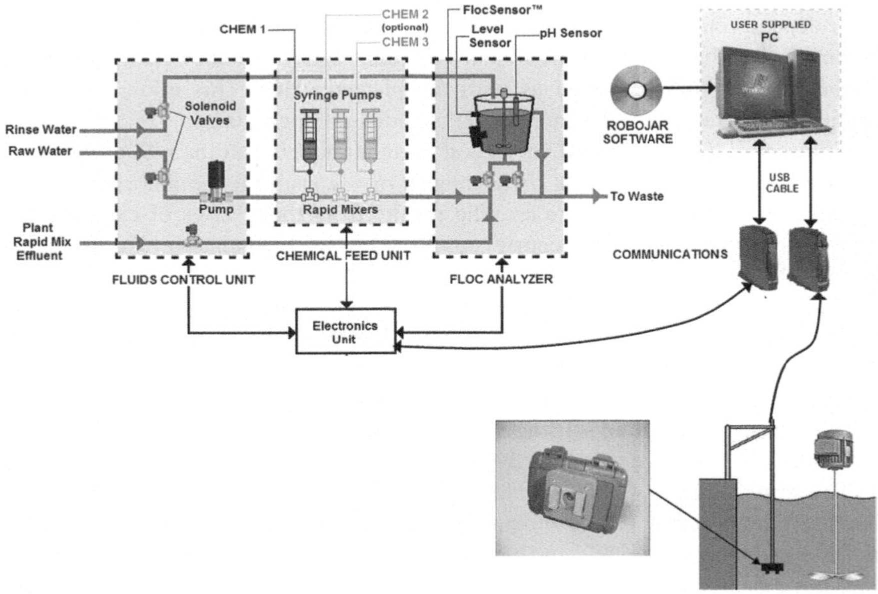

Figure 8.26 Schematic of floc control system. (Courtesy Clearcorp.)

A second component is an extension of the first, shown in Figure 8.26. Here, the system approaches an on-line instrument whereby raw water is automatically fed to the system for automated jar tests. The operator then has a constant feedback on the best coagulation conditions at any time.

A third component, also shown in Figure 8.26 is essentially an on-line monitor of the efficiency of plant applied chemical dosages. It would take coagulated samples after rapid mixing and analyze the efficiency of coagulation. The operator is able to compare the results from the automated jar test carried out on raw water, and compare with the results from the plant applied chemicals, and make adjustments if necessary. By practicing a technique of incremental small changes in chemical dosage in the operating plant, and monitoring the response in the floc control system, slight changes to the optimum coagulation dosage set-point can be determined. This technique is referred to as "dithering" (Jorden, 2005).

The floc monitor camera probe can also be immersed in the process stream – without pumping through a sample line. This avoids delays in transferring water from the process to the analyzer, and avoids the reactions that might occur in the sample lines. This approach also opens up the control system to a wider array of applications – including those in wastewater treatment, such as for chemically enhanced primary treatment (CEPT). Since samples are not passed through small sample lines, the issue of blockages of the lines becomes unimportant.

Although the floc monitor system may not be suitable as yet for on-line automatic dosage control, it could be combined with another system that is calibrated to the results of the floc monitor. One example is to use streaming current as the detector of setpoints optimized by the floc monitor. Another approach would be to apply the ongoing results of the automated jar test technique to a data driven control system.

8.7.4 Data driven control systems

Most of the models describing coagulation performance presented in this book are referred to as physically-based or knowledge-driven models. Examples include the model describing enhanced coagulation for organics removal by coagulation, and the model describing flocculation kinetics. Although such models attempt to explain processes based on our best understanding of the phenomena involved, and the parameters influencing a given outcome, a shortcoming of the models is that the processes are complex and are not fully described by the models. They also require careful calibration and quantification of various "constant" values, the values of which may vary under different conditions.

An alternative approach that is being increasingly developed, is to use data-driven models. These are "black-box" models that are not concerned with a rigorous scientific representation of the phenomena involved. Such models borrow heavily from artificial intelligence techniques. They are based on an incomplete knowledge of how a particular system works and rely on the data describing input and output characteristics.

A simple example of such an approach is to determine, from historical data, the coagulant dosage required for raw water turbidities. The Rio das Velhas water plant in Brazil, for example, experiences severe flash floods that cause instantaneous changes in raw water quality. There is no time to carry out jar tests when such incidents occur. Therefore, the operators developed a simple algorithm to help them respond rapidly to changing raw water conditions (Bratby, 1990):

$$\text{Dosage (mg/l } FeCl_3) = 2.75 \times (\text{raw water turbidity, NTU})^{0.30}$$

Another example is provided by Hudson (1965) where the results from three water plants in the United States and South America were correlated on a semi-log plot as follows:

$$\text{Dosage (mg/l Alum)} = 10.43 \times \log[(\text{raw water turbidity, NTU})/0.53]$$

Such algorithms are simple to use and may provide satisfactory treatment in some cases. However, the algorithm tells the operator nothing about the optimum pH and the required response to other constituents in the water. It is not uncommon during high rainfall runoffs that raw water chemistry parameters change significantly, in addition to turbidity.

A step further would be to associate a number of parameters with a particular output parameter, using a multiple regression analysis. One example is an expression relating suspended solids removal through primary clarifiers in a wastewater treatment plant (Bratby *et al.* 2001):

$$\text{TSS Removal (\%)} = \frac{278}{\exp\left(\dfrac{85}{\text{TSS}_{\text{inf}}} + 0.00047 \cdot \text{SOR} + \dfrac{8.94}{\text{T}}\right)}$$

where:

TSS_{inf} = influent suspended solids (mg/L)
SOR = surface overflow rate (gpd/ft^2)
T = temperature (°C)

Repeating the above exercise at the same plant, but at different times, showed that historical data was reasonably well predicted by the algorithm, but still only approximately. Values of the "constants" needed to be updated each time a significant amount of new plant performance data was available. This showed that although a reasonable approximation to reality was possible, the algorithm was still rather simplistic.

Logsdon *et al.* (2003) describe how a series of such algorithms or charts can be prepared, each one for a set of raw water conditions. Although the base correlation may relate to filtered turbidity, separate charts would be prepared for different temperature ranges, concentrations of TOC, and whether the season is one of rising turbidity, or falling turbidity. The charts also should not be used if the filtered water quality goals are more stringent than the background historical goals on which the charts were based.

The difficulty of predicting optimum coagulation conditions is evident in the following example: the ratio between organic content and turbidity, the humic fraction of the organic matter, and the water temperature, all govern optimum coagulation conditions. All of these parameters change independently of one another on a seasonal basis. For example, lower temperatures require slightly higher pH values or higher coagulant dosages, which is not indicated by the seasonal changes in the other parameters.

When several variables influence coagulation, such as wide seasonal temperature shifts, large changes in TOC and so on, the preparation of a suite of charts becomes impracticable. However, the overriding disadvantage of this approach is that automated coagulation control is not possible.

The increasingly stringent treated water quality requirements that operators are constantly faced with require more sophisticated and complete data driven models, if such an approach is to be adopted for control of treatment plants.

There are various such models. Currently, the most common of these are fuzzy logic based controllers, and artificial neural networks (ANN).

Fuzzy logic allows situations or problems to be described and processed in linguistic terms instead of precise mathematical models. An illustrative example is temperature: linguistic descriptions would be "warm", "very warm" or "cold". Such terms are referred to as fuzzy sets.

In contrast, the traditional method of characterizing temperature is to assign numeric values to the same quantities, such as 30°C, 40°C, or 10°C.

Human beings make decisions based on rules. Although decision rules are not always apparent, human decisions are based on if-then statements. Rules associate ideas and relate one event to another. Fuzzy logic tends to mimic the behavior of humans in decision-making. In fuzzy logic applications, the decision and the means of choosing that decision are replaced by fuzzy sets, and the rules are replaced by fuzzy rules. Fuzzy rules also operate using a series of if-then statements such as in the following example:

IF (water level is medium AND rising slowly) THEN (operate the pump at medium speed). The degree to which the actions are executed corresponds directly to the degree to which the respective conditions are true. A collection of these rules forms the rule base of the fuzzy logic system.

After all physical input values have been converted into fuzzy sets, conclusions are determined, or a hypothesis is generated, from the given input state. This process is known as fuzzy inference. In a fuzzy logic system, the rules define the dependencies between linguistically classified input and output values.

The symbolic control action that results from fuzzy inference cannot be used in a real plant. The linguistically manipulated variables must be defuzzified before control can be implemented. This process of defuzzification involves the calculation of a crisp numerical value based on the symbolic results. In most cases, several rules will apply and, therefore, different control actions will be activated. However, the actuator requires a single crisp value, one that can be calculated by different approaches.

A characteristic of fuzzy logic based systems is that they do require expert knowledge of the system to establish the numerous rule statements. This can be challenging because of a certain subjectivity in identifying cause-and-effect factors.

Liu and Wu (1997) describe the application of a fuzzy logic controller, coupled with streaming current inputs, to control coagulant dosage in a pilot plant system. Control rules were developed and modified based on the results of test pilot plant runs. There were a total of 29 rules based on coagulant dosage, and 43 control rules for sodium hydroxide dosage. Examples of the rules are as follows:

- If streaming current is *too small* then dose is *not large*
- If pH is *large* then dose is *not large*
- If pH is *medium* and streaming current is *not small* then dose is *small–small*
- If streaming current is *large* then NaOH is *medium*
- If pH is *medium* and streaming current is *large–large* then NaOH is *not small*

The control set-point for the streaming current detector was determined through jar tests. The control system was able to control feed turbidities of up to 110 NTU to less than 10 NTU through sedimentation. Further studies were recommended to test the controller under shock load conditions.

Systems based on artificial neural network (ANN) analysis are often preferred, because of the difficulties experienced with setting up the fuzzy logic rules. ANN essentially comprises a complex non-linear mathematical function with many parameters that are calibrated (or "trained").

Neural networks are a class of computational tools that approximate brain processes. Neural networks comprise a set of interconnected nodes, or processing units. The nodes are arranged in a series of layers that are interconnected. Figure 8.27 shows a typical three-layered feed-forward network (adapted from Brion *et al.* 2001).

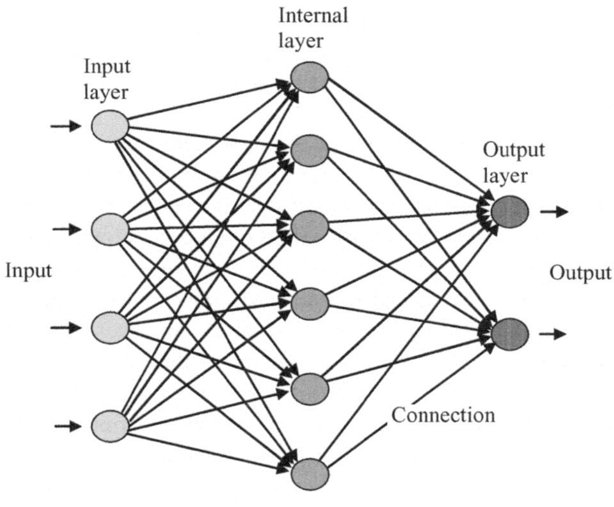

Figure 8.27 Schematic of a three-layered feed-forward neural network.

The number of nodes in the input layer is equal to the number of water quality parameters measured at a given site. Although Figure 8.27 shows four nodes, there may in reality be more. These may include parameters such as raw water turbidity, UV_{254}, temperature, pH, alkalinity, conductivity, and so on. Similarly, the number of nodes in the output layer is equal to the number of target outputs. These may include parameters such as coagulant dosage and pH.

A distinguishing characteristic of neural networks, when compared with simple multiple regression equations is their ability to learn complex systems and to organize themselves internally. This internal organization and the ability to develop generalized representations of the system is referred to as "training" and is crucial to the predictive phase (Brion *et al.* 2001). Obviously the validity and accuracy of the input data is crucial to the success of the training phase. Training comprises an iterative process of presenting the network with the data and adapting the weights of importance of each connection (Masters, 1993).

A common configuration of neural network applications is feed-forward control with back-propagation algorithms to train the network. Zhang and Stanley (1999) describe the coagulation control process shown in Figure 8.28. The neural network process reference model included inputs of raw water color, turbidity, pH, alkalinity, temperature, alum dosage, PAC dosage, clarifier overflow rate, and changes in color, turbidity and alkalinity compared with the immediate preceding sample. The sole output in their case was effluent turbidity from the clarifier. Given a set of these parameters, the model predicted a clarifier effluent turbidity. The model was trained with good operation cases (rejecting inadequate data) and also cases where under- or over-dosing of alum occurred.

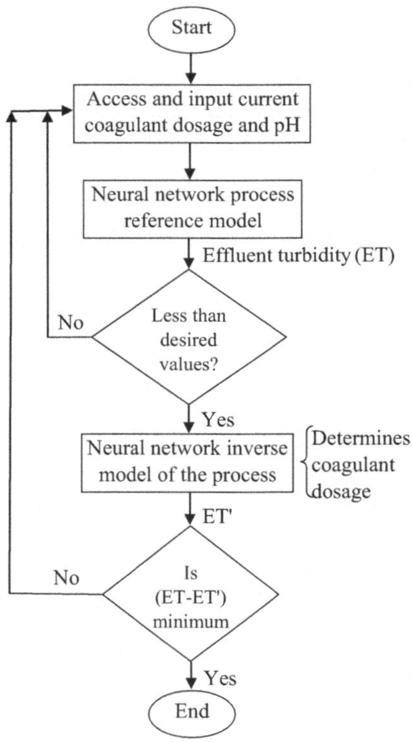

Figure 8.28 Optimization of coagulation using neural networks (adapted from Zhang and Stanley 1999).

The inverse neural network model then took all the same input parameters, with the exception that the effluent turbidity was also an input parameter, and the alum dosage was the sole output parameter. The inverse model was trained with good operating cases only, in terms of effluent turbidity.

An important step in model building is data validation. Invalid data needs to be rejected before calibration of the model. Typically the approach taken is to assign a confidence interval to each parameter data set. Validation comprises a comparison between each data point to a range of values typically obtained in the absence of erroneous measurements. There are sophisticated techniques for examining and rejecting data based on Kohonen maps (Kohonen, 1995).

Baxter *et al.* (2002) and Wierenga and Yavich (2005) describe successful applications of neural network control to pilot and full scale water plant operations. Wierenga and Yavich pointed out that the alternative of using streaming current control was unsatisfactory since coagulation was in the sweep-floc region where streaming current responses and control are poor.

Deveughèle and Do-Quang (2004) also described the control of coagulant dosage using neural networks at a large water plant in Viry, France treating water from the Seine. They described the main stages in setting up the control system as:

- Data processing and statistical data analysis
- Choice of relevant raw water parameters
- Selection of the neural network model
- Building of the model
- Implementation at the Viry water plant
- Comparison of the model predictions with the coagulant dosages set by the operator
- Continuous validation of the model over two years

Data processing used Kohonen mapping techniques. Statistical analyses of the data identified the relevant raw water parameters to be:

- Turbidity (2 to 104 NTU)
- Temperature (4 to 26°C)
- Conductivity (340 to 557 μS)
- pH (7.6 to 8.5)
- UV_{254} (2 to 12)
- Dissolved oxygen (7 to 15 mg/l)

Coagulant used was a basic polyaluminum chloride at dosages typically varying between 2 and 8.5 mg/l.

The historical data set available for building the model comprised about 43,000 data vectors. With the 6 parameters this comprised almost 260,000 data. After removing outliers, corrupt and missing data, about 3,200 vectors (or 19,200 data) remained for use in the model.

One of the benefits found by applying the model was a reduction in coagulant overdosing, which occurred because of over-reactions by the operator to rapidly changing raw water quality.

8.8 REFERENCES

Adin A. and Rebhun M. (1974). High rate contact flocculation-filtration with cationic polyelectrolytes. *Jour. AWWA*, 109–117.

Andreu-Villegas R. and Letterman R. D. (1976). Optimizing flocculator power input. *Jour. Env. Eng. Div., ASCE*, EE2, 251–263.

Argaman Y. and Kaufman W. J. (1968). Turbulence in orthokinetic flocculation. *SERL Report no. 68–5*, Berkeley: SERL, Univ. Calif., Berkeley, Calif., July.

Argaman Y. and Kaufman W. J. (1970). Turbulence and flocculation. *Jour. San. Eng. Div. ASCE*, 96, SA2, Proc. Paper 7201, 223–241.

Argaman Y. (1971). Pilot-Plant Studies of Flocculation. *Jour. AWWA*, 775–777.

AWWA (1992). *Operational control of coagulation and filtration processes.* AWWA Manual of Water Supply Practices, M37, Denver, CO.

Barthelme S. H. (1994). Modelling and optimisation of continuous clarifier operations from batch jar test data. Water SA, 20, 1, 35–48.

Baskerville R. C. and Gale R. S. (1968). A simple automatic instrument for determining the filtrability of sewage sludges. *Wat. Pollut. Control*, **67**, 233–241.

Baxter C. W., Shariff R., Stanley S. J., Smith D. W., Zhang Q. and Saumer E. D. (2002). Model-based advanced process control of coagulation. *Wat. Sci. Tech.*, **45**(4–5), 9–17.

Bean E. L., Campbell S. J. and Anspach F. R. (1964). Zeta potential measurements in the control of coagulation chemical doses. *Jour. AWWA*, **56**, 214–224.

Beard II J. D. and Tanaka T. S. (1977). A comparison of particle counting and nephelometry. *Jour. AWWA*, 533–538.

Benedek A. and Bancsi J. J. (1976). Laboratory evaluation of polymeric flocculants. *Jour. Env. Eng. Div. ASCE*, **102**, EEl, 17–28.

Benedek A. and Bancsi J. J. (1977). Comparative evaluation of commercial polyelectrolytes for flocculating alum precipitated domestic wastewater. *Prog. Wat. Tech.*, **9**, 33–42.

Black A. P., Buswell A. M., Eidsness F. A. and Black A. L. (1957). Review of the jar test. *Jour. AWWA*, 1414–1424.

Black A. P. and Walters J. V. (1964). Electrophoretic studies of turbidity removal with ferric sulphate. *Jour. AWWA*, **56**, 99–110.

Black A. P. and Christman R. F. (1961). Electrophoretic studies of sludge particles produced in lime-soda softening. *Jour. AWWA*, **53**, 737–747.

Black A. P. and Hannah S. A. (1961). Electrophoretic studies of turbidity removal by coagulation with aluminum sulphate. *Jour. AWWA*, **53**, 438–452.

Black A. P. and Willems D. G. (1961). Electrophoretic studies of coagulation for removal of organic color. *Jour. AWWA*, **53**, 589–604.

Black A. P., Birkner F. B. and Morgan J. J. (1965). Destabilization of dilute clay suspensions with labelled polymers. *Jour. AWWA*, **57**, 1547–1560.

Black A. P. and Chen C. (1965). Electrophoretic studies of coagulation and flocculation of river sediment suspensions with aluminum sulphate. *Jour. AWWA*, **57**, 354–362.

Black A. P. and Chen C. (1967). Electrokinetic behaviour of aluminum species in dilute dispersed kaolinite systems. *Jour. AWWA*, **59**, 1173–1183.

Black A. P. and Smith A. L. (1966). Suggested method for calibration of Briggs microelectrophoresis cells. *Jour. AWWA*, **58**, 445–454.

Bratby J. and Marais G. V. R. (1974). Dissolved air flotation. *Filtration and Separation*, 614–624.

Bratby J., Miller M. W. and Marais G. V. R. (1977). Design of flocculation systems from batch test data. *Water SA*, **3**, 173–182.

Bratby J. and Lutzow W. G. (1977). Flocculation characteristics of coloured water derived from Table Mountain catchment. Dept. Civil Eng., Univ. Cape Town, South Africa.

Bratby J. (1981). Optimizing coagulants and flocculant aids for settling. *Jour. AWWA*, 312–318. (Also Erratum: Jour. AWWA, Nov.)

Bratby J. (1981). Interpreting laboratory results for the design of rapid mixing and flocculation systems. *Jour. AWWA*, 318–325.

Bratby J. (1986). Optimizing direct filtration in Brasilia. *Jour. AWWA*, 106–115.

Bratby J. (1990). The importance of the sequence of chemical addition on the coagulation of manganese bearing waters. 21st Fine Particle Socitey Meeting, San Diego, Aug.

Bratby J., Hawkes D. and Tantone F. (1996). Operating experiences with filter press dewatering of water treatment plant sludge. Arizona Wat. & Pollut. Control Assoc., Annual conf., May, Tucson, AZ.

Bratby J., Johnson L., Parker D. and Van Derveer B. (2001). Are activated sludge simulation models useful to operators? Proc. WEFTEC conference, Atlanta, GA.

Briley D. S. and Knappe D. R. U. (2002). Optimizing ferric sulfate coagulation of algae with streaming current measurements. *Jour. AWWA*, 80–90.

Brink D. R., Choi S. I., Al-Ani M. and Hendricks D. W. (1988). Bench scale evaluation of coagulants for low turbidity water. *Jour. AWWA*, 199–204.

Brion G. M., Neelakantan T. R. and Lingireddy S. (2001). Using neural networks to predict peak cryptosporidium concentrations. *Jour. AWWA*, 99–105.

Camp T. R. (1946). Sedimentation and the design of settling tanks. *Trans. ASCE*, **III**, 909.

Camp T. R. (1955). Flocculation and flocculation basins. *Trans. ASCE*, **120**(1), 1–16.

Camp T. R. and Conklin G. F. (1970). Towards a rational jar test for coagulation. *Jour. NOWWA*, **84**(3), 325.

Ching H. W., Tanaka T. S. and Elimelech M. (1994). Dynamics of coagulation of kaolin particles with ferric chloride. *Wat. Res.*, **28**(3), 559–569.

Chou S., Lin S. and Chihpin H. (1997). Application of optical monitor to evaluate the coagulation of pulp wastewater. *IAWQ 7th Int. Workshop: Instrumentation, Control and Automation of Water and Wastewater Treatment and Transport Systems*, July, Brighton, UK.

Clark T. F. (1992). Effective coagulation to meet the new regulations. *AWWA Opflow*, **18**, 1.

Clark T. and Stephenson T. (1999). Development of a jar testing protocol for chemical phosphorus removal in activated sludge using statistical experimental design. *Wat. Res.*, **33**(7), 1730–1734.

Cleasby J. L. (1969). Approaches to a filtrability index for granular filters. *Jour. AWWA*, **61**, 372.

Cohen J. M. (1957). Improved jar test procedure. *Jour. AWWA*, 1425–1431.

Committee report (1971). State of the art of coagulation. *Jour. AWWA*, **63**, 99–108.

Conway R. A., Edwards V. H. (1961). How to design sedimentation systems from laboratory data. *Chem. Eng.*, **68**, 161–170.

Cornwell D. A. and Bishop M. M. (1983). Determining velocity gradients in laboratory and full-scale systems. *Jour. AWWA*, 470–475.

Critchley R. F., Smith E. O. and Pettit P. (1990). Automatic coagulation control at water treatment plants in the northwest region of England. *J. IWEM*, **4**, 535–543.

Dentel S. K., Resta J. J., Shetty P. V. and Bober T. A. (1988). Selecting coagualnt, filtration, and sludge conditioning aids. *Jour. AWWA*, 72–84.

Dentel S. K. and Kingery K. M. (1989). Using streaming current detectors in water treatment. *Jour. AWWA*, 85–94.

Dentel S. K., Thomas A. V. and Kingery K. M. (1989). Evaluation of the streaming current detector – I. Use in jar tests, and II. Continuous flow tests. *Wat. Res.*, **23**, 413–430.

Deveughèle S. and Do-Quang Z. (2004). Neural networks: an efficient approach to predict on-line the optimal coagulant dose. *Wat. Sci. Tech.: Water Supply*, **4**(5–6), 87–94.

Di Bernardo L. (1984). Comparação entre a filtração realizada em papel de filtro e em instalacao de filtração direta ascendente. XIX Congresso Interamericano de Engenharia Sanitária e Ambiental, Nov., Santiago, Chile.

Edwards D. M. and Monke E. J. (1967). Electrokinetic studies of slow sand filtration process. *Jour. AWWA*, **59**, 1310–1319.

Gale R. S. (1977). Optimizing the use of pre-treatment chemicals. Solid/Liquid Separation Equipment Scale-Up. Uplands Press, Croydon, England, U.K.

Ghosh M. M., Jordan J. A. and Porter R. L. (1975). Physicochemical approach to water and wastewater filtration. *Jour. Env. Eng. Div., ASCE*, EE1, 71–86.

Glaser H. T. and Edzwald J. K. (1979). Coagulation and direct filtration of humic substances with polyethylenimine. *Env. Sci. & Tech.*, **13**, 299–305.

Gollnitz W. D. and Kittle B. (1990). Operator evaluation of small systems. *Jour. AWWA*, 40–45.

Gregory J. and Nelson D. W. (1984). A new optical method for flocculation monitoring. In *Solid-Liquid Separation*, Ellis Horwood, UK, 172–182.

Gregory J. and Nelson D. W. (1986). Monitoring of aggregates in flowing suspensions. *Colloids and Surfaces*, **18**, 175–188.

Gregory J. (1988). A simple particle monitor for low-turbidity waters. AWWA Water Quality Technology Conference, Nov., St. Louis.

Gregory J. and Kayode T. O. (1988). On-line monitoring of polymer conditioning. Eng. Foundation Conf., Palm Coast, Florida, Jan.

Gregory J. (1990). Personal communication. University College, London.

Gregory J. (2004). Monitoring floc formation and breakage. *Wat. Sci. Tech.*, **50**(12), 163–170.

Griffith J. D. and Williams R. G. (1972). Application of jar-test analysis at Phoenix, Ariz. *Jour. AWWA*, 825–830.

Guibai L. and Gregory J. (1991). Flocculation and sedimentation of high-turbidity waters. *Wat. Res.*, **25**(9), 1137–1143.

Gupta V. S., Bhattacharjya S. K. and Dutta B. K. (1975). Zeta potential control for alum coagulation. *Jour. AWWA*, **67**, 21–23.

Habibian M. T. and O'Melia C. R. (1975). Particles, polymers and performance in filtration. *Jour. Env. Eng. Div. ASCE*, **101**, EE4, 567–583.

Hudson H. E. (1965). Physical aspects of flocculation. *Jour. AWWA*, **7**(7), 885–892.

Hudson H. E. (1973). Evaluation of plant operating and jar-test data. *Jour. AWWA*, 368–375.

Hudson H. E. and Wagner E. G. (1981). Conduct and uses of jar tests. *Jour. AWWA*, 218–224.

Ingels T., Drewes J., Yates G., McEncroe J., Crittenden R. and Jorden R. (2005). Optimizing filtration pre-treatment processes through state-of-the-art floc particle characterization. Proc. AWWA Water Quality Technology Conference, Quebec City.

Jin Y. (2005). Use of a high resolution photographic technique for studying coagulation/flocculation in water treatment. Masters thesis, Dept. Civil and Geological Eng., Univ. Saskatchewan, Saskatoon.

Jorden R. (2005). Personal communications, Longmont, CO.

Kawamura S., Hanna Jr. G. P. and Shumate K. S. (1967). Application of colloid titration technique to flocculation control. *Jour. AWWA*, **59**, 1003–1013.

Kayode T. O. and Gregory J. (1988). A new technique for monitoring alum sludge conditioning. *Wat. Res.*, **22**(1), 85–90.

Kerrison P. H., McEwen B., Phillips G. and Crook B. V. (1989). The use of redox potential to control ferric sulfate dosing during phosphate removal. *J. IWEM*, **3**, 397–403.

Kohonen T. (1995). *Self-organizing maps*. Heidelberg: Springer Verlag.

Lai R. J., Hudson H. E. and Singley J. E. (1975). Velocity gradient calibration of jar-test equipment. *Jour. AWWA*, 553–557.

Langelier W. F. (1921). Coagulation of water with alum by prolonged agitation. *Eng. News Record*, **86**, 924.

Lee N. (1988). Clarification improvement: reproducing plant mixing conditions in the beaker. *Wat. Sew. Eff.*, **8**(4), 22–27.

Letterman R. D. and Tanner R. D. (1974). Zeta potential controls direct filtration coagulation. *Wat. and Sew. Wks.*, 62–65.

Lind C. (1994). Coagulation control and optimization. Public Works.

Loganathan P. and Maier W. J. (1975). Some surface chemical aspects in turbidity removal by sand filtration. *Jour. AWWA*, **67**, 336–342.

Logsdon G., Ciemniecki D., Hess A. and Chipps M. (2003). Charting a course in coagulation control. *AWWA Opflow*, 8–10.

Liu J. C. and Wu M. D. (1997). Fuzzy control of coagulation reaction through streaming current monitoring. *Wat. Sci. Tech.*, **36**(4), 127–134.

Masters T. (1993). *Practical Neural Network Recipes in* C++. Academic Press, San Diego, CA.

Mhaisalkar V. A., Paramasivam R. and Bhole A. G. (1986). An innovative technique for determining velocity gradient in coagulation-flocculation process. Wat. Res., 20, 10, 1307– 1314.

Mical A. (1997). Jar testing simultaneous dosing device. AWWA Opflow, October, 10.

Mueller K. C. and Burbank Jr. N. C. (1964). Effects of polyelectrolyte coagulant aids on removal of missouri river turbidity with ferric sulphate. *Jour. AWWA*, **56**, 333–346.

Narkis N. and Rebhun M. (1977). Stoichiometric relationship between humic and fulvic acids and flocculants. *Jour. AWWA*, **69**, 325–328.

Ockershausen R. W. and Peterman C. J. (1964). Discussion to zeta potential measurements in the control of coagulation chemical doses. *Jour. AWWA*, **56**, 224–227.

Odegaard H. (1979). Orthokinetic flocculation of phosphate precipitates in a multi-component reactor with non-ideal flow. Prog. *Water. Tech.*, **11**(1), 61–88.

Oldshue J. Y. and Mady O. B. (1979). Flocculator impellers: a comparison. *Chem. Engrg. Progr.*, **75**(5), 72.

O'Melia C. R. (1967). Discussion to (*Orthokinetic flocculation in water purification, Harris et al.* 1966, Jour. San. Eng. Div., Proc. ASCE, 92, SA6, 95–111). Jour. San. Eng. Div. Proc. ASCE, 93, SA5, 120–125.

Packham R. F. (1960). The coagulation process Pt. III – The effect of pH on the precipitation of aluminium hydroxide. Tech. Paper No.17, Brit. Wat. Res. Asscn., U.K.

Packham R. F. (1972). The laboratory evaluation of polyelectrolyte flocculants. *Br. Polym. J.*, **4**, 305–315.

Packham R. F. and Sheiham I. (1977). Developments in the theory of coagulation and flocculation. *Jnl. Inst. Water Eng. and Sci.*, **31**(2), 96–108.

Posselt H. S., Reidies A. H. and Weber Jr. W. J. (1968). Coagulation of colloidal hydrous manganese dioxide. *Jour. AWWA*, **60**(1), 48–68.

Ratnaweera H., Lei L. and Lindholm O. (2002). Simulation program for wastewater coagulation. *Wat. Sci. Tech.*, **45**(4–5), 27–33.

Reed G. D. and Reece D. A. (1984). Sedimentation success from modified jar tests. *Jour. AWWA*, 101–105.

Riddick T. M. (1961). Zeta potential and its application to difficult waters. *Jour. AWWA*, **53**, 1007–1030.

Sharp E. L., Banks J., Billica J. A., Gertig K. R., Henderson R., Parsons S. A., Wilson D. and Jefferson B. (2005). Application of zeta potential measurements for coagulation control: pilot plant experiences from UK and US waters with elevated organics. *Water Sci. Tech.: Water Supply*, **5**(5), 49–56.

Shelton S. P. and Drewry W. A. (1973). Tests of coagulants for the reduction of viruses, turbidity and chemical oxygen demand. *Jour. AWWA*, **65**, 627–635.

Singley J. E. and Black A. P. (1967). Hydrolysis products of iron III. *Jour. AWWA*, **59**, 1549–1564.

Smith C. V. and Somerset I. J. (1971). Streaming current technique for optimum coagulant dose. Proc. 26th Ind. Waste Conf., Purdue Univ., 922–932.

Stanley S. J. and Smith D. W. (1995). Measurement of turbulent flow in standard jar test apparatus. *Jour. Env. Eng.*, 902–910.

Stumm W. and O'Melia C. R. (1968). Stoichiometry of coagulation. *Jour. AWWA*, **60**, 514–539.

Swanwick J. D. and Davidson M. F. (1961). Determination of specific resistance to filtration. *Water and Waste Treat. Jour.*, **8**, 386.

Swope H. G. (1977). Zeta potential measurement. *Water and Sew. Wks.*, Ref. No. R64–67.

Tambo N., Matsui Y., Kurotani K., Kubota M., Akiyama H., Ohto T., Zaitsu Y. and Itoh H. (1997). Control of coagualtion process by dual wavelength particle analyzer. *Wat. Sci. Tech.*, **36**(4), 135–142.

TeKippe R. J. and Ham R. K. (1970). Coagulation testing: a comparison of techniques – Part 1. *Jour. AWWA*, **62**, 594–602; – Part 2, *Jour. AWWA*, **62**, 620–628.

Verma S. and Chaudhuri M. (1978). Colloid-chemical parameters for coagulant dose control. *Jour. AWWA*, **70**, 102–105.

Wierenga J. and Yavich A. (2005). Calming the waters computer model predicts coagulant dosages. *AWWA Opflow*, **31**(9), 18–23.

Yao K. M., Habibian M. T. and O'Melia C. R. (1971). Water and wastewater filtration: concepts and applications. *Env. Sci. & Tech.*, **5**, 1105–1112.

Zhang Q. and Stanley S. J. (1999). Real-time treatment process control with artificial neural networks. *Jour. Env. Eng.*, 153–160.

Appendix A
Turbidity and its measurement

A.1 INTRODUCTION

In many municipal and industrial applications the concentration of particulate matter is important since suspended solids can shield microorganisms or create a disinfection demand that interferes with effective disinfection. Particulate matter can also adhere to heat transfer surfaces such as those found in heat exchangers, cooling ducts and radiators and interfere with the rate of heat exchange.

Particulate contaminants may vary from several nanometers to several millimeters in size. Consequently, cost-effective removal of these particles from the large volumes of water required by many communities represents a significant challenge for water treatment engineers. In the process of removing contaminants from water, certain particles may also be added to the water. Water treatment facilities must take into account the need to remove particles that are added or created during treatment as well as those present in the raw water. Natural weathering of minerals produces a variety of particles in water. Inorganic particles may consist of iron oxides such as goethite, haematite, and amorphous iron hydroxide; silica dioxide; calcites; clays such as kaolinite, montmorillonite, muscovite, or bentonite; pure and mixed oxides of aluminum; and many other minerals. Alone, these particles are objectionable in water because they affect the water's aesthetic qualities. However, in the form of small colloids, even small mass concentrations of these particles may present substantial surface areas for the adsorption of natural organic matter (NOM), pesticides and other synthetic organic chemicals (SOCs), metals, and other toxic substances. Bacteria and viruses may also attach to inorganic particles, and there is some concern that particulate materials (organic and inorganic alike) may "shield" pathogens from chemical disinfectants. In addition, asbestos fibers, which may come from natural or anthropogenic sources, are possible carcinogens when introduced into the gastrointestinal tract.

Clays, metal hydroxides, and other particles originating from mineral sources typically vary from several nanometers to several microns in diameter. The distribution of materials over this range is likely to be continuous. However, in surface waters most of these particles appear to be approximately 0.1 to 1 μm in diameter. Their small size renders them relatively stable in suspension. Moreover, particles in this size range scatter visible light efficiently and may create a cloudy appearance in water at very low concentrations.

Particulate matter is typically expressed as total suspended solids (TSS) measured by gravimetric analysis whereby particles are removed from a volume of water by dead-end filtration using a rinsed membrane of known dry weight. Membranes with a nominal pore size of 0.45 to 2 μm are often used. The membrane is then dried and weighed to determine the mass of particles deposited on the membrane. If the mass concentration of the suspension is low, relatively high volumes of water need to be filtered to deposit measurable amounts of solid material and minimize inherent errors in measurement. Measurements of the mass concentration of particles in the water by this method will be affected by the specific gravity of the particles, the effective pore size of the membrane, and the mass of material lost during drying. Some particles, such as algae or bacteria, may represent a relatively large particle volume concentration in the raw water. However, because of their high water content they may account for only a small mass of solids after drying. Similarly, organic compounds that volatilize during the drying process will reduce the estimate for the solids concentration in the sample.

The operational definition of particulate and dissolved materials is linked to the effective pore size of the membrane used to collect suspended materials. In addition, a cake of material that builds up on the membrane during filtration may become the effective membrane. In either case, significant concentrations of colloidal material may pass through the membrane and be excluded from the measurement of suspended material. In some cases this error may be important, particularly in estimating the quantities of colloid-bound pollutants.

A.2 THE ADVANTAGES OF TURBIDITY MEASUREMENTS

If rigorous laboratory procedures are adhered to, gravimetric measurements have the potential for accurate measurements. However, they are notorious for providing inconsistent results, particularly for more dilute suspensions. They are also time-consuming and are not convenient for real-time process-control work.

Turbidimetry on the other hand, is a more straightforward test and, although still not without its own challenges, is a reliable measure of source-water clarity and the effectiveness of subsequent treatment. In water treatment practice, turbidity is the most widely used indicator of the concentration of particles in water.

Turbidity is a measure of the scattering of light by particles suspended in a medium such as water. The deflection or scattering of light is related to a number of optical phenomena that in turn depend on the concentration, size, and shape of the particles as well as the wavelength of the incident light, the angle of observation, the optical properties of the particles, and the refractive index of the suspending medium (AWWARF/IWSA, 1997).

Originally, turbidity was a parameter used in limnological work to study the depth at which light can penetrate a body of water. Field measurements still use the Secchi disk, shown in Figure A.1. Figure A.2 shows a relationship between Secchi disk readings and standard turbidity readings.

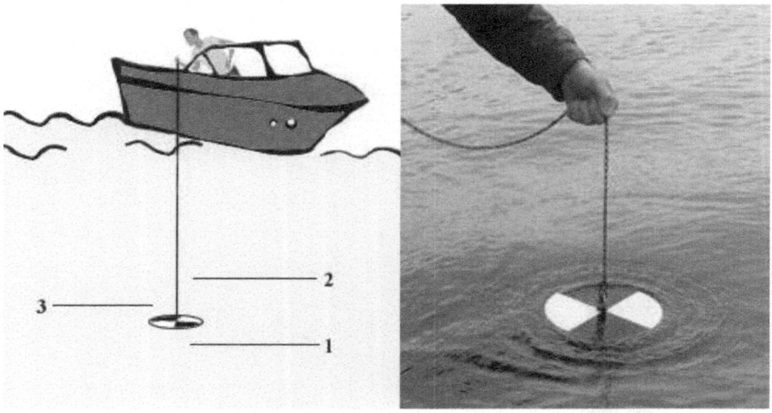

Figure A.1 Use of the Secchi disk: Lower the Secchi disk until it disappears and note the depth below the surface on the cord (1); Slowly raise the disk until it reappears and note the depth on the cord (2); Average the two depths and record as the Secchi depth (3).

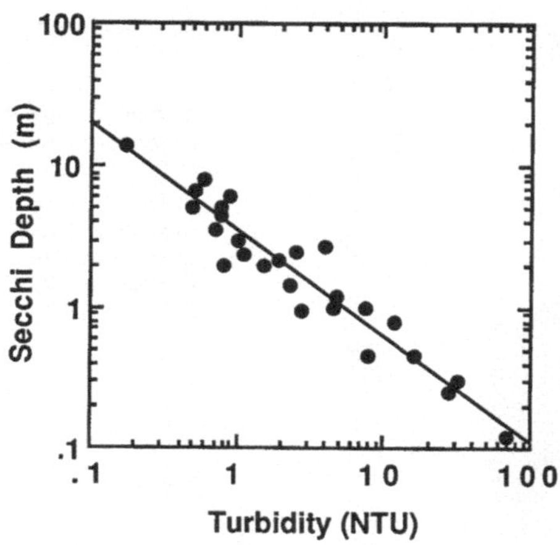

Figure A.2 Secchi disk depth versus turbidity in 27 New Zealand Lakes (from Davies-Colley, 1988).

In the past, the turbidity of water was measured based on the attenuation of light passing through a sample, as viewed through a Jackson candle turbidimeter, shown schematically in Figure A.3. This instrument could measure the clarity of water down to 25 Jackson Turbidity Units (JTU). The Jackson candle turbidimeter used a graded glass tube that was gradually filled with water until a person looking down through the tube could no longer discern the candle. This method took into account the effects of both light scattering and absorbance, and the reported value for turbidity was an apparent value.

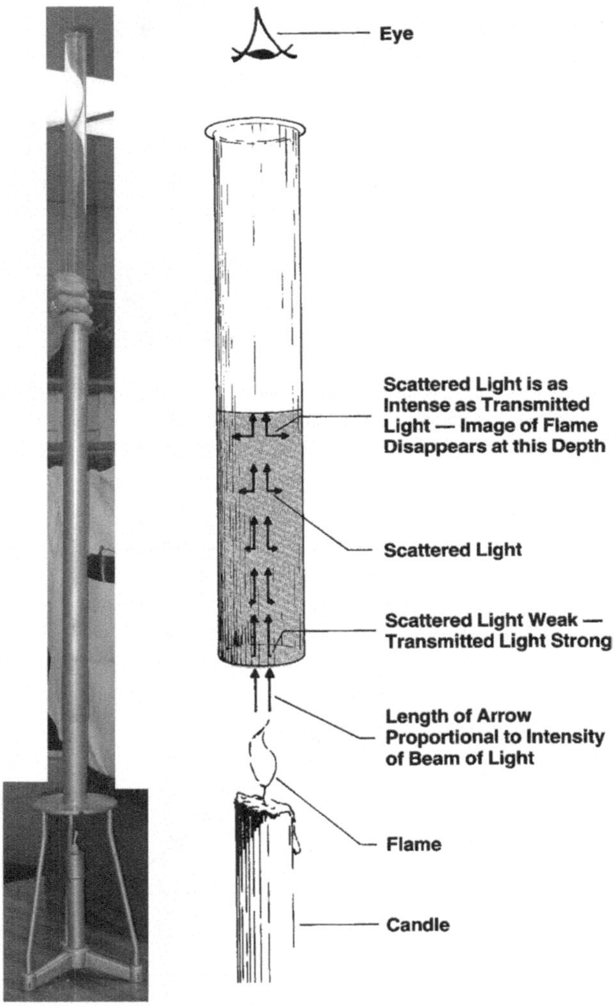

Figure A.3 The Jackson candle turbidimeter.

The appearance of a water sample is the product of the combined effects of scattering and absorption of light by dissolved and suspended material. It is a reasonable approximation that dissolved material absorbs light but particulate matter absorbs light and re-emits part of it as scattered light. Scattering of light is the cause of a hazy or milky appearance which is described by the turbidity, while color is generally regarded as being imparted by dissolved material. This separation of the two effects is not rigid because substances in solution do scatter light to some degree and particles in suspension, such as iron compounds, may be coloured. At high suspended solids concentrations the appearance of a sample may be due as much to attenuation of light by absorption as to the scattering of light. For practical purposes the contribution of particulate matter to appearance can be measured as turbidity directly on the water sample, but measurement of color generally requires the removal of suspended material by filtration.

There was a gradual shift from expressing turbidity as an aesthetic criterion, to stipulating turbidity goals for health reasons. This required accurate measurements of low levels of turbidity, far lower than was possible with the Jackson candle method. Typical ranges of required turbidity measurements at water treatment plants are presented in Table A.1 (Burlingame *et al.* 1998).

Table A.1 Example turbidity ranges at water treatment plants.

Location	Turbidity Range (NTU)
Surface water	1 to >100
Settled water applied to filters	0.5 to 1.5
Filtered water under steady state operation	0.05 to 0.15
Filtered water during ripening	0.05 to 0.50
Filter backwash water	10 to 500

As an example of the health-based role of low-level turbidity measurements, the Pennsylvania Department of Environmental Protection in 1996 pointed out the relationship between turbidity spikes and Giardia breakthrough in filtered water: a 10- to 50-fold increase in cyst breakthrough from filter media disturbance could occur with only a corresponding increase in turbidity of 0.1 NTU (Burlingame *et al.* 1998).

Monitoring turbidity at each step of the treatment process is performed at many water treatment plants. The Long Term 2 Enhanced Surface Water Treatment Rule (LT2ESWTR, USEPA, 2003) proposed a 0.5 log Cryptosporidium removal credit for a combined filter effluent turbidity less than 0.15 NTU in 95% of samples each month and (with certain conditions) a 1.0 log credit for filtered water turbidities of less than 0.10 NTU in 95% of daily maximum values from individual filters.

As more sensitive instruments developed, they adopted a tungsten filament light source at a color temperature of 2200 to 3000°K that impinged on a water sample held in a clear glass sample cell. A spectral peak response detector tuned between 300 and 400 nm was located at an angle to the incident light beam to measure the amount of light reflected towards the detector. Eventually this angle was fixed at 90°. This instrument is called a nephelometer and gives values in nephelometric turbidity units or NTU. However, in the past, a number of other turbidity units were used, based on the synthetic dispersions used to calibrate the instruments. The units are not inter-changeable and NTU is currently the preferred unit. Some of the different turbidity units that have been used are listed in Table A.2.

Table A.2 Different turbidity units (from Water Research Centre, 1979).

Nephelometric turbidity units	NTU
Formazin Turbidity Units (various)	FTU
Jackson Turbidity Units	JTU
Silica (Kieselgur) Units	$mgSiO^2/L$
Fullers Earth (Fulbent 570) Units	mgF.E./L
Absolute Units (Zeiss-Pulfrich Turbidity Unit)	A.E.
Mastic Units	drops of mastic
Langrohr Units (reciprocal length)	C*
Coleman Nephelos Units (TiO2)	CNU
Helm Units (BaSO4)	

A.3 TURBIDITY AS SURROGATE FOR PARTICLE CONCENTRATIONS

Turbidity is often assumed to be a surrogate for TSS (Allhands, 2007). It is generally true that the higher the TSS then the more particles are expected in suspension, and the higher should be the turbidity. However, suspended particles, particularly in natural waters contain a large variety of material that could include clays, silts, inorganic matter, organic matter, vegetation and living organisms that produce a large variety of particle sizes and optical characteristics. Therefore, there is no universal means of turbidity calibration that can define turbidity for all water sources.

Various relationships found between turbidity (NTU) and TSS (mg/L) for a number of raw water sources include: 0.74; 0.80; 1.0; 1.25; 1.40; and 1.60 NTU/TSS. Results using the overflow from water treatment sludge gravity thickeners have been reported as ranging from 2.0 to 2.27 NTU/TSS. These results demonstrate that turbidity is not a universal or inherent measure of suspended material.

Light scattering and light absorption depend on the range of optical properties of a suspension, which is dependent on the shape, size and refractive index of the particles. For example, particles less than 0.1 μm in diameter do not produce significant turbidity responses; particles less than 1 μm produce the highest turbidity responses; and particles greater than 1 μm can produce progressively lower turbidity readings (Burlingame, 1998).

The above is not to say that turbidity is not a good indicator of the presence of suspended particles, but that there is not a clear conversion factor between TSS and turbidity. However, for a given water or application, under a given set of circumstances, empirical correlations between TSS measurements and turbidity readings can be useful, although the relationship between the two parameters may vary over time and particularly seasonally for raw water sources, requiring periodic checks on the empirical relationship.

The true value of turbidity in modern water treatment plants is that it allows on-line monitoring of treated water quality, at extremely low levels of particle concentrations. In this case the conversion between such low levels of turbidity and TSS becomes almost meaningless. Concerns with cyst breakthroughs from water treatment filters are of concern to most water treatment utilities. The advent of modern high precision instruments, particularly those relying on laser technology have been shown to provide reliable indicators of particle breakthroughs from filters. Figure A.4 demonstrates one case where an online particle counter was unable to adequately detect some of the particle breakthroughs, whereas an online laser turbidimeter did detect the breakthroughs (Sadar, 2005). The major difference between the two instruments is that the particle counter detected particles larger than 2 μm, whereas the turbidimeter was able to detect sub-micron particles.

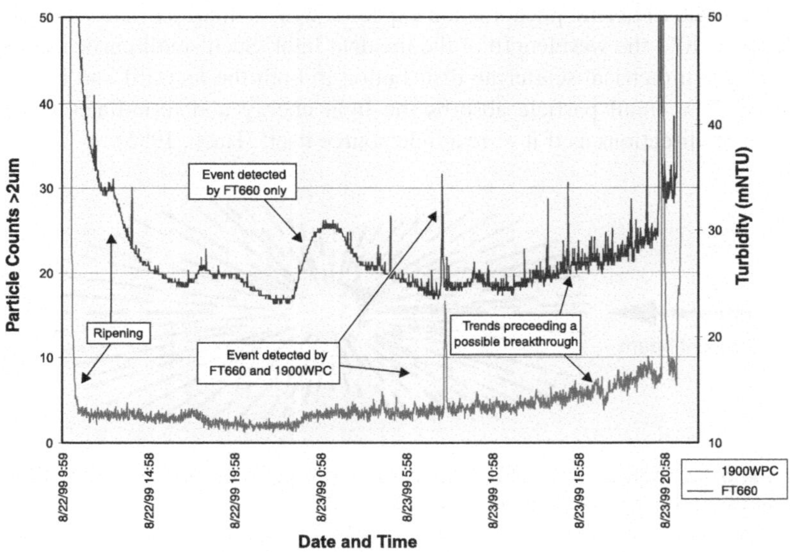

Figure A.4 Side-by-side filtered water quality using a laser turbidimeter (FT600; 1 mNTU = 0.001 NTU) and a particle counter (from Sadar, 2005).

A.4 PRINCIPLES OF TURBIDITY MEASUREMENT

When light is directed towards a suspension of particles, the intensity of light is reduced due to absorption and scattering of light. Absorption results in loss of energy from the light which is transferred to heat. Absorption of specific light wavelengths manifests as a characteristic color.

Unlike absorption, light scattering results in no net loss of energy from the light beam and results in radiation of energy in all directions at the same frequency as the incident light radiation. The intensity of the scattered light depends on the refractive index of the particles, and the difference between the refractive indices of the particles and the water, or other suspending medium. The refractive index of pure water is approximately 1.33, although the value changes with temperature and the concentration of dissolved organic and inorganic compounds. Temperature effects are relatively minor, within the ranges commonly encountered in water treatment. The effect of dissolved compounds is generally to increase the refractive index of water.

For a given particle size and wavelength of incident light, turbidity decreases with increasing refractive index of the suspending medium. Turbidity also increases as the particle size increases from zero up to a diameter equal to the wavelength of the incident light. As particle size increases further, turbidity then progressively decreases. These phenomena are mathematically complex and are described by the theory attributed to Gustav Mie, using Maxwell's equations to solve for regions inside and outside the particles (AWWARF/IWSA, 1997). Figure A.5 schematically shows the light scattering phenomenon expected with a small particle, of diameter smaller than 10% the wavelength of the incident light. Such small particles exhibit a relatively symmetrical scattering distribution in both the forward and backward directions. The small particle absorbs the light energy and re-radiates the light energy in all directions as if it were a light source itself (Hach, 1985).

Figure A.5 Scattered light pattern for very small particles (smaller than 10% the wavelength of the incident light (Hach, 1985).

Figure A.6 shows the scattered light pattern for a relatively large particle, with an assumed diameter approximately one-quarter the wavelength. Light scattered from different positions around the particle create interference patterns that are additive in the forward direction, resulting in forward-scattered light that is a higher intensity than light scattered in other directions.

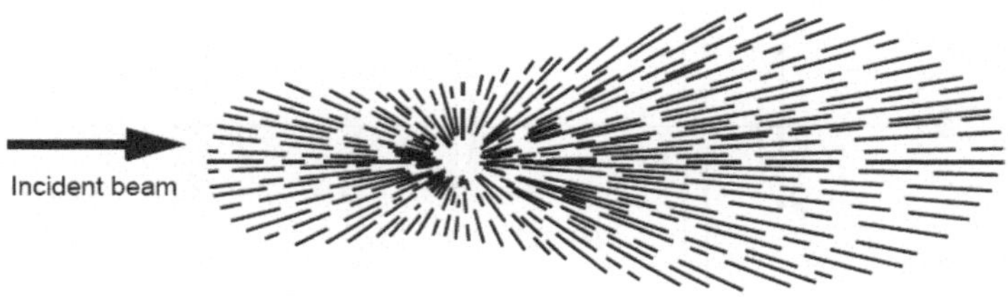

Figure A.6 Scattered light pattern for large particles (approximately 25% the wavelength of the incident light (Hach, 1985).

Figure A.7 shows the scattered light pattern for particles with diameters that are larger than the wavelength of the incident light. In this case the scattering in the forward direction is increased even further, with scattering also at other angles.

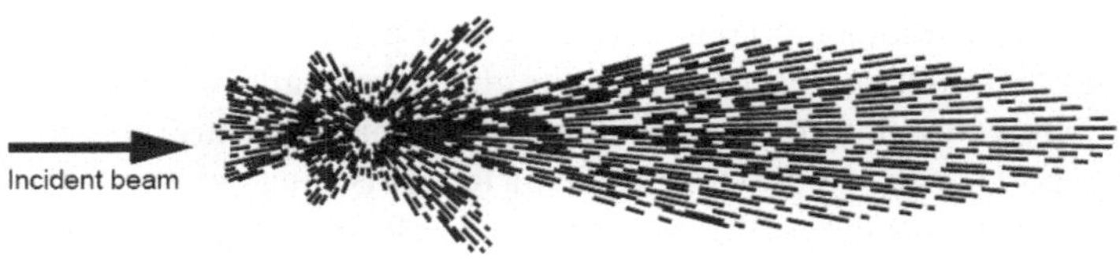

Figure A.7 Scattered light pattern for larger particles (larger than the wavelength of the incident light (Hach, 1985).

Since light scatters in all directions, there are essentially two methods of measuring turbidity: by measuring the reduction in intensity of the transmitted light, and by measuring the scattered light intensity, usually at 90° to the transmitted light beam (Gregory, 2006). Although the 90° angle is not the most sensitive to concentration it is probably least sensitive to variations in particle size.

Turbidity is calculated from the ratio of the light scattered by the sample to that produced by a suspension selected as a turbidity standard. However, different instruments can give different readings for the same sample, even when they have been similarly calibrated. Characteristics of the light source and detector determine the effective spectral distribution of the light source which could give rise to differences in the results between different nephelometer instruments.

A.5 TURBIDITY INSTRUMENTS

Light scattered by suspended particles is measured using a nephelometer, a schematic of which is shown in Figure A.8. Basically, a light source (tungsten filament or LED) directs light through a sample chamber and a detector measures scattered light, usually at 90° to the incident beam.

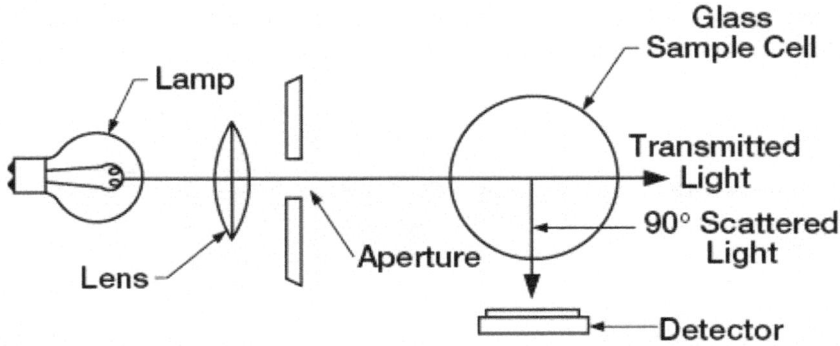

Figure A.8 Schematic of a nephelometric turbidimeter (Sadar, 1999).

There has been an attempt at standardization of turbidimeters so that measurements are relatively reproducible. There are two principal standards, summarized in Table A.3. USEPA 180.1 is widely used throughout the world. It calls for a tungsten-filament lamp and a spectral peak response for the detector between 400 and 600 nm corresponding to the wavelengths of visible light. The objective is that turbidity measurements should be most sensitive to particles that are 400 to 600 nm (0.4 to 0.6 μm) in diameter. This range of sensitivity should not be strongly dependent on the presence of dissolved compounds or the angle at which turbidity measurements are made.

Table A.3 Summary of design standards.

Requirement	USEPA 180.1	ISO 7027
Primary detector	90° ± 30°	90° ± 1.5°
Light source	Tungsten filament lamp with color temperature between 2200 and 3000 K.	Wavelength of 860 nm (LED or a combination of tungsten filament lamps with filters can be used)
Spectral response peak for the detector	400 to 600 nm (primary wavelengths of light)	Spectral bandwidth of the light must be within 860 nm ± 30 nm.
Measurement range	0 to 40 NTU (any sample above 40 NTU must be diluted)	0 to 40 NTU (any sample above 40 NTU must be diluted)

Method USEPA 180.1 is particularly applicable to clean water samples, below 1.0 NTU, which is the critical range for drinking water treatment. However, the method is sensitive to color that absorbs light in the 400 to 600 nm range of wavelength, and the tungsten filament lamps require lengthy warm-up times to achieve stability, and should be calibrated frequently. It is for this reason that laboratory instruments using this technology are often recommended to be permanently switched on for regular use.

Method ISO 7027 originated in the brewing industry and is commonly used in Europe. The method calls for a near monochromatic light source that is stable, has low absorbance interference with samples, and results in low stray light. However, the disadvantage of the long wavelength source adopted in this standard is a reduced sensitivity to small particle sizes. If the reduced sensitivity is electronically amplified this can result in increased background noise at low turbidity levels. Instruments using this method can have a disadvantage when measuring very low turbidities required in water treatment plants.

There are a number of turbidimeter configurations that have been designed to achieve increased accuracy or to reduce interferences. Figure A.9 shows a schematic of a ratio turbidimeter. In addition to a primary light detector positioned at 90° to the incident beam, it also includes secondary detectors situated (for example) at 0° and possibly at 15° to the transmitted light beam. The signals from the multiple light detectors combine electronically to give the turbidity reading. Such ratio-type instruments are useful when measuring the turbidity of colored liquid suspensions, such as beer, by minimizing the effect of light-absorbing soluble substances on the turbidity measurement. Some instruments are provided with a switch to activate or deactivate the secondary detectors thus changing the instrument from a ratio to standard design.

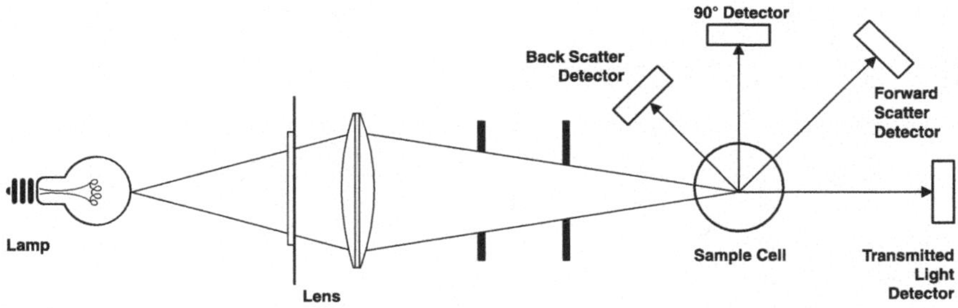

Figure A.9 Schematic of a ratio turbidimeter (Sadar, 1999).

A major benefit of turbidity measurements is that they are well suited to on-line process control. Figure A.10 shows a schematic of one such instrument which includes a network of baffled chambers exposed to atmosphere thereby minimizing the chance of false readings due to the presence of bubbles. This design also includes a “keyhole” design to reduce stray light and maximize the accuracy of the readings.

An alternative design of process instrumentation is shown in Figure A.11. This is an example of a surface scatter turbidimeter in which the light source and detector are mounted above the turbidimeter body thereby isolating optical components from the process water. The sample flow is introduced into the center of the body

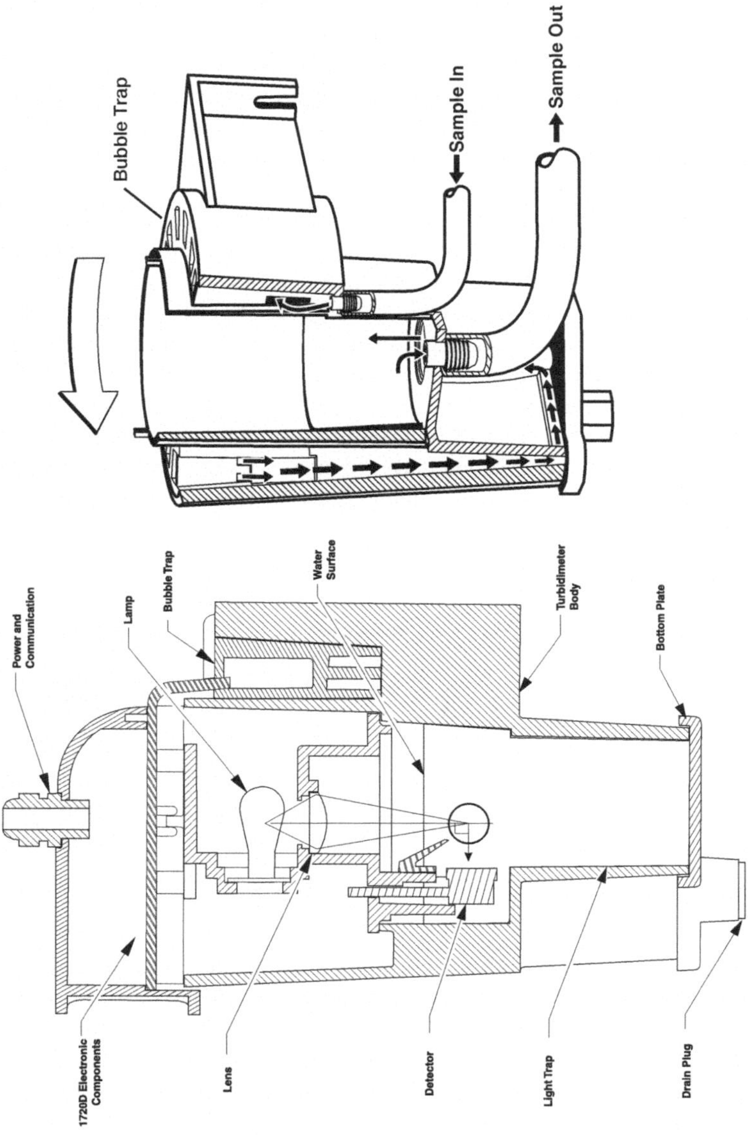

Figure A.10 Schematic of an inline process turbidimeter (Sadar, 1998).

and overflows a weir at the top forming a near perfect optical flat surface. The light beam is focused on the sample surface at an acute angle of about 15° and light scattered by the particles in the illuminated area is detected by the photodetector. The stray light in this instrument can be kept to a minimum by providing an entrance slit for the light beam, shielding baffles over the photocell, and providing an optical black surface to the sample container body. Stray light has been measured at an equivalent of less than 0.01 NTU. The instrument is particularly suited to wide-range measurements.

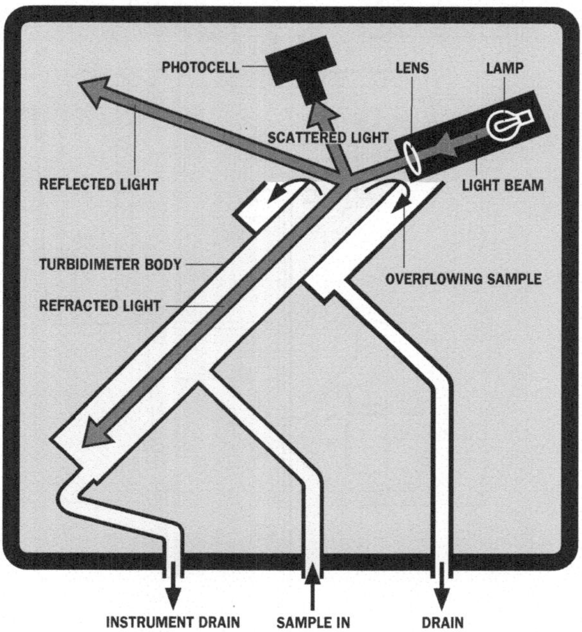

Figure A.11 Schematic of an inline process surface scatter turbidimeter (Sadar, 1998).

A turbidimeter design suited to very low in-line turbidity measurements is based on laser technology, as shown in Figure A.12. The light source in the instrument shown is a laser diode with a wavelength of 660 nm. The response system is optimized to detect this specific wavelength. The design produces a high beam power intensity that maximizes scattered light intensity from smaller particles. A light trap at the bottom of the nephelometer body reduces the stray light by about 50% compared with traditional instruments (Sadar, 2005). The instrument includes optical fibers to maximize the signal conveyed to the detector. The instrument is capable of detecting changes in turbidity down to 0.3 mNTU (0.0003 NTU). Some instruments include a Relative Standard Deviation (RSD) feature that quantifies the fluctuations in the laser turbidity measurements. This allows detection of precursors to turbidity events and impending turbidity spikes.

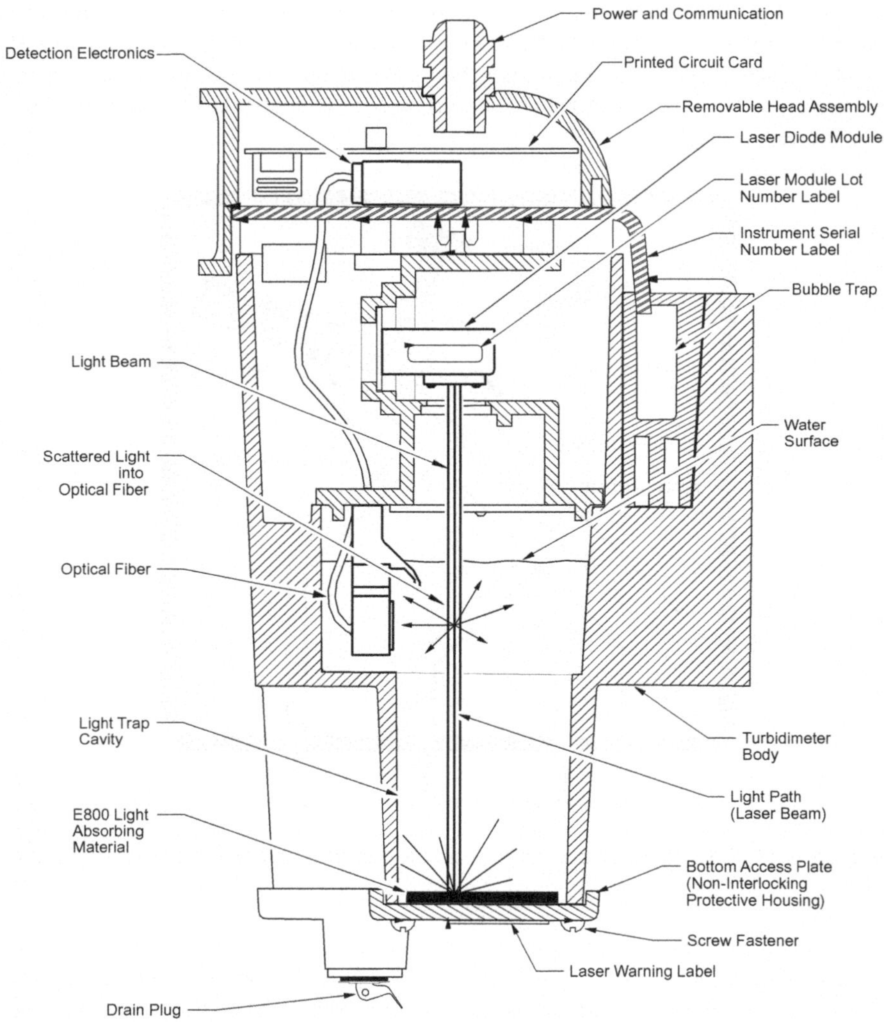

Figure A.12 Schematic of an inline process laser turbidimeter (Sadar, 2005).

Figure A.13 shows a turbidimeter suited to high solids concentrations. This type of instrument is used in filter backwash troughs or in activated sludge basins. The instrument shown includes a LED light source that transmits a beam of infrared light into the sample stream at an angle of 45° to the sensor face. A pair of photoreceptors detect scattered light at 90° to the transmitted beam. A backscatter photoreceptor can be mounted at 140° to the transmitted beam to detect light scattered in high solids sample streams. The sensor face can also include a self cleaning wiper device to minimize biological growth on the face of the device.

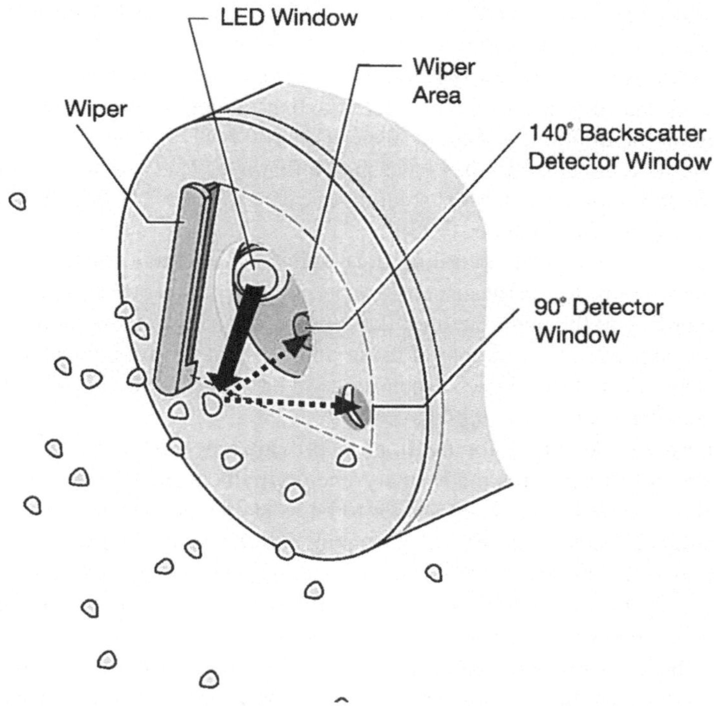

Figure A.13 Schematic of a high solids turbidimeter (Hach, 2005).

A.6 INSTRUMENT CALIBRATION

Turbidimeters should be regularly calibrated to ensure consistency between readings. Calibration includes adjusting the instrument to known turbidity standards. It is usually recommended that the instrument be calibrated to a standard suspension that is the closest to the expected range of turbidities of the samples to be measured. The primary standard for calibrating nephelometric turbidimeters has long been considered to be that based on formazin. It is prepared by accurately weighing 10 g of hydrazine sulfate and 100 g of hexamethylenetetramine in one liter of distilled water (method 180.1, USEPA, 1979). The solution develops a white turbidity after standing at 25°C for 48 hours. The polymer formed consists of random shapes and sizes, ranging from 0.01 to 10 μm and the suspension is highly reproducible. Care is required in handling and storing the chemicals due to their toxicity.

The turbidity of this suspension is 4,000 NTU and working suspensions of lower turbidity are prepared by diluting the 4,000 NTU suspension using particle-free water, typically less than 0.03 NTU.

Since formazin is considered a primary standard, it can be used in any instrument to calibrate against known turbidity values.

The 4,000 NTU stock suspension is generally stable for up to one year. However, diluted suspensions have a shelf life that ranges from minutes to weeks, depending on the dilution. A diluted suspension of 2 NTU is generally considered the minimum practicable. Studies have shown that a 40 NTU solution decayed to 28 NTU in 18 hours and a 1 NTU solution decayed to 0.4 NTU in 19 hours, even with refrigeration (Burlingame, 1998).

Because of issues with handling formazin suspensions, one manufacturer provides more stable formazin suspensions prepared in sealed cuvettes down to 0.10 NTU. According to the manufacturer, these preparations are stable for at least two years. The suspensions are prepared using an excess of hexamethylenetetramine. The characteristics of the stable suspensions are the same as the original formazin standard suspension (Sadar, 1999).

An alternative standard for calibration provided by one manufacturer uses suspensions of microspheres made of styrene divinylbenzene (SDVB) copolymer. Suspensions of SDVB have been shown to be as stable as concentrated formazin and much more stable than diluted formazin. Similar to the pre-prepared stable formazin suspensions, the SDVB suspensions are provided pre-diluted in sealed cuvettes, ready to use. SDVB suspensions are prepared to agree with formazin standards and are considered secondary standards.

Due to the mono-dispersed nature of the SDVB size distribution, incident light may over-scatter in the forward direction resulting in inaccurate calibration with some machines. This is the case with ratio turbidimeters, for example, where the manufacturer provides a special formulation to obtain acceptable readings when the instrument is tested against formazin standards.

Letterman *et al.* (2004) conducted comparative tests on a range of filtered, settled or disinfected only waters from different treatment technologies and water treatment plants from 10 geographically different water supply systems. They evaluated analyst techniques; 10 different instruments and 12 different instrument mode categories; sample turbidities ranging from 0 to 0.6 NTU; and four different calibration materials. The calibration materials evaluated were as follows:

- User-prepared formazin (following USEPA method 180.1). Subsequent dilutions used laboratory reverse osmosis (RO) water;
- Commercially supplied formazin stock solutions, subsequently diluted with RO water;

- Stabilized commercial formazin, prepared and sealed in a range of usable dilutions by the manufacturer;
- Styrene divinylbenzene microspheres (SDVB) supplied by the manufacturer in a range of pre-diluted suspensions.

Principal findings from this study are summarized as follows:

- The light source of the various instruments (tungsten filament, green LED, or infrared LED) cause minimal differences in readings. Instruments with LED light sources gave readings approximately 3% higher than tungsten filament instruments. For ratio instruments the discrepancy between the two light sources dropped to less than 0.5%.
- The calibration material used had minimal effect on the results of all instrument modes and water sources.
- A significant difference was noted between two groups of instruments when measuring low turbidity levels. The instruments that automatically set a pre-determined reading (for example, 0.02 NTU) when a deionized low particle water sample (for example, the laboratory RO water used for dilution of the calibration material) is placed in the instrument during calibration, subsequently gave significantly lower readings than instruments that do not operate this way. It was suggested that the differences between the two groups is related to the different electronic processes used in "zeroing" and calibrating the instruments (Letterman *et al.* 2004). Instruments that do not operate this way measure the turbidity of the dilution water and store the result in the instrument software. This value is then used to subtract the turbidity of the dilution water from the measured values of the other turbidity standards used to perform a calibration (Sadar, 1999).
- Analyst technique when calibrating and measuring low turbidities is a significant factor in the accuracy of the results.

A.7 TECHNIQUES FOR ACCURATE TURBIDITY MEASUREMENTS

A number of workers have pointed out techniques that are important for accurate and reproducible turbidity measurements. It is also important to follow the specific instrument operations manual. A number of the principal techniques are summarized below:

- It takes only a small amount of suspended matter, or a slightly dirty sensor, or a slightly dirty cell to cause significant inaccuracies when measuring low turbidities, for example at the 0.1 NTU range (Burlingame *et al.* 1998).

- The cleanliness of sample cells is an important issue for low-level turbidity measurements. One manufacturer recommends the following for ultra-low turbidity measurements:
  - Wash sample cells with soap and deionized water;
  - Soak the washed sample cells in 1:1 hydrochloric acid for at least 1 hour. Sample cells can also be placed in a sonic bath to release particles from the glass surfaces;
  - Rinse the cells with ultra-filtered deionized water (reverse osmosis or 0.2 μm filter) at least 15 times;
  - Immediately cap the cells to prevent contamination and to prevent drying out of the inner cell walls;
  - Polish the outside of the cell with silicone oil to prevent particle attachment and to fill in small imperfections in the outer glass.
- At low-level turbidity readings, it becomes more important that sample cells be indexed. Cells vary in optical quality; one side reflects more light, whereas another side absorbs more light. Burlingame et al. (1998) describe their practice whereby new cells are indexed and tested against the requirement that they not vary by more than 0.01 NTU. As many as 25 percent of new cells are discarded. The recommended procedure is as follows:
  - Once the sample cells have been cleaned, fill them with ultra-filtered low turbidity water;
  - Let samples stand to allow any bubbles to rise;
  - Then polish the cells with silicone oil;
  - Measure the turbidity at several points of rotation of the sample cell in the instrument;
  - Find the orientation where the turbidity reading is the lowest and mark the cell on the side in such a way that it can always be lined up the same way in the instrument's chamber for future measurements.
- If possible, use one sample cell for the low turbidity readings, particularly one that has been demonstrated to be of high optical quality and has been indexed.
- Dust inside the instrument can increase stray light, especially when it is stirred up when closing the instrument lid. Stray light can cause errors from less than 0.01 to greater than 0.04 NTU. It is important to follow the instrument operations manual in this regard.
- Condensation on the exterior glass window of the light source can interfere with low turbidity readings by causing noise and drift of the readings. Follow the instrument operations manual.
- Decay of the incandescent lamp or its variation over time can be significant at low-level measurements and can be a significant source of noise and drift in readings. Follow the instrument operations manual.
- It is important to remove bubbles from the sample before taking readings. Gas bubbles reflect light, produce light scatter and cause false turbidity responses

and erratic instrument behavior. The first measure should be to fill the sample cell slowly to avoid entraining bubbles. Then let the sample stand for several minutes to allow bubbles to vacate the sample. If the sample needs to be mixed, gently invert the sample several times to avoid introducing bubbles.

- If bubbles are still observed in the cell, for example attached to the cell wall, the capped cell can be rotated in the horizontal plane to dislodge the bubbles and allow them to rise to the headspace when returned to the upright position.
- A vacuum can be applied to the sample, although care must be taken to avoid contaminating the sample with the vacuum device. One method inserts a 50 mL syringe through a single hole in the cap of the sample cell. The plunger on the syringe is retracted to cause a partial vacuum. The vacuum is held until the formation of gas bubbles is no longer evident.
- Sonication can also be applied for 1 to 2 seconds to samples that are warming up to room temperature. Care should be taken with sonication since it could cause particles to fracture and change size, and cause particles attached to the sample cell wall to detach and cause higher turbidity of the sample.
- The use of in-line instruments is expected to reduce some of the errors caused by sampling. Grab sample collection can be affected by the cleanliness of the sample bottles; the sample holding time; sample storage conditions; particle breakup, clumping or settling; or particles sticking to the sample bottle. Some in-line process models include bubble traps or dissipators that reduce the effects of bubbles on the turbidity readings.

A.8 REFERENCES

Allhands M. (2007). For good measure. *Breaking Down Turbidity and Total Suspended Solids, WE&T,* 112–114.

AWWARF/IWSA (American Water Works Association Research Foundation and International Water Supply Association), (1997). Treatment Process Selection for Particle Removal, Cooperative Research Report.

Burlingame G., Pickel M., and Roman J. (1998). Practical applications of turbidity monitoring. *Jour. AWWA*, **90**(8), 57–69.

Davies-Colley R. (1988). Discussion of secchi disc transparency and turbidity, by S.W. Effler, *J. Env. Eng. Div. ASCE*, 114, 6.

Gregory J. (2006). Particles in Water. Properties and Processes. IWA Publishing/CRC Press, Taylor and Francis, London/ Boca Raton.

Hach C. (1985). Turbidity Standards. Technical Booklet 12 (also updated version by Sadar M.), Hach Company, U.S.A.

Letterman R., Johnson C. and Viswanathan S. (2004). Low-level turbidity measurements: A comparison of instruments. *Jour. AWWA*, **96**(8), 125–137.

Sadar M. (1998). Turbidity Science. Technical Information Series-11, Hach Company, U.S.A.

Sadar M. (1999). Turbidimeter Instrument Comparison: Low level Sample Measurements. Technical Information Series. Hach Company, U.S.A.

Sadar M. (2005). Introduction to Laser Nephelometry: An Alternative to Conventional Particulate Analysis Methods. Hach Company, U.S.A.

USEPA (1979). Methods for Chemical Analysis of Water and Wastes, 600/4/79/020, 180.1–1, Washington.

USEPA (2003). Long Term 2 Enhanced Surface Water Treatment Rule; Proposed Rule, Fed. Reg., 68:154, 47640, Washington.

Water Research Centre (1979). Measurement of Colour and Turbidity. Notes on Water Research, No. 22, August, U.K.

Appendix B

Processing and disposal of coagulant sludges

B.1 INTRODUCTION

It was estimated that in 1971 in the United States, approximately 2 million dry tons per year of alum sludge was produced in water treatment plants. For those plants using ferric salts approximately 0.3 million dry tons per year was estimated.

In Brazil, the state water company of São Paulo estimated that the seven principal water treatment plants in the metropolitan region, with a total treated water flow of 4,150 Ml/d (1,100 mgd), discharged a total of 90 dry tons solids per day to the environment. Total losses from unrecovered clarifier discharges and filter backwashes amounted to 120 Ml/d (32 mgd), equivalent to enough treated water to supply a population of over 450,000 people (SABESP, 1987).

In the past, the principal concern of potable water providers has been to produce water of the highest standard possible. Processing water plant residuals had not been of high concern. The common practice was to discharge the residuals to the same watercourse from which the water was abstracted. The arguments supporting this practice were based on the fact that the residuals are simply returned to the water body from where they were derived. However, this ignores the reduction in flow of the watercourse by the water plant, and the transformation of the substances originally in the water to what is essentially a chemical waste (Committee Report, 1972). Rigorous legislation is now in effect in several countries to limit the impacts of such residuals on the environment.

Water sludges are generally characterized by BOD concentrations in the hundreds, and COD concentrations in the thousands of mg/l. The high ratio between COD and BOD (generally around 15:1) suggests that the sludges are biologically relatively stable, somewhat resistant to decomposition and anaerobiosis, and do not generally present a high oxygen demand in the receiving body of water. Exceptions are those cases where organic discharges upstream of the point of abstraction in the body of water, do produce putrescible sludges. At times, volatile solids determinations of water sludges appear high, and suggest a high organic content. However, the apparently high volatile solids can be largely due to waters of hydration driven off at the high temperatures (600°C) used in the muffler furnace for the volatile solids test.

In general, waste sludge from water treatment plants may be described as a non-Newtonian, bulky, gelatinous substance typically comprising aluminum or ferric hydroxides, inorganic materials such as clay colloids, iron, and manganese, and organic material such as natural color, algae, bacteria and viruses, all removed from the treated water. Analyses on such residuals have identified over seventy distinct elements including, in order of predominance, silica, aluminum, iron, titanium, calcium, potassium, magnesium, and manganese.

Although the toxicity of metal hydroxides discharged to watercourses is generally low, some soluble complex species can present significant toxicity to aquatic organisms. For example, aluminate complexes, comprised of $Al(OH)_4^-$, are toxic to phytoplankton. Concentrations above 0.5 mg/l are acutely toxic to trout and also causes chronic injury, largely through gill hyperplasia (Committee Report, 1987). In general, toxicity effects of water plant residuals are more pronounced at lower and higher pH values, compared with neutral pH, as well as with softer waters (George, 1989). Obviously, the detrimental effects of water plant residual discharges are greater with rivers at low flow, and with rivers and lakes of low turbidity.

Apart from toxicity issues, other detrimental effects of water residuals in receiving waters arise from the general increase in turbidity, and the accumulation of sludge solids on the bed of the receiving body of water. These effects could impede reproductive processes of some higher species and reduce phytoplankton production.

In the United States, the 1972 Federal Water Pollution Control Act classified water plant residuals as an industrial waste. Today, the tendency when regulating these residuals is to completely avoid their discharge to bodies of water without adequate treatment, based on the enactment of the Clean Water Act of 1972 and the 1977 amendments.

In general, water treatment plant sludges are of a concentration too low for convenient handling and/or landfill (concentrations of at least 20 percent total solids concentration are generally required); compressible so that dewatering is hampered; and of such a composition that direct disposal on land gives rise to soil clogging.

Particularly problematical are sludges produced from water treatment plants using aluminum sulphate as coagulant. Lime sludges from plants practicing softening operations are generally easier to dewater. Ferric sludges have properties intermediate between the two.

Figure B.1 (adapted from Warden, 1982) shows the water plant sludge treatment and disposal routes in the United Kingdom, reported in 1977. Of the 341 plants reporting, 26 plants utilized mechanical dewatering, 38 reported drying beds, and 136 reported using lagoons. Almost 50-percent of the total sludge produced was processed and disposed of in lagoons. 175 plants reported disposing of liquid wastes. 35 of these disposed sludge to wastewater treatment plants, although the amount of sludge disposed of in this way represented less than 5-percent of the total.

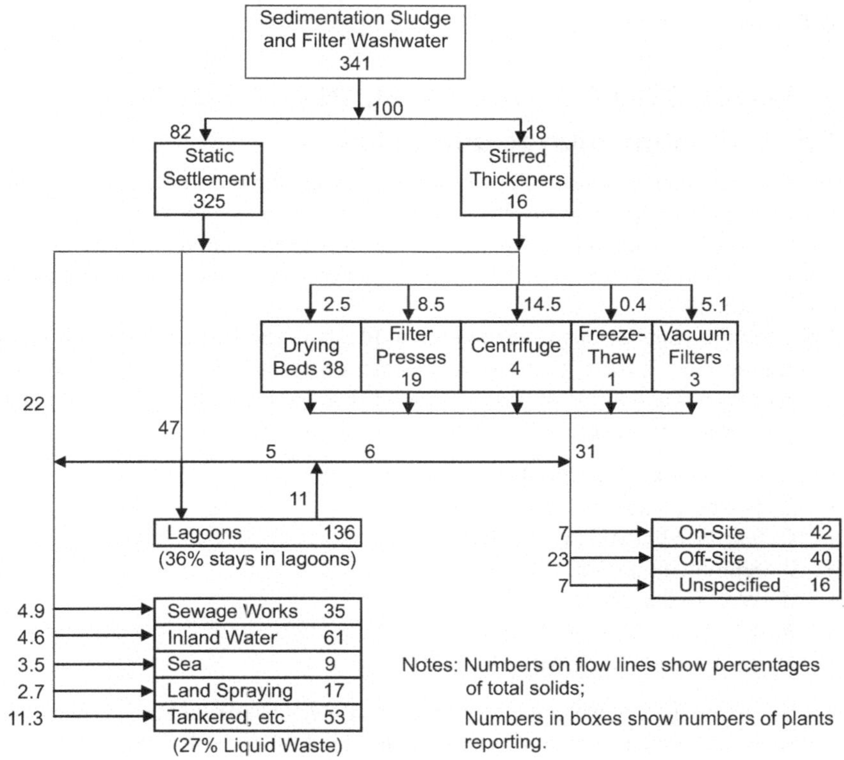

Figure B.1 Sludge processing routes for 341 water plants, reported in 1977 (adapted from Warden, 1982).

Water treatment plant wastes originate from two principal sources: the clarifier underflow and the filter backwash or washwater recovery process. The sludge deposited in clarifiers may be removed continuously or intermittently. Some installations still practice manual sludge removal, with sludge removed at intervals of 3 months or more. Filters are generally backwashed every 24 to 72 hours, with a relatively large volume of water used to displace retained solids during a relatively short time.

The solids concentration of sludges removed from clarifiers depends very much on the type of raw water and the chemicals used during treatment, as well as the mode of sludge removal. Typically the range of concentrations expected is from 0.1 to 3.5 percent (1,000 to 35,000 mg/l). Such sludges generally have BOD values ranging from 30 to 100 mg/1 and COD values from 500 to 10,000 mg/l.

Solids removed from rapid gravity filters during backwashing are very dilute, with concentrations ranging from 100 to 200 mg/l – but at times up to 800 mg/l – contained in a large flow of water of short duration. The concentration of backwash solids is typically high at the start of the wash, tapering down to low values at the end of the wash period.

B.2 PRODUCTION OF WATER PLANT RESIDUALS

B.2.1 Estimating sludge quantities

Water plant residuals comprise the raw water suspended material, coagulant metal hydroxides, and precipitation products of soluble substances removed from the raw water. Obviously the quantity of sludge produced per unit volume of water treated varies widely and depends largely on the characteristics of the raw water and the type of treatment.

Information from various water treatment plants in the United States, all using aluminum sulfate, showed that sludge production varied from 12 to 59 kg dry solids per Ml (100 to 490 lb/MG) water treated. The quantities for individual plants were as follows (Gruninger, 1974, 1975; Nielsen *et al.* 1973):

- Sobrante: 59 kg/Ml (490 lb/MG)
- Rochester: 25 kg/Ml (208 lb/MG)
- Kodak: 17.2 kg/Ml (143 lb/MG)
- Shoremont: 14 kg/ML (117 lb/MG)
- MCWA: 13.9 kg/Ml (116 lb/MG)
- ECWA: 12 kg/Ml (100 lb/MG)
- Sturgeon Point: 12 kg/Ml (100 lb/MG)

Tests done at the Chicago water treatment plant in 1939, related sludge solids produced to raw water turbidity, using alum or ferric chloride (Table B.1, Hudson, 1981):

Table B.1 Solids production related to turbidity (from Hudson, 1981).

Raw Water Turbidity NTU	Dry Solids Produced (Alum or Ferric Coagulants)	
	kg/Ml	lb/MG
2	2.8	23.3
5	4.5	37.5
10	6.7	56
20	12	100
50	36	300
100	86	716
200	240	1,998

The following expression was derived from data published in 1978 by AWWA (Committee Report, 1978) listing flocculation basin solids produced at different water treatment plants:

$$\text{Solids production} = 3.5 \times (\text{Raw Water Turbidity, NTU})^{0.66} \text{ kg dry solids/Ml}$$

$$= 29.1 \times (\text{Raw Water Turbidity, NTU})^{0.66} \text{ lb/MG}$$

Comparing the values calculated using the above expression with the values obtained by Hudson in the Table above, there are large discrepancies between the predicted values.

The above results illustrate the inaccuracies inherent when determining solids production using rule-of-thumb procedures. A more rational approach is to calculate sludge production based on raw water characteristics and the dosages of chemicals added during treatment.

Studies carried out by the Water Research Centre in the United Kingdom, at the Hodder water treatment plant, resulted in the following relationship (Burley *et al.* 1977):

Sludge production (mg/l of treated water)

$$= (\text{Raw water suspended solids, mg/l})$$

$$+ (0.07 \times \text{organic color, °H})$$

$$+ (\text{Coagulant hydroxide, mg/l } Al(OH)_3 \text{ or mg/l } Fe(OH)_3)$$

$$+ (\text{Other additives such as polyelectrolytes, carbon, etc., mg/l})$$

Warden (1982) also reported on Water Research Centre pilot work and presented basically the same expression, except for the organic color conversion. Instead of the 0.07 conversion factor, Warden found that a factor of 0.2 correlated with the pilot plant data.

In the expressions presented by Burley *et al.* (1977) and Warden (1982), correlations between raw water turbidity and suspended solids of 1.0 FTU: 2.0 to 2.27 mg/l were adopted. Such ratios are somewhat high for raw water. The likely reason is that the Water Research Centre work derived the correlations using data taken from gravity thickener overflows.

Similar correlations using raw water data from various treatment plants obtained the following ratios:

- 0.74 (Bishop *et al.* 1987)
- 0.80 (Burris *et al.* 1976)
- 1.0 (Croker, 1989)
- 1.25 (Acton, 1988)
- 1.40 (Bratby, 1988)
- 1.60 (Johnson *et al.* 1989)

In general, with raw water turbidities less than approximately 100 NTU, the ratio between suspended solids concentration and turbidity can be considered to be approximately 1.0.

An empirical expression for sludge production was presented by the Association Francaise pour L'Etude de Eaux (Levesque, 1982):

Sludge production (kg/Ml)

$$= (TSS + A \times {}^{\circ}H + X + F_1 + M_1 + C + K \times D)$$

where

TSS = raw water suspended solids (mg/l)
$^{\circ}H$ = organic color (°H)
A = coefficient to convert color to suspended solids after coagulation (= 0.05 to 0.07)
X = dosage of other chemicals (polymers, carbon, etc.), mg/l
F_1 = ferric hydroxide precipitate for iron removal (mg/l)
M_1 = Manganese dioxide precipitate for manganese removal (mg/l)
C = carbonate precipitates (mg/l)
D = coagulant dosage in terms of product (mg/l)
K = coefficient to convert coagulant dosage to hydroxide solids (mg/l)

K values derived from the original text are presented in Table B.2:

Table B.2 Corrected K coefficients for different coagulants.

Coagulant	Formula	S.G.	Strength	K	K[1]
Ferric Chloride (crystal)	$FeCl_3 \cdot 6H_2O$		60% $FeCl_3$	0.40	0.48
Ferric Chloride (anhydrous)	$FeCl_3$		99% $FeCl_3$	0.65	0.79
Ferric Chloride (40% soln.)	$FeCl_3$(40%)	1.41–1.45	596 g/l $FeCl_3$	0.27	0.33
Ferric Chloro-Sulfate (soln.)	$FeClSO_4 \cdot nH_2O$	1.6	594 g/l $FeCl_3$	0.24	0.29
PACl (solution)	$Al_n(OH)_nCl_3$	1.2	10% Al_2O_3	0.16	0.20
Alum (crystal)	$Al_2(SO_4)_3 \cdot 14H_2O$		17.2% Al_2O_3	0.27	0.34
Alum (solution)	$Al_2(SO_4)_3 \cdot 14H_2O$	1.3	8.3% Al_2O_3	0.13	0.16

S.G. = specific gravity
K[1] = corrected values assuming $1.25H_2O$

The original expression considers metal-hydroxide complexes in the form of $Me(OH)_3$. Theoretically, a more correct assumption would be $Me(OH)_3 \cdot 3H_2O$.

However, mass balances carried out at various water plants showed that the assumption of 1.25 water molecules fits the mass balances more consistently (Nielsen *et al.* 1973). Mass balances were conducted in terms of "dry" solids, analyzed by drying to 105°C in the laboratory. Therefore, the suggested format is $Me(OH)_3 \cdot 1.25H_2O$. The Table above includes adjusted values of K (K^1) to reflect the corrected values.

B.2.2 Alternative coagulants and dosage reduction

In general, sludges with a high percentage of metal-hydroxide are significantly more voluminous and difficult to process compared with sludges with lower percentages. Therefore, when evaluating sludge processing of water sludges, one of the first steps would be to investigate the possibility of reducing the amount of coagulant chemical added during water treatment.

Sometimes, changing the type of coagulant significantly reduces the amount of sludge produced. For example, the Union Hills plant in Phoenix, Arizona changed the primary coagulant from aluminum sulfate to ferric chloride and found a consequent reduction in water plant sludge from 4.6 kg/Ml (38 lb/MG) to 3.3 kg/Ml (27 lb/MG) – or a reduction of 28 percent (Thompson & Haas, 1972). An added benefit was that the ferric sludge produced fewer odors in the sludge lagoons and drying beds used at that time. In general, ferric sludges have been reported to have superior thickening and dewatering characteristics compared with alum sludges, although this may not always be the case – and the choice of primary coagulant will likely be based primarily on its efficacy during treatment, not only on the characteristics of the sludge produced.

Another alternative for reducing the amount of metal coagulant used for treatment is to substitute the coagulant wholly or partially for organic polymers. Some plants utilize polymers as the primary coagulant. However, since organic polymers seem to be somewhat specific to certain raw water constituents, many plants have found it preferable to use polymers as a coagulant aid, in conjunction with metal coagulants, thereby reducing the metal coagulant dosage.

One example is the Crown water plant in Ohio, where after changing from using alum alone as the primary coagulant, to a combination of alum and polymer, the sludge cake concentration obtained in the filter presses increased from an average of 32-percent to 45-percent. In this case the use of poly aluminum chloride as primary coagulant was unsuccessful (Heuer & Schwarzwalder, 1989).

The use of the family of polyaluminum hydroxychloride (PACl) coagulants has in many cases significantly reduced water plant residuals. These coagulants include polyaluminum chloride, aluminum chlorohydrate, aluminum polychloride, and others. Such products can be tailored to provide specific benefits for different waters. For example, sulfate substitutions produce the polyaluminum chloro-sulfate coagulants. Other substitutions are with phosphates and silicates, and also with cationic polymers which can impart enhanced coagulating properties for some waters.

The high cationic charge carried by these coagulants means that dosages can often be reduced significantly and the metal hydroxide complexes formed are much less. This means that the mass of sludge produced per unit volume of water treated is significantly reduced and the sludge thickening and dewatering properties of the sludges are improved, particularly for low turbidity waters where the hydroxide fraction of the sludge has more significance. With high turbidity waters the influence of the hydroxide portion of the sludge matrix is much less significant.

One example (Lind, 1993) is a plant treating Great Lakes water in a conventional sedimentation-filtration plant. Prior to switching from aluminum sulfate, alum dosages ranged from 10 to 15 mg/l, reaching more than 50 mg/l during storm conditions. When PACl was adopted, dosages were around 2.5 mg/l, reaching 12.5 mg/l during storm conditions. In general, sludge production reduced from the range 5 to 17 kg/Ml water treated (40 to 140 lb/MG), to less than 2.5 kg/Ml (20 lb/MG).

An important aspect related to sludge generation is to ensure that an excess of coagulant is not being applied. This can be the case particularly in conventional water plants using sedimentation upstream of the filters. In such cases, it could be the case that the coagulant dosage had been determined in terms of optimizing settling in the clarifiers. However, if chemical dosages are determined based on filtered water quality, using the filtered jar test for example, coagulant dosages have sometimes been reduced significantly. At the same time filter runs have not necessarily increased significantly.

A pertinent example is the Plano Piloto water plant in Brasilia (Bratby, 1986). Based on the results obtained in the *Pulsator* clarifiers, operators typically dosed approximately 15 mg/l alum with 10 mg/l lime to adjust pH. Filter runs were very short with clear hydroxide flocs passing from the clarifiers to the filters. By optimizing coagulant conditions in terms of filtration efficiency, the alum dosage could be reduced by half, to approximately 7.5 mg/l, with the added benefit of eliminating the lime. Filter runs increased markedly and sludge production was reduced by approximately half.

B.2.3 Sludge characteristics

Water treatment plant sludges are characterized as bulky, gelatinous and generally resistant to thickening and dewatering operations, particularly those sludges from plants treating low turbidity waters. The sludges are generally compressible such that increasing pressures applied during dewatering produce higher resistances to filtration. These sludges are thixotropic such that in a quiescent state they appear "solid", but on agitation become fluidized.

The concentrations of sludges discharged from clarifiers and filters are of low concentration and require thickening and dewatering before final disposal. Clarifier sludge concentrations will be lower when treating lower turbidity waters, and those raw waters with high algal contents.

Apart from the characteristics of the raw water treated in the plant, the concentration of sludge withdrawn from clarifiers depends also on the type of clarifier. For example, the concentration of settled alum sludge withdrawn from upflow blanket clarifiers is normally within the range 0.1 to 0.3% dry solids. With horizontal flow clarifiers, with continuous mechanical sludge removal, concentrations rarely exceed 1%, and are normally within the range 0.2 to 1%. With periodic sludge removal (by manual cleaning for example) concentrations of 4 to 6% are possible.

The density of sludges is a function of the sludge concentration. Density can be estimated on the basis of the density of dry solids after oven drying the sludge. For alum or ferric sludges, the density of dried solids is approximately 2,300 kg/m3 (19.2 lb/gal). Therefore sludge density is given by:

$$\text{Density} = \frac{2{,}300}{2.3 - 0.013 \times C}\ \text{kg/m}^3$$

where C = sludge concentration, % dry solids

Similarly, for lime softening sludges, the density of dry solids can be assumed to be 2,710 kg/m^3 and the sludge density is given by:

$$\text{Density} = \frac{2{,}710}{2.71 - 0.0171 \times C}\ \text{kg/m}^3$$

The volatile solids fraction of water sludges is generally in the range 28 to 35% (Johnson *et al.* 1989).

A useful indicator of dewatering characteristics is specific resistance to filtration. For alum sludges, reported specific resistance values have ranged from 500×10^{10} to $5{,}000 \times 10^{10}$ m/kg. Normally, specific resistance values reduce with increasing concentration, up to approximately 2%, beyond which they remain more-or-less constant.

At times, if sludge is left to stand for a time before dewatering, slightly higher specific resistance values are measured when compared with fresh sludge. However, the final dewatered concentrations achieved were actually slightly higher.

In general, sludges with specific resistance values higher than about 500×10^{10} m/kg are considered to be difficult to dewater. Values lower than 100×10^{10} m/kg are considered to be relatively easy to dewater.

For sludges from water treatment using lime, typical specific resistance values are from 1.2 to 8.2×10^{10} m/kg, reflecting the favorable dewatering characteristics of these sludges.

Tests carried out to determine the effects of the raw water organic color content, measured as TOC, on water sludge characteristics showed that thickened sludge concentrations could be 10 to 20-percent lower when compared with the same waters without organic color (Dulin & Knocke, 1989). Gravity thickening rates in the presence of organic matter were 15 to 30% the rates without organic material. The detrimental effect of organic matter on sludge characteristics was attributed to the incorporation of more water molecules in the sludge matrix and the production of smaller flocs which impeded the removal of interstitial water.

With ratios of (TOC removed)/(mg Al added) of 0; 0.25; 0.50; and 1.0, specific resistance values were measured at $1{,}200 \times 10^{10}$; $1{,}400 \times 10^{10}$; $2{,}100 \times 10^{10}$; and $3{,}400 \times 10^{10}$, respectively. Pre-oxidation with potassium permanganate improved dewatering characteristics. At a (TOC removed)/(mg Al added) ratio of 1.0, and $KMnO_4$ dosages of 0; 0.4; 0.8; 1.2 and 1.6 mg/l added to the raw water before coagulation, specific resistance values progressively decreased as follows: $3{,}400 \times 10^{10}$; $3{,}000 \times 10^{10}$; $2{,}200 \times 10^{10}$; $1{,}250 \times 10^{10}$; and $1{,}100 \times 10^{10}$, respectively. However, despite the improvements in dewatering characteristics, little improvement in thickening rates were observed with $KMnO_4$ addition.

The benefits on sludge characteristics of pre-oxidation before coagulation of organic matter has been observed elsewhere. For example, at the Newport News water plant in Virgina, pre-chlorination was interrupted because of disinfection byproduct formation. However, an undesirable consequence of this was to develop populations of micro-crustaceans and blue-green algae in the clarifiers. Clarified effluent quality deteriorated and filter runs reduced drastically to approximately 12 hours. The sludge removed from the clarifiers was black and odorous, indicating anaerobiosis in the sludge (Hoehn *et al.* 1987).

Tests conducted with chlorine dioxide (ClO_2) eliminated these problems without forming disinfection byproducts. In addition, sludge specific resistance values were approximately halved with ClO_2 addition. Improvements with pre-oxidation were attributed to an increase in floc sizes. Without ClO_2 the average floc size was 32 μm. After ClO_2 addition floc sizes increased to 55 μm. The increase in floc size was also accompanied by an increased floc density which improved settling characteristics.

B.2.4 Sludge conditioning

Because of the adverse nature of water plant sludges, particularly those generated using metal coagulants, some form of conditioning is essential for most thickening and dewatering applications. The most commonly used conditioning chemicals for these sludges are polyelectrolytes.

B.2.4.1 Polymer conditioning

In general, best results are achieved with high molecular weight polymers. The best type of polyelectrolyte for a given application is strongly influenced by pH.

Cationic polymers tend to function better at pH values less than 7.0. Nonionic and anionic products tend to function better within the pH range 6.5 to 8.5. At pH values higher than 8.5, anionic products with at least a 50% charge density are generally best (O'Brien & Novak, 1977; Warden, 1982).

When using polyelectrolytes the position and the type of device used to introduce the polymer is important. In many cases the use of orifice plates has produced good results with water sludges (Warden, 1982; Bratby *et al.* 1993, 1996).

After polymer dosing the turbulence regime should be as gentle as possible to avoid breaking the relatively fragile polymer-sludge bonds. Therefore, the dosage point should be as close as possible to the thickening device.

The intensity of mixing and the time allowed for rapid mixing both influence the polymer dosage necessary for satisfactory conditioning. Therefore, when evaluating polymers, tests should be done under conditions as close as possible to the thickening or dewatering device (Novak *et al.* 1975).

Novak and Bandak (1994) showed that the best polymer dosage for dewatering depends on both mixing time and intensity of shear. With alum sludges, the required polymer dosage increased as the shear intensity increased. For example, at a T of 30 seconds and a G of 1,400 s^{-1}, the optimal polymer dosage was 30 mg/l. At a T of 180 seconds and a G of 1,400 s^{-1}, the required polymer dosage increased to 60 mg/l.

Unconditioned alum sludges are easily fractured by dewatering processes. Polymer conditioning reduces, but does not eliminate sludge degradation by shear stresses. The dewatering rate of a sludge was found to be a function of G^nT, where n varied from 2.8 for unconditioned alum sludges, to 1.0 or less for well conditioned sludges. G and T are the root-mean-square velocity gradient ($second^{-1}$), and mixing time (seconds), respectively. Lime softening sludges do not degrade under shear, and a single polymer dosage will be adequate under a range of mixing and shear conditions.

Because of the influence of shear and mixing conditions on the effectiveness of polymer conditioning, the actual mixing regime is difficult to simulate during laboratory testing. Novak and Bandek (1994) found that GT values in filter presses and belt presses were 30,000 and 10,000, respectively. These values can be difficult to duplicate in the laboratory, particularly the intensity of mixing.

One alternative is to sample directly from the dewatering device and optimize polymer by trial-and-error during full scale operation. However, sampling the conditioned sludge can be an issue, particularly from the high pressure line feeding a filter press, for example. Here, the immediate drop in pressure at the sampling valve would cause excessive shear to the conditioned sludge sample.

Gale (1977) describes a device designed to overcome this difficulty (Figure B.2): The sampler is first filled with compressed air at the same pressure as the pipeline, by opening, then closing valve (1). The valve between the line and the sampler (3) is then opened fully. Air is then bled slowly from the sampler using valve (2), and the conditioned sludge suspension flows gently into the sampler. After closing valve

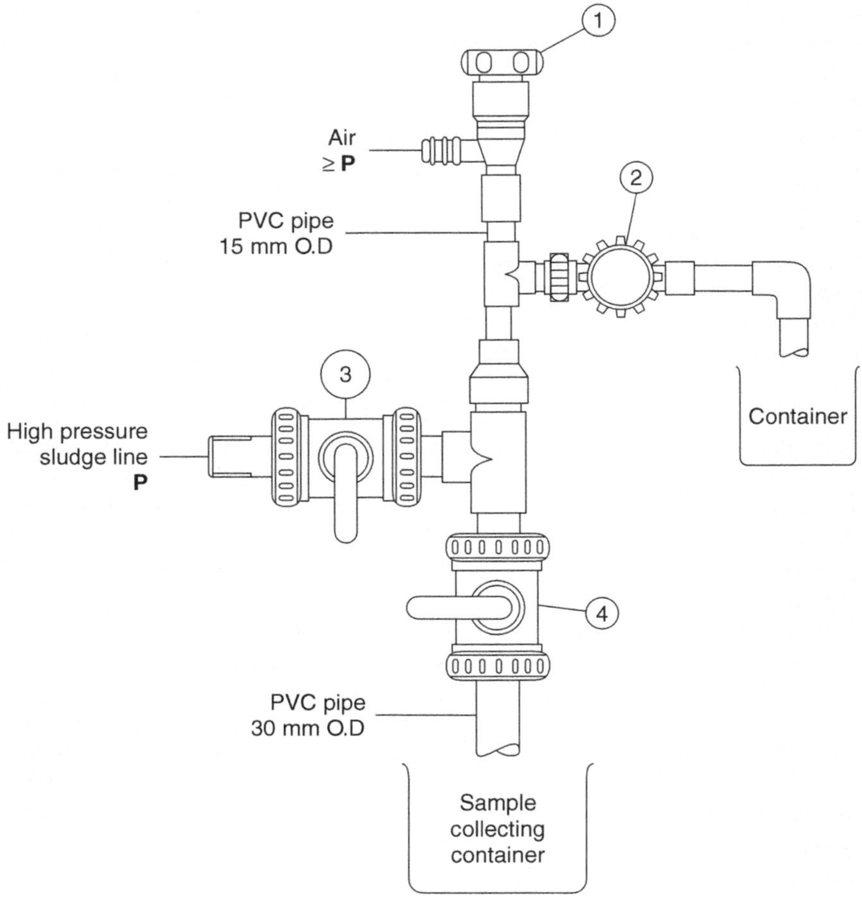

Figure B.2 Low-shear sampling device for high pressure sludge lines (after Gale, 1977).

(3) and releasing the air pressure using valve (2), the sample can be drained gently using valve (4) into a beaker for testing purposes.

The required polymer dosage depends on the particular application, but in general it is in the range 1.0 to 10 g/kg dry solids (2.0 to 20 lb/t dry solids). Figure B.3 shows the effect of polymer dosage on specific resistance to filtration for sludges from three plants in Brazil (Bratby, 1988).

For the Serra Azul plant, the pH was 9.5 and the best product was a high molecular weight anionic polymer at 4 g/kg (8 lb/t dry solids). For the Guaraú and ABV water plants in São Paulo, the best polymer was a high molecular weight non-ionic at a dosage of 4 g/kg (8 lb/t dry solids).

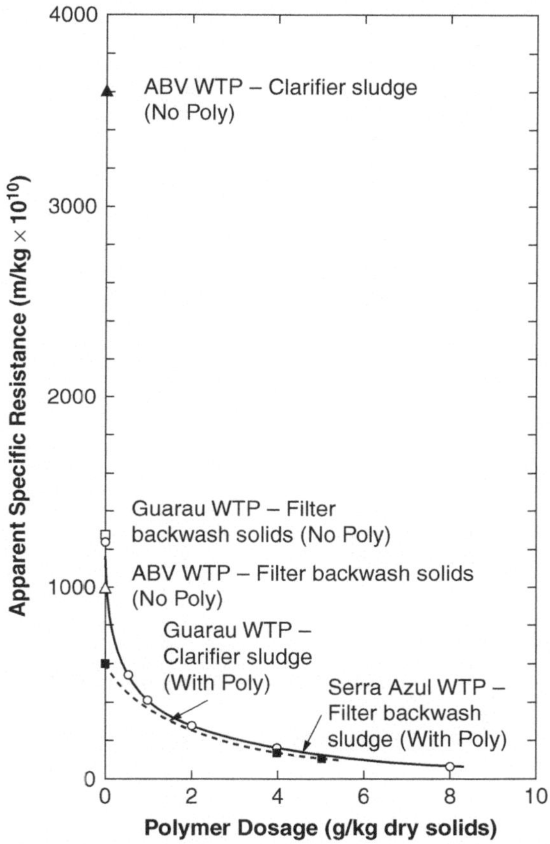

Figure B.3 Effect of polymer dose on specific resistance (Bratby, 1988).

One of the effects of polymer addition is to significantly increase the solids loading rate for thickening, although the final concentration achieved is often not higher (Knocke & Wakeland, 1983). However, Figure B.4 shows the results obtained at several plants in various catchments in different cities in Brazil. In general, for the same loading rates, increases in thickened concentrations range from zero to 5%, depending on the sludge. The very poor results obtained with the clarifier sludge from the ABV plant is due to the high degree of organic material in the raw water.

The use of two products in series sometimes provides superior results. For example, a cationic followed by an anionic product has given good results during dewatering operations of water sludges (Schwoyer & Luttinger, 1973). The addition of lime before polymer to increase cake concentrations is also well known.

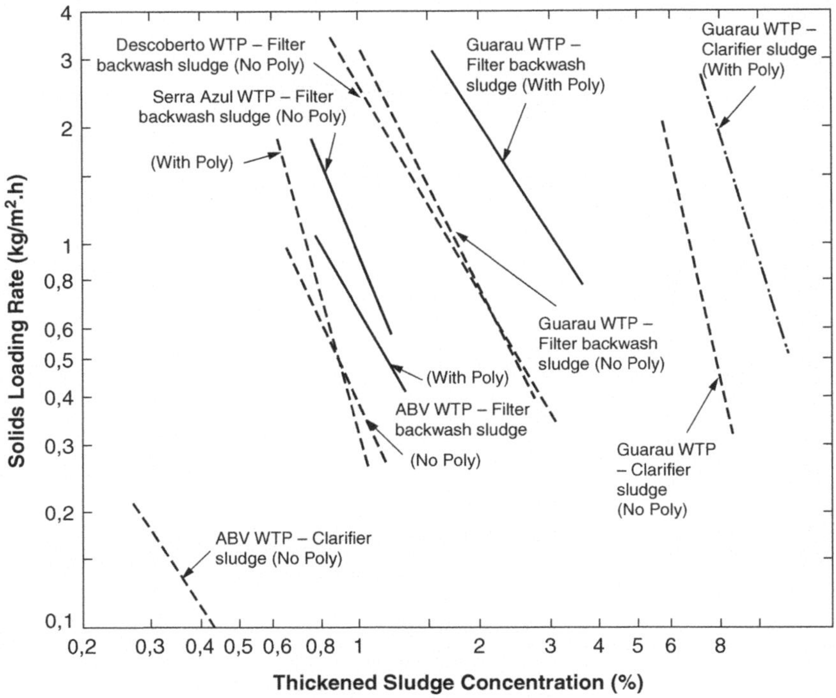

Figure B.4 Effect of polymer on solids loadings and thickened concentrations (Bratby, 1988).

Hydrogen peroxide has been tested (Drewry & Kelkar, 1989; Schafran *et al.* 1993) to reduce the gelatinous nature of alum sludges. With some of the sludges tested, the dewatering characteristics improved markedly, with reductions of specific resistance of 40 to 80% using H_2O_2 dosages of approximately 2.5 mg/l. However, some sludges showed little improvement with H_2O_2 addition.

Experiments were carried out with a variety of synthetic alum sludges, ranging from "pure" alum hydroxide sludge, to sludges with varying amounts of iron, natural organic matter and powdered activated carbon (PAC). The mechanism of conditioning with the addition of H_2O_2 is related to the partial oxidation of organic compounds, resulting in a shift towards lower molecular weight organics, an increase in floc density, and an increase in particle sizes. The authors also found that the relative amounts of iron, PAC and organic matter were important and influenced the effectiveness of the H_2O_2. It was believed that reactions between H_2O_2 and iron and PAC, catalyzed the rapid formation of hydroxyl radicals that were more effective at oxidation of the natural organic matter in the sludge.

Figure B.5 shows a flocculation device particularly appropriate for sludge conditioning. It includes a mechanized in-line rapid mixer for polymer (or conditioner) addition followed by a tapered floc formation device. The tapered cone arrangement progresses from a higher shear intensity (at the largest diameter) tapering down to a lower shear intensity (smaller diameter) before passing to the dewatering device.

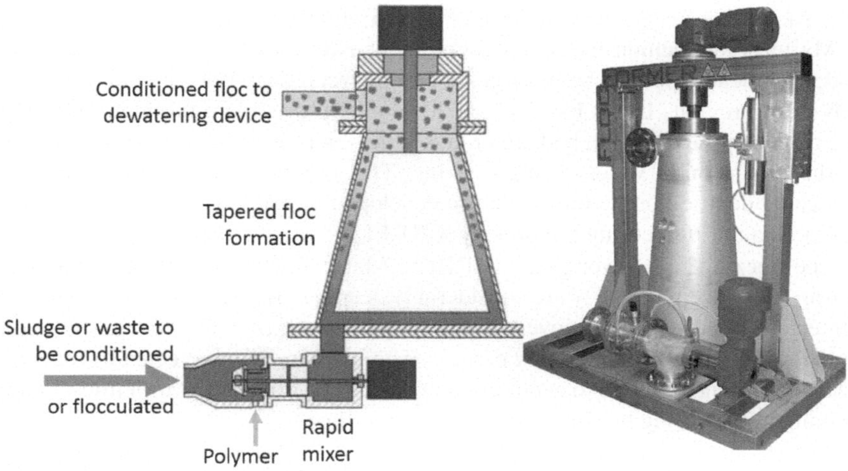

Figure B.5 In-line rapid mixing and tapered flocculation device for sludge conditioning (courtesy Aquen Aqua-Engineering GmbH, www.aquen.de).

Other forms of conditioning used with water sludges have been freezing, and thermal conditioning.

B.2.4.2 Freezing

The concept of freezing is that ice crystals extract hydration water from the hydroxide sludge thereby destroying the gelatinous structure of the flocs. The freeze treatment process gives rise to dehydration of sludge particles by freezing the water closely associated with them. As the sludge is chilled the particles are first concentrated by selective freezing of the liquid phase and finally dehydrated as the total mass is frozen. During the freezing process sludge particles and other impurities are forced to the ice crystal boundaries.

On thawing, water which has left the sludge matrix to form ice crystals has little tendency to return to the sludge particles. After freezing and thawing the sludge has a granular nature which is non-reversible. The process is so effective that sludge with a concentration of 2% can reach 20% just by sedimentation. After subsequent dewatering, concentrations up to 60% are possible (Wilhelm & Silverblatt, 1976).

The freeze-thaw process does not depend on the sludge characteristics, but only on the total volume of sludge. Evidently, thickening the sludge to reduce the volume as far as possible is beneficial. To date, it appears that the process is only viable when natural freezing is possible. Therefore, application of the process has generally been in lagoons in northern climates which go through natural freeze-thaw cycles. In Copenhagen, application of sludge to lagoons at the rate of 1.4 m/year (4.6 ft/year) accumulated only 40 mm (1.5-inches) after two years and after successive freeze-thaw cycles (AWWARF, 1969).

Martel (1989) demonstrated that for efficient natural freeze-thaw in sludge beds, sludge application should be in successive thin layers. Each layer should be applied only after previous layers have frozen. This maximizes the total depth of sludge that can be applied to each bed, and reduces the internal stresses in the bed caused by the expansion of water during freezing. This process can be automated using a freezing degree-hour counting device developed by the US Army Cold Regions Research and Engineering Laboratory (CRREL).

The freezing beds developed by CRREL (Martel & Diener, 1991) are similar to drying beds except that they are considerably deeper to freeze several layers during winter. The beds are also covered with a transparent roof to avoid snow cover, which would impede the freezing process. The sides are also protected by louvered sidewalls, or snow fences to avoid drifting snow. The transparent roof allows rapid thawing in the spring by solar radiation (Figure B.6).

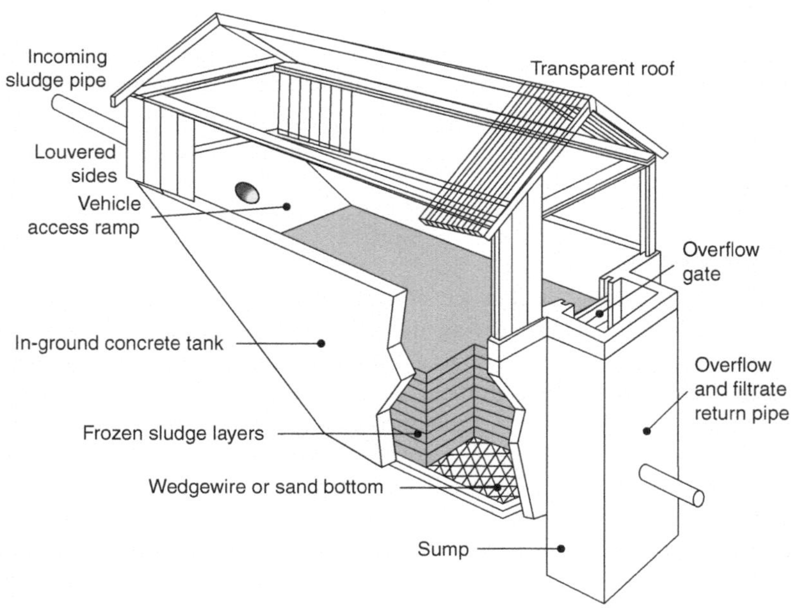

Figure B.6 Conceptual layout of sludge freezing bed (after Martel & Diener, 1991).

Martel (1989) developed equations that predicted design depths based on local climate conditions. He showed that for regions approximately north of the 50°-north latitude, design is based on the thawing depth. Designs for regions lower than this latidude are dependent on the freezing depth. Regions approximately south of the 40°-north latitude are not suitable for freezing beds.

Pilot studies showed that the volume of alum sludge can be reduced by 96-percent (Martel & Diener, 1991). The alum sludge, at an average concentration of 0.5% was applied in 20 to 100 mm (1 to 4-inch) layers. By the end of winter, 1 meter (3.2 ft) of sludge had been frozen. After thawing and drainage, the depth of sludge that remained was 30 to 50 mm. Further tests showed that the granular solids that remained could stay in the bed for several years, with successive sludge applications, before sludge removal was necessary.

In Japan, artificial freezing of sludge was studied extensively, but applied only rarely because of the high installation and operating costs (Committee Report, 1978). The most practicable temperatures for artificial freezing appear to be between -15°C and -4°C. Adequate freezing within this range is achieved in 20 to 60 minutes. Energy costs depend on sludge concentration (and hence volume) of sludge, as follows (Wilhelm & Silverblatt, 1976):

Sludge Concentration (% Dry Solids)	Energy Required (kWh/t Dry Solids)
1.5	2,300
2	1,700
4	1,000
6	700
10	400

Based on this data, a rule-of-thumb is that energy requirements are approximately 35 to 40 kWh/m^3 sludge (130 to 150 kWh/1,000 gallons).

Important factors for the artificial freezing process include the freezing rate, the heat exchange coefficient between the sludge and the heat exchanger, and the time during which the sludge is frozen (AWWARF & KIWA, 1990). If the speed of freezing is too high, particles remain dispersed and separated by the ice; in this case the dewaterability of the sludge after thawing may not be significantly improved. However, at lower freezing speeds, the hydrated sludge particles are concentrated in the unfrozen sludge mass; dehydration and compression act on the particles so that the dewaterability is greatly improved.

If the solids concentration of the sludge is too high, the heat exchangers do not function effectively and the sludge is not properly frozen.

Parker *et al.* (2000) studied the parameters affecting the mechanical freezing process. Residuals from five water plants were studied, and for all residuals under all conditions of freezing, dewaterability was improved by the freeze-thaw process. However, longer "curing" times (time that the samples are held frozen) resulted in improved filterability and cake solids content. Beyond 6-hours curing, improvements were marginal. Operating costs were minimized by thickening solids as far as practicable, while still maintaining flowability. Residuals should be frozen in thin layers to minimize energy costs.

B.2.4.3 Thermal conditioning

An additional form of sludge conditioning is heat treatment, where temperatures from 85°C to 200°C and pressures up to 1,300 kPa (190 psi) are applied to break down the gelatinous nature of the flocs to aid dewatering.

One application of this process was at the Sobrante water plant in California where the Porteous process was applied at 88°C. Although dewatering characteristics did improve, odors were an issue and filtrate was discolored. Because of these issues, the heat treatment process was discontinued at the Sobrante plant.

In general it has been found that thermal conditioning is not economically viable given the low calorific value of water sludge (Ventresque & Bablon, 1989). Conditioning effects are far less dramatic than with the freeze-thaw process.

B.3 FILTER BACKWASH

Water plant residual flows can either be discharged to watercourses, or recycled within the water plant. Direct discharge to watercourses is subject to pollution prevention regulations in most places, and is no longer a common practice. Discharge of residual flows that have been treated, typically by sedimentation is still common, although many States in the United States and elsewhere have included limitations on contaminants other than simply suspended solids and pH, that makes it difficult to discharge wastes, even after sedimentation treatment.

The internal recycle of in-plant residuals has also become regulated in many locations. For example, the Environmental Protection Agency in the United States has finalized a rule whereby recycled filter backwash water, sludge thickener supernatant, and liquids from dewatering processes must be returned to a location in the water plant such that all treatment processes in the plant are employed to treat the recycled flow, together with the raw water influent. Variances are possible for particular recycle flow treatment schemes.

The EPA Filter Backwash Recycling Rule has the prime objective of minimizing the incidences of pathogens such as *Cryptosporidium*, passing through to the final treated water. However, recovery of in-plant residual flows has the added benefit of minimizing wastage of water already treated by the water plant.

For example, water plant filter backwash rates are generally 2 to 20 times the filtration rate. The volume used for backwashing filters normally varies from 1 to 5 percent the volume of water treated. However, in some cases the volume of backwash water could be much higher. The Serra Azul direct filtration plant in Brazil, for example, during periods of lake turnover with high iron and manganese loads to the filters, experienced backwash volumes of 10 to 15% the volume treated (Bratby, 1988). This is a significant volume and, before washwater recovery was implemented, represented a significant wastage. The average treated water flow for the Serra Azul plant was 233 Ml/d (62 mgd). The backwash water which was wasted to a river downstream of the lake, represented up to 36 Ml/d (9.5 mgd) with 72 kg/d (160 lb/d) chlorine; 360 kg/d (800 lb/d) ferric chloride; and 550 kg/d (1,200 lb/d) lime.

The solids concentration contained in filter backwash is relatively low, normally between 100 and 200 mg/l. These suspended solids settle relatively fast. For example, tests carried out at the Serra Azul and Descoberto plants in Brazil showed that for solids removals of 90 percent from the backwash water, settling rates of 0.7 and 1.0 m/h (415 and 580 gpd/ft^2) were required, respectively (Figure B.7).

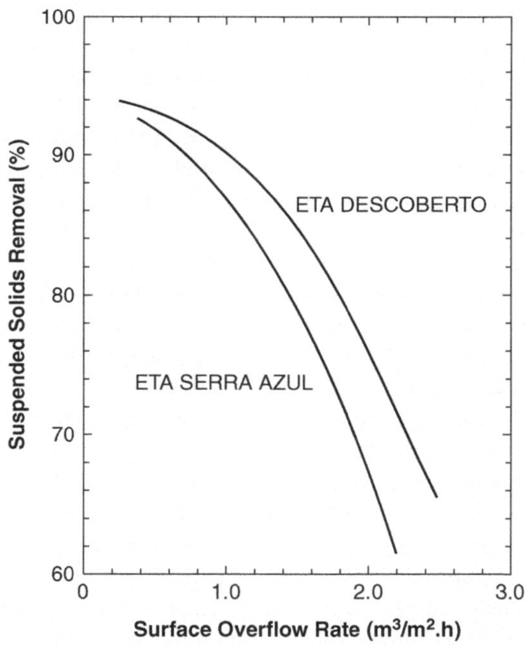

Figure B.7 Settling characteristics of filter backwash solids (Bratby, 1988).

Polymer addition can significantly increase the sedimentation rates of backwash solids. Column settling tests carried out in Connecticut at three water plants (Pines & Esponda, 2004), showed that settling rates without polymer ranged from approximately 0.4 to 1.0 m/h (240 to 565 gpd/ft^2). With 1 mg/l anionic polymer, corresponding settling rates increased to 1.8 to 2.7 m/h (1,100 to 1,600 gpd/ft^2).

Despite the fast settling rate of backwash solids, the thickening rate of these same solids is relatively slow. Without conditioning, solids loading rates of 0.5 and 1.0 $kg/m^2 \cdot h$ (2.5 and 5 $lb/ft^2 \cdot d$) for thickened concentrations of 1 and 2% respectively, are generally required – although actual rates depend on the raw water quality. With polyelectrolyte conditioning, thickening rates increase by as much as three times, but without necessarily increasing the thickened solids concentration significantly.

The specific resistance of thickened backwash solids varies, but is generally in the region of $1{,}200 \times 10^{10}$ m/kg. With polyelectrolyte addition, specific resistance reduces to approximately 150×10^{10} m/kg (see Figure B.3).

In conventional water treatment plants, by recycling filter backwash to the head of the plant after flow equalization, backwash solids are removed together with raw water suspended solids, in the sedimentation tanks. One advantage found with the recycle of backwash solids is improved flocculation because of the nucleating action of the recycled solids for floc growth. One example is the New Milford water plant in New Jersey, where the continuous recycle of backwash solids to the flocculation tanks and clarifiers resulted in lower head loss through the filters and longer filter runs. However, the plant also reported that filtered water turbidities were higher with the recycle of backwash solids. The alum dosage had to be increased by 15 percent to bring filtered water turbidities back to previous values (Wang *et al.* 1973).

Tobiason *et al.* (2003) surveyed backwash recycle practices at six water plants. All plants practiced some form of flow equalization, but only two of the six plants applied sedimentation to the backwash flow before recycle. None of the plants suffered deterioration in treated water quality based on filtered water turbidity and particle counts.

Other studies showed that when filter backwash solids were recycled back to the head of the plant, settled and filtered turbidity and *Cryptosporidium* oocysts were the same, or lower, than when using no recycle (Cornwell & Macphee, 2001).

Although not required by the EPA Filter Backwash Recycling Rule, removing suspended solids from recycle streams often does make sense because of the impacts of other contaminants contained in these residuals streams. Macphee *et al.* (2003) demonstrated that by adding polymer, recycle treatment clarification is much more efficient. Using an equalization tank, static mixer, flocculation and plate settlers to treat recycle streams, 0.1 mg/l nonionic polymer resulted in settled turbidities in the treated recycle of around 1 NTU, and particle counts lowered by approximately 50-percent compared with operation without polymer addition. The authors indicated that the key elements for recycle treatment were equalization, rapid mixing with polymer addition, and some form of solid-liquid separation.

McLane (2004) reported that at a water plant using lime for coagulation, by recycling fresh settled lime sludge back to the rapid mix stage of the coagulation-flocculation process, settled water turbidities reduced from 30 to 50 NTU, to approximately 3.0 NTU. Filter runs also increased from 30 to 50 hours, to over

150 hours. Other benefits included reduced disinfection byproducts due to the improved finished water turbidity and lowered chlorine demand.

Experiments carried out at the Rio das Velhas water plant in Brazil showed that by recycling equalized backwash at the rate of 6 percent the raw water flow, final filtered water turbidities were lower (Figure B.8). However, despite the improved settling characteristics of the flocculated solids, the settled water turbidity entering the filters was higher. However, with polyelectrolyte addition, settled water turbidities were significantly lowered.

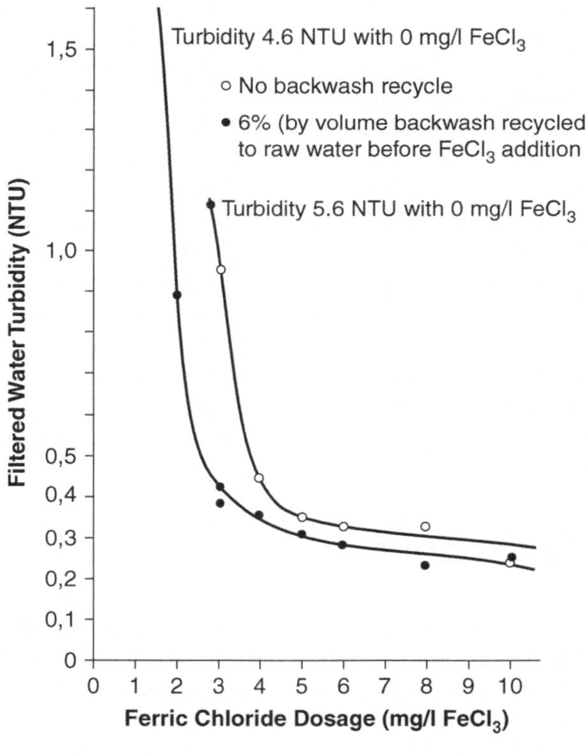

Figure B.8 Effect of backwash recycle on filtered water turbidity (Bratby, 1988).

The detrimental effects of recycled backwash solids on treated water quality sometimes observed, are due to breakup of flocs during backwash and recycle. These flocs don't necessarily re-form efficiently without using a polymer as flocculant aid. In some cases, increases in inert material and BOD, COD, and taste and odor compounds have also been observed (Faber & Nardozzi, 1972).

Increases in assimilable organic carbon (AOC), total organic carbon (TOC), manganese, preformed disinfection byproducts, and pathogens have also been observed when recycling residuals. In such cases, the removal of backwash solids prior to recycling back to the water plant may be advisable. Cornwell and Lee (1994) investigated the effects of recycle streams on treated water quality at a number of water plants. In general they found that by efficiently removing suspended solids from recycle streams, pathogens such as Cryptosporidium and Giardia, as well as contaminants such as TOC and manganese could be minimized in the recycle streams.

The mode of operation of unit processes within the plant could also reduce the concentrations of dissolved contaminants in recycle streams. For example, manually cleaned sedimentation basins and thickeners operated with deep blankets could exacerbate recycle quality issues. Cornwell and Lee also emphasized the importance of completely equalizing recycle flows – and recycling over a 24-hour period rather than intermittently to avoid spikes in treated water quality.

Backwash recovery tanks could be in the form of lagoons. However, in cases where the lagoons have the dual function of equalization and separation of solids, the accumulation of solids and storage for prolonged periods is not advisable. Decomposition of the solids and deterioration of recycled water quality is possible.

Sludge accumulated at the bottom of tanks or lagoons does not readily flow under water. Positive means for transporting the sludge should be provided. Mechanical means of sludge removal in backwash recovery tanks is one solution. Another is to design the tanks such that hydraulic sludge removal is possible. An example is the Serra Azul water plant in Brazil. The backwash recovery tanks were designed as long channels such that at each backwash discharge, sludge is transported to sludge hoppers at the ends of the channels by the first wave of backwash water. Recycled water is pumped from the head end of the channels where the water is the clearest (Bratby, 1988).

B.4 SLUDGE LAGOONS

Lagooning has been one of the most popular methods of sludge treatment due to its simplicity and low cost of operation. In areas where ample land is available, lagooning is quite economical. The general procedure in constructing a lagoon is to enclose a suitable area with dikes or excavation. Normally no attempt is made to provide additional underdrainage. The method of operation is to fill the lagoon with sludge and allow it to dry whereupon more sludge is added and the process repeated until the lagoon is full of sludge dry enough for removal.

Lagooning could have the objective of final disposal, or of dewatering. Lagoons used for disposal generally have depths of up to 4 m (13 ft). During the filling period, supernatant water is collected and often recycled back to the plant. Sludge is ideally accumulated in layers, so that previous layers are able to compact and dry to a certain extent before successive layers are applied. When the lagoon is full of sludge, a soil cover is applied and new lagoons utilized. In other cases, accumulated sludge is dredged and air dried, then carted to landfill or other final disposal.

Supernatant water removed from the lagoons is sometimes recycled back to the head of the water plant for treatment jointly with raw water, but if the quality of the supernatant is not adequate for recycling, it is sometimes permitted to be discharged to the water body, as long as suspended solids have been removed.

Concentrations of solids in lagoons depend principally on the method of operation of the lagoons. After various years of sludge accumulation, concentrations at the bottom of the sludge layers typically reach 10 to 20% dry solids, reducing significantly towards the top of the sludge layers, to as low as 2% dry solids. A dry crust is often formed on the surface of the sludge that impedes further atmospheric drying. In many cases, the average concentrations reached in such lagoons after various years of drying is only 10% dry solids (AWWARF, 1969–1970; Albrecht, 1972; Committee Report, 1978).

Better performance is achieved if the dried crust can be scraped from the surface of the lagoon and piled near the edges, allowing lower layers to also air dry. This methodology may not often be applied in practice because of the high operating costs.

In some cases, bottom drainage has been provided, but success has been limited because of the poor quality of the drainage water, and blockage of the drainage layers (Hubbs & Pavoni, 1974).

For relatively small water plants, sludge application in lagoons is probably the first choice if land area is available. Dixon (1985) reported that of 19 water plants in Pennsylvania, seven utilized lagoons for thickening sludge, and eight for dewatering sludge.

Thickening sludges in lagoons has been practiced successfully. Without polymer conditioning, thickened concentrations of 4 to 6% after 2 to 3 months of thickening are common, with constant supernatant withdrawal. Lime sludges can reach 30% dry solids after 2 to 3 months thickening.

Sludge at 4 to 8% dry solids concentration remains virtually vertical under water. In other words, such sludges do not flow of their own accord. Only when supernatant is completely withdrawn will such sludges flow – and then with difficulty. One solution adopted by the Lake Ontario water plant in New York was to modify a front-end loader which allowed it to enter the lagoons with sludge as deep as 1.5 m (5 ft). Sludge was thus pushed to suction sumps for removal by sludge pumps.

One issue with lagoons used for disposal is that eventually the lagoons are filled and are no longer available for sludge disposal. Utilities then need to find additional areas to dispose of the sludge – often a difficult undertaking – or to remove the wet lagoon sludge by dredging, mechanically dewater the sludge, haul the sludge to landfill or other means of disposal, and free up lagoon space for continued use. These alternatives are costly.

Hughes *et al.* (1993) reported on the North Jersey District Water Supply Commission's solution to this issue. The Wanaque Reservoir 530 Ml/d (140 mgd) water plant disposed of sludge in a 5 ha (12-acre) lagoon. The lagoon was filled to capacity with 85 Ml (26 MG) of 2 to 6% sludge. Additional room to expand the lagoon area was not available. The District had practiced lagoon dredging and mechanical dewatering using a belt press. The press only achieved 8 to 15% dry solids. This wet cake was hauled within the site and dumped on site.

A long term viable alternative implemented by the District was to install a gravity thickening and recessed chamber filter press dewatering system. The new facility processed new sludge generated at the plant, as well as lagoon sludge dredged from the lagoon. Sludge cake at 25 to 30% dry solids was further dried in a covered area to achieve 40 to 50% dry solids. The drying area required was 2,600 m^2 (28,000 ft^2). The cake was broken up by disc harrowing to accelerate drying and to reduce the final void volume.

The existing lagoon was partitioned by dikes formed using imported soil. One part was maintained for site drainage and emergency blowdown requirements – approximately 10-percent of the lagoon area, or 11,400 m^3 (9.2 acre-feet) of lagoon storage. The rest of the lagoon area was used for disposal of dried (40 to 50%) cake.

Once each area of lagoon was filled with dried sludge cake, to the dike road level, the area would then be covered by topsoil and vegetation. It was estimated that by following the new operating strategy, at the plant capacity of 530 Ml/d (140 mgd), a further twenty-five years of storage was available within the already filled lagoon volume.

When lagoons are used for sludge drying, applied sludge depths are normally 1 to 2 m (3 to 6.5 ft). After dewatering, sludge depths are usually 0.3 to 1 m (1 to 3 ft). During the sludge filling stage, supernatant is removed to allow partial drying of successive layers. When the lagoon is full, with concentrations of approximately 4 to 6% dry solids, sludge is allowed to stay in the lagoon for 4 to 12 months for further drying. Depending on the climatic conditions, concentrations of up to 40% are possible with alum or ferric sludges. With lime sludges, concentrations of 50% or more are possible.

For drying lagoons, at least three lagoons are required. Each lagoon is often sized for filling in three months, with a subsequent three-month drying cycle. In arid regions, a typical application rate is 80 kg dry solids/m^2 (16 lb DS/ft^2). In wet regions, applications are approximately halved (40 kgDS/m^2, or 8 lbDS/ft^2).

For example:

- Consider a sludge production of 1,000 kgDS/day (2,200 lbDS/d). Assume a filling time of 3 months and an application rate of 60 kg/m^2.
- The area required for each lagoon is $1{,}000 \times 3 \times 30/60 = 1{,}500$ m^2.
- Assume the average concentration after filling the lagoon is 5%. The sludge density is $2{,}300/(2.3 - 0.013 \times 5) = 1{,}029$ kg/m^3.
- The sludge volume would be $1{,}000 \times 3 \times 30/(0.05 \times 1{,}029) = 1{,}749$ m^3.
- The depth of sludge at the start of the drying cycle = 1.17 m (3.8 ft).

The Cornell University water plant treated 15 Ml/d (4 mgd) using alum. The sludge production was 314 kg/d (690 lb dry solids/day). Sludge removed from the clarifiers was discharged to two lagoons, each with 930 m^2 (10,000 ft^2) and 2 m

(6.5 ft) wetted depth. After 2 months of sludge accumulation, with supernatant withdrawal and under-drainage, the average concentration in the lagoon reached 25% dry solids. At this point, the sludge depth was 30 cm (1 ft).

In winter, sludge in the lagoon freezes and in spring, after thawing, the sludge concentration is 55% dry solids. Because of the drastic reduction in sludge accumulation by the freeze-thaw cycles, various years of successive sludge accumulations are possible. In this installation, sludge was only removed after 6 years of sludge accumulation, with a final sludge depth of approximately 80 cm (2.6 ft) (Schaefer & Clark, 1989).

Another example of freeze-thawing in sludge lagoons is the Lake Ontario direct filtration water plant in New York (Fitch & Elliott, 1986). This plant treated 260 Ml/d (68 mgd) using alum. Filter backwash water, with 250 to 300 mg/l suspended solids, was sent to two lagoons. After nine months of sludge accumulation, the sludge in the lagoons reached 8.5% dry solids. After freezing and subsequent thawing, the final concentration of the sludge was 25% dry solids.

B.5 SLUDGE DRYING BEDS

When suitable land areas are available, drying beds are used extensively for dewatering and drying sludges. The dewatering mechanisms in drying beds are a combination of decanting, drainage, and evaporation. The overall size of the beds is a function of the time it takes to evaporate to the desired dryness. Therefore, the amount of water removed by drainage and/or by decanting will reduce the overall bed size.

Drying beds can be constructed with bottom drainage – referred to as *sand drying beds*, or as shallow drying basins without bottom drainage – referred to as *solar drying beds* (Figure B.9).

Figure B.9 Removal of cake from drying bed.

Drying beds with drainage often include 15 to 25 cm (6 to 10-inches) 0.3 to 1.2 mm sand, overlying 30 cm (12-inches) gravel. Drains are often perforated PVC, clay or concrete pipe. After drying, sludge is removed either manually (particularly in small beds), or mechanically. With sand drying beds, to avoid removal of the sand layers with the dried sludge, perforated brick or concrete blocks, or some proprietary support media (see Figure B.10) are often installed to support a small front-end loader, or to facilitate the manual or mechanized removal of the sludge. Fouling of the sand layers within the block openings is an issue, and the sand within the openings is periodically removed and replaced with clean sand. In some larger installations, traveling sludge removal equipment could be used to more efficiently remove the dried sludge cake.

Figure B.10 Specialized sand drying bed support media (Courtesy Deskins).

Solar drying beds are normally constructed as shallow concrete basins in which decanting, then evaporation takes place. Sludge is removed manually in small installations, or with a small front-end loader in larger installations.

Since alum and ferric sludges generally drain poorly without polymer conditioning, drying beds without drainage are often employed when drying these sludges. However, polymer conditioning improves drainage significantly, as well as impedes the penetration of sludge into the sand layers. Polymer conditioning can significantly improve the yield of sand drying beds

(Dharmappa *et al.* 1997). Polymer could also be added before separate thickening, thereby applying a more concentrated sludge to the drying beds. Alternatively, thickening could occur within the bed itself, provided adequate drainage of supernatant is possible.

At least three or more beds are required to permit cake removal from one bed while the others are being charged with sludge, or drying. The layer of thickened sludge applied should be relatively shallow – preferably 15 cm (6-inches), but not more than 30 cm (12-inches).

Rainfall water is removed from the surface, in the same way as supernatant, by means of surface drawoff weirs or piping. Rainfall water is not readily incorporated back into thickened or drying sludge.

Sludge drying proceeds by means of evaporation. Sludge concentrations of 20 to 40% dry solids are typical, depending on the climatic conditions and the type of coagulant used during treatment. In the case of lime sludges, concentrations of 50% or more are possible.

Application rates often applied are in the region of 10 to 15 kg dry solids/m^2 (2 to 3 lb/ft^2). The Union Hills plant in Phoenix, Arizona reported that a treated water flow of 300 Ml/d (80 mgd) produced 1.4 t/day (1.5 ton dry solids/day). Sludge was thickened in two gravity thickeners then applied to solar drying basins with a total area of 15,000 m^2 (161,000 ft^2).

Sludges formed using alum and polyaluminum chloride drain poorly and, without conditioning, drainage beds release relatively little liquid. Sludges formed using ferric coagulants have somewhat better drainage characteristics, and lime sludges drain relatively well. Cornwell and Vandermeyden (1999) reported on tests carried out on the drainage characteristics of 57 different water plant sludges: 38 sludges from plants using aluminum sulfate, 5 using polyaluminum chloride, 9 using ferric chloride, and 9 using lime for softening. Mean specific resistance to filtration values for each of these sludge types were 15.7, 13.8, 6.4 and 0.54 × 10^{13} m/kg, for the alum, PACl, ferric and lime sludges respectively.

Based on the compiled data from pilot plant tests the drainage from sand drying beds could be predicted. Table B.3 (adapted from Vandermeyden & Cornwell, 1993) shows the compiled data for alum sludges. Without polymer conditioning, the data showed that the drainage volume through the sand and supporting layers varied from 8-percent to 64-percent the original volume. Lower solids loadings resulted in higher drainage volumes. Lower capillary suction times (CST), reflecting faster draining sludges, also resulted in higher drainage volumes.

Table B.3 shows that with unconditioned sludges, the drainage performance deteriorated markedly when the loading to the beds exceeded approximately 10 kg/m^2 (2 lb dry solids/ft^2), and applied sludge concentrations exceeded 2%.

Polymer conditioning of the applied sludge significantly improved drainage of water from the sludge. With polymer conditioning, higher solids loadings performed better than lower solids loadings.

Table B.3 Predicted sand bed drainage volumes.

Solids Loading kg/m² (lb/ft²)	Initial Solids Conc. (%)	Unconditioned CST Sec	(Solids Loading)/ (Solids Conc.) lb/ft² · (%)	Unconditioned Drainage Volume (%)	Conditioned Drainage Volume (%)
4.9 (1.0)	1	48	1.00	64	–
	2	104	0.50	37	–
	3	160	0.33	32	–
	4	270	0.25	27	–
9.8 (2.0)	1	48	2.00	54	65
	2	104	1.00	27	58
	3	160	0.67	25	52
	4	270	0.50	20	46
14.7 (3.0)	1	48	3.00	44	65
	2	104	1.50	20	73
	3	160	1.00	20	62
	4	270	0.75	15	50
19.6 (4.0)	1	48	4.00	–	65
	2	104	2.00	15	74
	3	160	1.33	13	68
	4	270	1.00	8	54

After decanting and drainage in sand drying beds, and after decanting in solar drying beds, the final step to produce a manageable cake for final disposal is evaporation. Calculations for this step of the process utilize local pan evaporation data available from meteorological stations. Cornwell and Vandermeyden (1999) found that from pilot and full scale tests at three water plants, the actual evaporation from drying beds was under-predicted using pan evaporation data when the drying sludge concentration was below a certain value. Beyond a certain cake concentration, when bound water is the predominant water content in the sludge, the evaporation rate in the bed is less than the pan evaporation rate. Table B.4 summarizes the full scale data obtained at three utilities (adapted from data in Cornwell & Vandermeyden, 1999).

Although it was shown that pan evaporation data underestimate the rate of actual evaporation occurring in drying beds until a certain concentration is reached (30 to 60% dry solids) it is probably adequate to calculate bed areas on the basis of pan evaporation data (Cornwell & Vandermeyden, 1999). The results will likely be somewhat conservative, but will compensate to a certain extent for other uncertainties in design.

Table B.4 Full scale drying bed evaporation results.

Solids Load kg/m² (lb/ft²)	Initial Drained/ Decanted Depth (mm) (in)	Initial Drained/ Decanted Conc. (%)	Evap. Rate mm/mth (in/mth)	Time to Attain Conc. (Days)				Cross-Over Conc. (%)
				25%		25%		
				Meas.	Calc.	Meas.	Calc.	
5.0 (1.0)	95 (3.75)	5.3	140 (5.6)	4	11	10	14	60
5.0 (1.0)	238 (9.5)	2.1	140 (5.6)	25	40	35	>50	–
17 (3.4)	303 (12)	5.6	160 (6.4)	15	26	30	30	40
15.5 (3.1)	250 (9.8)	6.2	303 (12.1)	6	8	12	9	30

Figure B.11 demonstrates the sequence of calculation to determine the drying times and number of beds required. For the particular example shown in Figure B.11, sludge deposited in a bed in January would reach a concentration adequate for removal and hauling within somewhere between 3 and 4 months. Sludge deposited in November would require between 4 and 5 months. The spreadsheet can also be set up to calculate on a weekly basis, although it is likely that evaporation and precipitation data would be input as monthly average data.

B.6 MECHANICAL THICKENING AND DEWATERING

Sludges generated in water plants are generally characterized as having relatively low concentrations. Consequently, in cases where lagoons or decanting beds are not feasible because of area or climatic restrictions, mechanical means of thickening and dewatering the sludge are often adopted.

Figure B.1 shows that in 1977, of 341 water plants in the U.K., 40% of the plants processed their sludge in lagoons; 11% in drying beds; and 7.6% dewatered the sludge by mechanical means. Of the latter mechanical dewatering group, 73% used filter presses, 15% centrifuges, and 12% vacuum filters. A similar study carried out in Europe in 1989 showed that of wastewater treatment plants that dewatered sludge by mechanical means, 35% used filter presses, 31% belt press, 27% centrifuges, and 7% vacuum filters. The decreasing popularity of vacuum filters is likely due to the lower cake concentrations achieved and the mechanical complexity of the machines. The advent of more efficient high-solids centrifuge machines has made this dewatering device more popular in the United States and elsewhere.

Prior to mechanical dewatering, thickening of water plant sludges is mandatory to reduce the cost of the high performance dewatering devices.

EVAPORATION IN DRYING BEDS

This program calculates the theoretical drying times for sludge applied to a series of drying beds given the climatic conditions, sludge characteristics, and applied sludge depth.

CLIMATIC CONDITIONS:

Month	Precip mm/mth	Free Water Evapn* mm/mth	(Ew-qr) mm per day
JAN	**55**	**59**	0.15
FEB	**30**	**74**	1.56
MAR	**38**	**116**	2.50
APR	**19**	**154**	4.51
MAY	**6**	**197**	6.15
JUN	**30**	**220**	6.33
JUL	**145**	**165**	0.66
AUG	**165**	**137**	–0.90
SEP	**57**	**133**	2.55
OCT	**36**	**118**	2.67
NOV	**17**	**74**	1.90
DEC	**37**	**59**	0.70

*Assumed100%of free pan evap.rate

Sludge concentration after settling and decanting, Sd =	**3**	%
Initial Sludge Depth =	**150**	mm
(Initial dry solids load =	4.5	kg/m2)

The spreadsheets below calculate the sludge concentrations achieved after each month drying period, up to 12 months, for sludge applied in each month.

qi = initial water content
qr = precipitation
fee = evaporation factor relative to free wat
Ew = free water evaporation
qd = final water content

Month beginning in January:

End of	qi (kg/m2)	qr (kg/m2)	Ew (kg/m2)	qd (kg/m2)	Conc (%)
JAN	146	55	59	140.9	3.1
FEB	141	30	74	97.4	4.4
MAR	97	38	116	20.0	18.4
APR	20	19	154	0.0	100.0

Month beginning in February:

End of	qi (kg/m2)	qr (kg/m2)	Ew (kg/m2)	qd (kg/m2)	Conc (%)
FEB	146	30	74	102.0	4.2
MAR	102	38	116	24.5	15.5
APR	25	19	154	0.0	100.0

Month beginning in November:

End of	qi (kg/m2)	qr (kg/m2)	Ew (kg/m2)	qd (kg/m2)	Conc (%)
NOV	146	17	74	88.6	4.8
DEC	89	37	59	66.7	6.3
JAN	67	55	59	62.2	6.7
FEB	62	30	74	18.6	19.4
MAR	19	38	116	0.0	100.0

Month beginning in December:

End of	qi (kg/m2)	qr (kg/m2)	Ew (kg/m2)	qd (kg/m2)	Conc (%)
DEC	146	37	59	123.7	3.5
JAN	124	55	59	119.1	3.6
FEB	119	30	74	75.6	5.6
MAR	76	38	116	0.0	100.0

Figure B.11 Example of spreadsheet calculations for evaporation in drying beds.

B.6.1 Sludge thickening

B.6.1.1 Gravity thickeners

Water plant sludges thickened in gravity thickeners can reach dry solids concentrations of 0.3 to 10 percent, when metal coagulants are used during water treatment. The actual concentration achieved depends largely on the raw water quality, coagulants used during treatment, and the organic fraction of the raw water.

Without conditioning, solids loading rates to the thickeners are generally between 0.5 and 1.0 $kg/m^2 \cdot h$ (2.5 to 5 $lb/ft^2 \cdot d$). With conditioning, rates can be increased to 1.0 to 4.0 $kg/m^2 \cdot h$ (5 to 20 $lb/ft^2 \cdot d$).

Gravity thickener performance is generally improved when sludge is fed continuously with essentially constant properties. In cases where significant changes in sludge quality are anticipated, equalization of the sludge is advantageous (Warden, 1982).

The thickened sludge concentration achieved depends primarily on raw water quality. High turbidity waters produce sludges with higher thickened concentrations compared with lower turbidity waters. The density of flocs generally increase when raw water turbidity increases (Calkins & Novak, 1973).

Occasionally lime is used in a second thickening stage, achieving concentrations of up to 8% with hard to thicken sludges (Levesque, 1982). Lime conditioning requires a contact time of 15 to 30 minutes. pH values after lime addition are typically 9.5 to 11.5.

When lime is used during water treatment, gravity thickened concentrations of up to 30% are possible.

The Choisy-le-Roi water plant in France had a treated water flow of 800 Ml/d (211 mgd) and sludge production of 20 t/day (22 U.S. ton dry solids/day). Alum was used as primary coagulant during treatment. The clarifiers were split into four levels, with mechanical sludge removal at each level. The average concentration of sludge removed from the clarifiers was approximately 0.5% (Ventresque & Bablon, 1989).

A high molecular weight cationic polymer was used for sludge conditioning, prior to gravity thickening. The polymer accelerated the rate of thickening, but did not significantly increase the thickened concentrations achieved. Polymer was dosed at 1.5 to 2 g/kg (3 to 4 lb/U.S. ton dry solids). The best point of polymer addition was immediately upstream of the thickener using an orifice plate in the sludge piping that provided a G value of approximately 10,000 s^{-1}.

The gravity thickeners were operated at average loading rates of 0.75 $m^3/m^2 \cdot h$ and 1.4 $kg/m^2 \cdot h$ (440 gpd/ft^2 and 6.9 $lb/ft^2 \cdot d$). The sludge removal rate was 0.2 m/h (118 gpd/ft^2).

Feed sludge concentrations were at 0.1 to 0.5% and the thickened sludge 5 to 10% dry solids. The supernatant solids concentration was less than 30 mg/l.

A two-stage thickening installation was utilized at the Nishi-Nagasawa water plant in Japan (Bishop, 1978). In the first thickener concentrations were up to 6 to

9%. The second thickener attained concentrations of 7 to 10%. Sludge conditioning was by sodium silicate and polymer. Thickened sludge was dewatered in a rotating drum device ("dehydrum") which was a part of the pelletization process. Concentrations of 25 to 30% were achieved. Sludge was then dried in oil-fired dryers to attain cake concentrations of 65 to 70% solids.

Studies were carried out by the Water Research Centre (WRC) in the U.K. to determine the best form of sludge scraper mechanism in gravity thickeners. Warden (1982) pointed out that a sludge with a concentration of 10% would have a density of approximately 1,065 g/l. However, when submerged, the relative density is only 65 g/l. This is the reason sludges do not flow readily under water – at the bottom of thickeners, and require a positive means of mechanical transport.

The WRC design has a central spiral "scoop" scraper and echelon scrapers situated towards the outside. The echelon scrapers gradually push the sludge towards the center at each rotation of the mechanism. At a certain radius the spiral scraper scoops the sludge towards the central hopper.

From pilot plant results, Warden (1982) presented the following correlation for the thickened sludge concentration obtained with the sludges tested:

$$\text{Thickened sludge (\%)} = \frac{\text{Total sludge solids (mg/l raw water)}}{0.25 \times \text{coagulant hydroxide (mg/l)} + 0.02 \times \text{color (°H)}}$$

Bratby *et al.* (1993) applied the principles of the WRC design to the sludge processing facility at the Papago water plant in Tempe, Arizona. The treatment plant comprises presedimentation, coagulation and flocculation, sedimentation, filtration, disinfection, and fluoridation. Minimum, average and maximum design influent raw water flow rates were 75, 150 and 215 Ml/d (20, 40 and 57 mgd) respectively.

Chemicals used at the plant included aluminum sulfate for coagulation, polymers for flocculant aid and for filter aid, chlorine for disinfection, potassium permanganate for pre-oxidation, powdered activated carbon for taste and odor control, lime for pH control, and hydrofluosilicic acid for fluoridation. Alum is the chemical that most contributes to sludge production; the other chemicals have a relatively minor impact.

Sludge is produced from three major sources in the plant: presedimentation, sedimentation, and filter backwash. Filter backwash water is recycled, after flow equalization, to the presedimentation basins. Suspended solids contained in the backwash water settle out principally in the presedimentation basins and, to a certain extent, in the sedimentation basins.

During a large part of the year, raw water turbidities are relatively low, from about 2 to 10 NTU. However, during the rainy season, the silt load in the raw water is high and turbidities peak up to about 1,000 NTU. Figure B.12 presents typical raw water turbidities and alum dosages at the plant. During high turbidity periods, the alum-to-turbidity ratio is relatively low and the sludge produced at the plant is typically easy to thicken and dewater. Conversely, at low raw water turbidity periods, the alum-to-turbidity ratio is high and the sludge is correspondingly more difficult to thicken and dewater.

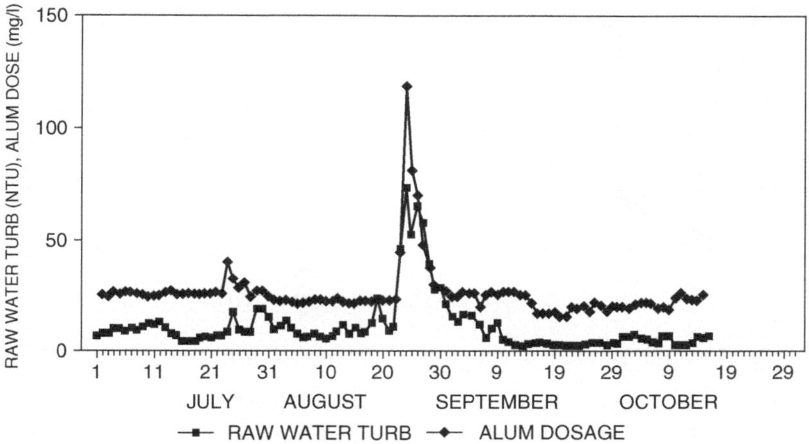

Figure B.12 Raw water turbidity and alum dosage (Bratby *et al.* 1993, 1996).

Figure B.13 shows the general relationship between raw water turbidity and alum dosage. Figure B.14 presents the total quantity of sludge produced per mgd of raw water flow, based on the raw water turbidity and aluminum sulfate and other chemical dosages.

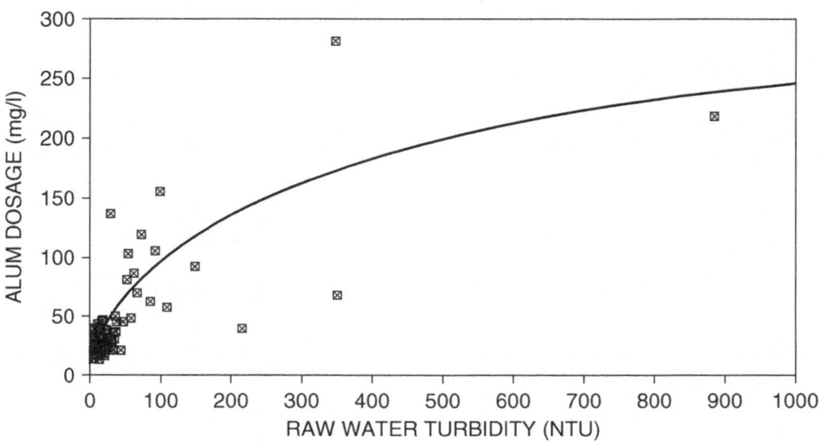

Figure B.13 Raw water turbidity and alum dosage (Bratby *et al.* 1993, 1996).

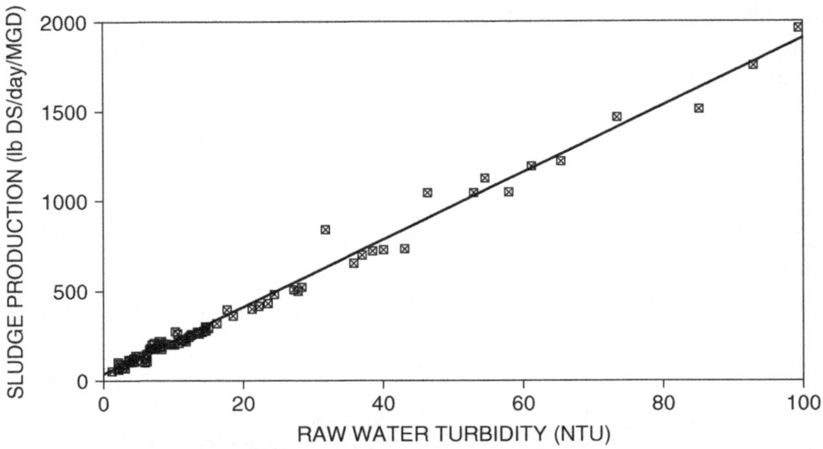

Figure B.14 Raw water turbidity and sludge produced (Bratby *et al.* 1993, 1996).

Figure B.15 presents typical sludge concentration versus time profiles for both the presedimentation and the sedimentation tanks. In both cases, the high initial concentration in the sludge hoppers rapidly decreases with time during sludge withdrawal.

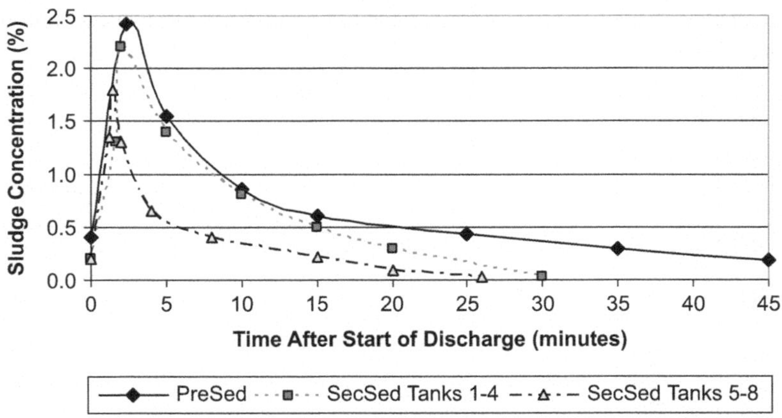

Figure B.15 Sedimentation tank sludge discharge concentrations with time (Bratby *et al.* 1993, 1996).

Figure B.16 presents typical combined sludge concentrations from the presedimentation and sedimentation tanks, as a function of raw water turbidity.

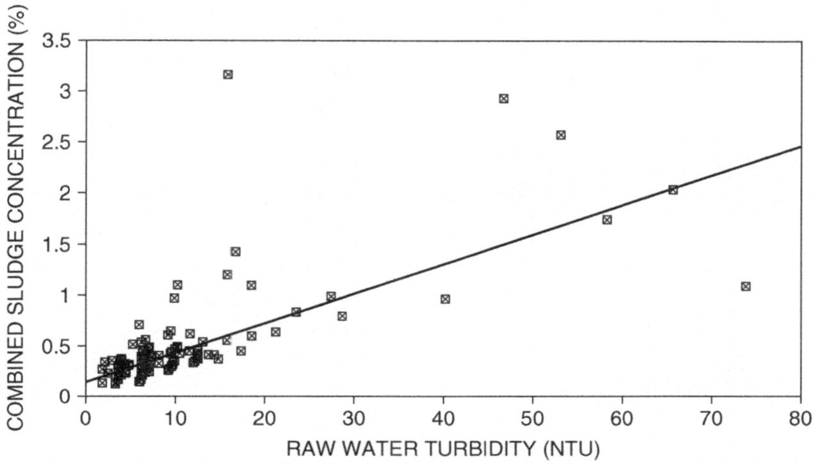

Figure B.16 Combined presedimentation and secondary sludge discharge concentrations (Bratby *et al.* 1993, 1996).

From these analyses of sludge production characteristics, the design of the mechanical sludge processing facility was based on a maximum sludge production month and an 8-hour day, 5-day per week operating regime. The design basis was 10,200 kg/d (22,400 lb dry solids per day). The 7-day sludge discharge volume was 1.0 Ml/d (260,000 gallons/day). The peak daily sludge production reaches 41,000 kg/d (90,000 lb dry solids/day) but the peaks are of relatively short duration, and can be processed by running extra shifts during the day or running the system into the weekend.

Figure B.17 presents a schematic of the facility. Sludge was discharged to a 64 m^3 (17,000 gallon) balancing tank. Constant speed pumps provided a constant sludge flow to the gravity thickener. Two pumps were provided, each with separate discharge piping. Each pump was rated at 1.4 Ml/d (250 gpm). Dual pipes entered the gravity thickener at a horizontal shallow trough extending from the periphery to the central feedwell of the thickener. The shallow trough served to gently introduce the flocculated solids to the thickener, and to provide an effective visual evaluation of the efficacy of polymer conditioning.

The constant speed pump rate provided optimal polymer conditioning. An in-line orifice plate on each line provided a constant G value for mixing. The orifice plate was interchangeable so that the best G could be determined by trial and error. Orifice plates of diameter 57 to 110 mm (2.25 to 4.31 inches) provided G values ranging from 1,000 to 10,000 s^{-1} (assuming a mixing length of 7.5 pipe diameters downstream of the orifice plate, and a velocity that is the average of the velocity through the orifice and the velocity in the pipe downstream of the orifice).

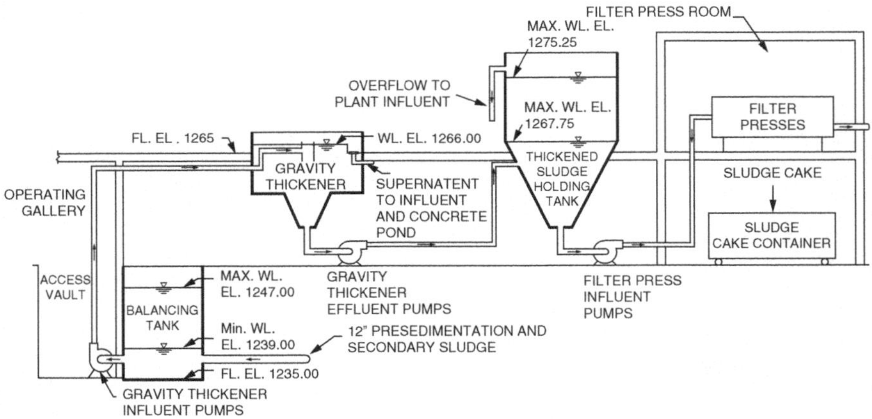

Figure B.17 Schematic of the Papago WTP Sludge Processing Facility (Bratby *et al.* 1993, 1996).

The gravity thickener was 7 m (23 ft) diameter and 2.1 m (7 ft) wetted sidewall depth, with a flat floor. The constant hydraulic loading rate was 1.5 m/h (866 gpd/ft^2). The expected design thickened sludge concentration under maximum month loading conditions was 8% dry solids.

Thickened sludge pumps were also constant speed rated at 0.38 Ml/d (70 gpm). Therefore, the sludge withdrawal rate was at 0.41 m/h (240 gpd/ft^2). Thickened sludge was pumped to a 265 m^3 (70,000 gallon) thickened sludge storage tank. Sludge was then batch fed to two recessed chamber filter presses.

The concentration of sludge collected in the balancing tank ranged from approximately 0.2 to 3.8%, with a mean value of 1.5% dry solids. Sludge from the balancing tank was pumped to the gravity thickener, after polymer addition. Thickened sludge concentrations attained in the gravity thickener were above 10%. However, there is a practical limit to the maximum concentration which is desirable in the gravity thickener. If the thickened sludge concentration and the depth of sludge in the gravity thickener are allowed to rise to excessive levels, a phenomenon occurs whereby sludge effectively "gels" and is not transported to the center of the thickener, but tends to simply rotate with the scraper mechanism, thereby preventing the sludge from being pumped to the sludge holding tank. At this point the only recourse was to drain the thickener and "squeegee" the sludge to the hopper. Two other plants in the United States that employed this thickening technology were contacted and basically the same problem was found.

The key to successful operation of the gravity thickener was to rigorously prevent polymer overdosing, limit the total depth of sludge to approximately 2 feet, and limit the thickened sludge concentration to 5% dry solids. These guidelines are somewhat contrary to optimal operation of the filter press. However, sludge does continue to thicken in the sludge holding tank. Thickening in the sludge holding tank was achieved by including decant piping in the holding tank. Therefore, thickened sludge at approximately 8% dry solids was still fed to the filter presses.

Evidently, success of operation of the WRC design depends on the sludge type. Sludges produced from turbid waters appear to be subject to the above limitations if the thickened sludge concentration is allowed to increase beyond approximately 5% dry solids.

B.6.1.2 Belt thickeners

Belt thickeners have gained popularity because of the relatively low cost and compact area of the installation (Figure B.18). Sludge is conditioned using polymers– or sometimes metal coagulant and polymer, and applied evenly to the belt of the thickener. The sludge then moves with the belt. Fixed guide vanes or plows located just above the surface of the moving belt cut through the sludge and create clear zones through which liquid drains freely through the belt fabric. Thickened solids, scraped from the belt, are collected in a hopper and pumped to further processing stages.

Figure B.18 Belt thickener (Courtesy Eimco).

Pilot tests conducted at the Papago water plant in Arizona (Bratby *et al.* 1993) tested recessed chamber filter presses from different manufacturers. One of the suppliers used a belt thickener to thicken sludge prior to dewatering. This application was very successful and produced thickened sludge concentrations of 11 to 18% dry solids during low raw water turbidity (2 to 7 NTU) periods when thickening and dewatering of the sludge was typically the most difficult.

B.6.1.3 Flotation thickeners

Dissolved-air flotation has been used to thicken water plant sludges. Thickened concentrations of at least 2.5 to 4.5% have been reported on alum sludges using polymer. Application rates are typically significantly higher than with gravity thickeners. Loading rates of 5 to 10 $kg/m^2 \cdot h$ (25 to 50 $lb/ft^2 \cdot d$) are typical with flotation thickeners applied to water sludges (Levesque, 1982).

One example is the application of flotation to sludges produced from the treatment of highly colored, low turbidity waters. These sludges are very difficult to thicken. One example is the Kloofnek water plant in Cape Town, South Africa (Bratby & Marais, 1977) where water treatment consisted of 10 mg/l sodium aluminate, followed by lime, then by 50 mg/l alum, with final pretreatment pH 5.5. Even after one week of gravity thickening in the settler hoppers, sludge concentrations never exceeded 0.25%. Flotation thickening using a cationic polyelectrolyte at a dosage of 0.5 g/kg (1.0 lb/t) dry solids, achieved thickened sludge concentrations of up to 12% dry solids.

Flotation separation and thickening has also been applied successfully for the recovery of filter backwash residuals. For example, tests carried out at the City of Boulder Betasso water plant showed that, compared with ballasted sedimentation and plate settlers, dissolved-air flotation was the best suited technology for spent filter backwash recovery (Patawaran *et al.* 1999). Flotation effluent particle counts were less than 100 and turbidities less than 1 NTU. The flotation system also provided robust treatment with good effluent quality even under stressed conditions. A cationic polymer at 0.2 mg/l was found optimal for flotation of the spent backwash water.

B.6.1.4 Centrifuge thickening

Centrifuges have been most commonly applied for dewatering water plant sludges, after thickening by other means (such as gravity thickening). However, just as with wastewater sludges, where centrifuge thickening has made successful inroads, there may be applications where centrifuge thickening of water plant sludges is a viable option.

Peck *et al.* (1993) studied different thickening alternatives for processing water plant residuals at the 140 mgd Val Vista water plant in Mesa, Arizona. On the basis of continuous operation, the project cost for implementing a centrifuge thickening system would be U.S.$8.8 million. Belt or gravity thickener options would each be $2.7 million. For this application gravity thickeners were installed. However, centrifuges were adopted for dewatering the gravity thickened sludge.

B.6.1.5 Other thickening devices

There are other mechanical thickening devices that could be effective in coagulant sludge thickening applications. These include rotating drum, disc and screw thickeners.

The rotating drum thickener, RDT, an example of which is shown in Figure B.19, comprises polymer addition; a tangential flow flocculation tank; a dewatering drum with augers and woven wire mesh, or wedge wire, or perforated plate screens for water-sludge separation; and wash water sprays to clean the drum screens. The drum rotates slowly and the augurs transport the thickening sludge to the discharge end. One manufacturer provides nominal unit sizes ranging from 5 to 90 m^3/h (25 GPM to 400 GPM).

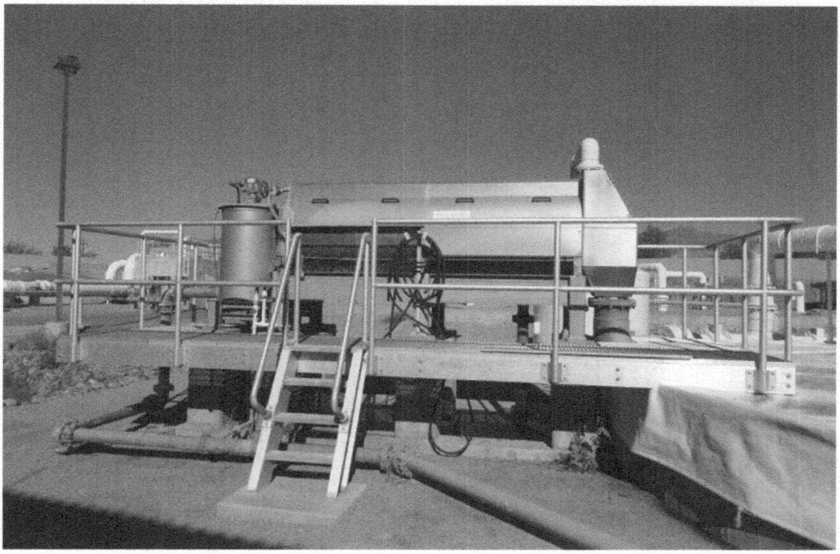

Figure B.19 Rotating drum thickener (Courtesy Parkson: www.parkson.com).

The disc thickener, shown in Figure B.20, comprises polymer addition; a flocculation tank with stirrer; an inclined rotating filter disc; and wash water sprays to clean the stainless steel filter cloth. The filter disc rotational speed is adjustable depending on the thickening conditions, but is generally less than 5 rpm. The flocculated sludge as it enters accumulates at the lower end of the disc providing a small hydrostatic head for drainage. Plows help with further drainage as the thickening sludge is transported up the inclined screen. A curved baffle plate directs the rotating thickened sludge to the upper and outer edge for discharge to a chute. Filtrate is discharged from the opposite lower end. One manufacturer provides nominal unit sizes ranging from 20 to 40 m^3/h (90 GPM to 180 GPM).

The screw thickener, shown in Figure B.21, comprises polymer addition; a flocculation tank with stirrer; an inclined cylindrical wedge-wire screen/basket; a slowly rotating screw that conveys the thickening sludge up the inclined basket; and wash water sprays to periodically clean the wedge-wire screen. The screw speed and the sludge discharge weir height are adjustable depending on the sludge and thickening conditions. One manufacturer provides nominal unit sizes ranging from 20 to 100 m^3/h (90 GPM to 450 GPM).

Figure B.20 Disc thickener (Courtesy Huber technology: www.huber.de).

Figure B.21 Screw Thickener (Courtesy Huber technology: www.huber.de).

B.6.2 Sludge dewatering

B.6.2.1 Vacuum filters

Vacuum filters (Figure B.22) are still used in some industries, including mining and paper processing. In the case of water sludges, some lime softening plants still use this dewatering device. However, it appears that plants processing alum or ferric sludges have opted for alternative means of sludge dewatering.

Figure B.22 Vacuum filter (Courtesy Eimco).

Rotary vacuum filters basically comprise a cylindrical drum that rotates partially submerged in the sludge to be dewatered. The drum is divided in various compartments. When compartments are immersed in the sludge, vacuum is applied and water passes through the filtering medium and forms a cake on the drum surface. After almost a full rotation, the vacuum is released and the cake is released from the surface of the drum. Operation passes through three zones: firstly a pick-up or form section where a vacuum is applied to draw liquid through the filter media and to form a covering of partially dewatered sludge. Suction is maintained through the cake drying zone to continue dewatering. After the final zone, vacuum is released and a concentrated sludge cake is discharged.

Vacuum filters provide a continuous process dependent on a filtering pressure of less than 1 atmosphere. The vacuum is applied by a dry vacuum pump on the downstream side of the filter septum. The filter medium is either a rotary fabric belt which is continuously and automatically washed or a series of overlapping coils.

A large number of factors affect the actual yield from a vacuum filter such as depth of submergence, speed of rotation, solids content and filterability of the sludge. A large variety of sizes are available ranging from 0.9 to 3.6 m (3 to 12 feet) diameter and 0.3 to 0.6 m (1 to 2 feet) in length giving filter areas of from 1 to 75 m^2 (10 to 800 ft^2). Fabric belts are often preferred for sludge dewatering service, generally using polypropylene monofilament fabrics.

Lime sludges dewater successfully with vacuum filtration although the coil types are subject to incrustation of the coils. Belt type filters yield a crumbly cake of 45 to 65 percent solids which is sufficiently dry for landfill or road stabilization. Loading rates with lime sludges range from 50 to 200 $kg/m^2 \cdot h$ (245 to 980 $lb/ft^2 \cdot d$).

With alum sludges (at about 1.5 percent solids feed), solids appear to penetrate the fabric readily and a solution in many cases is to use some form of pre-coat, for example, diatomaceous earth. The pre-coat is applied to the drum of the rotary vacuum filter to a depth of about 50 mm (2-inches) before alum sludge filtration (Gruninger, 1975). Cake solids with alum (pre-coat) filtration range from 20 to 30 percent. Pre-coat requirements range from 30 percent for concentrated sludges (at approximately 5 percent) to 50 percent for dilute sludges (less than 1 percent).

Filtration rates with alum sludges are of the order 0.1 to 0.2 $m^3/m^2/h$ (60 to 120 gpd/ft^2). The filtrate solids are typically 10 to 15 mg/l. A disadvantage of pre-coat operation is that associated operating costs may be excessive. An alternative is polymer preconditioning. However, in some cases the applied dosage is very critical (Nielsen *et al.* 1973) and may give rise to operational problems. Reported issues include adhesion of the cake to the filter medium when using polymers (Levesque, 1982).

At loading rates of approximately 25 $kg/m^2 \cdot h$ (122 $lb/ft^2 \cdot d$) alum sludge cake concentrations of only 9% have been reported. Lower loading rates of 2 to 5 $kg/m^2 \cdot h$ (10 to 25 $lb/ft^2 \cdot d$) have achieved up to 15% cake concentrations with alum sludges (Nielsen *et al.* 1973). With lime conditioning of alum sludges, at 20 to 25% by dry weight ratio, cake concentrations of 18 to 25% have been reported at loading rates of 15 to 25 $kg/m^2 \cdot h$ (74 to 122 $lb/ft^2 \cdot d$) (Levesque, 1982).

Although favorable results have been reported for dewatering of lime sludges, this is not always the case, especially where high magnesium concentrations are present which may result in dewatering characteristics similar to those of alum sludges. Furthermore, since dewatering of dilute sludges (i.e. without prior thickening) is precluded with vacuum filtration, in some cases vacuum filtration of alum sludges may not be economically and technologically feasible and other methods may be preferable.

B.6.2.2 Centrifuges

Centrifuges have become popular dewatering devices for both wastewater and water plant sludges. Advantages of centrifugation are the small space requirements, complete process automation and ability to handle dilute (settled) or thickened sludges (Figure B.23).

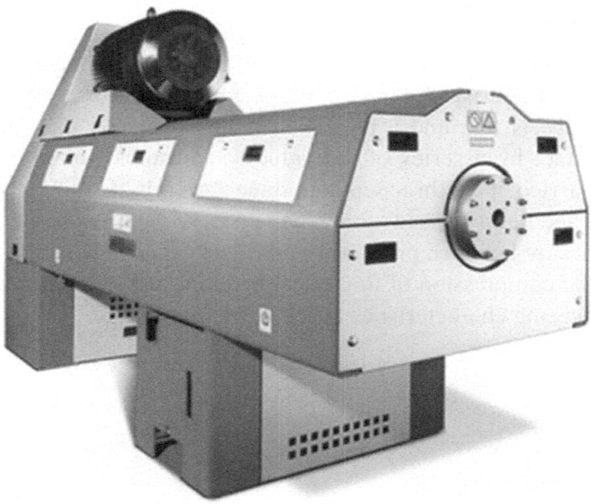

Figure B.23 Centrifuge (Courtesy Westfalia).

The scroll centrifuge is basically a rotating cylindrical bowl with a conical end to allow solids discharge. The opposite end of the bowl has effluent ports of varying distance from the imperforate bowl wall, thus allowing different "pond depths" for liquid discharge. Inside the imperforate bowl is a screw conveyor that rotates at a slight differential speed relative to the bowl. This screw conveyor moves the thickened solids to the conical end of the bowl for discharge from the centrifuge.

For alum sludges conditioned with polyelectrolytes a very short sedimentation time in the centrifuge (less than 1 minute) is required for solids thickening. With influent solids concentrations of approximately 3 percent, the scroll centrifuge may be expected to produce a cake discharge of 24 to 28 percent with approximately 1.5 g/kg polymer addition. With filter backwash waters (at 1 to 2 percent solids) cake discharges of 18 percent have been reported (Gruninger, 1975). As with other dewatering devices, sludges from the treatment of high turbidity waters produce the highest cake concentrations. Waters with low turbidities may produce cake concentrations below 15 to 20% dry solids.

A disadvantage sometimes reported with the scroll centrifuge is the high centrate solids concentration. For example, the Thrum Hall water plant in England applied centrifuges to dewater alum sludge derived from the treatment of a highly colored water. The centrifuge cake was typically around 18% and centrate solids about 320 mg/l (Kent *et al.* 1987).

Variables influencing solids recovery and cake concentration with the scroll centrifuge are the centrifugal acceleration produced by the bowl's rotational speed, the shape of the bowl, the pond depth, the pitch of the screw conveyor and the differential speed between screw conveyor and bowl.

B.6.2.3 Belt filter press

Many belt filter press variations have been evolved by machinery suppliers. In essence, they all consist of a horizontal perforated or woven belt on which the conditioned sludge is fed, and a press belt which bears on the other. The belts are kept in contact by a series of individually adjustable rollers (Figure B.24). Dewatering is carried out in three separate stages: An initial draining zone followed by a press zone and finally a shear zone. In the press zone, the sludge is subjected to mechanical pressure between press belt and filter belt. The belt speed is infinitely variable as is the compression of the rollers enabling the belt press to be adjusted to suit the dewatering characteristics of the particular sludge.

Figure B.24 Belt press (Courtesy Ashbrook).

Using alum sludges (of approximately 2.5 percent feed concentration) pre-conditioned with 5 to 7 g/kg polyelectrolyte, filter cake concentrations of only 8 to 12 percent were reported in the past (Gruninger, 1975; Westerhoff & Daly, 1974). However, newer technologies have considerably improved performance.

The Robert A. Skinner water plant in Los Angeles treated 665 Ml/d (175 mgd) using alum. Raw water turbidities were 0.6 to 2.5 NTU and suspended solids 1 to 4 mg/l. The alum dosage typically varied from 4 to 8 mg/l with 0.5 to 2.0 mg/l cationic polymer. Sludge production was reported at approximately 13.3 t/day (14.6 U.S. ton dry solids/day).

Technical and economic studies carried out on two occasions, comparing filter presses, belt presses, centrifuges, vacuum filters, drying beds, drying basins, and sludge lagoons, concluded that the belt press option was the most viable (Johnson *et al.* 1989).

Filter backwash water with solids concentrations of 200 to 800 mg/l, was recovered and thickened to 4 to 10% dry solids concentration. Sludge from the clarifiers, at 0.07 to 1.5% dry solids, was mixed directly with the thickened backwash sludge, to produce a blended sludge of 0.7 to 4% concentration, and fed to two 1.5 meter (5 ft) belt presses.

A cationic polymer was used for dewatering at a concentration of 1.5 g/kg (3 lb/ton dry solids). The application rate to the presses was 185 kg/h · m (124 lb/h · ft) and 7.6 m^3/h · m (10 gpm/ft) at a mean thickened concentration of 3.4%. The belt velocity was approximately 1.8 m/minute (5.9 ft/min). The mean concentration of the dewatered sludge was 24%, and the cake density 1,124 kg/m^3 (70 lb/ft^3).

The dewatering system operated without operator attention 85 to 90-percent of the time. Operator attention was approximately 45 to 60 minutes per operator shift.

The Newcastle water plant in Pennsylvania (Croker, 1989) treated a plant flow of 32 Ml/d (8.4 mgd) using alum. Sludge production was 736 kg/d (1,620 lb dry solids/day). Filter backwash water was collected in equalization tanks and recycled back to the raw water influent channel. Backwash solids settled jointly in the clarifiers. Sludge removed from the clarifiers, at a mean concentration of 0.5%, was sent to gravity thickeners at a loading rate of 0.17 kg/m^2 · h (0.83 lb/ft^2 · d). Polymer was added immediately upstream of the thickener at a dosage of 0.9 to 1.5 g/kg (1.8 to 3 lb/ton dry solids).

Thickened sludge was sent to a 2-meter (6.5 ft) belt press at a loading of 75 kg/h · m (50 lb/h · ft). Cake concentrations were typically 25% dry solids. The sludge dewatering operation was typically less than 8 hours per day.

The City of Longmont, Colorado used a belt press to dewater alum sludge generated at two water treatment plants. The plants had a collective design capacity of 95 Ml/d (25 mgd) and generated approximately 0.9 t/d (1.0 ton dry solids/day) of alum sludge. Residuals from the plant including filter backwash and sedimentation tank blowdowns were sent to two washwater recovery ponds. Sludge in the ponds was dredged and pumped to the dewatering facility.

The City evaluated different types of drying beds, as well as the belt press option. Principally because of limited land availability and costs, the belt press was selected for dewatering. Dewatered sludge was used for land application, including the recovery of mined land.

The feed rate to the press averaged 0.33 Ml/d (61 gpm) with a feed concentration of 3.1%. Polymer was fed at 2 to 3.8 kg/t (4 to 7.5 lb/dry ton), depending on the sludge characteristics. The press achieved cake concentrations averaging 18.7% with a solids capture of 98-percent.

The Metropolitan Water District of Southern California reported on alternative processing methods for water plant sludges (Clark & Anderson, 1997). On-site pilot tests were conducted at the Jensen water plant using a high-solids (14-roll) belt press, a tower belt press, a centrifuge, and a membrane filter press. The results obtained are summarized as follows:

- The high solids belt press achieved 30 to 32% dry solids cake at polymer dosages of 2 to 3 kg/t (4 to 6 lb/ton dry solids) and application rate of 140 to 230 kg/h · m (300 to 500 lb/h · m) belt width. Solids capture was 90 to 97-percent.

- The tower press achieved 29 to 31% dry solids cake at polymer dosages of 5.5 to 13.5 kg/t (11 to 27 lb/ton dry solids) and application rate of 150 to 263 kg/h · m (330 to 580 lb/h·m) belt width. Solids capture was 96 to 98-percent.
- The centrifuge achieved 28 to 32% dry solids cake at polymer dosages of 2.7 to 4.0 kg/t (5.4 to 9.0 lb/ton dry solids) and solids capture of 96 to 97-percent.
- The membrane filter press achieved 32 to 42% dry solids cake at polymer dosages of 0.5 to 1.0 kg/t (1.0 to 2.0 lb/ton dry solids) and cycle times of 2.5 to 4 hours. A gravity belt filter was used to pre-thicken sludge before the press

The District selected the belt press option (either the high solids or tower press) from life cycle costs, including final sludge disposal costs. Dewatered sludge would be further air dried on concrete pads to 50% or more. The dried sludge would be transported off site for use as potting soil amendment.

B.6.2.4 Filter presses

Pressure filtration is essentially a batch process. The press consists basically of a series of substantial plates supported by a robust structure. Filter cloths are mounted between the plates. The format of the plates is such that when pressed together, they form recesses within which the sludge cake forms (Figure B.25). The principal advantage of the filter press is that it produces the highest cake concentrations when compared with other mechanical dewatering systems. Filtrate quality is also very good. However, initial costs can be high, although operating costs (including transport and disposal costs) can be lower than other methods.

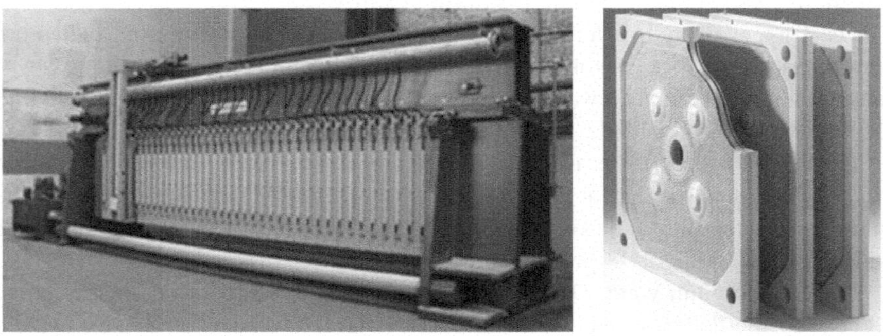

Figure B.25 Filter press (Courtesy Andritz Netzsch) and filter plates (Courtesy Menardi).

During the press cycle, the plates are pressed together, hydraulically or electromechanically, and preconditioned sludge is pumped into the cloth lined recesses between the plates. Sludge is introduced through holes in the centre or at the edges of the plates into the space between the filter cloths at pressures of 700 to 7,000 kPa (100 to 1,000 psi). Filtrate is forced through the cloths and the retained solids form a cake approximately 25 to 35 mm thick. Pressing time varies considerably with sludge characteristics, and can be from 2 hours to 3 days. Discharge of the pressed cake can be manually or automatically controlled by releasing pressure and separating the plates.

Press plates used to be made of cast iron, which often gave rise to problems with cracking due to unequal feed distribution. However, these have largely been superseded by polypropylene or moulded rubber units. Two disadvantages often cited in the past with filter presses are the short life of the filter cloths, and the lack of automation. However, modern synthetic fibres greatly extend cloth life, and increased use is made of automatic feeding and pressing cycles. It is only at the end of the pressing cycle that no fully unmanned process for discharging the sludge cake has as yet generally been used. There are automatic cake discharge devices, but some operator attention is generally preferred to ensure that no cake remains between the plates before the press is closed and subject to high pressure for the next press cycle.

With polyelectrolyte conditioning and after thickening to 2 to 10 percent solids, filter cake concentrations of at least 30%, and up to 60% dry solids are attainable with alum sludges, depending on the raw water characteristics.

With lime conditioning of alum sludges, filter cake concentrations of at least 40 to 50 percent solids are obtainable. Lime requirements in such cases may amount to 25 percent on a mass basis of the waste sludge. Typical cycle times are of the order 90 to 120 minutes but depend very much on the sludge conditions. Sometimes polymer addition with lime gives best results. The polymer dosage is typically between 2 and 5 g/kg (4 to 10 lb/ton dry solids) and lime 5 to 40-percent by weight.

Normally the lime is dosed at a 7.5-percent solution to bring the pH to 11. The reaction time of the lime with the sludge is generally at least 30 minutes. At times lime is added in stages to reduce the amount required. First, part of the lime is added followed by polymer. Sludge is thickened and supernatant or filtrate recycled to the water plant. Lime is then added to the thickened sludge to bring the pH to 11. By this means the overall quantity of lime required is less.

At the Choisy-le-Roi water plant in France (Ventresque & Bablon, 1989) alum sludge was dewatered in filter presses after gravity thickening using polymer for conditioning. For dewatering, lime was added at 25 to 40-percent by weight. The operating cycle of the presses comprised 700 kPa (100 psi) for 4 minutes, followed by 7,000 kPa (1,015 psi) for 7 minutes, and compressed air injection for 45 seconds. Each press cycle was for 15 minutes, including time required for cake discharge. The loading rate to the presses was 10 to 12 $kg/m^2 \cdot h$ (49 to 58 $lb/ft^2 \cdot d$). The presses operated for 14 hours per day and produced cake concentrations of at least 30% dry solids.

The Crown water plant in Ohio reported a treatment flow of 190 Ml/d (50 mgd) and a mean raw water turbidity of 5 NTU (Heuer & Schwarzwalder, 1989).

Alum was used as primary coagulant for water treatment. Filter backwash water at 300 to 400 mg/l solids was equalized and returned to the raw water channel. Settled sludge from the clarifiers was removed continuously by scrapers at a mean concentration of 0.5%. Settled sludge was thickened in gravity thickeners to 2.5 to 5% dry solids without polymer addition. Thickened sludge was pumped to a reaction tank and lime added to approximately 20-percent by weight.

Dewatering was achieved in two filter presses each with 95 plates and 4.6 m^3 (162 ft^3) capacity. Cake concentrations achieved were 30 to 40% dry solids. No filter pre-coat was added. Operation of the presses was for 9 hours per day, 5 days per week. The dewatering facility had only one operator.

The Langsett water plant in the United Kingdom treated 47.5 Ml/d (12.5 mgd) of highly colored water, using ferric sulfate (Briens *et al.* 1989). Sludge withdrawn from the clarifiers varied in concentration from 0.2 to 1%. Filter backwash water was equalized and recycled to the head of the plant. Settled sludge was thickened in two gravity thickeners using polymer for conditioning. Sludge was thickened to approximately 4% dry solids and dewatered in two filter presses, each with 120 plates. Sludge production was reported as 2.5 t/day (2.75 ton dry solids/day).

The cake concentration achieved was 20 to 25% with press cycles of 8 hours. No further conditioning chemical was added, apart from the polymer added at the thickening stage. Subsequent air drying of the cake produced concentrations of 35 to 50%.

The Papago water plant in Tempe, Arizona treated up to 215 Ml/d (57 mgd) using alum as primary coagulant, as described in Section B.6.1 (Bratby *et al.* 1993, 1996). Raw water turbidities ranged from 2 to 10 NTU in the dry season, to peaks of 1,000 NTU in the wet season. Sludge was derived from the presedimentation and sedimentation tanks. Filter backwash was equalized and recycled back to the presedimentation tanks. The combined sludge concentration discharged from the sedimentation and presedimentation tanks varied from approximately 0.2% at low turbidity periods, to higher than 2% at high turbidity periods.

Sludges discharged from the sedimentation tanks were blended and equalized in a balancing tank. Sludge was then pumped to a gravity thickener, after polymer conditioning. Sludge was thickened to approximately 5%, then pumped to a thickened sludge holding tank where further thickening and decanting took place. Supernatant overflowed the peripheral weirs of the thickener and flowed by gravity to the plant influent and existing ponds. The sludge holding tank allowed sludge to be pumped out of the thickener at a constant rate, thereby optimizing gravity thickener performance. Thickened sludge storage also allowed for periods of press inactivity, such as nights, weekends, and maintenance.

Conditioned and thickened sludge was pumped from the sludge holding tank by the feed pumps, to the filter press. No further polymer addition was required after the thickening process although, for flexibility, an alternative polymer feed at the suction side of the filter press pumps was provided. The concentration at the bottom of the sludge holding tank and fed to the filter presses varied from 6 to 15% and averaged at approximately 8% dry solids.

Two recessed chamber filter presses were provided, each press with seventy 1,500 × 1,500 mm (4.9 ft) recessed chamber polypropylene plates with a 30 mm (1.2-inches) cake, giving a total cake volume per press of 3.8 m^3 (135 ft^3). The filtration pressure was 1,550 kPa (225 psig). The plates were center feed, and filter cloths used were polypropylene sateen finish monofilament. The cloths have an average life of up to 3,000 press cycles, or up to 2 years operation.

The filter press feed pumps were variable speed single plunger pumps. The press fill period was at a high flow rate of 6.5 l/s (100 gpm) per press, gradually reducing to 1.25 l/s (20 gpm) at the end of the compression period, at 1,550 kPa (225 psig).

The dewatering system was normally operated eight hours per day, five days per week. However, maximum daily sludge production peaks, which were of relatively short duration, could be processed by running extra shifts during the day or running the system into the weekend.

With raw water turbidities ranging from approximately 6 to 13 NTU, cake concentrations ranged from 30 to 46% dry solids. Press cycle run times were less than 1.5 hours and the polymer dosage approximately 1.5 g/kg (3 lb/ton dry solids). The press cycle time included the following operations:

- Time to fill the press
- Time the press is under pressure
- Time to blow the core back to the sludge holding tank
- Time to open the press
- Time to discharge the cake
- Time to spray wash the cloths (when required)
- Time to close the press

At higher raw water turbidities, from 50 to 75 NTU, cake concentrations of approximately 55% dry solids were obtained, with press cycle times less than 1.5 hours.

Filtrate solid concentrations were consistently low, with values ranging from 1.5 to 5 mg/l.

B.6.2.5 Other dewatering devices

There are other mechanical dewatering devices that could be effective in coagulant sludge dewatering applications. These include the hydraulic piston press, and screw press.

The Dehydris™ Twist press, shown in Figure B.26, is based on the Bucher Unipektin piston press. The press comprises a rotating drum, hydraulic piston, and numerous flexible drainage elements. During sludge introduction, the cylinder-piston is closed and rotation started. The piston moves back and forth and at each return stroke additional volumes of pre-conditioned and thickened sludge are injected. This continues until the press has been filled. The piston continues to move back and forth and the cylinder-piston unit continues to rotate. This action continuously exposes sludge to the drainage elements thereby facilitating drainage of the interstitial water. When pressing is complete, part of the cylinder casing slides open, the cylinder/piston rotates while the piston moves forward, the sludge cake is dislodged and is discharged from the bottom.

Pilot testing was carried out with an alum sludge from the Seneca WTP in South Carolina. Existing filter presses typically achieved approximately 23% cake with this sludge. After conditioning and thickening, the Dehydris™ press achieved cake concentrations above 40% using polymer dosages of approximately 5 lb/ton active polymer.

Nominal unit sizes of this press range from 50 to 120 m^3/h (220 GPM to 530 GPM).

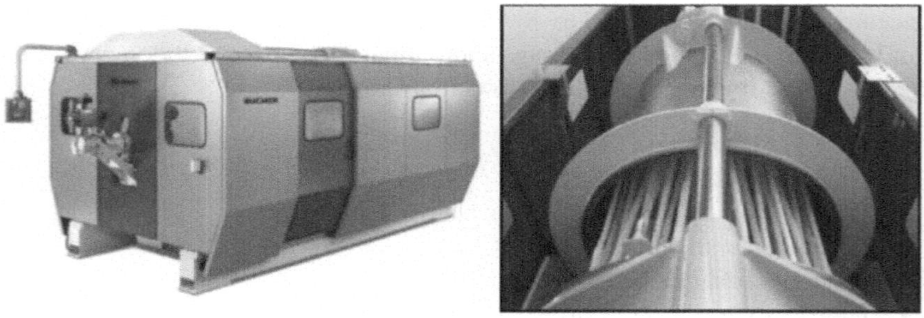

Figure B.26 Dehydris™ twist press (Courtesy Infilco Degremont Inc.: www.degremont-technologies.com).

The screw press, shown in Figure B.27, comprises a conical screw shaft and cylindrical wedge-wire sieves. The sludge residence time within the press can be regulated by adjusting the speed of the screw conveyor. Typically, the speed of the screw is 2 to 3 rpm which conveys sludge up the screens towards the discharge at the top. The decreasing screw pitch, drum diameter and screen apertures progressively increase the dewatering pressure on the sludge. A pneumatic cone at the press discharge maintains a counter-pressure to the dewatering cake within the press. Wash water sprays are used to clean the wedge-wire screen. Nominal sludge feed capacities to this press range from 20 to 100 m^3/h (90 GPM to 450 GPM).

Figure B.27 Screw press (Courtesy Huber technology: www.huber.de).

B.7 COAGULANT RECOVERY

The recovery of aluminum sulfate from water plant sludges became relatively popular in Japan in the 1960s. It appears that approximately fifteen water plants operated some form of coagulant recovery process. However, it also appears that since that time, no additional coagulant recovery plants were implemented.

It was also reported in 1990 that three full scale plants in the United States utilized coagulant recovery. All three used acidification: two for alum recovery, and one for iron recovery. At that time one additional plant was under construction in Durham, North Carolina (AWWARF & KIWA, 1990). This plant initiated operation in 1992 (Anon, 1994).

Apart from alum recovery, the recovery of lime by recalcination was also practiced in various plants in the United States. One disadvantage of this practice was related to the high energy costs.

The recovery of metal coagulants is also possible by increasing pH to 10 or 11, but the efficiency of re-dissolution of aluminum or iron is relatively low. For this reason, research has tended to acidification and other means to recover the metal from the hydroxide precipitates.

Original research was directed towards sludge conditioning by acidification. It was observed that as the pH is reduced to 1.5 to 2.5, thickening of the sludges was markedly improved. After acidification, thickened sludge concentrations of 10% or more were possible, with sludge volume reductions of 50 to 80-percent. By dissolving the hydroxides, approximately 15-percent solids destruction was achieved.

Acidification also improved dewatering of the sludges. Using filter presses, cake concentrations of 40 to 55% were possible. However, although the specific resistance values were lowered, the sludge cake formed was often found to be sticky and tended to clog the filter cloths.

One alternative has been to add diatomaceous earth as a pre-coat before sludge filtration. Almost at the end of the filter press fill cycle, lime is added to the sludge to neutralize the cake. Control of the lime is by monitoring the pH of the filtrate. When the pH is neutral, the press cycle is halted and the cake released. While lime is added, the filtrate is not used for coagulant recovery.

With the acidification process, conditioning tanks, press feed pumps, press plates and other equipment and piping are made of corrosion resistant materials.

Figure B.28 shows a general scheme of the acid recovery process.

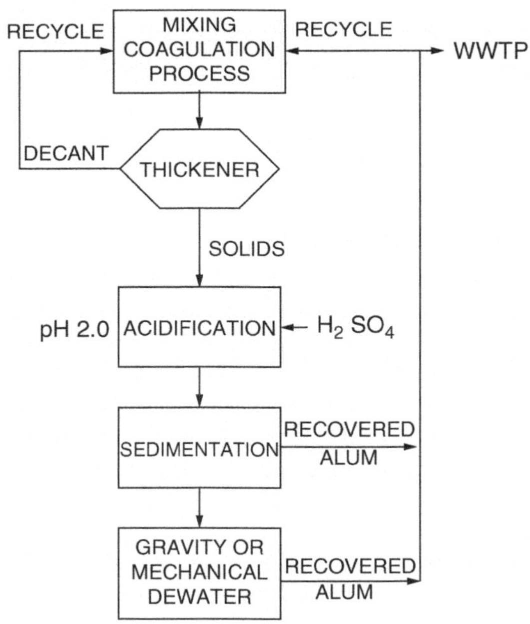

Figure B.28 Acid recovery schematic.

In general, coagulant recovery by acidification follows three steps:

(1) Thickening of the waste sludge is required to facilitate subsequent dewatering, to reduce the quantity of acid used, and to ensure a sufficiently

high concentration of recovered alum of at least 2 percent (Fulton, 1974). Lower coagulant concentrations result in lowered coagulation efficiencies during water treatment.

(2) The thickened sludge is acidified with sulphuric acid to dissolve the hydroxides. Acid, in excess of stoichiometric quantities is required to ensure a low pH (approximately 2.0). Acidification first releases waters of hydration, then dissolves the hydroxides. Reactions with aluminum and ferric hydroxide are as follows:

$$2Al(OH)_3 + 3H_2SO_4 \rightarrow Al_2(SO_4)_3 + 6H_2O$$

$$2Fe(OH)_3 + 3H_2SO_4 \rightarrow Fe_2(SO_4)_3 + 6H_2O$$

Theoretically, 1.5 moles of H_2SO_4 are required per mole of aluminum. This equates to 1.9 kg (lb) of sulfuric acid (H_2SO_4) per kg (lb) of aluminum hydroxide ($Al(OH)_3$). An excess of approximately 8 percent over the stoichiometric amount is usually required. In practice, approximately 23 kg H_2SO_4 per m^3 hydroxide sludge at 1.5% dry solids content are required (1.4 lb/ft^3). In other words, 0.5 kg H_2SO_4 per kg aluminum sulfate added during water treatment (0.5 lb/lb). The reaction time required, between acid and sludge is approximately 10 to 25 minutes. Figure B.29 shows the relationship between acid dosage and recovered alum (adapted from Rencken *et al.* 1989).

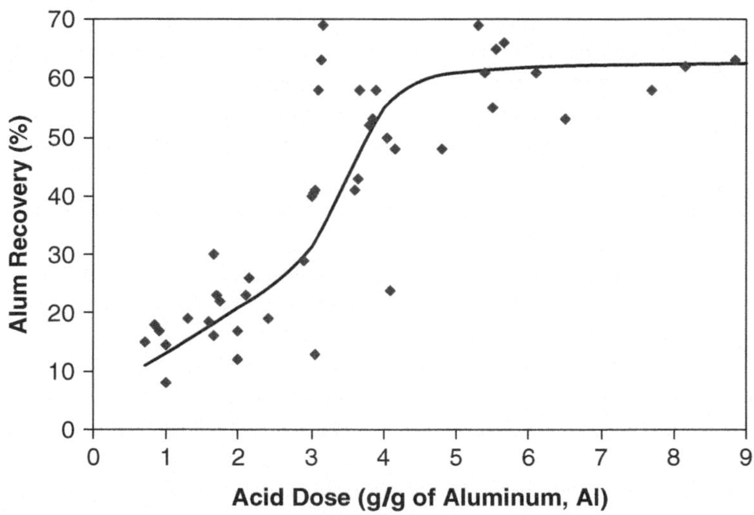

Figure B.29 Sulfuric acid dosage and alum recovery.

(3) The acidified sludge passes to a solid-liquid separation phase. The liquid fraction, containing the recovered coagulant is used for water treatment coagulation, and the solid phase is neutralized to avoid corrosion of sludge handling equipment, and to facilitate sludge disposal.

Undesirable suspended matter remaining in the liquid fraction is separated from the alum solution. Such separation may be accomplished either by settling out the solids in a basin or by removing the solids with a filter press.

The sludge fraction is usually mixed with lime for pH neutralization and conditioning prior to dewatering, although in some cases the sludge is passed directly to a dewatering facility. This has the advantage of improving the percentage of recovered alum and reducing the subsequent lime dosage necessary to neutralize the waste sludge prior to disposal. With this procedure cake solid concentrations, using pressure filtration, have been reported to be of the order 40 to 55 percent (AWWA, 1972). The recovery of reusable alum is of the order 50 to 75 percent with an acidulator retention time of 10 to 25 minutes.

With sludges derived from high raw water turbidities, the recovery rates are generally lower. Higher hydroxide fractions in the sludge generally provide higher coagulant recoveries. In general, the process becomes impractical when the aluminum or iron fraction of the sludge is less than approximately 5-percent.

For example, raw waters with a high organic color fraction, and low turbidity, would presumably be ideal for coagulant recovery by this means. Such waters tend to require high coagulant dosages and, when acidified, release almost all of the coagulant back into solution. However, the organic color is also released together with the coagulant, making the coagulant almost useless for recovery.

In some cases the recovered coagulant produces better results than the virgin coagulant. This has been found to be the case with low turbidity waters. It is possible that the impurities present in the recovered coagulant serve as nucleation sites during coagulation. The inverse appears to be the case with high turbidity waters, where the performance of the recovered alum is inferior to the virgin coagulant.

It has been reported that the acidification recovery process is more viable when coagulant dosages during treatment are relatively high and turbidities low. One example is the Tampa water plant in Florida (Dunkle, 1975). The alum dosage in this case was 50 to 100 mg/l and the coagulant was effectively recovered 8 to 10 times without affecting the quality of the treated water.

The Williams water plant in Durham, North Carolina is a relatively recent coagulant recovery installation that started operation in 1992 (Anon, 1994; Bishop *et al.* 1987; Bishop *et al.* 1991; Cornwell *et al.* 1993). The plant had a design capacity of 83 Ml/d (22 mgd) using alum at an average dosage of 30 mg/l. Raw water turbidities averaged 30 NTU and color 30 color units. The treatment process produced approximately 730 t/year (800 tons dry solids/year of alum residuals. Sludge withdrawn from the clarifiers was thickened in batch thickening tanks to concentrations between 2 and 4% dry solids. After acidifying the thickened sludge, the sludge is allowed to settle. The decant liquor (recovered alum) is pumped to a storage tank, and the settled residuals dosed with polymer and defoamer and dewatered using a centrifuge. Centrate is also recovered as alum. The sulfuric acid required to adjust pH to 2.0 was approximately 0.67 kg/kg dissolved aluminum sulfate.

The reduction of sludge solids due to dissolution of the hydroxide sludge was approximately 35 to 40-percent.

Previous work at this plant used drying beds for dewatering (Bishop *et al.* 1987). The characteristics of the acidified sludge during dewatering on the drying beds was such that only 40-percent the total bed area was required with acidification, compared with sludge without acidification. Without acidification, drainage through the drying beds was approximately 50-percent the original volume applied to the beds, using polymers. With acidification, the drainage volume increased to approximately 60-percent, using polymers.

The cake concentration taken from the beds was approximately 25% dry solids. The average application rate to the beds without acidification was approximately 115 $kg/m^2 \cdot year$ (23.5 $lb/ft^2 \cdot year$). With acidification, the application rates increased to 160 $kg/m^2 \cdot year$ (33 $lb/ft^2 \cdot year$). The percentage recovery of aluminum sulfate was 74 to 79-percent. The efficiency of recovery depended largely on the dewatering efficiency. The recovered aluminum sulfate was dilute, at approximately 2-percent strength.

It was recommended that the recycled alum be limited to 50 to 65-percent of the total used at the plant to avoid the build up of heavy metals and organic compounds.

Later tests at the same plant (Bishop *et al.* 1989) tested centrifuge and filter press dewatering to complement the drying beds. Thickened sludge was acidified and then settled again in gravity thickeners. Without acidification the thickened sludge concentration was 1.0 to 1.5% dry solids. After acidification concentrations of 3.0 to 3.5% were achieved. Using the filter press without acidification, total press cycles were approximately 65 minutes with cake concentrations of 26 to 31% using 20 to 30% CaO and 1.5 to 2.5 g/kg (3 to 5 lb/dry ton) polymer. The overall production rate in the press was approximately 0.45 $kg/m^2 \cdot h$ (2.2 $lb/ft^2 \cdot d$).

With acidification, cycles were 55 minutes with a production rate of 0.6 $kg/m^2 \cdot h$ (2.9 $lb/ft^2 \cdot d$). Cake concentrations were 35 to 40% using 5 to 10% $FeCl_3$ and 1.5 to 2.5 g/kg (3 to 5 lb/dry ton) polymer.

Using the centrifuge, the polymer concentration was 1 to 2 g/kg (2 to 4 lb/dry ton). Without acidification, cake concentrations were 22 to 25%. With acidification, cake concentrations were above 30% dry solids.

The final configuration at this plant using gravity thickening of the acidified residuals, and centrifuge dewatering of the decanted sludge, achieved a 35-percent reduction in residuals by acidification of the hydroxide sludge. The centrifuge dewatered solids concentration was 35% with 98-percent solids recovery during dewatering. The alum recovery has been virtually 100-percent.

The recovered alum was shown to be as effective as fresh alum during treatment. The project achieved an overall reduction of 70-percent in the consumption of new alum. The recovered alum is used in both the water plant, and at the wastewater plant for phosphorus precipitation and removal. The decision was made to reuse the alum at the water plant once or twice then use the recovered alum at the wastewater plant to avoid the build-up of impurities from repeated recycling.

Operating costs, including hauling and disposal costs were monitored closely by the City. It was determined that the alum recovery facility produced annual savings of approximately $100,000 which represented a payback period for the facility of 2 years.

The Hazelmere water plant in South Africa (Rencken *et al.* 1989) treated a flow of 15 Ml/d (4 mgd) using aluminum sulfate and polyelectrolyte as flocculant aid. Raw water turbidities were 15 to 80 NTU. Sludge produced was 750 kg/d (0.83 ton dry solids/d). Sludge processing included a gravity thickener operated in batch mode, addition of sulfuric acid, reaction tank, polyelectrolyte addition, and vacuum assisted drying beds.

After thickening, supernatant was decanted from the thickener and the sludge homogenized with diffused air. The sludge was then pumped to the acid reaction tank with a reaction time of 15 minutes. pH was maintained at 2.0. After acid reaction, a non-ionic polymer was added at a dosage of 1.25 g/kg (2.5 lb/ton dry solids). Conditioned sludge was sent to a 125 m^2 (1,345 ft^2) drying bed. Supernatant was withdrawn from the surface of the bed during the first 3 hours, then vacuum was applied to the bottom of the bed for 12 hours until the surface of the bed cracked and the vacuum broken. After dewatering, sludge cake was removed at a concentration of 25% using a bob-cat loader. Sludge was stockpiled on a patio and air dried to approximately 50% dry solids.

Cake removal from the drying bed took approximately 1.5 hours. After removing the cake, the drainage plates were cleaned with water jets. This part of the operation took approximately 4 to 6 hours.

The application rate to the beds was approximately 14.7 kg/m^2 (3 lb/ft^2 dry solids). The optimized sulfuric acid dosage was found to be 6 g/g of aluminum as Al. The aluminum sulfate recovery was 55 to 60%, and the consumption of virgin alum was reduced by about 70-percent. Analyses did not indicate any significant build-up of heavy metals or organic compounds using the recycled alum.

In some cases of using recycled alum, increases in final water turbidity have been noted. One example is the Sturgeon Point water plant in New York (Westerhoff, 1973). Final treated water turbidity increased from 0.14 to 0.28 NTU. To maintain the same treated water quality, the alum dosage had to be increased.

The principal problem with coagulant recovery is the release of unwanted constituents from the sludge, together with the coagulant. When raw waters contain significant concentrations of iron, manganese, organic color and other constituents, these substances would be recycled. For example, if a raw water contained 0.05 mg/l of iron, and assuming a coagulant recovery efficiency of 80-percent, the concentration of iron would increase to 1.0 mg/l within 25 days, and to 3.0 mg/l within 75 days (Fulton, 1974).

Figure B.30 (adapted from Sengupta & Shi, 1992) shows the release of dissolved organic carbon (DOC) as well as the recovered aluminum during acidification.

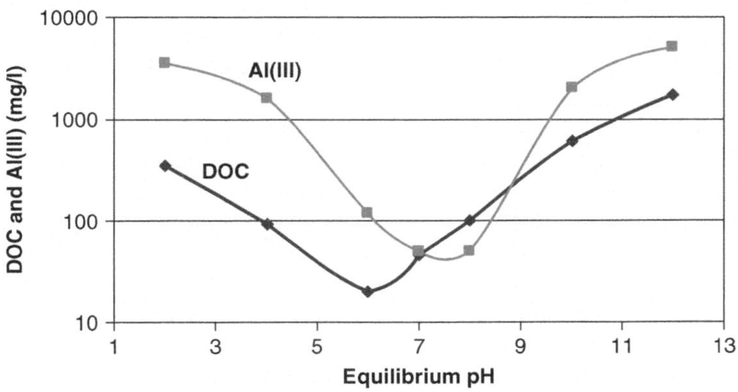

Figure B.30 DOC and aluminum (III) released from sludge at different pH values.

An operational problem noted during acidic alum recovery processes is the generation of gas bubbles (probably CO_2) which serves to impede subsequent gravity thickening and develop a troublesome froth layer on the surface of gravity thickeners (Westerhoff & Daly, 1974).

Although in some cases acid alum recovery and recycling within the water treatment plant may not be feasible, particularly where impurity build-up is likely, the process may be economical where reuse of recovered alum at a wastewater treatment plant for phosphorus precipitation is feasible. In this case the presence of impurities in the recovered alum may not be a disadvantage. Advantages would be, apart from phosphorus precipitation, improved COD and solid removals during primary sedimentation. It may also be feasible to use the acidified sludge without solid-liquid separation directly in the wastewater plant. However, the wastewater treatment plant must have sufficient capacity to handle the increased solids load.

One development to resolve the issue of coagulant impurities was to use a liquid ion-exchange process (Cornwell & Zoltek, 1977; Cornwell, 1979; Cornwell & Lemunyon, 1980; Cornwell *et al.* 1981; Westerhoff & Cornwell, 1978). In this process a solvent is mixed with the sludge to form a dispersion of small water drops. By a process of ion exchange (aluminum for hydrogen) the aluminum ions are concentrated at the interface between the solvent and the water drops. Coalescence of the water drops then takes place in a separation tank and the aluminum is concentrated in the solvent. After separation of the aluminum from the water, the aluminum is recovered from the solvent by the addition of sulfuric acid. During this phase, aluminum sulfate is formed, and the solvent is recycled for further aluminum extraction.

The solvent used in this process was a mixture of equal parts of mono- and di-(2-ethylhexyl) phosphoric acid, diluted in a hydrocarbon such as kerosene. The recovery efficiency of aluminum sulfate was reported to be as high as 90-percent.

Despite the high recovery efficiency, the process does not appear to have been implemented because of the high costs of reagents which tended to be higher than the cost of the virgin alum itself.

Another novel alum recovery process evaluated at laboratory scale was reported by Sengupta and Shi (1992). The process uses a special class of composite polymeric membranes. The membranes are composed of sheets of polytetrafluoroethylene (PTFE) within which are enmeshed 100μm diameter beads of chelating exchangers. These chelating ion exchange beads have a high affinity for aluminum, in preference to other contaminants present in the sludge matrix. Therefore the aluminum recovered has a high purity. The aluminum is then released by immersing the membrane material in a sulfuric acid bath, thereby forming aluminum sulfate. Sludge is not passed through the membranes; rather, the membrane is immersed in the sludge until exchange has occurred.

The recovery procedure involves acidifying the sludge to a pH of approximately 3.5. The membrane is immersed in the acidified sludge until exchange equilibrium is reached. The membrane is then withdrawn from the sludge and immersed in a 2 to 10 percent sulfuric acid solution. The regenerated membranes are then returned to a new batch of sludge to continue the exchange reactions. Although the recovered alum has a high purity, the liquid fraction from the first acidification step contains impurities that would require treatment for disposal (see also Figure B.30).

A further recovery process of importance for water softening plants is lime recalcination. Essentially the lime recovery process comprises the following steps (AWWA, 1969–1970):

(1) Sludge thickening and/or blending (in some cases recarbonation is utilized in this step);
(2) Dewatering by centrifuge or vacuum filter;
(3) Flash drying of dewatered sludge by use of off-gases;
(4) Recalcining at high temperature, i.e. converting calcium carbonate to calcium oxide;
(5) Product cooling, where required.

The two predominant types of recalciners which have been used are the rotary kiln and fluidized bed furnaces. Fluidized bed furnaces appear to have gained favor; the system operates more economically, produces less dust, produces a harder CaO product, is less susceptible to air-slaking, requires less installation space and is capable of faster start-up and shutdown. Other types have been variations of multiple-hearth furnaces.

Successful recalcination relies on the production of an end product of minimal impurities. For example the concentration of magnesium should not exceed 10 percent. Because of this limitation, many water treatment sludges are not amenable to recalcination. However, in cases where recalcination is feasible, several plants have shown a definite profit; the quantity of lime reclaimed had in some cases exceeded that required for originally softening the water.

A typical recovery process included thickening of the calcium carbonate sludge to 20 to 30 percent solids; dewatering by centrifuge to about 67 percent solids; calcination at temperatures of 1,000 to 1,200°C in an oil fired rotary kiln.

Separation of magnesium from the sludge prior to recalcination may in some cases be accomplished at the dewatering (centrifugation) stage by suitably adjusting the centrifuge operating conditions. Recovery of $Mg(OH)_2$ may lead to a useful end product. Carbonation of softening sludge produces $MgCO_3$ (which can be physically separated from $CaCO_3$ precipitates), which may be reused as a water treatment coagulant in place of alum in certain cases (Thompson *et al.* 1972).

When tastes, odors or color are present in the raw water, the centrate is usually wasted rather than recycled to the plant. The effect of silt and clay particles, if present in excessive amounts, may give rise to a recalcined product with the characteristics of cement rather than lime. In certain cases removal of such particulate material prior to softening operations has been a solution.

A useful end product of the recalcining process is the production of CO_2 which is present in the furnace stack gas (15 to 27 percent CO_2). The CO_2 liberated from $CaCO_3$ on reduction to CaO may be used for recarbonation during water conditioning in the plant, or it may be recovered and marketed as "liquid or dry ice".

B.8 SLUDGE DISPOSAL

B.8.1 Introduction

There are a number of disposal routes for water plant sludges. These include lagooning, land application, landfill, disposal to sewers, co-disposal with wastewater plant sludges, and disposal to water bodies. The last option is to a large extent defunct due to environmental legislation in many countries. The remaining options are used on a case specific basis.

In the United States there are a number of federal, state and local laws that control the management, transportation, disposal and recycling of wastes generated by water plants. Figure B.31 provides a schematic overview of the regulations governing residuals management (adapted from Koorse, 1993).

In the United Kingdom Warden (1982) reported that approximately 40-percent of total water plant sludge solids was disposed of in lagoons; 5-percent to municipal sewers; 5-percent to inland waters; 3.5-percent to sea; 3-percent sprayed on land; and 7-percent disposed of on-site. The remainder went to some form of landfill or land disposal.

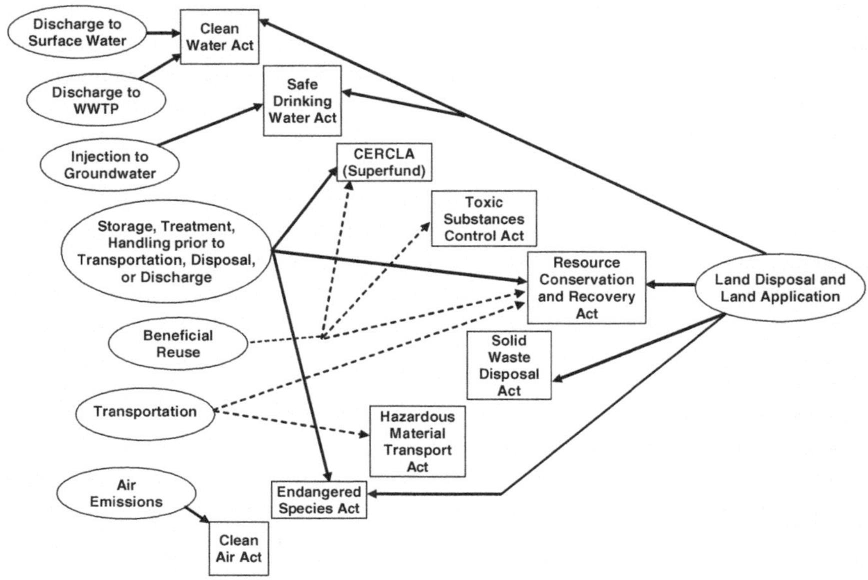

Figure B.31 Overview of water sludge regulations (United States).

In the United States, approximately 19-percent of water treatment plants utilize lime softening. Of 132 lime softening plants surveyed around the country in 1995, 76 disposed of the lime sludge in lagoons; 53 by land application; 47 by landfill; 20 discharged to wastewater collection systems; and 9 disposed of the lime sludge jointly with wastewater sludge (Alfaro *et al.* 1995).

A survey was carried out on 19 water plants in Pennsylvania using alum and iron coagulants (Dixon & Lee, 1989). Of these plants, the existing practices were as follows: one discharged liquid sludge to a river, two sent liquid sludge to incineration centers, eight used landfills for the disposal of dewatered sludges, and eight used liquid sludge for land application. General conclusions drawn from the studies were as follows:

- Despite the simplicity of the solution, direct discharge to sewers is subject to acceptance by wastewater treatment entities and may not be an acceptable form of disposal in the long term.
- Because of increasing costs of tipping and disposal fees for municipal landfill disposal, this method of disposal was considered to be unacceptable in the long term.
- Because of the high costs associated with the transport of liquid sludge, land disposal of liquid sludge was not viable in the long term. This form of disposal represented the highest life cycle cost amongst the alternatives.

- The final recommendation from the studies was to acquire facility owned land for dedicated land disposal. This solution required that those plants that disposed of liquid sludge invest in thickening and dewatering of the sludge.
- Only one of the plants among the nineteen studied would have issues with dedicated landfill operations: the sludge from this plant had excessive barium concentrations that required an impermeable layer and capture of the leachate.

For dedicated landfill of water plant sludge, it is generally required that the sludge cake is at least 20% dry solids (Croker, 1989). The sludge is often mixed with soil in the proportion of 0.25 to 1.0 parts soil to 1.0 parts of sludge. The actual proportion depends on the sludge cake concentration and the characteristics of the sludge. With cake concentrations greater than 35%, blending with soil can often be avoided.

Dedicated sludge landfill application rates are normally between 4,000 and 25,000 m^3/ha (57,000 to 360,000 ft^3/acre). The methodology typically adopted is to construct dikes then apply the sludge, or a mixture of sludge and soil, within the dikes. Soil cover is often used to depths ranging from 0.6 to 1.5 m (2 to 5 feet).

In-situ soil conditions, groundwater contamination with aluminum, iron and other contaminants is case specific and would influence the design of the landfill in particular cases.

A related application of water plant sludges is as landfill cover. Studies carried out on the physical and engineering properties of alum water sludges from two water plants found that, in general, the sludges exhibited fine particle size distributions, high specific resistance, high plasticity index, low dry unit weight, and low shear strength (Aldeeb *et al.* 2003). These characteristics made the residuals unsuitable as landfill cover. The low shear strength of the residuals would result in instability, particularly in sloped areas. However, when the residuals were blended with natural topsoil, the physical properties were improved such that the material could be used as cover. A blend of 20-percent residuals and 80-percent natural topsoil was found to be suitable.

Another option for water sludge disposal is co-disposal with wastewater residuals. The Vest water plant, owned by the Charlotte-Mecklenburg Utility Department (CMUD) in North Carolina, dewatered thickened water plant alum sludge with wastewater sludge at the Irwin Creek wastewater treatment plant. Thickened sludge from the Vest water plant was transported to the Irwin Creek plant by tanker truck. Dewatered solids from Irwin Creek were disposed of by land application (Mendenhall *et al.* 1993).

An indirect co-disposal option is to discharge water plant residuals to municipal sewers. This is discussed in more detail in the following section.

B.8.2 Disposal to municipal sewers

A solution to the water treatment sludge problem in many cases is to discharge the waste stream into sanitary sewers and thereby effectively treat the sludge at the wastewater treatment plant. With such a practice care must be taken that optimum pH conditions for biological efficiency are maintained. Apart from this, however, there does not appear to be much evidence indicating a detrimental effect on biological activity. For plants with primary sedimentation the presence of a coagulated sludge may well result in improved sedimentation and decreased loads to subsequent biological processes (Salotto *et al.* 1973).

Russell and Russell (1955) reported that the discharge of water softening lime sludges to municipal sewers may remove 45 percent of the BOD and 75 percent of suspended solids during primary (in their case, upflow) sedimentation. It has been stated that although alum sludge is beneficial during raw sewage primary sedimentation, this is especially so if flocculation of the combined streams occurs before sedimentation.

When considering direct discharge to a sewer, the following considerations have been recommended (Doe, 1966; Smith, 1948):

- Amenability of the waste sludge to existing treatment processes;
- Hydraulic capacity of sewage treatment facilities;
- The effect of the waste on the final plant effluent;
- Cross connections must be precluded, possibly by a water break, to ensure that if the sewer backs up the raw sewage passes elsewhere;
- Water treatment sludge should be discharged on a continuous basis rather than as slug flows;
- The discharge of gelatinous metal hydroxide sludges may choke trickling filters, if this is the treatment process used;
- Possibility of silting and plugging of pipelines. Krasauskas (1969) recommended a minimum flow velocity of 0.75 m/s (2.5 ft/s).

In cases where the wastewater plant is large compared to the water treatment plant, discharge to sewers may offer a satisfactory solution, provided that sludge handling facilities at the wastewater plant are adequate to cope with the greater sludge volume. Lime sludges may preclude such a method of disposal since deposition problems may be overwhelming and settling problems in the wastewater plant, for example in anaerobic digesters, may be serious.

The principal disadvantage with disposal of water plant sludges to wastewater plants that include primary clarifiers is related principally to anaerobic digesters. Although volatile solids loadings are not significantly increased, the increase in sludge volume can significantly reduce the hydraulic residence time for digestion. Another issue is the potential deposition of inert material in the digesters.

The Alvorado water plant in San Diego, California treated a flow of 458 Ml/d (120 mgd) and determined the following scheme to be the most viable for disposal (Price *et al.* 1989):

- Filter backwash water was equalized in a balancing and settling tank and recycled back to the head of the water plant;
- Sludge from the clarifiers was thickened then discharged by gravity to the wastewater treatment plant.

Sending filter backwash water directly to the wastewater treatment plant was not an option in this case due to the large flows involved.

The pipe diameter required for the combined thickened sludge, at a flow of 0.115 Ml/d (30,000 gpd) and a concentration of around 0.8%, was 200 mm (8-inches).

The wastewater treatment plant provided primary treatment, with primary sedimentation and anaerobic digestion. Ferric chloride was added to improve suspended solids and biochemical oxygen demand (BOD) removals through the primary clarifiers. Tests before and after sending water sludge to the wastewater plant showed that there was no detrimental effect due to the water plant sludge.

The Hillandale water plant in North Carolina carried out tests to evaluate sending water sludge to the Northside wastewater plant. The Hillandale water plant was a conventional plant using alum for coagulation. The Northside wastewater plant included pretreatment with trickling filters, primary sedimentation, activated sludge, anaerobic digesters, and drying beds.

The introduction of the water plant sludge represented an increase of 78 mg/l suspended solids in the raw wastewater influent. The principal effects noted at the wastewater plant were a decrease in the digester retention time of 23-percent, and an increase in the solids loading to the drying beds of 46-percent. Despite the increase in solids load to the drying beds, the drainage characteristics of the sludge were improved.

Primary effluent suspended solids increased from 100 mg/l to 120 mg/l, although primary effluent BOD decreased by approximately 4-percent. Overall phosphorus removal increased from 0.2 mg/l to 1.3 mg/l. On average, for every 100 mg/l water plant sludge added to the raw wastewater, approximately 1 mg/l of phosphorus was removed. Nitrogen removal was not affected by water sludge addition.

The major negative impact of water sludge addition was an increase in final effluent suspended solids, when the quantity of water plant sludge exceeded the equivalent of approximately 35 mg/l in the wastewater influent. Quantities of water sludge less than this equivalent amount did not affect effluent quality.

Guan *et al.* (2004) evaluated the beneficial effects of water plant alum sludge discharges on the performance of a primary treatment wastewater plant in Hong Kong. They found that TSS removals increased from the range 52 to 66-percent, to 72 to 86-percent through the primary clarifiers. Similarly, COD removals increased from the range 28 to 38-percent, to 43 to 53-percent. The authors concluded that sweep floc and physical adsorption onto $Al(OH)_3(s)$ were predominant mechanisms of removal. The authors found that to achieve these removal efficiencies, the alum sludge dosage needed to be at least 18 mg/l as Al. If one assumes that a water plant produces from 10 to 100 kg alum sludge per Ml water treated, and the Al content of the water sludge reported by the authors was 0.135 kg Al/kg dry solids, then the sludge from water plants treating 1.3 to 13.3 Ml/d would need to be discharged for every 1.0 Ml/d wastewater plant capacity, to reach a concentration of 18 mg/l as Al. Under normal circumstances this might be impracticable, and require the transportation of sludges from numerous water plants to achieve this concentration in the wastewater plant.

The City of Boulder wastewater treatment plant found a number of benefits with the discharge of iron sludge from the water treatment plants (Edwards *et al.* 1997). The authors concluded that, due to the benefits of water plant ferric sludge addition, there would be a net economic benefit of US$530/year per Ml/d (US$2,000/year per mgd wastewater plant capacity). However, wastewater treatment plant personnel found that treatment problems occurred at the wastewater plant if discharges of water plant sludge were not equalized carefully.

Adding iron sludge to the wastewater collection system virtually eliminated hydrogen sulfide from anaerobic digester gas. Between 2.6 and 4.2 mg sulfide were removed for every mg of iron added, or a molar ratio of 4.5 to 7.2 mole sulfide/mole Fe added. The authors concluded that the mechanism of sulfide removal was a combination of direct oxidation to sulfate, and precipitation.

Severe struvite formation was also eliminated – within a week after discharging the ferric sludge. It was estimated that the equivalent iron content due to the ferric sludge was approximately 30 kg Fe/t digester TS (60 lb Fe/ton digester dry solids).

Jar test experiments determined that the ferric sludge removed from 0.65 to 4.8 mole phosphorus per mole iron added. Similar experiments using alum sludge showed a lower removal, of approximately 0.023 mole P/mole Al. The phosphorus removal by alum sludge follows the stoichiometry expected by phosphorus sorption to aluminum hydroxide flocs. However, removals using ferric sludge exceeded the stoichiometry expected for the precipitation of vivianite (ferrous phosphate) (Singer, 1972).

One important issue with the discharge of coagulant sludges to wastewater plants is an increase in heavy metal concentrations in the wastewater effluent. Edwards *et al.* (1997) showed that manganese and copper concentrations increased in the wastewater samples after adding water plant sludges. Dosages of 0.4 mg/l iron sludge, and 19.7 mg/l alum sludge added 25 μg/l and 176 μg/l of manganese to the wastewater, respectively. The same coagulant sludge dosages also added 52.4 to 57.2 μg/l total copper to the samples. With the iron sludge, soluble copper increased up to a dosage of 0.2 mg/l Fe, then decreased. At a dosage of 0.4 mg/l Fe the soluble copper remaining was equal to the initial concentration before sludge addition. There appeared to be an interaction between the copper added from the sludge and the copper precipitated by the iron.

B.8.3 Land application of water plant sludge

In general, water plant sludges do not present major issues regarding pathogens, organic compounds or heavy metals, particularly when compared with wastewater plant sludges.

Analyses carried out on eight sludges (Elliott *et al.* 1989) showed that concentrations of cadmium, copper, chromium, nickel, lead and zinc were well below the maximum allowable levels. Other tests carried out in Connecticut (Grabarek & Krug, 1987) showed that sludges from water treatment with aluminum sulfate did not cause toxicity to growing plants, even with those planted directly in the water plant sludge before mixing with soil. Other workers also found that chemical and physical properties of water plant sludges were adequate for plant growth (Dayton & Basta, 2001).

Sludge from the Hays Mine water plant in Pittsburgh has been successfully applied as a topsoil, and as topsoil amendment. The water source for treated water contains algae, silts, sands and sediments that, after the water treatment process, results in a nutrient rich substance that is effective as a topsoil substitute, but considerably cheaper – estimated at a third the cost of commercial topsoil (Watson & Cynar, 2002).

One of the major problems when using water plant sludges as a soil amendment is the potential for fixation of phosphorus in the soil. However, if the application rate is limited to 25 to 50 t/ha · year (11 to 22 ton dry solids/acre · year) this often minimizes this issue (Dempsey *et al.* 1989; Elliott *et al.* 1989). Higher application rates could be used if supplemental fertilization is used (Lucas *et al.* 1994).

Extensive studies carried out in Colorado on the joint application of wastewater biosolids and water plant residuals found that co-mixing at the dry solids ratio of 8 parts water sludge: 1 part biosolids was effective at adsorbing all soluble phosphorus contained in the biosolids (Ippolito *et al.* 1999). At higher water sludge application rates, not only all the biosolids P but some plant available soil P was adsorbed. This finding was important in allowing higher biosolids application rates without exceeding the maximum P application rate. Biosolids application rates can be determined based on the agronomic nitrogen requirements. The water sludge to be mixed then could be determined based on the amount of available P required in the applied mixture.

The phosphorus appeared to be fixed via adsorption to the surface of the aluminum hydroxide solids, or via aluminum-phosphate precipitation (Ippolito *et al.* 2003). The ability of the water sludge to adsorb and fix phosphorus is an advantage, particularly where runoff and eutrophication are concerns.

Co-application of water sludges and biosolids also presents other advantages (Ippolito *et al.* 2002). Without water plant residuals, biosolids land application rates may be limited not only by the P load, but also by the molybdenum (Mo) load. Water plant residuals are able to reduce plant-available Mo concentrations by means of dilution, or by sorption reactions, allowing the biosolids to be applied on the basis of agronomic nitrogen limits.

The application of water plant sludges in silviculture, particularly with trees that have a low phosphorus demand, can be successful with minimal detrimental effects. Tests carried out in Connecticut (Grabarek & Krug, 1987) showed that sludge applied at a concentration of approximately 1.5% at depths of up to 10 cm (4-inches) had no detrimental effect on tree growth.

Geertsema *et al.* (1994) also showed that alum sludge could be applied to forest land at loadings of at least 1.5 to 2.5 percent by dry weight without any adverse effects. Soil, groundwater, and pine growth analyses showed no statistically significant differences between sites with and without alum sludge additions. There was no evidence of significant metals migration through the soil or groundwater. No effects on plant available phosphorus or pine tree growth were observed after alum sludge addition.

Phosphorus runoff from animal waste applied to agricultural land is a significant contributor to non-point source nutrient pollution. Meals *et al.* (2008) showed that by applying water plant residuals to liquid dairy manure at rates ranging from 5 to 10% by dry weight, reductions in the soluble phosphorus content of 20% to 30% were achieved. As an example, at a 5% dose by weight, 175 000 kg/year (385 000 lb/year) could treat the manure generated by 2200 cows, fixing over 2500 kg (5500 lb) of soluble P.

Lime sludges have been widely used for land application. Ohio has a large number of softening plants and the application of the lime sludges produced is widely used (Che *et al.* 1988).

B.9 REFERENCES

Acton N. R. (1988). Chemical reactions in the purification of water by the Rand water board. *Wat. Sew. & Eff.*, **8**(2), 22–24.

Albrecht A. E. (1972). Disposal of alum sludges. *Jour. AWWA*, 46–52.

Aldeeb A. A., Qasim S. R., Puppala A. J. and Anderson C. F. (2003). Physical and engineering properties of treatment plant residuals and disposal. *Jour. AWWA*, **95**(8), 127–137.

Alfaro N., Dubcak T. and Hengst R. (1995). Waking a sleeping giant–water treatment sludge management. WEF/AWWA Joint Residuals/Biosolids Conf., Kansas City.

Amirhor P. and Engelbrecht R. S. (1975). Virus removal by polyelectrolyte-aided filtration. *Jour. AWWA*, 187–192.

Anon (1994). Coagulant recovery system wins big award. *Water Eng. & Mgmt*, 12–13.

AWWA and ASCE (1997). Water Treatment Plant Design – Third Edition. American Water Works Association and American Society of Civil Engineers, McGraw-Hill.

AWWARF (1969–1970). Disposal of wastes from water treatment plants. *American Water Works Association Research Foundation*, Vol. 61, No. 10–12 and Vol. 62, No. 1.

AWWARF and KIWA Ltd. (1990). Slib, Schlamm, Sludge. American Water Works Association Research Foundation (AWWARF) and Keuringsinstituut voor Waterleidingartikelen (KIWA), AWWARF, Denver, CO.

Benn D. and Doe P. W. (1969). The disposal of sludge by the freezing and thawing process. *Filtration and Separation*, **6**.

Bishop S. L. (1978). Alternate processes for treatment of water plant wastes. *Jour. AWWA*, 503–506.

Bishop M. M., Rolan A. T., Bailey T. L. and Cornwell D. A. (1987). Testing of alum recovery for solids reduction and reuse. *Jour. AWWA*, 76–83.

Bishop M. M., Cornwell D. A. and Bailey T. L. (1989). Dewatering evaluation of alum solids and acidified solids with mechanical equipment, AWWA/WPCF Residuals Mgmt Conf., San Diego.

Bishop M. M., Cornwell D. A., Rolan A. T. and Bailey T. L. (1991). Mechanical dewatering of alum solids and acidified solids: an evaluation. *Jour. AWWA*, 50–55.

Bratby J. and Marais G. vR. (1977). Thickening of brown water sludges by dissolved–air flotation. *Water SA*, **3**.

Bratby J. (1986). Optimizing direct filtration in Brasilia. *Jour. AWWA*, 106–115.

Bratby J. (1988). Optimizing manganese removal and washwater recovery at a direct filtration plant in Brazil. *Jour. AWWA*, 71–81.

Bratby J. (1988). Internal Report SABESP, Brazil – Experimental results on water treatment wastes processing.

Bratby J. (1988). Internal Report CAESB, Brazil – Pilot plant results on Descoberto WTP wastes processing.

Bratby J., Baugh C. and Ambrose W. A. (1993). Water plant sludge dewatering by recessed chamber filter press. WEF/AWWA Joint Residuals/Biosolids Conf., Phoenix, Arizona.

Bratby J., Hawkes D. and Tantone F. (1996). Operating experiences with filter press dewatering of water treatment plant sludge. Ariz. Water & Pollut. Control Assoc., May 1–3, Tucson, Arizona.

Briens P. J., Lyon D. K. and Baggaley R. (1989). Operating experiences at Langsett water treatment works. *J. IWEM*, **3**, 50–59.

Burley M. J., Butcher R. A. F., Chappell T. J. W. and Warden J. H. (1977) Thickening and dewatering of waterworks sludges, Water Research Centre, Tech. Rep. TR62.

Burris M. A., Cosens K. W. and Mair D. M. (1976). Softening and coagulation sludge – disposal studies for a surface water supply. *Jour. AWWA*, 247–257.

Calkins R. J. and Novak J. T. (1973). Characterization of chemical sludges. *Jour. AWWA*, 423–428.

Che M. D., Logan T. J., Traina S. J. and Bigham J. M. (1988). Properties of water treatment lime sludges and their effectiveness as agricultural limestone substitutes. *Jour. WPCF*, **60**(5), 674–680.

Chen B. H. H., King P. H. and Randall C. W. (1976). Alum recovery from representative water–treatment plant sludges. *Jour. AWWA*, 204–207.

Clark D. K. and Anderson C. A. (1997). Shared residuals processing will benefit adjacent agencies. *Jour. AWWA*, **89**(6), 48–59.

Collins A. G., Mitra S. and Pavlostathis S. G. (1991). Microwave heating for sludge dewatering and drying. *Res. Jour. WPCF*, **63**(6), 921–924.

Committee Report (1972). Disposal of water treatment plant wastes. *Jour. AWWA*.

Committee Report (1978). Water treatment plant sludges – an update of the state of the art: I & II. *Jour. AWWA*, 498–502 & 548–554.

Committee Report (1981). Lime softening sludge treatment and disposal. *Jour. AWWA*, 600–608.

Committee Report (1987). Research needs for alum sludge discharge. *Jour. AWWA*, 99–104.

Cornwell D. A. and Zoltek J. (1977). Recycling of alum used for phosphorus removal in domestic wastewater treatment. *Jour. WPCF*, 600–612.

Cornwell D. A. (1979). An overview of liquid ion exchange with emphasis on alum recovery. *Jour. AWWA*, 741–744.

Cornwell D. A. and Susan J. A. (1979). Characteristics of acid–treated sludges. *Jour. AWWA*, 604–608.

Cornwell D. A. and Lemunyon R. M. (1980). Feasibility studies on liquid ion exchange for alum recovery from water treatment plant sludges. *Jour. AWWA*, 64–68.

Cornwell D. A., Cline G. C., Przbyla J. M. and Tippin D. (1981). Demonstration testing of alum recovery by liquid ion exchange. *Jour. AWWA*, 326–332.

Cornwell D. A., Bailey T. and Rolan A. T. (1993). The benefits and performance of a full-scale alum recovery facility. WEF/AWWA Joint Residuals/Biosolids Conf., Phoenix, Arizona.

Cornwell D. A. and Lee R. G. (1994). Waste stream recycling: its effect on water quality. *Jour. AWWA*, 50–63.

Cornwell D. A. and Vandermeyden C. (1999). Sizing residuals drying beds. *Jour. AWWA*, 94–105.

Cornwell D. A. and Macphee M. J. (2001). Effects of spent filter backwash recycle on Cryptosporidium removal. *Jour. AWWA*, 153–162.

Croker R. A. (1989). Field experiences in optimization & solids inventory of a water plant belt filter press. AWWA/WPCF Residuals Mgmt Conf., San Diego.

Dayton E. A. and Basta N. T. (2001). Characterization of drinking water treatment residuals for use as a soil substitute. *Wat. Environ. Res.*, **73**(1), 52–57.

DeFilippi J. A. (1972). Designing filtration plant waste–disposal systems. *Jour. AWWA*, 185–187.

Dempsey B. A., Elliott H. A., DeWolfe J., Hamilton D., Lee Y. and Liebowitz R. (1989). Land application of water plant sludges. AWWA/WPCF Residuals Mgmt Conf., San Diego.

Dharmappa H. B., Hasia A. and Hagare P. (1997). Water treatment plant residuals management. *Wat. Sci. Tech.*, **35**(8), 45–56.

Dixon K. L. and Lee R. G. (1989). A residuals management strategy for the Pennsylvania–American Water Company. AWWA/WPCF Residuals Mgmt Conf., San Diego.

Doe P. W., Benn D. and Bays L. R. (1965). The disposal of washwater sludge by freezing. *J. Indus. Water Eng.*, **6**, 251.

Doe P. W. (1966–67) A report on the disposal of sludge from water treatment plants, British Water-Works Asscn., Jubilee Travelling Scholarship.

Drewry W. A. and Kelkar U. (1989). Effects of hydrogen peroxide addition on alum sludge dewatering. AWWA/WPCF Resid. Mgmt Conf., San Diego.

Dulin B. E. and Knocke W. R. (1989). The impact of incorporated organic matter on the dewatering characteristics of aluminum hydroxide sludges. *Jour. AWWA*, 74–79.

Dunkle C. G. (1975). Coagulant recovery takes planning. *Wat. & Wastes Eng.*, **104**, 47–48.

Edwards M., Courtney B., Heppler P. S. and Hernandez M. (1997). Beneficial discharge of iron coagulation sludge to sewers. *Jour. Env. Eng. Div. ASCE*, 1027–1032.

Elliott H. A., Dempsey B. A. and DeWolfe J. (1989). Soil metal loading and phosphorus fertility effects from land application of water treatment sludges. AWWA/WPCF Residuals Mgmt Conf., San Diego.

Elliott H. A., Dempsey B. A. and DeWolfe J. (1988). Land disposal of drinking water sludge. Amer. Soc. Ag. Engnrs Metting, Chicago, 13–16.

Elliott H. A. and Dempsey B. A. (1991). Agronomic effects of land application of water treatment sludges. *Jour. AWWA*, 126–131.

Faber H. A. and Nardozzi A. D. (1972). Water treatment plant waste disposal – action now! *Jour. AWWA*, 674–680.

Faber H. A. and Taras M. J. (1973). Water treatment plant sludge. *Jour. AWWA*, 381–382.

Fitch D. E. and Elliott C. M. (1986). Implementing direct filtration and natural freezing of alum sludge. *Jour. AWWA*, 52–56.

Fulton G. P. (1973). Sludge handling: what's the best method? *Wat. and Wastes Eng.*, **42**, 43–47.

Fulton G. P. (1974). Recover alum to reduce waste disposal costs. *Jour. AWWA*, 312–318.

Gale R. S. (1977). Optimizing the use of pretreatment chemicals. In Solid/Liquid Separation Equipment Scale-up. Uplands Press, London.

Geertsema W. S., Knocke W. R., Novak J. T. and Dove D. (1994). Long-term effects of sludge application to land. *Jour. AWWA*, 64–74.

George D. B., Berk S. G. and Adams V. D. (1989). Toxicity of alum sludge discharge on the aquatic environment. AWWA/WPCF Residuals Mgmt Conf., San Diego.

Glenn R. W., Judkins Jr. J. F. and Morgan J. M. (1973). Filtrability of water treatment plant sludge. *Jour. AWWA*, 414–417.

Goldman M., Varma M. M. and Watson F. (1977). Feasibility of alum sludge reclamation. Conf. 31st Ind. Waste Conf., Purdue Univ.

Gotoh K. et al (1990). "Dewatering characteristics of polymer conditioning alum sludge under low compressive pressure by a constant rate strain compression test", Water Supply, 8, Jonkoping Conf, 195–203.

Grabarek R. J. and Krug E. C. (1987). Silvicultural application of alum sludge. *Jour. AWWA*, 84–87.

Gruninger R. M. (1974). Chemical treatment for surface water. *Wat. & Sew. Wks*, 110–114.

Gruninger R. M. (1975). Disposal of waste alum sludge from water treatment plants. *Jour. WPCF*, **47**(3), 543–552.

Guan X. H., Shang C., Yu S. M. and Chen G. H. (2004). Exploratory study on reusing water treatment works sludge to enhance primary sewage treatment. *Wat. Sci. and Tech.: Water Supply*, **4**(1), 159–164.

Hawkins F. C., Judkins J. F. and Morgan J. M. (1974). Water treatment sludge filtration studies. *Jour. AWWA*, 653–657.

Heuer D. J. and Schwarzwalder R. O. (1989). Alum sludge press operations at Cleveland, Ohio, AWWA/WPCF Resid. Mgmt. Conf., San Diego.

Hoehn R. C., Novak J. T. and Cumbie W. E. (1987). Effects of storage and preoxidation on sludge and water quality. *Jour. AWWA*, 67–75.

Huang J. C. and Nguyen H. Q. (1979). Codisposal of water softening and wastewater sludges. *Jour. WPCF*, **51**(10), 2413–2418.

Hubbs S. A. and Pavoni J. L. (1974). Optimization of sludge dewaterability in sludge disposal lagoons. *Jour. AWWA*, 658–663.

Hudson H. E. Jr. Water Clarification Processes Practical Design and Evaluation, Van Nostrand Reinhold Environmental Engineering Series, 1981.

Hughes D. M., Bassell S. E. and Notte J. J. (1993). Reprocessing residuals new life for an old sludge lagoon. WEF/AWWA Joint Residuals/Biosolids Conf., Phoenix, Arizona.

Inhoffer W. R. and Doe P. W. (1973). Design of wash–water and alum sludge disposal facilities. *Jour. AWWA*, 404–409.

Ippolito J. A., Barbarick K. A. and Redente E. F. (1999). Co-application of water treatment residuals and biosolids on two range grasses. *J. Environ. Qual.*, **28**, 1644–1650.

Ippolito J. A., Barbarick K. A. and Redente E. F. (2002). Combinations of water treatment residuals and biosolids affect two range grasses. *Commun. Soil Sci. Plant Anal.*, **33**(5–6), 831–844.

Ippolito J. A., Barbarick K. A., Heil D. M., Chandler J. P. and Redente E. F. (2003). Phosphorus retention mechanisms of a water treatment residual. *J. Environ. Qual.*, **32**, 1857–1864.

James C. R. and O'Melia C. R. (1982). Considering sludge production in the selection of coagulants. *Jour. AWWA*, 148–151.

Johnson G., Buchanan G. G. and Newkirk D. D. (1989). Case history of a belt filter press dewatering operation at the Robert A. Skinner filtration plant. AWWA/WPCF Residuals Management Conference, San Diego.

Kawczynski E. (1989) "AWWA Research Foundation update on residual management research", AWWA/WPCF Resid Mgmt Conf, San Diego.

Kent A., Wilson D. and Pickles D. M. (1987). Colour removal in a reservoir and the dredging and dewatering of the deposited waterworks sludge. *J. IWEM.*

King P. H., Medding W. S. and Randall C. W. (1970). Lagoon disposal of water treatment plant wastes. *J. San. Eng. Div. ASCE*, **96**(SA4), 1031–1033.

Knocke W. R. and Wakeland D. L. (1983). Fundamental characteristics of water treatment plant sludges. *Jour. AWWA*, 516–523.

Knocke W. R., Hamon J. R. and Dulin B. E. (1987). Effects of coagulation on sludge thickening and dewatering. *Jour. AWWA*, 89–98.

Koorse S. J. (1993). The role of residuals disposal laws in treatment plant design. *Jour. AWWA*, 57–62.

Kos P. (1977). Gravity thickening of water treatment plant sludges. *Jour. AWWA*, 272–282.

LaFargue A. A. and Henriques C. V. (1989). A second and successful attempt at sludge management. AWWA/WPCF Residuals Mgmt Conf., San Diego.

Levesque L. (1982). *Traitement des boues de stations de production d'eau potable (Drinking water plant sludge treatment)*. Association Francaise pour l'Etude des Eaux, Paris.

Levis C. E. and Smith W. B. (1972). Lancaster's new water treatment plant. *Jour. AWWA*, 25–28.

Lind C. B. (1993). Reduction of water treatment residuals by polyaluminum hydroxychloride coagulants. WEF/AWWA Joint Residuals/Biosolids Conf., Phoenix, AZ.

Lucas J. B., Dillaha T. A., Reneau R. B., Novak J. T. and Knocke W. R. (1994). Alum sludge land application and its effect on plant growth. *Jour. AWWA*, 75–83.

Martel C. J. (1989). Development and design of sludge freezing beds. *Jour. Env. Eng. Div. ASCE*, **115**(4), 799.

Martel C. J. and Diener C. J. (1991). A pilot scale study of alum sludge dewatering in a freezing bed. *Jour. AWWA*, 51–55.

McLane J. (2004). Water quality improves by recycling settled sludge. *AWWA Opflow*, 15–18.

McTigue N. E., Cornwell D. A. and Rolan A. T. (1989). Impact of water plant waste discharge on wastewater plants. AWWA/WPCF Residuals Mgmt Conf., San Diego.

Meals D., Fay J. and Barsotti M. (2008). Effects of water treatment residual addition on phosphorus in liquid dairy manure. *Jour. AWWA*, **100**(4), 140–150.

Nicholson J. and Goldbach J. C. (1991). Belt press dewatering of alum sludge proves effective. *Water Eng. & Mgmt*, 28–31.

Nielsen H. L., Carns K. E. and DeBoice J. N. (1973). Alum sludge thickening and disposal. *Jour. AWWA*, 385–394.

Nielsen H. L., Carns K. E. and DeBoice J. N. (1973). Scroll centrifuge for dewatering wins at one plant. *Wat. and Wastes Eng.*, 44–47.

Nielsen H. L. (1977). Alum sludge disposal– problems and success. *Jour. AWWA*, 335–341.

Novak J. T. and Calkins D. C. (1975). Sludge dewatering and its physical properties. *Jour. AWWA*, 42–45.

Novak J. T., Bandak N., Knocke W. R. and Zoccola G. (1989). The effect of shear on sludge dewatering. AWWA/WPCF Resid. Mgmt Conf., San Diego.

Novak J. T. and Bandak N. (1994). The effect of shear on the dewatering of water treatment residuals. *Jour. AWWA*, 84–91.

O'Brien J. H. and Novak J. T. (1977). Effects of pH and mixing on polymer conditioning of chemical sludge. *Jour. AWWA*, 600–605.

O'Connor J. T. (1971). Management of water treatment plant residues. In Water Quality and Treatment, McGraw Hill, NY, 625–646.

Olson R. L. (1976). Alum sludge drying with basic extractive treatment. *Jour. AWWA*, 321–323.

Parker P. J., Collins A. G. and DeWolfe J. R. (2000). Freeze-thaw residuals conditioning. *Jour. AWWA*, **92**(4), 168–181.

Patawaran R., Lew J. and Phillips C. (1999). City of Boulder – Public Works/Utilities – Spent backwash water pilot study. Report on study results, City of Boulder.

Peck B., Russell J., Ardizzone R. and Hayes J. (1993). A case history at the Val Vista water treatment plant implementing a used water recovery system. WEF/AWWA Joint Residuals/Biosolids Conf., Phoenix, Arizona.

Peters G. H., Baumann E. R. and Larson M. A. (1989). Effects of various parameters on the thickening of softening plant sludges. *Jour. AWWA*, 74–84.

Pigeon P. E., Linstedt K. D. and Bennett E. R. (1978). Recovery and reuse of iron coagulants in water treatment. *Jour. AWWA*, 397–403.

Pines D. S. and Esponda R. (2004). Comparison of an anionic polymer to no polymer addition for treating filter backwash water. AWWA Water Technology Conference, San Antonio, TX, Nov. 14–18.

Price S. D., Owen D. M., Cline G. C. and Spiker K. E. (1989). Water treatment residuals management for San Diego– the feasibility of co–disposal. AWWA/WPCF Residuals Mgmt Conf., San Diego.

Reh C. W. (1980). Disposal and handling of water treatment plant sludge. *Jour. AWWA*, 115–118.

Rencken G. E., Hodgkinson D. and Pillay J. (1989). The coagulant recovery sludge dewatering system at Hazelmere water treatment works. WISA Conference, South Africa, March.

Ruiz G., Sancha A. and Carabias A. (1984). *Estudio sobre barro producido en decantadores de plantas de tratamiento de agua potable* (*Study of sludge production in sedimentation tanks in drinking water plants*). Congresso AIDIS, Santiago.

Russell G. D. and Russell G. S. (1955). The disposal of sludge from a lime-soda softening plant as industrial waste. Proc. 9th Ind. Waste Conf., Purdue Univ.

Russell G. A. (1975). From lagooning to farmland application: the next step in lime sludge disposal. *Jour. AWWA*, 585–588.

SABESP (1987). *Recuperacao das aguas de lavagens, tratamento e disposicao de residuos solidos das ETAs da RMSP* (*Washwater recovery, treatment and disposal of residuals from water plants in the metropolitan region of Sao Paulo*). DAE, 47, 150, 216–219.

Salotto B. V., Farrell J. B. and Dean R. B. (1973). The effect of water utility sludge on the activated sludge process. *Jour. AWWA*, 428–431.

Saunders F. M. and Shaw R. C. (1988). Aluminum reclamation by acidic extraction of aluminum–anodizing sludges. *Jour. WPCF*, 369–378.

Schaefer J. K. and Clark D. (1989). Freeze/thaw dewatering for alum sludge. AWWA/WPCF Residuals Mgmt Conf., San Diego.

Schaefer J. K. and Clark (1990). Freeze/thaw dewatering: a proven solution for alum sludge disposal. *Waterworld News*, 15–18.

Schafran G. C., Kelkar U. G. and Drewry W. A. (1993). Dewatering alum sludges with hydrogen peroxide: deciphering the mechanism. WEF/AWWA Joint Residuals/Biosolids Conf., Phoenix, Arizona.

Schwoyer W. L. and Luttinger L. B. (1973). Dewatering of water plant sludges. *Jour. AWWA*, 399–403.

Sengupta A. K. and Shi B. (1992). Selective alum recovery from clarifier sludge. *Jour. AWWA*, 96–103.

Singer P. C. (1972). Anaerobic control of phosphate by ferrous iron. *Jour. Wat. Pollut. Control Fed.*, **44**, 663.

Smith F. E. (1948). A study of sludge disposal of water purification plants in new England. *Jour. NEWWA*, **62**, 265.

Thompson C. G., Singley J. E. and Black A. P. (1972). Magnesium Carbonate – a recycled coagulant, I & II. *Jour. AWWA*, 11–19 & 93–99.

Thompson C. K. and Haas C. A. (1989). Where do all the solids go? A mass balance study. AWWA/WPCF Residuals Mgmt Conf., San Diego.

Treffry–Goatley K., Rencken G. E. and Buckley C. A. (1988). The application of a newly developed tubular filter press to the dewatering of a waterworks sludge. *J. IWEM*, **2**, 376–382.

Treffry–Goatley K., Rencken G. E. and Buckley C. A. (1988). Sizing and optimization of the tubular filter press process for dewatering municipal and industrial sludges. *Explochem Water Treatment.*

Vadovic J. P. (1974). Practical reduction of water treatment wastes. *Wat. & Sew. Wks*, 73–76.

Vandermeyden C. and Cornwell D. A. (1993). Characterizing and optimizing alum residuals dewatering on sand drying beds. WEF/AWWA Joint Residuals/Biosolids Conf., Phoenix, Arizona.

Ventresque C. and Bablon G. (1989). Optimization of sludge treatment in a large scale potable water works. AWWA/WPCF Resid. Mgmt Conf., San Diego.

Wang L. K., Pallo P. E., Schwartz B. J. and Kown B. T. (1973). Continuous pilot plant study of recycling of filter backwash water. *Jour. AWWA*, 355–358.

Warden J. H. (1982). Polymer treatment of waterworks coagulant sludges. *Chem. Eng.*, 460–463.

Watson J. A. and Cynar P. J. (2002). Everything's coming up green. *AWWA Opflow*, **28**(10).

Westerhoff G. P. (1973). Alum recycling: an idea whose time has come? *Wat. & Wastes Eng.*, **10**, 28–31 & 48.

Westerhoff G. P. and Daly M. P. (1974). Water treatment plant wastes disposal – I, II & III. *Jour. AWWA*, 319–324, 379–384 & 441–444.

Westerhoff G. P. and Cornwell D. A. (1978). A new approach to alum recovery. *Jour. AWWA*, 709–714.

Westerhoff G. P. and Cline G. C. (1980). Planned processing beats back water plant sludge disposal problems. *Wat. & Sew. Wks*, 32–34, 36, 58–59.

Wilhelm J. H. and Silverblatt C. E. (1976). Freeze treatment of alum sludge. *Jour. AWWA*, 312–314.

Index

IWA Publishing's authorised EU representative for General Product Safety Regulations is Diane D'Arras, 15 rue Duret, 75116 Paris, France, e-mail: safety@iwap.co.uk.

Printed and bound by CPI Group (UK) Ltd, Croydon, CR0 4YY

16/06/2026

02140349-0003